**Genotyping by Sequencing for Crop Improvement**

# Genotyping by Sequencing for Crop Improvement

*Edited by*

*Humira Sonah*
*National Agri-Food Biotechnology Institute*
*Punjab, India*

*Vinod Goyal*
*CCS Haryana Agriculture University*
*Hisar, India*

*S.M. Shivaraj*
*Laval University*
*Quebec City, QC, Canada*

*Rupesh K. Deshmukh*
*National Agri-Food Biotechnology Institute*
*Punjab, India*

WILEY

This edition first published 2022


*Registered Offices*
John Wiley & Sons, Inc., 111 River Street, Hoboken, NJ 07030, USA
John Wiley & Sons Ltd, The Atrium, Southern Gate, Chichester, West Sussex, PO19 8SQ, UK

*Editorial Office*
The Atrium, Southern Gate, Chichester, West Sussex, PO19 8SQ, UK

For details of our global editorial offices, customer services, and more information about Wiley products visit us at www.wiley.com.

Wiley also publishes its books in a variety of electronic formats and by print-on-demand. Some content that appears in standard print versions of this book may not be available in other formats.



*Library of Congress Cataloging-in-Publication Data*

Names: Sonah, Humira, editor. | Goyal, Vinod, editor. | Shivaraj, S. M., editor. | Deshmukh, Rupesh editor.
Title: Genotyping by sequencing for crop improvement / edited by Humira Sonah, Vinod Goyal, S. M. Shivaraj, Rupesh K. Deshmukh.
Description: First edition. | Hoboken, NJ, USA : John Wiley & Sons, Inc., [2022] | Includes bibliographical references and index.
Identifiers: LCCN 2021046855 (print) | LCCN 2021046856 (ebook) | ISBN 9781119745655 (hardback) | ISBN 9781119745662 (adobe pdf) | ISBN 9781119745679 (epub)
Subjects: LCSH: Genetics–Technique. | Gene mapping. | Genomics. | Plant genomes.
Classification: LCC QK981.45 .G46 2021 (print) | LCC QK981.45 (ebook) | DDC 572.8/62–dc23/eng/20211001
LC record available at https://lccn.loc.gov/2021046855
LC ebook record available at https://lccn.loc.gov/2021046856

Cover Design: Wiley
Cover Image: © Billion Photos/Shutterstock

Set in 9.5/12.5pt STIXTwoText by Straive, Pondicherry, India

Printed and bound by CPI Group (UK) Ltd, Croydon CR0 4YY

C9781119745655_160322

*Dedicated to the two most eminent agricultural scientists of Canada whose work in plant genomics and breeding helped in food security and inspired many young scientists worldwide.*

Prof. Richard Bélanger
Département de phytologie
Université Laval, Canada

Prof. François Belzile
Département de phytologie
Université Laval, Canada

Dr. Humira Sonah
Dr. Vinod Goyal
Dr. S.M. Shivaraj
Dr. Rupesh K. Deshmukh

# Contents

# List of Contributors

**Alish Alisha**, Department of Gene Expression, Faculty of Biology, Adam Mickiewicz University, Poznań, Poland

**Gagandeep Singh Bajwa**, Department of Plant Breeding and Genetics, Punjab Agricultural University, Ludhiana, Punjab, India

**Vitthal T. Barvkar**, Department of Botany, Savitribai Phule Pune University, Pune, Maharashtra, India

**Shubham Bhardwaj**, National Agri-Food Biotechnology Institute (NABI), Mohali, Punjab, India

National Institute of Plant Genome Research (NIPGR), New Delhi, India

**Dharminder Bhatia**, Department of Plant Breeding and Genetics, Punjab Agricultural University, Ludhiana, Punjab, India

**Bharat Char**, Mahyco Research Centre, Mahyco Private Limited, Jalna, Maharashtra, India

**Viswanathan Chinnusamy**, Division of Plant Physiology, ICAR-IARI, New Delhi, India

**Shalu Choudhary**, Mahyco Research Centre, Mahyco Private Limited, Jalna, Maharashtra, India

**Rupesh K. Deshmukh**, Agricultural Biotechnology, National Agri-Food Biotechnology Institute (NABI), Mohali, Punjab, India

**Vikas Devkar**, Department of Plant and Soil Science, Institute of Genomics for Crop Abiotic Stress Tolerance (IGCAST), Texas Tech University, Lubbock, TX, USA

**Pallavi Dhiman**, Department of Agriculture Biotechnology, National Agri-Food Biotechnology Institute (NABI), Mohali, Punjab, India

**Kishor Gaikwad**, ICAR – National Institute for Biotechnology, New Delhi, India

**Naina Garewal**, Department of Biotechnology, Panjab University, Chandigarh, India

**Dhananjay Narayanrao Gotarkar**, International Rice Research Institute, Los Baños, Philippines

**Md Aminul Islam**, Department of Botany, Majuli College, Majuli, Assam, India

**Priyanka Jain**, ICAR – National Institute for Biotechnology, New Delhi, India

**Riya Joon**, Department of Biotechnology, Panjab University, Chandigarh, India

**Swapnil B. Kadam**, Department of Botany, Savitribai Phule Pune University, Pune, Maharashtra, India

**Ravindra Ramrao Kale**, ICAR-Indian Institute of Rice Research, Hyderabad, Telangana, India

**Ravneet Kaur**, Department of Biotechnology, Panjab University, Chandigarh, India

**Suneetha Kota**, ICAR-IIRR, Hyderabad, Telangana, India

**Amit Kumar**, National Agri-Food Biotechnology Institute (NABI), Mohali, Punjab, India

**Kuldeep Kumar**, ICAR – Indian Institute of Pulses Research, Kanpur, Uttar Pradesh, India

**Manish Kumar**, Department of Seed Science and Technology, Dr. Yashwant Singh Parmar University of Horticulture and Forestry, Solan, Himachal Pradesh, India

**Sandeep Kumar**, Xcelris Lab Pvt Ltd., Ahmedabad, Gujarat, India

**Virender Kumar**, Department of Agriculture Biotechnology, National Agri-Food Biotechnology Institute (NABI), Mohali, Punjab, India

**Surbhi Kumawat**, Department of Agriculture Biotechnology, National Agri-Food Biotechnology Institute (NABI), Mohali, Punjab, India

**Brij Kishore Kushwaha**, Department of Molecular Biology and Genetic Engineering, Bihar Agricultural University, Sabour Bhagalpur, Bihar, India

**Omkar Maharudra Limbalkar**, ICAR-Division of Genetics, Indian Agriculture Research Institute, New Delhi, India

**Rushil Mandlik**, Department of Agriculture Biotechnology, National Agri-Food Biotechnology Institute (NABI), Mohali, Punjab, India

**Venugopal Mikkilineni**, Mahyco Research Centre, Mahyco Private Limited, Jalna, Maharashtra, India

**Pankaj S. Mundada**, Department of Botany, Savitribai Phule Pune University, Pune, Maharashtra, India

Department of Biotechnology, Yashavantrao Chavan Institute of Science, Satara Maharashtra, India

**Narender Negi**, Department of Fruit Science, ICAR-NBPGR Regional Station, Shimla, Himachal Pradesh, India

**Anupama A. Pable**, Department of Microbiology, Savitribai Phule Pune University, Maharashtra, India

**Gunashri Padalkar**, Department of Agriculture Biotechnology, National Agri-Food Biotechnology Institute (NABI), Mohali, Punjab, India

**Jayendra Padiya**, Mahyco Research Centre, Mahyco Private Limited, Jalna, Maharashtra, India

**Arushi Padiyal**, Department of Seed Science and Technology, Dr. Yashwant Singh Parmar University of Horticulture and Forestry, Solan, Himachal Pradesh, India

**Brajendra Parmar**, ICAR-IIRR, Hyderabad, Telangana, India

**Gunvant B. Patil**, Department of Plant and Soil Science, Institute of Genomics for Crop Abiotic Stress Tolerance (IGCAST), Texas Tech University, Lubbock, TX, USA

**Vinaykumar Rachappanavar**, Department of Seed Science and Technology, Dr. Yashwant Singh Parmar University of Horticulture and Forestry, Solan, Himachal Pradesh, India

Department of Agriculture, MS Swaminathan School of Agriculture, Shoolini University, Solan, Himachal Pradesh, India

**Nitika Rajora**, Department of Agriculture Biotechnology, National Agri-Food Biotechnology Institute (NABI), Mohali, Punjab, India

**Santosh Rathod**, ICAR-IIRR, Hyderabad, Telangana, India

**Gaurav Raturi**, Department of Agriculture Biotechnology, National Agri-Food Biotechnology Institute (NABI), Mohali, Punjab, India

**Rita**, ICAR – National Institute for Biotechnology, New Delhi, India

**Akshay S. Sakhare**, Division of Plant Physiology, ICAR-IARI, New Delhi, India

**Swati Saxena**, ICAR – National Institute for Biotechnology, New Delhi, India

**Senthilkumar Shanmugavel**, Crop Improvement Division, ICAR – Sugarcane Breeding Institute, Coimbatore, Tamil Nadu, India

**Jitender Kumar Sharma**, Department of Agriculture, School of Agriculture, Baddi University of Emerging Sciences & Technology, Baddi, Himachal Pradesh, India

**Sandhya Sharma**, ICAR – National Institute for Biotechnology, New Delhi, India

**Shivani Sharma**, National Agri-Food Biotechnology Institute (NABI), Mohali, Punjab, India

**Yogesh Sharma**, Department of Agriculture Biotechnology, National Agri-Food Biotechnology Institute (NABI), Mohali, Punjab, India

**Prashant Raghunath Shingote**, Vasantrao Naik College of Agricultural Biotechnology, Dr. Panjabrao Deshmukh Krishi Vidyapeeth, Akola, Maharashtra, India

**Harsha Srivastava**, ICAR – National Institute for Biotechnology, New Delhi, India

**Anuradha Singh**, Department of Genomics, ICAR – National Institute on Plant Biotechnology, New Delhi, India

**Kashmir Singh**, Department of Biotechnology, Panjab University, Chandigarh, India

**Manipal Singh**, Department of Agriculture Biotechnology, National Agri-Food Biotechnology Institute (NABI), Mohali, Punjab, India

**Nisha Singh**, Department of Genomics, ICAR – National Institute on Plant Biotechnology, New Delhi, India

**Avinash Singode**, ICAR – Indian Institute of Millets Research, Hyderabad, Telangana, India

**Sweta Sinha**, Department of Molecular Biology and Genetic Engineering, Bihar Agricultural University, Sabour Bhagalpur, Bihar, India

**Sreeja Sudhakaran**, Department of Agriculture Biotechnology, National Agri-Food Biotechnology Institute (NABI), Mohali, Punjab, India

**Lakshmipathy Thalambedu**, Crop Improvement Division, ICAR – Sugarcane Breeding Institute, Coimbatore, Tamil Nadu, India

**Vandana Thakral**, Department of Agriculture Biotechnology, National Agri-Food Biotechnology Institute (NABI), Mohali, Punjab, India

**Prathima P. Thirugnanasambandam**, Crop Improvement Division, ICAR – Sugarcane Breeding Institute, Coimbatore, Tamil Nadu, India

**Anshuman Tiwari**, Mahyco Research Centre, Mahyco Private Limited, Jalna, Maharashtra, India

**Kishor Tribhuvan**, ICAR – Indian Institute of Agricultural Biotechnology, Ranchi, Jharkhand, India

**Abhijit Ubale**, Mahyco Research Centre, Mahyco Private Limited, Jalna, Maharashtra, India

**Sanskriti Vats**, Agricultural Biotechnology, National Agri-Food Biotechnology Institute (NABI), Mohali, Punjab, India

Regional Centre for Biotechnology, Faridabad, Haryana (NCR Delhi), India

**Joshita Vijayan**, ICAR – National Institute for Biotechnology, New Delhi, India

**Dhiraj Lalji Wasule**, Vasantrao Naik College of Agricultural Biotechnology, Dr. Panjabrao Deshmukh Krishi Vidyapeeth, Akola, Maharashtra, India

**Himanshu Yadav**, Department of Agriculture Biotechnology, National Agri-Food Biotechnology Institute (NABI), Mohali, Punjab, India

# Preface

Recent advances in sequencing technology and computational resources have accelerated genomics and translational research in crop science. The technological advances have provided many opportunities in genomics-assisted plant breeding to address issues related to food security. Among the several applications, genotyping-by-sequencing (GBS) technology has evolved as one of the frontier areas facilitating high-throughput plant genotyping. The GBS approaches have proved effective for the utilization in genotyping-based applications like quantitative trait loci (QTL) mapping, genome-wide association study (GWAS), genomic selection (GS), and marker-assisted breeding (MAB). Considering the current affairs in plant breeding, we decided to compile the advances in GBS methods, statistical approaches to analyze the GBS data, and its applications including QTL mapping, GWAS, and GS in crop improvement.

Presently, the food produced around the world is adequate for the existing population. However, the constantly increasing population mounting pressure on a food production system. Hence efficient utilization of technological advances and existing knowledge is essential to enhance food production to match the growing food demand. In this direction, most of the countries around the globe have adopted advanced genomic methodologies to breed superior plant genotypes. Among such technological advances, the high-throughput genotyping using GBS has shown promising results in different crop plants. The GBS has predominantly been used for germplasm evaluation, evolutionary studies, development of dense linkage map, QTL mapping, GWAS, GS, and MAB. The cost-effectiveness and whole-genome coverage make GBS more reliable than other next-generation sequencing (NGS) techniques.

This book describes advanced molecular markers, high-throughput genotyping platforms, whole-genome resequencing (WGR), QTL mapping using advanced mapping populations, analytical pipelines for the GBS analysis, advances in GWAS, advances in GS, application of GBS, GWAS, and GS in different crop plants. The different marker types including traditional and advanced markers used in plant genotyping have been presented in great detail. DNA extraction directly from seeds without germination can save time and effort. Several modified and crop-specific nondestructive seed DNA extraction protocols have been compiled and presented. Many advanced genotyping platforms are now available which cater to specific research purposes because of the differences in terms of reaction chemistry involved, cost, method of signal detection, and flexibility in the protocols. Such advanced platforms along with their principles have been discussed. The WGR methodology and available resources have been covered in detail. The WGR has emerged as a powerful method to identify genetic variation among individuals. The recent advancement in WGR includes pool-Seq which provides an alternative to individual sequencing and a cost-effective method for GWAS. Compared to biparental populations the multiparental population provides an opportunity to interrogate multiple alleles and to provide an increased level of recombination and mapping resolution of QTLs. The use of such improved populations in the era of high-throughput genotyping has been presented in one of the chapters. The dedicated section focused on the basic principle of GWAS, the efficiency of different markers, candidate gene identification, meta-GWAS, and statistical methods involved in GWAS analysis has been included. For genetic mapping, and marker-assisted selection, rapid and quality DNA isolation is mandatory to accelerate the whole process. A focused section about GS has been included which gives an account of the basic concept, advances, applicability, and challenges of GS. Similarly, a separate chapter is included which discusses the analytical pipelines used for GBS data. Application of technologies such as GBS, GWAS, and GS in different crop categories like cereals, pulses, oilseeds, and commercial crops has been discussed in different chapters.

Here, we have tried to compile basic aspects and recent advances in GBS, GWAS, and GS in plant breeding. We believe that the book will be helpful to researchers and scientists to understand and plan future experiments. This book will enable plant scientists to explore GBS application more efficiently for basic research as well as applied aspects in various crops improvement projects.

Editors
Dr. Humira Sonah
Dr. Vinod Goyal
Dr. S. M. Shivaraj
Dr. R. K. Deshmukh

# 1

# Molecular Marker Techniques and Recent Advancements

*Dharminder Bhatia and Gagandeep Singh Bajwa*

*Department of Plant Breeding and Genetics, Punjab Agricultural University, Ludhiana, Punjab, India*

## 1.1 Introduction

Plant selection and systematic breeding efforts led to the development of present-day improved cultivars of crop plants. From a historical perspective, increased crop yield is the result of genetic improvement (Fehr 1984). Markers play an important role in the selection of traits of interest. Markers can be morphological, biochemical, or molecular in nature. Morphological markers are visual phenotypic characters such as growth habit of the plant, seed shape, seed color, flower color etc. Biochemical markers are the isozyme-based markers characterized by variation in molecular form of enzyme showing a difference in mobility on an electrophoresis gel. Very few morphological and biochemical markers are available in plants, and they are influenced by developmental stage and environmental factors. Since a large number of economically important traits are quantitative in nature, which are affected by both genetic and environmental factors, the morphological and biochemical markers-based selection of traits may not be much reliable. The subsequent discovery of abundantly available DNA-based markers made possible the selection of almost any trait of interest. DNA-based markers are not affected by the environment. Besides, these markers are highly reproducible across labs and show high polymorphism to distinguish between two genetically different individuals or species.

In the last four decades, DNA-based molecular marker technology has witnessed several advances from low throughput hybridization-based markers to high-throughput sequencing-based markers. These advances have been possible due to critical discoveries such as polymerase chain reaction (PCR) (Mullis et al. 1986), Sanger sequencing method (Sanger et al. 1977), automation of Sanger sequencing (Shendure et al. 2011), next-generation sequencing (NGS) technologies (Mardis 2008), and development of bioinformatics tools. This chapter will briefly discuss different types of molecular markers while particularly focusing on recent developments in molecular marker technologies. These

*Genotyping by Sequencing for Crop Improvement*, First Edition.
Edited by Humira Sonah, Vinod Goyal, S.M. Shivaraj, and Rupesh K. Deshmukh.

developments have expedited the mapping and cloning of several loci governing important traits, precise trait selection, and transfer into elite germplasm.

## 1.2 What is a Molecular Marker?

DNA or molecular marker is a fragment of the DNA that is associated with a particular trait in an individual. These molecular markers aid in determining the location of genes that control key traits.

Generally, molecular markers do not represent the gene of interest but act as "flags" or "signs." Similar to genes, all the molecular markers occupy a specific position within the chromosomes. Molecular markers located close to genes (i.e. tightly linked) are referred to as "gene tags."

DNA-based molecular markers are the most widely used markers predominantly due to their abundance. They arise from different classes of DNA mutations such as substitution mutations (point mutations), rearrangements (insertions or deletions), or errors in replication of tandemly repeated DNA. These markers are selectively neutral because they are usually located in noncoding regions of DNA. Unlike morphological and biochemical markers, DNA markers are practically unlimited in number and are not affected by environmental factors and/or the developmental stage of the plant.

DNA markers show genetic differences that can be visualized by using a gel electrophoresis technique and staining ethidium bromide or hybridization with radioactive or colorimetric probes. Markers that can identify the difference between two individuals are referred to as polymorphic markers, whereas those that do not distinguish the individuals are called monomorphic markers. Based on how polymorphic markers can discriminate between individuals, they are described as codominant or dominant. Codominant markers indicate differences in size whereas dominant markers reveal differences based on their presence or absence. The different forms of a DNA marker in the form of band size on gels are known as marker "alleles." Dominant marker has only two alleles whereas codominant markers may have many alleles.

## 1.3 Classes of Molecular Markers

Based on the method of their detection, DNA markers are broadly classified into three groups: (i) hybridization-based, (ii) PCR-based, and (iii) DNA sequence-based molecular markers. Molecular markers have been discussed earlier in several reviews (Collard et al. 2005; Semagn et al. 2006; Gupta and Rustgi 2004) and book chapters (Mir et al. 2013; Singh and Singh 2015), which readers can also consult for more details. However, a brief description of each of these markers has been presented below.

### 1.3.1 Hybridization-based Markers

#### 1.3.1.1 Restriction Fragment Length Polymorphism (RFLP)

These are the first molecular markers used by Grodzicker et al. (1975) in adenovirus and Botstein et al. (1980) in human genome mapping. These were first used in plants by Helentjaris et al. (1986). In this type of marker, polymorphism is detected by cutting DNA into fragments by the use of restriction enzymes followed by hybridization of radioactively labeled DNA probes which are single or low copy DNA fragments and visualized by autoradiography. DNA probes could be genomic clones, cDNA clones, or even cloned genes. The RFLP markers show co-dominance and are highly reliable in linkage analysis and breeding (Semagn et al. 2006). However, this technique requires a large quantity of DNA, labor-intensive, relatively expensive, and hazardous. RFLP shows polymorphism in two different species if they differ due to point mutations, insertion/deletion, inversion, translocation, and duplication.

#### 1.3.1.2 Diversity Array Technology (DArT™)

This is a high-throughput DNA polymorphism analysis method which combines microarray and restriction-based PCR methods. It is similar to AFLP where hybridization is used for the detection of polymorphism. It can able to provide a comprehensive genome coverage even in those organisms not having genome sequence information (Jaccoud et al. 2001). Diversity array technology (DArT) is a solid-state open platform method for analyzing DNA polymorphism. DArT procedure includes (i) Generating a diversity panel and (ii) Genotyping using a diversity panel. The diversity panel is generated using a set of lines representing the breadth of variability in germplasm (~10 lines). An equal quantity of DNA from each representative line is pooled followed by restriction with two to three restriction endonucleases (REs) and ligation of RE-specific adaptors. Later DNA fragments are amplified using adaptor complementary primers. The representation fragments are ligated to vector and transformed into *Escherichia coli* cells. The transformed cells with recombinant DNA are selected and amplified using M13 forward and reverse primer. The amplified DNA is isolated and purified. The purified DNA is coated onto polylysine-coated glass slides to generate a diversity array.

For genotyping, the representation fragments of the target genotypes are prepared in the same as in the diversity panel. The DNA fragments are column purified and fluorescently labeled with two different dyes (Cy3 or Cy5). The labeled DNA fragments are used for hybridization onto the diversity array. Two representative panels – one labeled with Cy3 and another with Cy5 – can be hybridized simultaneously and hybridization signal intensities are measured for each spot. DArT, thus detects DNA polymorphism at several hundred genomic loci in a single array without relying on sequence information.

### 1.3.2 Polymerase Chain Reaction (PCR)-based Markers

#### 1.3.2.1 Simple-Sequence Repeats (SSRs)

Simple-sequence repeats (SSRs) (Litt and Luty 1989) are also known as microsatellites or short tandem repeats (STRs) or simple sequence length polymorphism (SSLP). These are widely used markers and are also referred to as the mother of all the markers. These are STRs, generally of one to eight nucleotide length. These are found dispersed throughout the genome and are hypervariable. These repeat regions are flanked with unique sequences that are highly conserved. The flanking unique sequences are used to design complementary primers which can be assayed with PCR. SSRs are highly polymorphic and codominant markers. These show polymorphism as a result of the variable number of repeat units. Before the era of genome sequencing, it was difficult to develop SSRs due to the extensive cost and labor involved in the identification of repeat regions and flanking unique sequences. However, with the availability of genome sequences of several organisms, the development of SSR has become very easy which involves in silico identification of STRs, designing of SSR from flanking unique sequences, and validation through experimentation. SSR markers have shown immense application in population genetic analysis, gene mapping, and cloning due to their abundance in the genome and high polymorphism, and very high reproducibility across labs. SSR-based linkage maps have been developed in several important crop plants such as rice (Temnykh et al. 2000; McCouch et al. 2002; Orjuela et al. 2010), wheat (Roder et al. 1998), maize (Sharopova et al. 2002), potato (Milbourne et al. 1998), etc.

#### 1.3.2.2 Sequence-Tagged Sites (STSs)

Sequence-tagged sites (STSs) were first developed for physical mapping of the human genome by Olsen et al. (1989). STS is the short unique sequences developed from polymorphic RFLP probe or AFLP fragment which is linked to desirable traits. The RFLP probes or AFLP fragments showing polymorphism are end-sequenced and primers are designed to specifically amplify these fragments. STS markers are co-dominant and highly reproducible. For example, STS markers have been developed for RFLP markers linked with bacterial blight resistance genes *xa5*, *xa13*, and *Xa21* (Huang et al. 1997). One major limitation of these types of markers is the reduced polymorphism than the corresponding RFLP probe.

#### 1.3.2.3 Randomly Amplified Polymorphic DNAs (RAPDs)

Williams et al. (1990) first developed these markers to amplify DNA without prior sequence information. In this type of marker, the arbitrary decamer sequences are used as primers at low annealing temperatures for DNA amplification. These markers are referred to as dominant markers because the polymorphism is determined based on the presence or absence of a particular amplified fragment. Polymorphism may also be due to varying brightness of bands at a particular locus due to copy number differences. These markers have been used for constructing linkage maps in several species (Hunt 1997; Laucou et al. 1998) and also for tagging genes of economic importance. However, due to the dominant nature, these may not be appropriate for genetic mapping and marker-assisted selection (MAS). One major limitation of these markers is the lack of repeatability in certain cases. Variations of RAPD include AP-PCR (arbitrarily primed PCR) and DAF (DNA amplification fingerprinting (Table 1.1).

**Table 1.1** Details of the other important molecular markers.

| Marker | Description |
|---|---|
| Variable number tandem repeat (VNTR) or minisatellites | A short DNA sequence (10–100 bp) is present as tandem repeats and is a highly variable copy number |
| DNA amplification fingerprinting (DAF) | A variation of RAPD, where 4–5 bp single and arbitrary primer is used to detect polymorphism |
| Arbitrary-primed PCR (AP-PCR) | A variation of RAPD, where 18–32 bp long single and arbitrary primer is used to detect polymorphism |

**Table 1.1** (Continued)

| Marker | Description |
|---|---|
| Inter-simple sequence repeat (ISSR) | Primers are designed based on the repeat region of microsatellites. These primers are used to amplify the region between two microsatellites. The stretches of unique DNA in between or flanking the SSRs are amplified. A single SSR-based primer is used to prime PCR |
| Selective amplification of microsatellite polymorphic loci (SAMPL) | A modification of ISSR, where SSR-based primer is used along with AFLP primer. The template is identical to the AFLP template and the rare cutter primer is replaced by SSR-based primer |
| Cleaved amplified polymorphic sequences (CAPS) | These markers are also called PCR-RFLP, where amplified PCR product is digested with endonucleases to reveal polymorphism. These are used when PCR product does not show polymorphism and restriction enzyme site present in amplified PCR product may detect polymorphism |
| Derived cleaved amplified polymorphic sequences (dCAPS) | A variation of CAPS, where a primer containing one or more mismatches to template DNA is used to create a restriction enzyme recognition site in one allele but not in another due to the presence of SNP. Thus, obtained PCR product is subjected to restriction enzyme digestion to find the presence or absence of the SNP |
| Single-strand conformational polymorphism (SSCP) | DNA fragments of size ranging from 200 to 800 bp were amplified by PCR using specific primers (20–25 bp), followed by gel-electrophoresis of single-strand DNA to detect nucleotide sequence variation. The method is based on a principle that the secondary structure of single-strand DNA molecule changes significantly if it harbors mutation. This method detects nucleotide variation without sequencing a DNA sample |
| Denaturing/temperature gradient gel electrophoresis (DGGE, TGGE) | These methods reveal polymorphism due to differential movement of the same genomic double-stranded region with different base-pair composition. As an example, the AT-rich region would have a lower melting temperature than the GC-rich region |
| Target region amplification polymorphism (TRAP) | This method employs primers designed from the EST database for detecting polymorphism around a selected candidate gene. This includes two primers of 18 bp, one of which is designed from targeted EST and the other is an arbitrary primer |

#### 1.3.2.4 Sequence Characterized Amplified Regions (SCARs)

These markers overcome the limitation of RAPDs. In this case, the RAPD fragments that are linked to a gene of interest are cloned and sequenced. Based on the terminal sequences, longer primers (20 mer) are designed. These SCAR primers more specifically amplify a particular locus. These are similar to STS markers in design and application. The presence or absence of the band indicates variation in sequences. The SCAR markers thus are dominant in nature. These, however, can be converted to codominant markers in certain cases by digesting the amplified fragment with tetranucleotide recognizing restriction enzymes. There are several examples where the RAPD markers linked to the gene of importance have been converted to SCAR markers (Joshi et al. 1999; Liu et al. 1999; Kasai et al. 2000; Akkurt et al. 2007; Chao et al. 2018).

#### 1.3.2.5 Amplified Fragment Length Polymorphism (AFLP)

This marker technique was developed by Vos et al. (1995) and is patented by Keygene (www.keygene.com). In this technique, DNA is cut into fragments by a combination of restriction enzymes which are frequent (four bases) and rare (six bases) cutters that generate restriction overhangs on both sides of fragments. This is followed by the annealing of double-stranded oligonucleotide adapters of a few oligonucleotide bases with respective restriction overhangs. The oligonucleotide adapters are designed in such a way that the original restriction sites are not reinstated and also provide the PCR amplification sites. The fragments are PCR amplified and visualized on agarose gel. This method produces many restriction fragments enabling the polymorphism detection. The number of amplified DNA fragments can be controlled by selecting different number or composition of bases in the adapters. The stringent reaction conditions used for primer annealing make this technique more reliable. This method is a combination of both RFLP and PCR techniques and is extremely useful in the detection of polymorphism between closely related genotypes. Like RAPD, AFLP is a dominant marker and is not preferred for genetic mapping studies and MAS. AFLP maps have been constructed in several species and integrated into already existing RFLP maps e.g. tomato (Haanstra et al. 1999), rice (Cho et al. 1997), and wheat (Lotti et al. 2000).

#### 1.3.2.6 Expressed Sequence Tags (ESTs)

These markers are developed by end sequencing (generally 200–300 bp) of random cDNA clones. The sequence thus obtained is referred to as expressed sequence tags (ESTs). A large number of ESTs have been synthesized in several crop plants and are available in the EST database at NCBI (https://www.ncbi.nlm. nih.gov/dbEST/). These markers were originally developed to identify gene transcripts and have played important role in the identification of several genes and the development of markers such as RFLP, SSR, SNPs, CAPS, etc. (Semagn et al. 2006). However, EST-based SSRs show less polymorphism as compared to genomic DNA-based SSRs. Since EST markers are from expressed sequence regions, these are highly conserved among the species and can be used for synteny mapping. Most of these could also be functional genes. A large number of EST markers have been used in rice for developing a high-density linkage map (Harushima et al. 1998) and for chromosome bin mapping in wheat using deletion stocks (Qi et al. 2003). In addition to these, several other molecular marker variants have been developed. The description of those markers is presented in Table 1.1.

## 1.4 Sequencing-based Markers

### 1.4.1 Single-Nucleotide Polymorphisms (SNPs)

Single-nucleotide polymorphisms (SNPs) are more abundant resulted from single-base pair variations. These are evenly distributed in a whole genome that can tag almost any gene or locus of a genome (Brookes 1999). However, the distribution of SNPs varies among species with 1 SNP per 60–120 bp in maize (Ching et al. 2002) and 1 SNP per 1000 bp in humans (Sachidanandam et al. 2001). SNPs are more prevalent in the noncoding region. In the coding region, SNPs could be synonymous or nonsynonymous. In synonymous SNPs, there is no change in the amino acid resulting in no phenotypic differences. However, phenotypic differences could be produced due to modified mRNA splicing (Richard and Beckman 1995). In nonsynonymous SNPs, change in amino acid results in phenotypic differences. SNPs are mostly bi-allelic and cause polymorphism due to nucleotide base substitution. The two types of nucleotide base substitutions result in SNPs. A transition substitution occurs between purines (A, G) or between pyrimidines (C, T). This type of substitution constitutes two-thirds of all SNPs. A transversion substitution occurs between a purine and pyrimidine. SNPs can be detected by the alignment of the similar genomic region of two different species. The SNPs have only two alleles compared to typical multiallele SSLP; however, this disadvantage can be compensated by using the high density of SNPs.

### 1.4.2 Identification of SNP in a Pregenomic Era

Initially, identification of SNP markers was laborious and expensive and involved allele-specific sequencing (Ganal et al. 2009). This includes sequencing of unigene-derived amplicons using Sanger's method from two or more than two lines. In an experiment, about 350 bp of the RFLP clone, A-519 was end sequenced in soybean and the flanking amplification primers were designed (Coryell et al. 1999). Primers were used to screen for allele diversity using PCR from ten genotypes and the amplicons were sequenced followed by sequence comparison to identify SNP. SNPs were also identified through mining a large number of EST sequences in EST databases, which are generated through improved sequencing technologies (Soleimani et al. 2003). These SNPs are further validated using PCR (Batley et al. 2003). These approaches allowed the identification of mainly gene-based SNPs, but their frequency is generally low. Additionally, SNPs located in low-copy noncoding regions and intergenic spaces could not be identified.

Several assays have been developed for genotyping based on identified SNPs which include, allele-specific hybridization, primer extension, oligonucleotide ligation, and invasive cleavage (Sobrino et al. 2005). Besides, DNA chips, allele-specific PCR, and primer extension were also attractive options since these are suitable for automation and can be used for the development of dense genetic maps. Allele-specific hybridization was used for the identification of polymorphism in 570 genotypes of soybean (Coryell et al. 1999).

## 1.5 Recent Advances in Molecular Marker Technologies

The improvement of Sanger sequencing technology in the 1990s combined with the beginning of EST and genome sequencing projects in model plants led to the spurt in the identification of variation at the single-base resolution (Wang et al. 1998). From 2005 onward, the emergence of NGS platforms such as Roche 454, Illumina HiSeq2500, ABI 5500xl SOLiD, Ion Torrent, PacBio RS, Oxford Nanopore, and advances in bioinformatics tools simplified the process of identification of genome-wide SNPs and changed the face of molecular marker technology. NGS-based genotyping platforms such as genotyping-by-sequencing (GBS), whole-genome resequencing (WGR), and high-density SNP arrays helped to type thousands of SNPs in a single reaction in hundreds of individuals.

### 1.5.1 Genotyping-by-Sequencing (GBS)

GBS is an NGS-based reduced representation sequencing technique for the identification of genome-wide SNPs and genotyping large populations (Bhatia et al. 2013). GBS is a one-step approach for the identification and utilization of markers in a single reaction. It is a complexity reduction procedure where a combination of restriction enzymes is used to separate low copy sequences from high copy repetitive regions. In general, GBS involves the sequencing of fragments generated through restriction digestion of the genome on the NGS platform. In this process, the DNA of the population is digested with RE followed by ligation of RE-specific adaptors containing genotype-specific barcode sequences and sites for binding PCR and sequencing primers (Figure 1.1). The fragments thus generated can be PCR amplified and an equal volume of PCR product from different individuals are pooled in a tube. The fragments in the pool can be selected based on their size and sequenced on

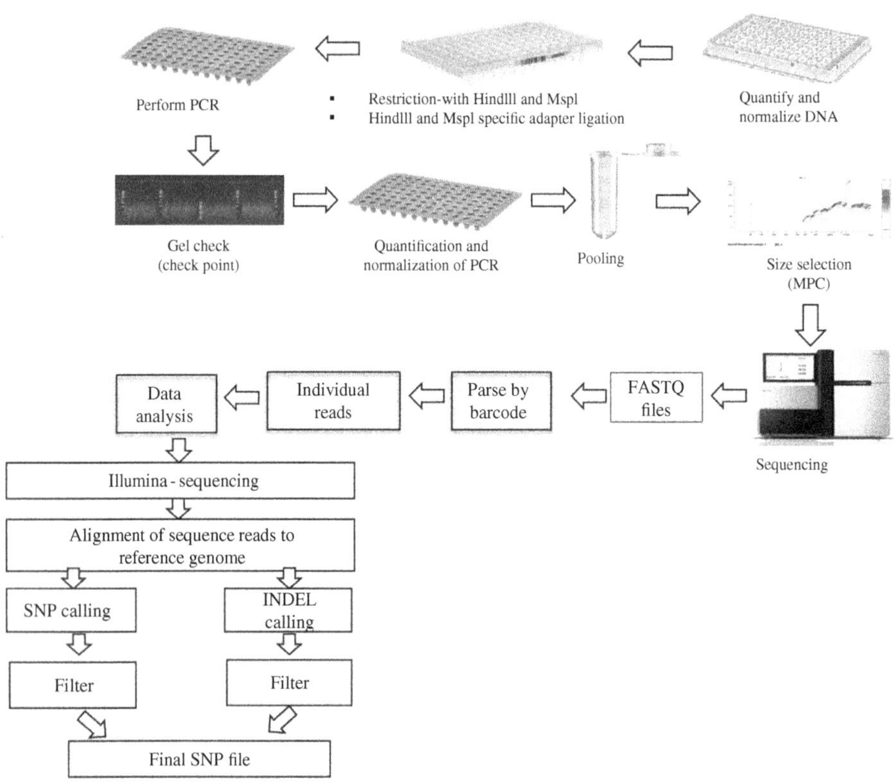


**Figure 1.1** An example of GBS and GBS data analysis workflow for identification of SNP markers.

the NGS platform. The choice of restriction enzymes depends upon the complexity and size of the genome. Presently, different versions of GBS are available, which includes RAD-seq (restriction associated DNA sequencing), ddRAD-seq (double-digest restriction associated sequencing), SLAF-seq (specific-locus amplified fragment sequencing), Rest-seq (restriction DNA sequencing), Skim GBS (skim-based GBS) (Bhatia 2020). These versions differ with respect to fragment size selection, the extent of complexity reduction, and genome coverage. Since GBS is a population-dependent genotyping method, to make it cost-effective a low-depth sequencing is adopted which caused a high rate of missing data. The low-depth sequencing makes it an ineffective genotyping approach in heterozygous populations. GBS has low genome coverage due to reduced representation sequencing.

GBS is being widely used to capture SNPs and other marker variations by NGS. GBS overtook the conventional genotyping procedures involving the use of traditional markers such as RAPD, AFLP, SSR, and many others in terms of time, labor, and cost involved. As an example, GBS can generate data of thousands of markers in a large population in a week, which can be analyzed in a month (Bhatia et al. 2018). The approach has been utilized in the mapping of several economically important traits in a number of crop plants (Poland and Rife 2012). Most of the developing countries have in-house computational facilities that are being used for GBS analysis. Few online servers are also available, where GBS analysis can be done using in-built pipelines such as cyverse (www.cyverse.org); however, these are unable to analyze the large dataset. Further speed of analysis depends upon the internet speed. Alignment of NGS-based reads and calling SNPs and Indels are the two major steps in GBS analysis, for which several pipelines are available publically such as Stacks, IGST, GB-eaSY, TASSEL-GBS, FAST-GBS, UNEAK, etc. (Wickland et al. 2017).

Another important pipeline widely used for NGS data analysis is dDocent pipeline (www.dDocent.com) which is a simple bash wrapper to quality analysis, assemble, map, and call SNPs from almost any kind of RAD sequencing (Puritz et al. 2014). However, most of these pipelines are hard to code for a student with little bioinformatics background. Most of these pipelines vary with respect to the complexity of the genome and computational space required. Besides there are several bioinformatics tools such as BWA, Bowtie2, SAM tools, GATK, BCFtools including a set of Perl utility scripts (Kagale et al. 2016) that can be used for GBS data analysis. However, there should be knowledge of the installation and usage of these tools for proper utilization in data analysis. With the advancements in NGS approaches, GBS has become a widely used approach in plant breeding and genetics, particularly for understanding complex quantitative traits.

DArT-seq GBS (https://www.diversityarrays.com/technology-and-resources/dartseq/) somehow overcomes the limitation of the missing data point. The technique is an extension of traditional DArT technology where DArT representations are sequenced on the NGS platform. The fragment sequencing enables a dramatic increase in the number of genomic fragments analyzed and an increase in the number of reported markers thus making it a cost-effective technology than the initial DArT method.

### 1.5.2 Whole-Genome Resequencing (WGR)

WGR with high coverage and depth overcomes the limitations of GBS due to missing data points and heterozygous calls. In general, WGR involves the sequencing of enough DNA fragments (>5×–20×) to cover the whole genome of an organism. Due to sequencing cost, the technique is suitable in crop plants having smaller genome sizes such as rice. In such cases, GBS can be replaced by resequencing of a larger size population at 5–6× depth. However, WGR for few samples can be done at a much higher read depth of 10–20× as in the case of the BSA-seq approach (Nguyen et al. 2019). One of the important BSA-seq-based approaches is quantitative trait loci (QTL)-seq developed by Takagi et al. (2015) in rice. Later this technique has been widely used in several crop plants. Takagi et al. (2015) developed a pipeline for analysis of the whole genome sequence of bulks and identification of causative variants. WGR has been used in several studies for identification of genome-wide SNPs, genotyping mapping populations for construction of high-density linkage maps and QTL mapping, linkage and genome-wide association studies (GWASs), of reference genome improvement, and genomic selection (Poland and Rife 2012; Bhatia et al. 2013; Chung et al. 2017; Nguyen et al. 2019).

### 1.5.3 SNP Arrays

Along with GBS, high-density DNA array-based SNP chips or SNP arrays have become a widely used SNP detection platform for high multiplex genotyping. SNP arrays work by hybridization of DNA fragments with allele-specific oligonucleotide probes (SNP probes) and fluorescence-based detection of signals. In general, SNP arrays can be roughly categorized into two types based on SNP detection methods: (i) nonenzymatic differential hybridization including allele-specific hybridization, (ii) enzymatic reactions including primer extension, and mini-sequencing (Ding and Jin 2009). For making SNP arrays, the first step is the identification of genome-wide SNPs by sequencing (preferably WGR) of a large diverse panel. The SNPs arrays may include SNPs from coding (genic) regions only and/or genome-wide SNPs from other noncoding regions. SNPs are in silico validated with several custom tools and final filtered SNPs are identified. The oligonucleotide probes containing SNP alleles are designed and bound on a solid glass plate surface. SNP chips can be custom designed commercially from two widely used platforms: Affymetrics (www.affymetrics.com) as Axiom Affymetrics SNP Chips (Affymetrix/Thermo Fisher Axiom®) or Illumina (https://www.illumina.com/science/technology/microarray.html) as Immunia Infinium assay (Illumina Infinium®). Affymetrics SNP array relies on differential hybridization due to different melting temperatures for matched and mismatched SNPs binding to target DNA sequence. On the other hand, Illumina Infinium assay uses Illumina BeadArray technology that relies on primer extension to distinguish two SNP alleles. The Affymetrix SNP array uses 25-mer for SNP calling while the Illumina BeadArray uses 50-mer for target capture. In rice, a high-resolution 44K Affymetrix array, 50K Infinium array, and 700K high-density rice array are available for rice SNP genotyping (McCouch et al. 2010; Tung et al. 2010; Chen et al. 2013; McCouch et al. 2015). Additionally, high-density SNP arrays have been developed for other crop plants such as maize (Ganal et al. 2011) and sunflower (Bachlava et al. 2012) as well as domestic animal species, including cattle (Gibbs et al. 2009; Matukumalli et al. 2009) and pig (Ramos et al. 2009). One major advantage of SNP arrays is the reproducibility of data points where GBS does have some shortcomings. However, the disadvantage is the less polymorphism as compared to GBS and WGR and detection of only alleles present in the array (Table 1.2).

**Table 1.2** Comparison between different marker techniques commonly used in plant research.

| | SSR | GBS | WGR | SNP array | KASP™ |
|---|---|---|---|---|---|
| DNA quality | Moderate | High | High | High | High |
| PCR-based | Yes | Yes | No | No | No |
| Allele detection | High | High | High | Low | Low |
| Polymorphism | High | High | High | Low | Low |
| Ease to use | Easy | Not easy | Not easy | Easy | Easy |
| Reproducibility | High | Low | High | High | High |
| Cost | Moderate | Low to moderate | High | High | moderate |
| Cost for analysis | High | High | High | Low | Low |
| *Suitability for different approaches* | | | | | |
| Genetic diversity analysis | High | Moderate | High (cost concerns) | High | High |
| Bi-parental QTL mapping | High | High | High | High | High |
| Genome wide association analysis | Moderate | High | High | High | Low |
| Genomic selection | Low | Moderate | High (cost concerns) | High | Low |

### 1.5.4 Kompetitive Allele-Specific PCR (KASP™)

KASP™ is a trademark technology of KBiosciences (http://www.kbioscience.co.uk/) or LGC genomics (http://www.lgcgenomics.com) initially developed for in-house genotyping, thereafter evolving as a benchmark technology for SNP genotyping. Any candidate SNP identified through GBS or WGR and associated with any important traits can be validated through KASP assay. Furthermore, any identified candidate SNP associated with a trait of interest can be readily converted into KASP assay to serve as a robust and cost-effective marker to be used as a MAS tool for crop improvement. It works on the principle of competitive allele-specific PCR permitting bi-allelic scoring of SNP, insertion, and deletions (InDels) at a specific location in the genome (Figure 1.2). KASP genotyping reaction consists of DNA sample, KASP assay mix, and universal KASP master mix. Allele-specific two forward primers and common reverse primer all unlabelled constitute KASP assay mix. Allele-specific primers have a unique tail sequence complementary to FRET (fluorescence resonant energy transfer) cassette. Each allele harbors a tail sequence linked to different dyes (FAM™ and HEX™ dyes). KASP master mix has FRET cassettes in the quenched state, a passive reference dye (ROX™), and other components for PCR reaction. During the first round of reaction allele-specific primer binds to template incorporating tail sequence in newly synthesized strands. In the next round of PCR, a complementary strand of the allele-specific tail sequence is generated allowing the FRET cassette to bind enabling an unquenched state, and generating a fluorescent allele-specific signal. In the case of a homozygous DNA sample, only one signal specific to the allele is seen and mixed signal is generated in the case of the heterozygous individual.

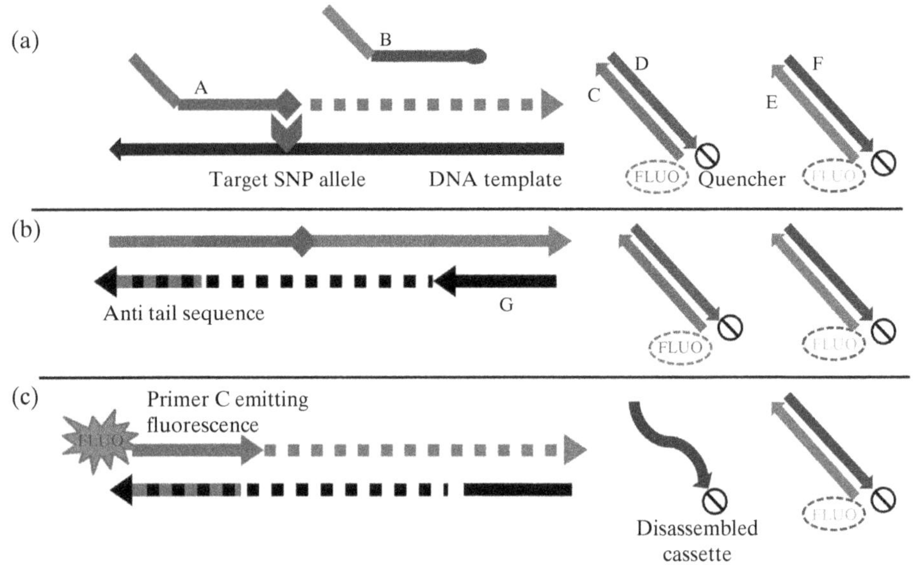


**Figure 1.2** Steps in KASP reaction: (a) annealing: allele-specific primer binds to target SNP, (b) extension: anti tail sequence generation leading to disassembly of allele-specific FRET cassette, and (c) fluorescent emission: sample emitting fluorescence on exposing to a specific wavelength. *Source:* The figure is reproduced from Rosas et al. (2014) available with CC-BY-4.0.

It can be carried out in 96- to 1536-well plate format. Application of KASP includes QC (quality control) analysis, QTL mapping, allele mining (Semagn et al. 2013), and MAS. However, it may not be a suitable platform for genome-wide association mapping and genomic selection due to fewer data points. KASP markers have been utilized extensively for MAS in major crops like rice (Yang et al. 2019; Kang et al. 2019), wheat (Makhoul et al. 2020; Fang et al. 2020; Grewal et al. 2020), and maize (Zhang et al. 2017; Su et al. 2016).

## 1.6 SNP Databases

SNP databases correspond to a publically available archive of genetic information of economically important species. Recent developments in genome sequencing technologies have ushered the era of cost-effective, high-throughput genomics resulting in the creation of huge datasets of sequence information. In a similar manner, developments in computing facilities and data sciences had enabled us to compare, categorize, and compute relationship matrices among and between species leading to the creation of databases. As numerous labs are working on the same species, collaborative consortiums were established to avoid redundancy leading to high-quality SNP databases containing a range of molecular variations constituting SNP's, insertions, and deletions (InDels), trait-specific characterized SNP's and called variants. Table 1.3 enlists some important SNP databases corresponding to humans, model genetic organisms, and important crop species. These public databases serve as an important resource for crop improvement for genetic diversity analysis, establishing a genetic association and linkage disequilibrium studies.

**Table 1.3** List of important online SNP databases.

| SNP database | Organism | URL |
|---|---|---|
| dbSNP | Human | http://www.ncbi.nlm.nih.gov/sites/entrez?db=snp |
| Ensembl | Human | http://www.ensembl.org/Homo_sapiens/index.html |
| 1001 Genomes | Arabidopsis | https://1001genomes.org/ |
| CropSNPdb | Brassica crops, wheat | http://snpdb.appliedbioinformatics.com.au |
| SNP-Seek Database | Rice 3K panel | https://snp-seek.irri.org/ |
| MaizeSNPDB | 1210 Maize inbred lines | http://150.109.59.144:3838/MaizeSNPDB/ |
| CerealDB | Wheat | http://www.cerealsdb.uk.net/cerealgenomics/CerealsDB/indexNEW.php |

## 1.7 Application of Molecular Markers

### 1.7.1 Application of Molecular Markers in Crop Improvement

Molecular markers have several applications in genetic studies and crop improvement programs. These have been used in the development of saturated linkage maps, gene/QTL mapping, map-based cloning of genes, orthologous gene mapping, and marker-assisted transfer of targeted genes/QTLs in the background of different cultivars/lines. Saturation of linkage maps refers to increased marker density to cover the entire chromosomal region. In general, when molecular markers are arranged on a linkage map with less than 1 cM distance apart, is considered as saturated linkage map. The development of saturated linkage maps could only be possible with the availability of molecular markers. These maps are prerequisite for gene/QTL mapping, map-based cloning of genes, and MAS. Several molecular markers-based saturated linkage maps have been developed in crop plants including rice (Harushima et al. 1998; McCouch et al. 2002; IRGSP 2005; Zhu et al. 2017; Kumar et al. 2018), wheat (Somers et al. 2004; Song et al. 2005; Poland et al. 2012; Li et al. 2015; Hussain et al. 2017), maize (Sharopova et al. 2002; Zhou et al. 2016; Su et al. 2017), and tomato (Tanksley et al. 1992; Haanstra et al. 1999; Sim et al. 2012). In one of the studies, a rice genetic map helped to enrich the genetic region of Ph1 locus of wheat and facilitated the identification of candidate genes governing the locus (Sidhu et al. 2008).

Plant breeders have relied heavily on generating new gene combinations and selecting these new gene combinations empirically. Though phenotype-based selection has largely been successful, MAS has improved the efficiency and precision of selection. MAS can be practiced more efficiently for characters whose phenotypic selection is difficult. As an example, selecting a fertility restorer gene in segregating generations need test crossing before subsequent backcrossing. However, if such genes are tagged with molecular markers, desirable plants with fertility restorer genes (in heterozygous condition), can be identified and backcrossed. Similarly, if desirable genes conferring tolerance to abiotic stresses are tagged, these can be selected easily in segregating generations. Also, genetic markers can be assayed in nontarget areas such as growth chambers, greenhouses, or off-season nurseries, thus permitting more rapid progress. The efficiencies of scale and time accorded by DNA markers are valuable in breeding horticultural plants where fewer individuals might save several hectares and fewer generations may save several decades (Paterson et al. 1991). MAS has been used efficiently in (i) gene pyramiding (Huang et al. 1997; Singh et al. 2001; Bhatia et al. 2011; Kumar et al. 2013; Yasuda et al. 2015), (ii) marker-assisted alien introgressions (Jena et al. 1992; Brar and Dhaliwal 1997; Elkot et al. 2015), and (iii) simultaneous identification and pyramiding of QTLs from primitive cultivars and wild species (Tanksley and Nelson 1996; Tanksley et al. 1996; Fulton et al. 1997; Xiao et al. 1996, 1998).

### 1.7.2 Role of Molecular Markers in Germplasm Characterization

Molecular markers are also used in DNA fingerprinting for varietal identification, germplasm evaluation, phylogenetic and evolutionary studies, etc. The molecular marker-based DNA fingerprinting data are useful for the characterization of plant germplasm accessions, quantification of genetic diversity, and protection of proprietary germplasm (Smith and Smith 1992). Molecular markers have been utilized to distinguish closely related crop cultivars (Melchinger et al. 1991; Paull et al. 1998), in sex identification of dioecious plants (Parasnis et al. 1999). They are also used to understand evolutionary relationships within and between species, genera, or higher taxonomic groups. Such studies involve large number of markers to study similarities and differences among taxa (Paterson et al. 1991). Although phylogeny has been established for many plant species based on morphological markers, biochemical markers, and chromosome homology, the genetic markers have enhanced our understanding of phylogeny. In one important study, DNA-based markers enabled the designation of GG for *Oryza granulata* and HHJJ for *Oryza ridleyi* (Aggarwal et al. 1997).

### 1.7.3 Deployment of Molecular Markers in Plant Variety Protection and Registration

The current system of plant variety protection and registration using assessment of DUS (Distinctness, Uniformity, and Stability) characteristics predominantly relies upon morphological traits which are quite laborious, time-consuming, requires skills, expertise, and evaluation under special designs for most quantitative traits. With the advent of novel breeding technologies, new varieties differ only for few traits or at few loci which make the process of detecting distinctness in varieties, a challenging task. Even with increasing numbers of plant varieties, DUS testing is becoming quite expensive. In a review by Jamali et al. (2019), DNA-based molecular markers have been proposed to be a reliable alternative for conducting DUS testing. Molecular markers not only cut cost, time and labor but will also help in the proper sharing of Plant Breeder's Rights with an assessment of few distinct traits, particularly in essentially derived varieties.

## 1.8 Summary

The DNA-based molecular markers are widely recognized for their enormous potential in plant breeding and genetic studies. The past few years have seen remarkable developments in the field of molecular markers technology particularly with the emergence of NGS technologies. SNPs have become the choice of markers of present and future based on their genome-wide abundance, high polymorphism, amenability to high-throughput automation, and easier analysis. SNPs can be utilized in different genotyping platforms such as GBS, DArT, WGR, SNP arrays, and KASP. Any genotyping platform can be chosen based on the objective of the study and cost concerns. As an example, GBS, WGR, SNP arrays, and KASP can generate similar type of results for genetic diversity analysis; however, KASP and GBS could be cheaper than others. While the development of KASP will depend upon the availability of SNPs particularly in SNP databases, but GBS can be done without any previous information. It has been observed that these recent marker genotyping technologies have accelerated the crop improvement programs particularly in the identification and utilization of novel genes and QTLs.

## References

Aggarwal, R.K., Brar, D.S., and Khush, G.S. (1997). Two new genomes in the *oryza* complex identified on the basis of molecular divergence analysis using total genomic hybridization. *Molecular & General Genetics* 254: 1–12.

Akkurt, M., Welter, L., Maul, E. et al. (2007). Development of SCAR markers linked to powdery mildew (*Uncinula necator*) resistance in grapevine (*Vitis vinifera* L. and *Vitis* sp.). *Molecular Breeding* 19: 103–111.

Bachlava, E., Taylor, C.A., Tang, S. et al. (2012). SNP discovery and development of a high-density genotyping array for sunflower. *PLoS One* 7 (1): e29814.

Batley, J., Barker, G., O'Sullivan, H. et al. (2003). Mining for single nucleotide polymorphisms and insertions/deletions in maize expressed sequence tag data. *Plant Physiology* 132 (1): 84–91.

Bhatia, D. (2020). Advanced quantitative genetics technologies for accelerating plant breeding. In: *Accelerated Plant Breeding*, vol. I (eds. S.S. Gosal and S.H. Wani). Springer Verlag https://doi.org/10.1007/978-3-030-41866-3_5.

Bhatia, D., Sharma, R., Vikal, Y. et al. (2011). Marker assisted development of bacterial blight resistant, dwarf and high yielding versions of two traditional Basmati rice cultivars. *Crop Science* 51: 759–770.

Bhatia, D., Wing, R.A., and Singh, K. (2013). Genotyping by sequencing, its implications and benefits. *Crop Improvement* 40: 101–111.

Bhatia, D., Wing, R.A., Yu, Y. et al. (2018). Genotyping by sequencing of rice interspecific backcross inbred lines identifies QTLs for grain weight and grain length. *Euphytica* 214: 41. https://doi.org/10.1007/s10681-018-2119-1.

Botstein, D., White, R.L., Skolnick, M., and Davis, R.W. (1980). Construction of a genetic linkage map in man using restriction fragment length polymorphisms. *American Journal of Human Genetics* 32 (314): 331.

Brar, D.S. and Dhaliwal, H.S. (1997). Molecular markers and their application in crop improvement. In: *Proceedings 3rd Agricultural Science Congress* (eds. M.S. Bajwa et al.). New Delhi, India: Natl. Acad. Agr. Sci.

Brookes, A.J. (1999). The essence of SNPs. *Gene* 234: 177–186.

Chao, W.Z., Tang, C.H., Zhang, J.S. et al. (2018). Development of a stable SCAR marker for rapid identification of *Ganoderma lucidum* Hunong 5 cultivar using DNA pooling method and inter-simple sequence repeat markers. *Journal of Integrative Agriculture* 17 (1): 130–138.

Chen, H., Xie, W., He, H. et al. (2013). A high density SNP genotyping array for rice biology and molecular breeding. *Molecular Plant* 7: 541–553.

Ching, A., Caldwell, K.S., Jung, M.T., and Dolan, M. (2002). SNP frequency, haplotype structure and linkage disequilibrium in elite maize inbred lines. *BMC Genetics* 3 (1): 19. https://doi.org/10.1186/1471-2156-3-19.

Cho, Y.G., McCouch, S.R., Kuiper, M. et al. (1997). Integration of AFLP markers into an RFLP and SSLP map of rice (*Oryza sativa* L.). Rice Genet. *Newsletter* 14: 106–109.

Chung, Y.S., Choi, S.C., Jun, T.-H., and Kim, C. (2017). Genotyping by sequencing: a promising tool for plant genetics research and breeding. *Horticulture, Environment and Biotechnology* 58 (5): 425–431.

Collard, B.C.Y., Jahufer, M.Z.Z., Brouwer, J.B., and Pang, E.C.K. (2005). An introduction to markers, quantitative trait loci (QTL) mapping and marker-assisted selection for crop improvement: the basic concepts. *Euphytica* 142: 169–196.

Coryell, V.H., Jessen, H., Schupp, J.M. et al. (1999). Allele-specific hybridization markers for soybean. *Theoretical and Applied Genetics* 98: 690–696.

Ding, C. and Jin, S. (2009, 2003). High-throughput methods for SNP genotyping. In: *Single Nucleotide Polymorphisms, Methods in Molecular Biology*, vol. 578 (ed. A.A. Komar). Humana Press, A Part of Springer Science+Business Media, LLC https://doi.org/10.1007/978-1-60327-411-1_16.

Elkot, A.F.A., Chhuneja, P., Kaur, S. et al. (2015). Marker assisted transfer of two powdery mildew resistance genes *PmTb7A.1* and *PmTb7A.2* from *Triticum boeoticum* (Boiss.) to *Triticum aestivum* (L.). *PLoS One* 10 (6): e128297.

Fang, T., Lei, L., Li, G. et al. (2020). Development and deployment of KASP markers for multiple alleles of *Lr34* in wheat. *Theoretical and Applied Genetics* 133: 2183–2195.

Fehr, W. (1984). *Genetic Contributions to Yield Gains of Five Major Crop Plants.* Madison, Wisconsin: Special publication No. 7 Crop Science Society of America.

Fulton, T.M., Nelson, J.C., and Tanksley, S.D. (1997). Introgression and DNA marker analysis of *Lycopersicon peruvianum*, a wild relative of cultivated tomato into *L. esculentum* followed through three successive backcross generations. *Theoretical and Applied Genetics* 95: 895–902.

Ganal, M.W., Altmann, T., and Roder, M.S. (2009). SNP identification in crop plants. *Current Opinion in Plant Biology* 12 (2): 211–217.

Ganal, M.W., Durstewitz, G., Polley, A. et al. (2011). A large maize (*Zea mays* L.) SNP genotyping array: development and germplasm genotyping, and genetic mapping to compare with the B73 reference genome. *PLoS One* 6.

Gibbs, R.A. et al. (2009). Genome-wide survey of SNP variation uncovers the genetic structure of cattle breeds. *Science* 324: 528–532.

Grewal, S., Othmeni, M., Walker, J. et al. (2020). Development of Wheat-Aegilops caudata introgression lines and their characterization using genome-specific KASP markers. *Frontiers in Plant Science* 11: 606.

Grodzicker, T., Williams, J., Sharp, P., and Sambrook, J. (1975). Physical mapping of temperature sensitive mutants of adenovirus. *Cold Spring Harbor Symposia on Quantitative Biology* 39: 439–446.

Gupta, P.K. and Rustgi, S. (2004). Molecular markers from the transcribed/expressed region of the genome in higher plants. *Functional & Integrative Genomics* 4: 139–162.

Haanstra, J.P.W., Wye, C., Odinot, P. et al. (1999). An integrated high density RFLP-AFLP map of tomato based on two *Lycopersicon esculentum* x *L. pennellii* $F_2$ populations. *Theoretical and Applied Genetics* 99: 254–271.

Harushima, Y., Yano, M., Shomura, A. et al. (1998). A high density rice genetic linkage map with 2275 markers using a single $F_2$ population. *Genetics* 148: 479–494.

Helentjaris, T., Slocum, M., Wright, S. et al. (1986). Construction of genetic linkage maps in maize and tomato using restriction fragment length polymorphisms. *Theoretical and Applied Genetics* 61: 650–658.

Huang, N., Angeles, E.R., Domingo, J. et al. (1997). Pyramiding bacterial blight resistance gene in rice: marker assisted selection using RFLP and PCR. *Theoretical and Applied Genetics* 95: 313–320.

Hunt, G.J. (1997). Construction of linkage maps with RAPD markers. In: *Fingerprinting Methods Based on Arbitrarily Primed PCR* (eds. M.R. Micheli and R. Bova). Berlin, Heidelberg: Springer Lab Manuals. Springer https://doi.org/10.1007/978-3-642-60441-6_22.

Hussain, W., Baenziger, P.S., Belamkar, V. et al. (2017). Genotyping-by-Sequencing derived high-density linkage map and its application to QTL mapping of flag leaf traits in bread wheat. *Scientific Reports* 7: 16394.

IRGSP (2005). The map-based sequence of the rice genome. *Nature* 436: 793–800. (Suppl. Table 18).

Jaccoud, D., Peng, K., Feinstein, D., and Kilian, A. (2001). Diversity arrays: a solid state technology for sequence information independent genotyping. *Nucleic Acids Research* 29: e25.

Jamali, S.H., Cockram, J., and Hickey, L.T. (2019). Insights into deployment of DNA markers in plant variety protection and registration. *Theoretical and Applied Genetics* 132: 1911–1929.

Jena, K.K., Khush, G.S., and Kochert, G. (1992). RFLP analysis of rice introgression lines. *Theoretical and Applied Genetics* 84: 608–616.

Joshi, S.P., Ranjekar, P.K., and Gupta, V.S. (1999). Molecular markers in plant genome analysis. *Current Science* 77: 230–240.

Kagale, S., Koh, C., Clarke, W.E. et al. (2016). Analysis of genotyping by sequencing data. *Methods in Molecular Biology* 1374: 269–284.

Kang, J.-W., Shin, D., Cho, J.H. et al. (2019). Accelerated development of rice stripe virus-resistant, near-isogenic rice lines through marker-assisted backcrossing. *PLoS One* 14 (12): e0225974.

Kasai, K., Morikawa, Y., Sorri, V.A. et al. (2000). Development of SCAR markers to the PVY resistance gene *Ryadg* based on a common feature of plant disease resistance genes. *Genome* 43: 1–8.

Kumar, A., Dixit, S., and Henry, A. (2013). Marker-assisted introgression of major QTLs for grain yield under drought in rice. In: *Translational Genomics for Crop Breeding Volume 2: Abiotic Stresses, Yield and Quality* (eds. R.K. Varshney and R. Tuberosa). Wiley.

Kumar, K., Sarao, P.S., Bhatia, D. et al. (2018). High-resolution genetic mapping of novel brown planthopper resistance locus, *Bph34* in *Oryza sativa* X *Oryza nivara* (Sharma &

Shastry) derived interspecific $F_2$ population. *Theoretical and Applied Genetics* 131: 1163–1171.
Laucou, V., Haurogne, K., Ellis, N., and Rameau, C. (1998). Genetic mapping in pea. 1. RAPD-based genetic linkage map of *Pisum sativum*. *Theoretical and Applied Genetics* 97: 905–915.
Li, H., Vikram, P., Singh, R.P. et al. (2015). A high density GBS map of bread wheat and its application for dissecting complex disease resistance traits. *BMC Genomics* 16: 216.
Litt, M. and Luty, J.M. (1989). A hypervariable microsatellite revealed by in vitro amplification of a dinucleotide repeat within the cardiac muscle actin gene. *American Journal of Human Genetics* 44: 397–401.
Liu, Z., Sun, Q., Ni, Z., and Yang, T. (1999). Development of SCAR markers linked to the *Pm21* gene conferring resistance to powdery mildew in common wheat. *Plant Breeding* 118: 215–219.
Lotti, C., Salvi, S., Pasqualone, A. et al. (2000). Integration of AFLP markers into an RFLP-based map of durum wheat. *Plant Breeding* 119 (5): 393–401.
Makhoul, M., Rambla, C., Voss-Fels, K.P. et al. (2020). Overcoming polyploidy pitfalls: a user guide for effective SNP conversion into KASP markers in wheat. *Theoretical and Applied Genetics* 133 (8): 2413–2430.
Mardis, E.R. (2008). Next generation DNA sequencing methods. *Annual Review of Genomics and Human Genetics* 9: 387–402.
Matukumalli, L.K., Lawley, C.T., Schnabel, R.D. et al. (2009). Development and characterization of a high density SNP genotyping assay for cattle. *PLoS One* 4: e5350.
McCouch, S.R., Teytelman, L., Xu, Y. et al. (2002). Development and mapping of 2240 new SSR markers for rice (*Oryza sativa* L.). *DNA Research* 9: 199–207.
McCouch, S.R., Zhao, K., Wright, M. et al. (2010). Development of genome-wide SNP assays for rice. *Breeding Science* 60: 524–535.
McCouch, S.R., Wright, M.H., Tung, C.-W. et al. (2015). Open access resources for genome-wide association mapping in rice. *Nature Communications* 7: 10532.
Melchinger, A.E., Messemer, M.M., Lee, M. et al. (1991). Diversity and relationship among US maize inbreds revealed by RFLPs. *Crop Science* 31: 669–678.
Milbourne, D., Meyer, R.C., Collins, A.J. et al. (1998). Isolation, characterization and mapping of simple sequence repeat loci in potato. *Molecular & General Genetics* 259: 233–245.
Mir, R.R., Hiremath, P.J., Lizarazu-Riera, O., and Varshney, R.K. (2013). Evolving molecular marker technologies in plants: from RFLPs to GBS. In: *Diagnostics in Plant Breeding*, https://doi.org/10.1007/978-94-007-5687-8__11 (eds. T. Lübberstedt and R.K. Varshney). Dordrecht: Springer Science Business Media.
Mullis, K., Faloona, F., Scharf, S. et al. (1986). Specific enzymatic amplication of DNA in vitro: the polymerase chain reaction. *Cold Spring Harbor Symposium on Quantitative Biology.* 51: 263–273.
Nguyen, K.L., Grondin, A., Courtois, B., and Gantet, P. (2019). Next-Generation sequencing accelerates crop genome discovery. *Trends in Plant Science* 24: 263–274.
Olsen, M., Hood, L., Cantor, C., and Botstein, D. (1989). A common language for physical mapping of the human genome. *Science* 245: 1434–1435.
Orjuela, J., Garavito, A., Bouniol, M. et al. (2010). A universal core genetic map for rice. *Theoretical and Applied Genetics* 120: 563–572.

Parasnis, A.S., Ramakrishna, W., Chowdari, K.V. et al. (1999). Microsatellite (GATA)n reveals sex-specific differences in papaya. *Theoretical and Applied Genetics* 99: 1047–1052.

Paterson, A.H., Tanksley, S.D., and Sorrels, M.E. (1991). DNA markers in crop improvement. *Advances in Agronomy* 46: 39–90.

Paull, J.G., Chalmers, K.J., Karakousis, A. et al. (1998). Genetic diversity in Australian wheat varieties and breeding material based on RFLP data. *Theoretical and Applied Genetics* 96: 435–446.

Poland, J. and Rife, T.W. (2012). Genotyping-by-sequencing for plant breeding and genetics. *Plant Genitics* 5: 92–102.

Poland, J.A., Brown, P.J., Sorrells, M.E., and Jannink, J.-L. (2012). Development of high-density genetic maps for barley and wheat using a novel two-enzyme genotyping-by-sequencing approach. *PLoS One* 7 (2): e32253.

Puritz, J.B., Hollenbeck, C.M., and Gold, J.R. (2014). dDocent: a RADseq, variant-calling pipeline designed for population genomics of non-model organisms. *Peer Journal* 2: e431.

Qi, L., Echalier, B., Friebe, B., and Gill, B.S. (2003). Molecular characterization of a set of wheat deletion stocks for use in chromosome bin mapping of ESTs. *Functional & Integrative Genomics* 3: 39–55. https://doi.org/10.1007/s10142-002-0063-5.

Ramos, A.M., Crooijmans, R.P.M.A., Affara, N.A. et al. (2009). Design of a high density SNP genotyping assay in the pig using SNPs identified and characterized by next generation sequencing technology. *PLoS One* 4: e6524.

Richard, I. and Beckman, J.S. (1995). How neutral are synonymous codon mutations? *Nature Genetics* 10: 259.

Roder, M.S., Korzun, V., Wendehake, K. et al. (1998). A microsatellite map of wheat. *Genetics* 149: 2007–2023.

Rosas, J.E., Bonnecarrère, V., and Pérez de Vida, V.F. (2014). One-step, codominant detection of imidazolinone resistance mutations in weedy rice (*Oryza sativa* L.). *Electronic Journal of Biotechnology* 17: 95–101.

Sachidanandam, R., Weissman, D., Schmidt, S.C. et al. (2001). A map of human genome sequence variation containing 1.42 million single nucleotide polymorphisms. *Nature* 409 (6822): 928–933.

Sanger, F., Nicklen, S., and Coulson, A.R. (1977). DNA sequencing with chain chain-terminating inhibitors. *Proceedings of the National Academy Sciences of the United States of America* 74: 5463–5467.

Semagn, K., Bjornstad, A., and Ndjiondjop, M.N. (2006). An overview of molecular marker methods for plants. *African Journal of Biotechnology* 5 (25): 2540–2568.

Semagn, K., Babu, R., Hearne, S., and Olsen, M. (2013). Single nucleotide polymorphism genotyping using *Kometitive allele* specific PCR (KASP): overview of the technology and its application in crop improvement. *Molecular Breeding* 33: 1–14.

Sharopova, N., McMullen, M.D., Schultz, L. et al. (2002). Development and mapping of SSR markers for maize. *Plant Molecular Biology* 48: 463–481.

Shendure, J.A., Porreca, G.J., Church, G.M. et al. (2011). Overview of DNA sequencing strategies. *Current Protocols in Molecular Biology* 7 (1): 1–7.1.23. https://doi.org/10.1002/0471142727.mb0701s96.

Sidhu, G.K., Rustgi, S., Shafqat, M.N. et al. (2008). Fine structure mapping of a gene-rich region ofwheat carrying *Ph1*, a suppressor of crossing over between homoeologous chromosomes. *Proceedings of the National Academy Sciences of the United States of America* 15: 5815–5820.

Sim, S.-C., Durstewitz, G., Plieske, J. et al. (2012). Development of a large SNP genotyping array and generation of high-density genetic maps in tomato. *PLoS One* 7 (7): e40563. https://doi.org/10.1371/journal.pone.0040563.

Singh, B.D. and Singh, A.K. (2015). *Marker-Assisted Plant Breeding: Principles and Practices*. https://doi.org/10.1007/978-81-322-2316-0. Springer (India) Pvt. Ltd.

Singh, S., Sidhu, J.S., Huang, N. et al. (2001). Pyramiding three bacterial blight resistance genes (*xa*5, *xa*13 and *Xa*21) using marker assisted selection into indica cultivar PR106. *Theoretical and Applied Genetics* 102: 1011–1015.

Smith, J.S.C. and Smith, O.S. (1992). Fingerprinting crop varieties. *Advances in Agronomy* 47: 85–140.

Sobrino, B., Briona, M., and Carracedoa, A. (2005). SNPs in forensic genetics: a review on SNP typing methodologies. *Forensic Science International* 154: 181–194.

Soleimani, V.D., Baum, B.R., and Johnson, D.A. (2003). Efficient validation of single nucleotide polymorphisms in plants by allele-specific PCR, with an example from barley. *Plant Molecular Biology Reporter* 21: 281–288.

Somers, D.J., Isaac, P., and Edwards, K. (2004). A high-density microsatellite consensus map for bread wheat (*Triticum aestivum* L.). *Theoretical and Applied Genetics* 109: 1105–1114.

Song, Q.J., Shi, J.R., Singh, S. et al. (2005). Development and mapping of microsatellite (SSR) markers in wheat. *Theoretical and Applied Genetics* 110: 550–560.

Su, A., Song, W., Xing, J. et al. (2016). Identification of genes potentially associated with the fertility instability of S-Type cytoplasmic male sterility in maize via bulked segregant RNA-Seq. *PLoS One* 11 (9): e0163489.

Su, C., Wang, W., Gong, S. et al. (2017). High density linkage map construction and mapping of yield trait QTLs in maize (*Zea mays*) using the genotyping-by-sequencing (GBS) technology. *Frontiers in Plant Science* 8: 706. https://doi.org/10.3389/fpls.2017.00706.

Takagi, H., Tamiru, M., Abe, A. et al. (2015). MutMap accelerates breeding of a salt-tolerant rice cultivar. *Nature Biotechnology* 33: 445–449.

Tanksley, S.D. and Nelson, J.C. (1996). Advanced backcross QTL analysis: a method for simultaneous discovery and transfer of valuable QTLs from unadapted germplasm into elite breeding lines. *Theoretical and Applied Genetics* 92: 191–203.

Tanksley, S.D., Ganal, M.W., Prince, J.P. et al. (1992). High density molecular linkage maps of tomato and potato genomes. *Genetics* 132: 1141–1160.

Tanksley, S.D., Grandillo, S., Fulton, T.M. et al. (1996). Advanced backcross QTL analysis in a cross between an elite processing line of tomato and its wild relative *L. pimpinellifolium*. *Theoretical and Applied Genetics* 92: 213–224.

Temnykh, S., Park, W.D., Ayres, N. et al. (2000). Mapping and genome organization of mirosatellite sequences in rice (*Oryza sativa*) Theor. *Applied Genetics* 100: 697–712.

Tung, C.W., Zhao, K., Wright, M.H. et al. (2010). Development of a research platform for dissecting phenotype genotype associations in rice (*Oryza* spp.). *Rice* 3: 205–217.

Vos, P., Hogers, R., Bleeker, M. et al. (1995). AFLP: a new technique for DNA fingerprinting. *Nucleic Acids Research* 23: 4407–4414.

Wang, D.G., Fan, J.B., Siao, C.J. et al. (1998). Large-scale identification, mapping, and genotyping of single-nucleotide polymorphisms in the human genome. *Science* 280: 1077–1082.

Wickland, D.P., Battu, G., Hudson, K.A. et al. (2017). A comparison of genotyping-by-sequencing analysis methods on low-coverage crop datasets shows advantages of a new workflow, GB-eaSy. *BMC Bioinformatics* 18: 586.

Williams, J.G.K., Kublelik, A.R., Livak, K.J. et al. (1990). DNA polymorphism's amplified by arbitrary primers are useful as genetic markers. *Nucleic Acids Research* 18: 6531–6535.

Xiao, J., Grandillo, S., Ahn, S.N. et al. (1996). Genes from wild rice improved yield. *Nature* 384: 223–224.

Xiao, J., Li, J., Grandillo, S. et al. (1998). Identification of trait improving QTL alleles from a wild rice relative *Oryza rufipogon*. *Genetics* 150: 899–909.

Yang, G., Chen, S., Chen, L. et al. (2019). Development of a core SNP arrays based on the KASP method for molecular breeding of rice. *Rice* 12: 21.

Yasuda, N., Mitsunaga, T., Hayashi, K. et al. (2015). Effects of pyramiding quantitative resistance genes *pi21, Pi34*, and *Pi35* on rice leaf blast disease. *Plant Disease* 99: 904–909.

Zhang, X., Yang, Q., Rucker, E. et al. (2017). Fine mapping of a quantitative resistance gene for gray leaf spot of maize (*Zea mays* L.) derived from teosinte (*Z. mays* ssp. *parviglumis*). *Theoretical and Applied Genetics* 130: 1285–1295.

Zhou, Z., Zhang, C., Zhou, Y. et al. (2016). Genetic dissection of maize plant architecture with an ultra-high density bin map based on recombinant inbred lines. *BMC Genomics* 17: 178. https://doi.org/10.1186/s12864-016-2555-z.

Zhu, M., Liu, D., Liu, W. et al. (2017). QTL mapping using an ultra-high-density SNP map reveals a major locus for grain yield in an elite rice restorer R998. *Scientific Reports* 7: 10914. https://doi.org/10.1038/s41598-017-10666-7.

# 2

# High-throughput Genotyping Platforms

*Sandhya Sharma[1], Kuldeep Kumar[2], Kishor Tribhuvan[3], Rita[1], Sandeep Kumar[4], Priyanka Jain[1], Swati Saxena[1], Joshita Vijayan[1], Harsha Srivastava[1], and Kishor Gaikwad[1]*

[1] ICAR – National Institute for Biotechnology, New Delhi, India
[2] ICAR – Indian Institute of Pulses Research, Kanpur, Uttar Pradesh, India
[3] ICAR – Indian Institute of Agricultural Biotechnology, Ranchi, Jharkand, India
[4] Xcelris Lab Pvt Ltd., Ahmedabad, Gujarat, India

## 2.1 Introduction

Single-nucleotide polymorphisms (SNPs) are the most common genetic variations present in the genome. These are the genetic loci where variation exists for a single nucleotide. Even for variation to be called as SNPs, it should be present in more than 1% of the individuals within a population. SNPs may be present as change or InDel of a single base which has originated mainly due to mutation or replication error. SNPs may fall within coding, non-coding as well as intergenic regions. Due to degeneracy of the genetic code, it is not necessary that SNPs within a coding region will change the amino acid sequence of the protein. A synonymous (often silent mutation) SNP is one in which both types of alleles result in the same polypeptide sequence, whereas nonsynonymous SNPs create a distinct polypeptide sequence. A nonsynonymous change further may be missense or nonsense. The missense mutation results in a different amino acid, while a nonsense mutation will result in a premature stop codon, ultimately forming a truncated peptide. SNPs that are not in protein-coding regions may still affect the gene splicing, transcription factor binding resulting in a change in expression dynamics of the gene. As SNPs can occur throughout the genome, they offer a good advantage of genome-wide coverage and higher frequency which assist high-resolution mapping compared to other marker systems. Automated genotyping techniques, high reproducibility, easy cost of genotyping, presence within a genic region, and gel-free methodology are some advantages of SNP which other markers lack. Besides several advantages, SNPs also offer certain disadvantages, such as detection of SNPs mostly requires sequencing of the DNA, high-throughput analysis is required in postsequencing steps, also the SNP-based genotyping experiments are always associated with high equipment cost.

*Genotyping by Sequencing for Crop Improvement*, First Edition.
Edited by Humira Sonah, Vinod Goyal, S.M. Shivaraj, and Rupesh K. Deshmukh.

In the last couple of years, the detection and exploitation of DNA polymorphisms with molecular markers and their utilization in breeding programs have transformed the pace and accuracy of these experiments. Ultimately, a complete shift from phenotype-based breeding to molecular breeding has been evident. With the advancement of NGS technology, there is a rapid rise in the application of SNP-based genotyping. The popularity of these programs depends on the cost incurred as well as the ease of their application. Also, these SNP-based genotyping platforms offer many vital advantages over other marker systems like reproducibility, high marker density, automation, and many more. Many modern breeding techniques, such as genome-wide association studies (GWAS) and genomic selection (GS), necessitate a large number of markers, which is difficult to achieve using PCR-based markers like SSR.

## 2.2 SNP Genotyping Platforms

Accurate detection of SNP is the first and foremost step for its mainstream application in breeding experiments. Besides the sequencing techniques used, reads length, coverage, and robust assembly approach are the control points for sequencing errors. Sequencing technologies with long reads length (in kb's) like PacBio single-molecule real-time (SMRT) sequencing (Eid et al. 2009), Illumina TruSeq synthetic long-read technology (McCoy et al. 2014), and Oxford nanopore sequencing (Branton et al. 2010) are available to develop the complete and correct reference genome assembly of any organism, which sort out many of the problems in SNP discovery. Not only in genome sequencing, these long-read methodologies have been successfully deployed for SNP mining and provide a more precise understanding of the complexity of isoforms for organisms lacking a reference (Piriyapongsa et al. 2018). Once a reference genome is available, the application of SNPs in its breeding program becomes comparatively easier. A concise layout for SNP-based breeding experiment is provided (Figure 2.1).

### 2.2.1 SNP Genotyping Versus SNP Discovery

First, a diverse set of genotypes are selected based on phenotype for the creation of SNP panel with the concept that as these genotypes are hypervariable in several phenotypes, the

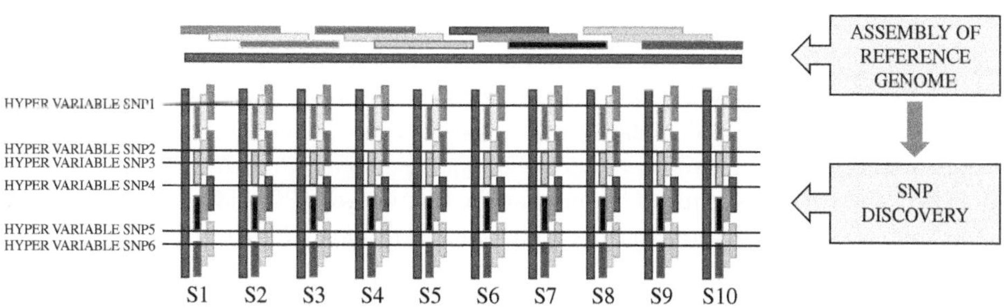


**Figure 2.1** A pipeline for SNP discovery (S1–S10 are different diverse accession of the same species to which the reference belongs).

same should reflect from their sequences. The reference genome sequence is used for reference-based assembly of the reads generated from sequencing of these genotypes (S1–S10 in Figure 2.1). A set of hypervariable loci are filtered through some statistical analysis. Once discovered, these can be genotyped in any other accession, which is known as SNP genotyping, as genotyping is the process of knowing the allelic status of an organism at a particular locus. The process till the identification of hypervariable loci is known as SNP discovery. The discovery of SNPs in crops prompted the development of the simple and low-cost genotyping platforms. A flowchart related to different SNP genotyping technologies has been provided (Figure 2.2).

### 2.2.2 Types of SNP Genotyping Platforms

Although multiple classification systems are based on the reaction, fluorescence used, PCR usage, etc., but broadly all SNP genotyping methods can be categorized into two groups based on the detection mechanism, i.e. Allelic discrimination and Allelic detection. Both of these groups can be further classified based on the reaction chemistry and other variables.

#### 2.2.2.1 Allelic Discrimination

Allelic discrimination depends on allele-specific biochemical reactions where the alternate alleles are discriminated based on their extension (primer extension methods), hybridization (array for both alleles), and differential enzymatic cleavage pattern. A broader classification for different allelic discrimination method is mentioned below:

##### *2.2.2.1.1 PCR-Free Genotyping Technology*

Among the conventional molecular markers, RFLP is based on the detection of mutation at the restriction site which is usually SNP. Except for RFLP, which relies on restriction digestion followed by detection of digested DNA via southern blotting, other majority of the SNP genotyping methods require amplification of the SNP containing genomic region prior to polymerase chain reaction (PCR). This preamplification step is inevitable from most of the genotyping techniques, despite the fact that the PCR process is relatively expensive.

**Invader Assay** Several PCR-free genotyping methods have been developed to date, one of such methods is invader assay (Lyamichev et al. 1999). Invader assay is based on nucleotide-specific cleavage by a structure-specific "flap" endonuclease, in the presence of an invading oligonucleotide. This reaction is followed by a subsequent secondary reaction that generates allele-specific signals using fluorescence resonance energy transfer (FRET) oligonucleotide cassettes. Both of the reactions in invader assay are a "single vessel reaction" as well as isothermal reaction, hence are easily automatable. Although the whole process is highly accurate and automatable, the need of a large amount of DNA is one of its major drawbacks which can be omitted by coupling the reaction with PCR. Many other PCR-free methodologies have been proposed like padlock probe ligation (Nilsson et al. 1994) the rolling circle DNA amplification (RCA) process (Baner et al. 1998). Although these gel-free methods can be automated, and are highly accurate, their application is limited to only a small number of SNPs, hence genotyping of large numbers of SNPs covering the whole genome for genomic assisted breeding is very difficult.

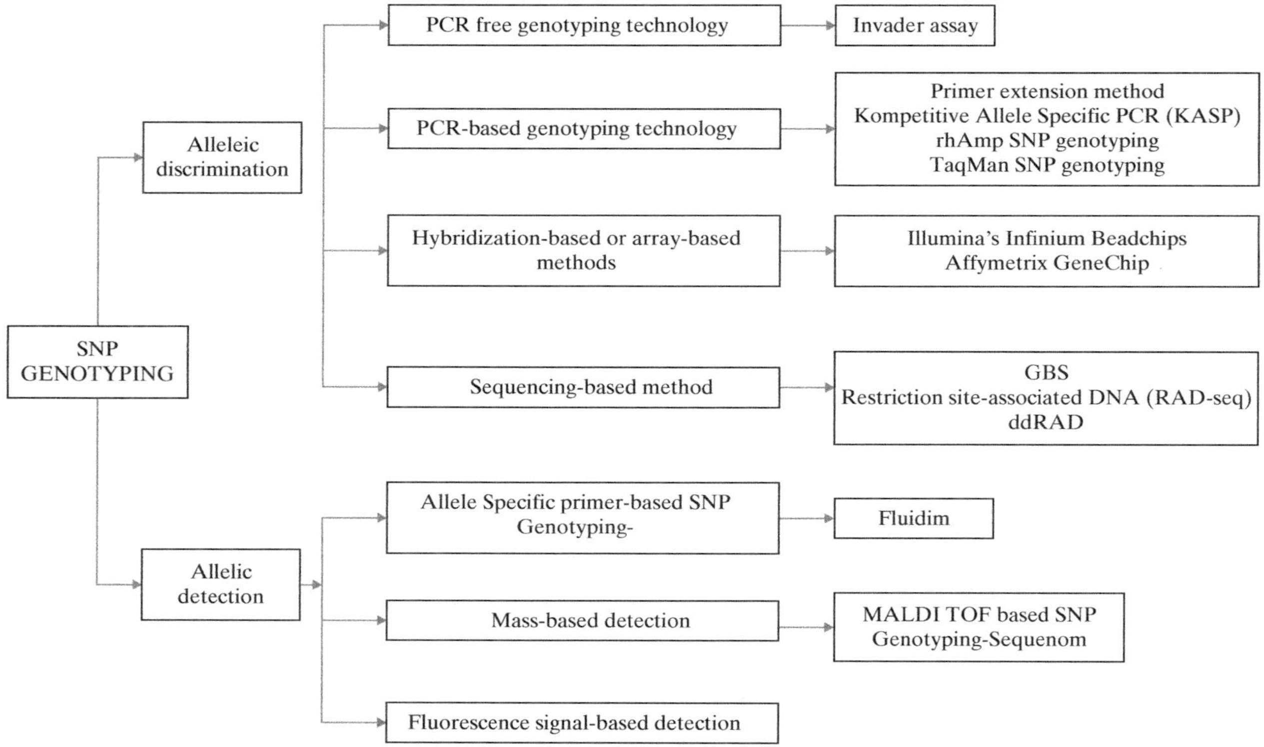


**Figure 2.2** A schematic representation of different SNP genotyping technologies.

The fluorescence-based detection systems were further advanced and were designed in such a manner to detect change in polarization of fluorescence which is caused by the decreased mobility of a fluorophore due to an increase in molecular mass. During SNP genotyping with this method, the allele-specific incorporation of fluorescently labeled ddNTPs can be detected as an increase in polarized fluorescence (Chen et al. 1999). As this amendment does not require fluorescently labeled primer and also a simple detection platform for genotyping is required, so the cost incurred is decreased. The addition of one extra step (PCR and post-PCR processing steps) compared to single-step methods makes it more complicated. Further, these fluorescence polarization-based methods are usually unsuitable for high-throughput genotyping because of the lack of multiplexing capability.

#### *2.2.2.1.2 PCR-based Detection System*

**Primer Extension Method** PCR-based primer discrimination method simply exploits the accuracy of polymerase to exclude nonspecific nucleotides during extension. In this method, two forward primers are designed from the locus differing in the genotypes while a common reverse primer is used. Both forward primers are specific for their respective alleles. Two different reactions are set for a sample where each reaction contains a different forward primer. Amplification will take place only in any one of these reactions otherwise both DNA is homozygous for the concerned loci.

Taqman (Livak 1999), molecular beacon (Thelwell et al. 2000), and scorpion assay (Whitcombe et al. 1999) are detection systems which are performed in a microtiter plate and based on fluorescent detection systems. Taqman and molecular beacon, both rely on allele-specific hybridization of oligonucleotide probes to DNA during PCR for allelic discrimination, while scorpion assay can use either allele-specific PCR or allele-specific hybridization chemistry for allelic discrimination. All of these microtiter reactions are end point assays and all reagents and genomic DNA are mixed at the beginning, and the fluorescent signal is detected. They are straightforward to accomplish since they do not require a separate preamplification step or intermediary processing.

**TaqMan SNP Genotyping** TaqMan SNP genotyping assay utilizes PCR amplification principle for discriminating the samples based on the presence of the SNPs where forward and reverse primers along with Taq polymerase and two SNP specific fluorescent-labeled probes are used for the amplification. The first probe, specific to allele 1 is labeled with VIC green fluorescent dye, while the second probe is specific to allele 2 labeled with FAM blue fluorescent dye. During the amplification process, probe specific to allele hybridize with template DNA and 5′ to 3′ exonuclease activity of *Taq* polymerase cleave the probe and remove the fluorescent dye. The dye was further detected by the detector which gives allele-specific signals. If the sample is homozygous for allele 1, green fluorescence signals from VIC dye are detected by the detector, while the sample is homozygous for allele 2, blue fluorescence signals from FAM dye are detected by the detector. Under the heterozygous condition, there should be a roughly equal signal from both the dyes that produce cyan color signals.

**Kompetitive Allele-Specific PCR (KASP)** KASP assay is a fluorescence-based genotyping variant of PCR. It utilized allele-specific primer for amplification and generation fluorescence signals utilizing dsDNA template, two allele-specific forward primers, common reverse primer where the forward primers are fluorescently labeled having sequences identical to that of 5′ tail of allele-specific primers. The first round of PCR amplification is carried out using allele-specific forward primer and common reverse primer. It allows the extension of only perfectly matched primes to result in an amplified product, while mismatched primers remain unamplified. In the third PCR cycle, fluorescence-labeled forward primer and common reverse primer bind to PCR-amplified products and extend the amplification reaction. During the primer annealing process quencher molecules present in complementary strands goes away and fluorescence starts emitting which are further detected by the detector.

**rhAmp SNP Genotyping** rhAmp SNP genotyping is a PCR assay that uses a unique two-enzyme system coupled with RNA–DNA hybrid primers to detect the target SNPs. It required a double-stranded DNA template, two allele-specific forward primers containing RNA base and blocking moiety at 3′ end, common reverse primer containing RNA base and blocking moiety at 3′ end, RNase H2 enzyme, universal forward primer and universal probe having fluorescence-labeled at 5′ end and quencher molecule at 3′ end. It involves an allele-specific amplification process, where allele 1 specific forward primer 1 perfectly binds with SNP site and common reverse primer binds at the 5′ end of the complementary strand. RNase H2 binds cleaves at the RNA base of primer and removes blocking moiety, which allows extension of primers and amplification of both the strand. In the subsequent amplification cycle, the universal primer and universal probe 1 site merged into an amplified product. In PCR cycle 3, universal probe 1 and universal primer 1 bind to the PCR cycle 2 amplified product. Universal forward primer start extension, 5′ to 3′ exonuclease activity of Taq polymerase cleaves the probe and removes fluorescence which was detected by the detector. When allele 2 specific primer 2 binds to the template DNA which generates a universal prob 2-binding sites labeled with different fluorescence molecules. In this way, different genotypes were clustered based on fluorescence signals detected.

#### *2.2.2.1.3 Hybridization-based Platforms or Array-based Methods*

Hybridization methods exploit differences in dsDNA thermal stability to discriminate between perfectly matched and mismatched target-probe combinations in order to achieve allelic discrimination. In the array-based system, customized probes from the flanking sequence of the SNP are used for genotyping. For a particular locus, the number of probes to be used depends on the number of alleles present. SNPs are usually biallelic in nature, hence mostly two probes distinguishing both alleles are used for each locus. The probes in the SNP array are hybridized with DNA samples to determine the specific alleles of all SNPs on the array for the hybridized DNA sample (LaFramboise 2009). This kind of array has been successfully used in several crops like rice (Thomson et al. 2017; McCouch et al. 2017) maize (Unterseer et al. 2014), wheat (Wang et al. 2014), and barley (Bayer et al. 2017). The number of SNP genotyped at one go has been scaled up to 700 k in rice (Mccouch et al. 2016) and 820 k in case of wheat (Winfield et al. 2016).

There are several well-established genotyping platforms which can be used for customizing a new array, two of which, i.e., Illumina Infinium Beadchip and Affymetrix GeneChip are most widely used. Further, a new concept of customized array which has a highly specific set of probes used to validate known SNPs in subsets of population is emerging nowadays. Both of the above mention genotyping technology are briefly described below

a) **Illumina's Infinium Beadchips**: Infinium array was capable of genotyping only 10k–100k SNPs initially which later has been increased up to 1000k SNPs. It includes first a whole-genome amplification step, followed by hybridization to bead arrays of 50 bp long capture probes. The locus-specific sequences include an allele-specific 3′ terminal base. An allele-specific primer extension reaction is used to incorporate biotin-labeled nucleotide-based or single-base primer extension step detection systems for positive detection.
b) **Affymetrix GeneChip**: The GeneChip assays are based on allelic discrimination by direct hybridization of genomic DNA to allele-specific oligonucleotides (25 mers probes). These oligonucleotides represent all allelic combinations for the particular loci, i.e. perfect match or mismatch probes to each SNP. The range of SNPs to be genotyped ranges from 10k to 2000k SNPs. As gene chip requires high-quality DNA, so for GeneChip assays to work efficiently, the complexity of the genomic DNA must be reduced through digestion with restriction endonucleases and fractionation. In previous years, researchers used a high-density oligonucleotide probes array to genotype the entire genome, paving the way for genome-wide association studies (GWAS) (Kennedy et al. 2003; Matsuzaki et al. 2004; McGall and Christians 2002). Affymetrix GeneChip array contains a large number of synthetic fragments (25-mer) probes immobilized on a solid substrate. First, the denatured ssDNA are hybridized with these probes, which takes place in a highly specific manner as noncomplementary alleles will not hybridize. Subsequently, the fragments of noncomplementary strands are washed away. The hybrid strands containing probes are stained and then the gene chip is subjected to the CCD imaging device for scanning. Each SNP is represented by a probe set that contains multiple probe pairs. The probe pairs differ in the location of SNP within the oligonucleotide sequence (5 location's probes are selected). For each position, probes are included from the sense and antisense strand. Hence, the total probes for each allele are 40 probes per SNP.

Although DNA array-based genotyping has completely revolutionized the field of RNA sequencing and expression analysis, but comparably it has lesser use for SNP genotyping mainly because of the difficulty in obtaining a good signal/noise ratio in allele-specific hybridization. The resolution of discrimination between completely matched/mismatched oligonucleotides in the array is totally based on high-throughput fluorescence detection system which is further complicated with an increase in the number of SNPs to be genotyped under a single condition. This specificity compromises genotype call rates and accuracy. Generally, effectiveness in allele differentiation depends on length and sequence of the probe, location of SNP in the probe, and hybridization conditions (Table 2.1).

**Table 2.1** Customized SNP array details in plant species.

| Crop | Species | Array fsize | References | Trait |
|---|---|---|---|---|
| Pigeonpea | (*Cajanus cajan*) | 62K | Singh et al. (2020) | Genetic diversity |
| Wheat | (*Triticum*) | 55K | Zhang et al. (2019) | Adult-plant resistance to leaf and stripe rust |
| Apple | (*Malus domestica*) | 480K | Bianco et al. (2016) | Phenology, fruit quality, disease resistance, or drought tolerance |
| Apple | (*Malus domestica*) | 8K | Chagné et al. (2012) | Quantitative traits |
| Rice | (*Oryza sativa*) | 50K | Singh et al. (2015) | Genetic diversity |
| Rice | (*Oryza sativa*) | 44 K | Zhao et al. (2011) | Plant morphology, grain quality, plant development |
| Rye | (*Secale cereale*) | 5 K | Li et al. (2011) | Diversity analysis |
| Barley | (*Hordeum vulgare*) | 50 K | Bayer et al. (2017) | Evaluation and use of barley genetic resources |
| Sweet Cherry | (*Prunus avium*) | 6K | Peace et al. (2012) | Fruit taste |
| Pear | (*Prunus*) | 70K | Montanari et al. (2019) | Genetic diversity studies |
| Potato | (*Solanum tuberosum*) | 22K | Khlestkin et al. (2019) | Starch phosphorylation |
| Pear | (*Pyrus*) | 200K | Li et al. (2019) | Flowering time and candidate genes linked to the size of fruit |
| Walnut | (*Juglans regia*) | 700K | Marrano et al. (2019) | Genetic diversity of germplasm |
| Cotton | (*Gossypium barbadense*) | 63K | Kumar et al. (2019) | Fiber quality |

#### *2.2.2.1.4 Sequencing-based Platforms*

The simplest method of genotyping is whole-genome resequencing of genotypes followed by the reference-based assembly for variant calling. But as the majority of the genome is noncoding, these noncoding regions are also reflected in the variant file generated by the mapping of the reads which in turn increases the complexity of analysis as well as interpretation. Reduced representation approaches like RAD-seq and GBS methods have solved this issue as in spite of sequencing the whole genome, these methods are confined to coding/genic regions only. For this, we first digest the DNA with methylation-sensitive restriction enzyme with the hypothesis that DNA is more methylated in the noncoding region hence the digestion will be limited to coding regions only. The restricted ends generated are then ligated with adapters and sequenced. Other widely used sequencing-based technologies are exome sequencing (Exom-seq) and double digest RAD-seq (ddRAD-seq).

**Restriction Site-Associated DNA (RAD-seq)** Restriction site-associated DNA (RAD) is a next-generation sequencing-based SNP genotyping method, which allows simultaneous discovery and scoring of thousands of SNP markers in hundreds of individuals. The basic principle behind RAD is the identification of genomic SNPs adjacent to enzyme restriction sites through sequencing. It requires genomic DNA, restriction enzymes, P1 adaptor containing Illumina sequencing forward primer and barcode information, and P2 adaptor containing reverse complement of the reverse amplification primer site. The RAD sequencing library is prepared by digesting genomic DNA of genotypes using restriction enzymes and P1 adaptor is ligated to the compatible ends of the fragments. The adaptor-ligated fragments are subsequently pooled, randomly sheared, and a specific size fraction is selected following electrophoresis. The DNA fragments are then ligated with P2 adapter and sequenced using Illumina platform (Figure 2.3). The obtained RAD data are analyzed using bioinformatics tools like stack, PyRAD, dDocent, GATK HaplotypeCaller, GATK unifieldGenotyper, samtools mpileup, etc.

One enzyme digestion followed by random sharing causes high DNA loss during RAD sequencing. For organisms without a reference genome, a significant portion of the RAD-Seq data has been discarded due to sequence read errors and the presence of variable sites. To overcome this problem, slight modification in the RAD sequencing protocol has been carried out. Use of two restriction enzymes instead of a single restriction enzyme and removal of the random sharing steps is adopted. The modified method is renamed as double digest RAD (ddRAD).

**Genotyping-by-Sequencing (GBS)** Genotyping-by-sequencing (GBS) is also a next-generation sequencing-based SNP genotyping method that uses restriction enzymes to reduce the genome complexity. GBS only targets as little as 2.3% of a genome. This reduced representation can be achieved by digestion of genomic DNA with methylation-sensitive restriction enzymes (REs). Methylation-sensitive REs do not cut repetitive DNA because of its highly methylated in nature and hence digested fragments generated from low-copy genomic regions. This helps in avoiding the repetitive regions and targeting low copy regions of the genome (Elshire et al. 2011). In GBS, genomic DNA is extracted and digested with methylation-sensitive restriction enzymes. Illumina sequencing forward and reverse adaptors ligated to both the end of fragments. The forward adapter is enriched with barcode sequences which is helpful for multiplexing of the samples. Now the PCR amplification is performed using stranded pair-end Illumina primers. The PCR is used to amplify the GBS library and add the Illumina sequencing regions to the fragment. The prepared GBS library was sequenced using Illumina platform and generated data were analyzed using bioinformatics software's like Next Generation Sequencing Eclipse Plugin (NGSEP), TASSEL-GBS, R package msgbsR, etc. (Perea et al. 2016; Wickland et al. 2017).

#### 2.2.2.2 Allelic Detection

##### *2.2.2.2.1 Allele-Specific Primer-based SNP Genotyping-Fluidigm*

Fluidigm EP1 is the fast, cost-effective, and efficient system for medium-throughput SNP genotyping. This system works on a principle of locus-specific primer-based PCR amplification and genotyping (similar to Taqman chemistry). The system incorporates a nanofluidic chip, also known as integrated fluid circuit (IFC) which consists of a network of fluid lines from sample and assay inlet, nanoflex valves, and reaction chambers. This technology has the potential to perform up to 9216 independent TaqMan reactions in a single experiment (Maurice Chan et al. 2011). The nanoflex valve, which is composed of elastomeric rubber, is used to regulate the flow of fluids in the IFC by deflecting under pressure and creating a tight seal is the cornerstone of the IFC.

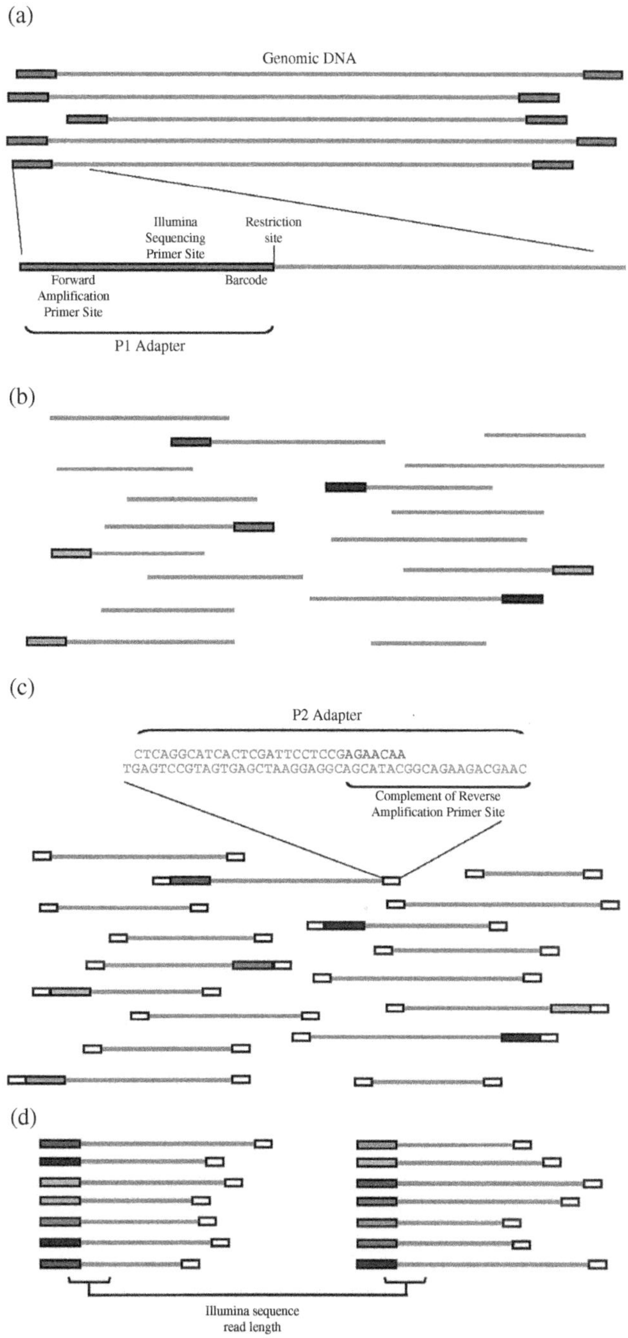


**Figure 2.3** Illustration of various steps involved in the generation of RAD-based SNP marker generation. (a) Ligate P1 adopter to digested genomic DNA. (b) Poor barcoded sample and shear. (c) Ligate P2 adopter to sheared fragments. (d) selective amplification of RAD tags. Source: The figure is reproduced from Baird et al. (2008) available with Creative Commons Attribution License (CC BY).

An IFC is primed in the IFC Controller before loading of samples and assays. The IFC controller pressurizes the control lines and closes the interface valves. After priming of IFC, samples and Taqman assays are loaded into the IFC and then again put into the controller for the mixing of sample and assays into the reaction chambers. In the loading process, on application of pressure, fluid from the sample inlets and assay inlets are pushed into the corresponding fluid lines. The chip utilizes carry over slug design (CS) which allows precise metering and mixing of fluids in a reaction (Unger et al. 2000) chamber (nanolitres of samples and assays are used in a reaction). The loading process takes 45 minutes in the controller and after that, the IFC is being loaded into the standalone PCR and the reaction starts in reaction chambers by thermal cycling in presence of fluorophore-containing probes. After completion of PCR, the data are acquired in the EP1 fluorescence system and the data analysis is done using bioinformatic tools and Software.

The processing of plant samples, i.e. DNA requires preamplification, which is done using specific-target amplification and locus-specific primers. After diluting the preamplified products with distilled water, PCR amplification with allele-specific primers is carried out (Kishora et al. 2020). During PCR amplification, two fluorogenic probes corresponding to one of the two alleles in a biallelic SNP are used to probe the SNP site. The probes are made up of a fluorescent reporter dye attached to the probe's 5′ end and a nonfluorescent quencher attached to the probe's 3′ end. The reporter dye's fluorescence is muted when the probe is intact because the reporter dye is close to the quencher. The probe anneals selectively to a complementary sequence during PCR. This causes the probe to cleave, releasing the reporter dye, resulting in increased fluorescence with each PCR cycle. This increase in fluorescence occurs only if the target sequence is complementary to the probe and is amplified during PCR (Seeb et al. 2009).

**MALDI TOF-based SNP Genotyping-Sequenom** MALDI MS (matrix-assisted laser desorption/ionization mass spectrometry) allows for the quick study of high molecular weight biomolecules (e.g. DNA, proteins, peptides, polysaccharides, and other organic compounds) and other chemical substances (e.g. polymers, macromolecules, and hybrid materials). It produces single-charged ions and has a high detection accuracy and sensitivity. Sequenom incorporates matrix-assisted laser desorption ionization-time of flight (MALDI-TOF) mass spectrometry technique for highly accurate, low-throughput SNP genotyping. MALDI-TOF-based SNP genotyping platforms take the advantage of differences in mass between allele-specific primer and extension products. During the MALDI process, analyte molecules are mixed with low molecular weight organic molecules like α-cyano-4-hydroxycinnamic acid (CHCA) and 2,5-dihydroxybenzoic acid (DHB) called matrix and the mixture is crystallized when dried over SpectroCHIP. Then the crystallized matrix-analyte mixture (a large quantity of matrix is used) is irradiated with a laser beam with an emitting wavelength range in UV or mid IR range. The irradiation causes desorption and ionization of analyte molecules in the form of singly protonated ions. The system uses a TOF mass analyzer. The protonated ions are then accelerated with uniform translational energy on application of the electric field. Then the charged ions are moved to the drift region which is free of electric field. The velocity of charged ions in the drift region is determined by the mass to charge ratio of the ions which in turn is calculated by measuring time required by the ions to travel through the flight tube (Figure 2.4).

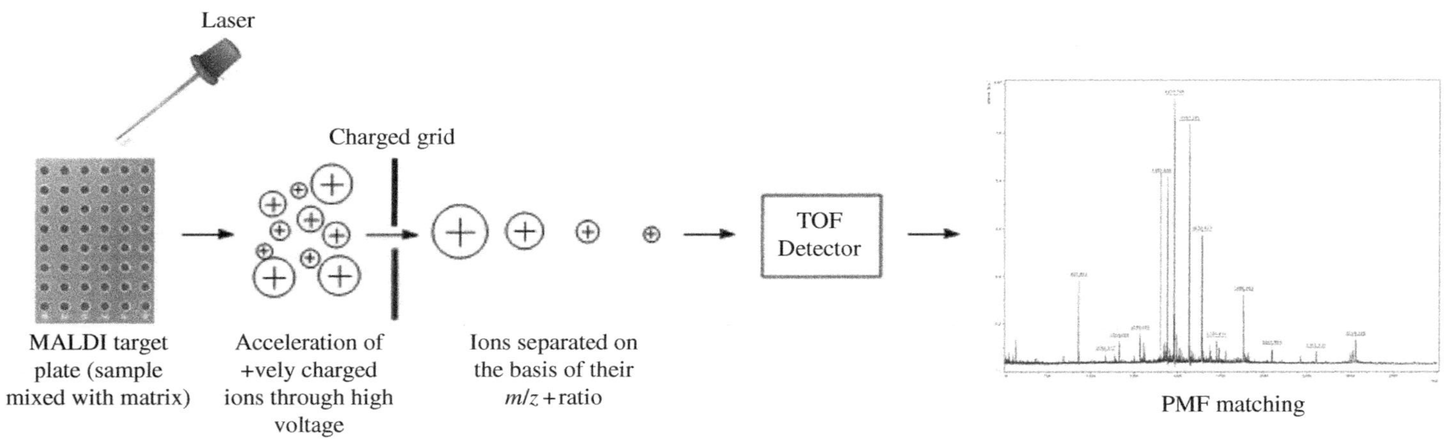


**Figure 2.4** Schematic illustration of work-flow in a MALDI-TOF MS *Source:* The figure is reproduced from Singhal et al., 2015, figure 01(p.03), available with Creative Commons 4.0

For SNP genotyping, the target regions of the sample (5–10 ng) are amplified for each assay within the multiplex. The PCR product is mixed with the shrimp alkaline phosphatase (SAP) which dephosphorylates any residual dNTPs, which might interfere with the allele-specific termination at the later stage. The amplification process is followed by a post-PCR primer extension where an allele-specific primer is annealed at the adjacent position of the SNP site. The extension reaction cocktail contains four terminating nucleotides (ddNTPs). The extension involves the addition of a single nucleotide and hence alleles are differentiated by the molecular mass of the extension product. The extension products are then conditioned with the ion-exchange resin and a nano dispensing device loads a few nanoliters of sample product to the prefabricated matrix loaded chip array (384-pad spectroCHIP) (Oath et al. 2009). The SpectroCHIP is then loaded to the MALDI-TOF MS analyzer for the analysis.

### 2.2.3 Custom Assay Technologies

Once any significant SNP with a vital role in gene regulation, phenotype change, disease response, and so on is discovered, the next step is to validate it in a larger population. Arrays designed with few vital SNPs, mainly for validation purposes are called custom-designed SNP assays. These assays must meet at least majority of the following criteria:

1) should be able to analyze large amounts of samples quickly,
2) should have a high design to assay conversion rate for custom SNPs, and
3) should be reliable and cost-effective.

To achieve these requirements, these technologies have incorporated a variety of sophisticated molecular biology techniques. The advantage of customized array is that SNPs from genomic regions of interest can be specifically added to the array and the number of interested SNPs to be fabricated on the array can be adjusted to the customer's needs. Customized SNP array cost per sample is low to moderate approximately $28–$90 (USD) while the NGS method have approximately $35 per sample price for GBS (Peng et al. 2017) which will increase drastically from diploid to polyploid as much higher coverage is required for accurate SNP calling. The major disadvantage of customized SNP array is that it requires prior genomic information and location (Vos et al. 2015). Furthermore, its design and further optimization can take a long time. Another issue is that SNP discovery requires fewer samples, which will primarily remove rare alleles while capturing common alleles (Gravel et al. 2011).

### 2.2.4 Summary

SNPs constitute one of the most popular and significant genetic markers in studying disease development and progression, breeding crops for improved traits etc. Two major strategies are involved in SNP genotyping viz. allele discrimination and allele detection. The high-throughput platforms used for SNP genotyping have generated huge amount of data in many crops. Further utilization of this information in plant breeding is important for crop improvement.

## References


Baird, N.A., Etter, P.D., Atwood, T.S. et al. (2008). Rapid SNP discovery and genetic mapping using sequenced RAD markers. *PLoS One* 3 (10): e3376.

Baner, J., Nilsson, M., Mendel-Hartvig, M. et al. (1998). Signal amplification of padlock probes by rolling circle replication. *Nucleic Acids Research* 26 (22): 5073–5078.

Bayer, M.M., Rapazote-Flores, P., Ganal, M. et al. (2017). Development and evaluation of a barley 50k iSelect SNP array. *Frontiers in Plant Science* 8: 1792.

Bianco, L., Cestaro, A., Linsmith, G. et al. (2016). Development and validation of the Axiom® Apple480K SNP genotyping array. *Plant Journal* 86: 62–74. https://doi.org/10.1111/tpj.13145.

Branton, D., Deamer, D.W., Marziali, A. et al. (2010). The potential and challenges of nanopore sequencing. *Nanoscience and Technology: A Collection of Reviews from Nature Journals*: 261–268.

Chagné, D., Crowhurst, R.N., Troggio, M. et al. (2012). Genome-wide SNP detection, validation, and development of an 8K SNP array for apple. *PLoS One* 7: e31745. https://doi.org/10.1371/journal.pone.0031745.

Chan, M., Chan, M.W., Loh, T.W. et al. (2011). Evaluation of nanofluidics technology for highthroughput SNP genotyping in a clinical setting. *The Journal of Molecular Diagnostics* 13 (3): 305–312. https://doi.org/10.1016/j.jmoldx.2010.12.001.

Chen, X., Levine, L., and Kwok, P.Y. (1999). Fluorescence polarization in homogeneous nucleic acid analysis. *Genome Research* 9 (5): 492–498.

Eid, J., Fehr, A., Gray, J. et al. (2009). Real-time DNA sequencing from single polymerase molecules. *Science 323* (5910): 133–138.

Elshire, R.J., Glaubitz, J.C., Sun, Q. et al. (2011). A robust, simple genotyping-by-sequencing (GBS) approach for high diversity species. *PLoS One* 6 (5): e19379.

Gravel, S., Henn, B.M., Gutenkunst, R.N. et al. (2011). Demographic history and rare allele sharing among human populations. *Proceedings. National Academy of Sciences. United States of America* 108: 11983–11988. https://doi.org/10.1073/pnas.1019276108.

Kennedy, G.C., Matsuzaki, H., Dong, S. et al. (2003). Large-scale genotyping of complex DNA. *Nature Biotechnology* 21: 10.

Khlestkin, V.K., Rozanova, I.V., Efimov, V.M. et al. (2019). Starch phosphorylation associated SNPs found by genome-wide association studies in the potato (*Solanum tuberosum* L.). *BMC Genetics* 20 (1): 45–53.

Kishora, D.S., Songa, W.-H., Noha, Y. et al. (2020). Development of SNP markers and validation assays in commercial Korean melon cultivars, using genotyping-by-sequencing and Fluidigm analyses. *Scientia Horticulturae* 263: 109113.

Kumar, N.M., Katageri, I.S., Gowda, S.A. et al. (2019). 63K SNP chip based linkage mapping and QTL analysis for fibre quality and yield component traits in *Gossypium barbadense* L. cotton. *Euphytica* 215 (1): 1–16.

LaFramboise, T. (2009). Single nucleotide polymorphism arrays: a decade of biological, computational and technological advances. *Nucleic Acids Research* 37 (13): 4181–4193.

Li, Y., Haseneyer, G., Schön, C.C. et al. (2011). High levels of nucleotide diversity and fast decline of linkage disequilibrium in rye (*Secale cereale* L.) genes involved in frost response. *BMC Plant Biology* 11: 6.

Li, X., Singh, J., Qin, M. et al. (2019). Development of an integrated 200K SNP genotyping array and application for genetic mapping, genome assembly improvement and genome wide association studies in pear (Pyrus). *Plant Biotechnology Journal* 17 (8): 1582–1594.

Livak, K.J. (1999). Allelic discrimination using fluorogenic probes and the 5′ nuclease assay. *Genetic Analysis - Biomolecular Engineering* 14 (5-6): 143–149.

Lyamichev, V., Mast, A.L., Hall, J.G. et al. (1999). Polymorphism identification and quantitative detection of genomic DNA by invasive cleavage of oligonucleotide probes. *Nature Biotechnology* 17 (3): 292–296.

Marrano, A., Martínez-García, P.J., Bianco, L. et al. (2019). A new genomic tool for walnut (*Juglans regia* L.): development and validation of the high-density Axiom™ J. regia 700K SNP genotyping array. *Plant Biotechnology Journal* 17 (6): 1027–1036.

Matsuzaki, H., Dong, S., Loi, H. et al. (2004). Genotyping over 100,000 SNPs on a pair of oligonucleotide arrays. *Nature Biotechnology* 21: 10.

McCouch, S.R., Wright, M.H., Tung, C.W. et al. (2016). Open access resources for genome-wide association mapping in rice. *Nature Communications* 7 (1): 1–14.

McCouch, S.R., Wright, M.H., Tung, C.-W. et al. (2017). Open access resources for genome-wide association mapping in rice. *Nature Communications* 7: 10532. https://doi.org/10.1038/ncomms10532.

McCoy, R.C., Taylor, R.W., Blauwkamp, T.A. et al. (2014). Illumina TruSeq synthetic long-reads empower de novo assembly and resolve complex, highly-repetitive transposable elements. *PLoS One* 9 (9): e106689.

Montanari, S., Bianco, L., Allen, B.J. et al. (2019). Development of a highly efficient Axiom™ 70 K SNP array for Pyrus and evaluation for high-density mapping and germplasm characterization. *BMC Genomics* 20 (1): 1–18.

Nilsson, M., Malmgren, H., Samiotaki, M. et al. (1994). Padlock probes: circularizing oligonucleotides for localized DNA detection. *Science* 265 (5181): 2085–2088.

Oath, P., del Mistro, G., Marnellos, G. et al. (2009). *Single Nucleotide Polymorphisms, Methods in Molecular Biology*, vol. 578 (ed. A.A. Komar). Totowa: Humana Press Inc. https://doi.org/10.1007/978-1-60327-411-1_20.

Peace, C., Bassil, N., Main, D. et al. (2012). Development and evaluation of a genome-wide 6K SNP array for diploid sweet cherry and tetraploid sour cherry. *PLoS One* 7 (12): e48305.

Peng, Z., Fan, W., Wang, L. et al. (2017). Target enrichment sequencing in cultivated peanut (*Arachis hypogaea* L.) using probes designed from transcript sequences. *Molecular Genetics and Genomics* 292: 955–965. https://doi.org/10.1007/s00438-017-1327-z.

Perea, C., De La Hoz, J.F., Cruz, D.F. et al. (2016). Bioinformatic analysis of genotype by sequencing (GBS) data with NGSEP. *BMC Genomics* 17: 498. https://doi.org/10.1186/s12864-016-2827-7.

Piriyapongsa, J., Kaewprommal, P., Vaiwsri, S. et al. (2018). Uncovering full-length transcript isoforms of sugarcane cultivar Khon Kaen 3 using single-molecule long-read sequencing. *PeerJ* 6: e5818.

Seeb, J.E., Pascal, C.E., Ramakrishnan, R. et al. (2009). *Single Nucleotide Polymorphisms, Methods in Molecular Biology*, vol. 578 (ed. A.A. Komar). Totowa: Humana Press Inc. https://doi.org/10.1007/978-1-60327-411-1_18.

Singh, N., Jayaswal, P.K., Panda, K. et al. (2015). Single-copy gene based 50 K SNP chip for genetic studies and molecular breeding in rice. *Scientific Reports* 5 (1): 1–9.

Singh, S., Mahato, A.K., Jayaswal, P.K. et al. (2020). A 62K genic-SNP chip array for genetic studies and breeding applications in pigeonpea (*Cajanus cajan* L. Millsp). *Scientific Reports* 10 (1): 1–14.

Singhal, N., Kumar, M., Kanaujia, P.K. et al. (2015). MALDI-TOF mass spectrometry: an emerging technology for microbial identification and diagnosis. *Frontiers in Microbiology* 6: 791. https://doi.org/10.3389/fmicb.2015.00791.

Thelwell, N., Millington, S., Solinas, A. et al. (2000). Mode of action and application of Scorpion primers to mutation detection. *Nucleic Acids Research* 28 (19): 3752–3761.

Thomson, M.J., Singh, N., Dwiyanti, M.S. et al. (2017). Large-scale deployment of a rice 6 K SNP array for genetics and breeding applications. *Rice* 10 (1): 1–13.

Unger, M.A., Chou, H.-P., Thorsen, T. et al. (2000). Monolithic microfabricated valves and pumps by multilayer soft lithography. *Science* 288: 113–116.

Unterseer, S., Bauer, E., Haberer, G. et al. (2014). A powerful tool for genome analysis in maize: development and evaluation of the high density 600 k SNP genotyping array. *BMC Genomics* 15 (1): 1–15.

Vos, P.G., Uitdewilligen, J.G., Voorrips, R.E. et al. (2015). Development and analysis of a 20K SNP array for potato (*Solanum tuberosum*): an insight into the breeding history. *Theoretical and Applied Genetics* 128: 2387–2401. https://doi.org/10.1007/s00122-015-2593-y.

Wang, S., Wong, D., Forrest, K. et al. (2014). International Wheat Genome Sequencing Consortium. Characterization of polyploid wheat genomic diversity using a high-density 90 000 single nucleotide polymorphism array. *Plant Biotechnology Journal* 12: 787–796.

Whitcombe, D., Theaker, J., Guy, S.P. et al. (1999). Detection of PCR products using self-probing amplicons and fluorescence. *Nature Biotechnology* 17 (8): 804–807.

Wickland, D.P., Battu, G., Hudson, K.A. et al. (2017). A comparison of genotyping-by-sequencing analysis methods on low-coverage crop datasets shows advantages of a new workflow, GB-eaSy. *BMC Bioinformatics* 18 (1): 586. https://doi.org/10.1186/s12859-017-2000-6.

Winfield, M.O., Allen, A.M., Burridge, A.J. et al. (2016). High-density SNP genotyping array for hexaploid wheat and its secondary and tertiary gene pool. *Plant Biotechnology Journal* 14 (5): 1195–1206.

Zhang, P., Li, X., Gebrewahid, T.W. et al. (2019). QTL mapping of adult-plant resistance to leaf and stripe rust in wheat cross SW 8588/Thatcher using the wheat 55K SNP array. *Plant Disease* 103 (12): 3041–3049.

Zhao, K., Tung, C.W., Eizenga, G.C. et al. (2011). Genome-wide association mapping reveals a rich genetic architecture of complex traits in *Oryza sativa*. *Nature Communications* 2 (1): 1–10.

# 3

# Opportunity and Challenges for Whole-Genome Resequencing-based Genotyping in Plants

*Surbhi Kumawat[1], Gaurav Raturi[1], Pallavi Dhiman[1], Sreeja Sudhakarn[1], Nitika Rajora[1], Vandana Thakral[1], Himanshu Yadav[1], Gunashri Padalkar[1], Yogesh Sharma[1], Vinaykumar Rachappanavar[2], and Manish Kumar[2]*

[1] *Department of Agriculture Biotechnology, National Agri-Food Biotechnology Institute (NABI), Mohali, Punjab, India*
[2] *Department of Seed Science and Technology, Dr. Yashwant Singh Parmar University of Horticulture and Forestry, Solan, Himachal Pradesh, India*

## 3.1 Introduction

Recent advancements in next-generation sequencing (NGS) have speed up genomics and transcriptome research across the board, especially in plant science (Rana et al. 2019; Ratnaparkhe et al. 2020). Genomics is a key for the better understanding of whole-genome involving the DNA sequences determined in the order in which they are present in a genome which indeed in the sequential exploration of the evolution of genome structure and interpretation of the molecular phylogeny. It involves the combination of various technologies including recombinant DNA, sequencing methods of DNA, and bioinformatics to assemble, analyze and assign the structure and function of the plant genome. Earlier, the Sanger method was commonly used for nucleic acid sequencing in genomics but in 2005 the development of NGS brought sequencing platforms that are fast and cost-effective enabling the excavating of characteristics unique to particular species in terms of gene structure as shown in Figure 3.1 (Unamba et al. 2015; Loera-Sánchez et al. 2019). It also enables us to understand the mechanism lying behind the process of gene expression as well as secondary metabolism. Further recent advances in the NGS platforms bring the gratifying efforts in de-novo whole-genome/transcriptomic sequencing and also for the developments of markers based on genome-wide sequences (Vasupalli et al. 2020). The most common platforms of NGS include Illumina/Solexa, ABI/SOLiD, Helicos, and 454/Roche (Unamba et al. 2015).

Earlier techniques for gene expression analysis include northern blotting in which specific RNA are detected and quantified (Goda and Minton 1995). Another technique involves the hybridization of antisense RNA with the complementary known target sequence thereby preventing the digestion of target sequence by single-strand RNAase resulting in the specific

*Genotyping by Sequencing for Crop Improvement*, First Edition.
Edited by Humira Sonah, Vinod Goyal, S.M. Shivaraj, and Rupesh K. Deshmukh.

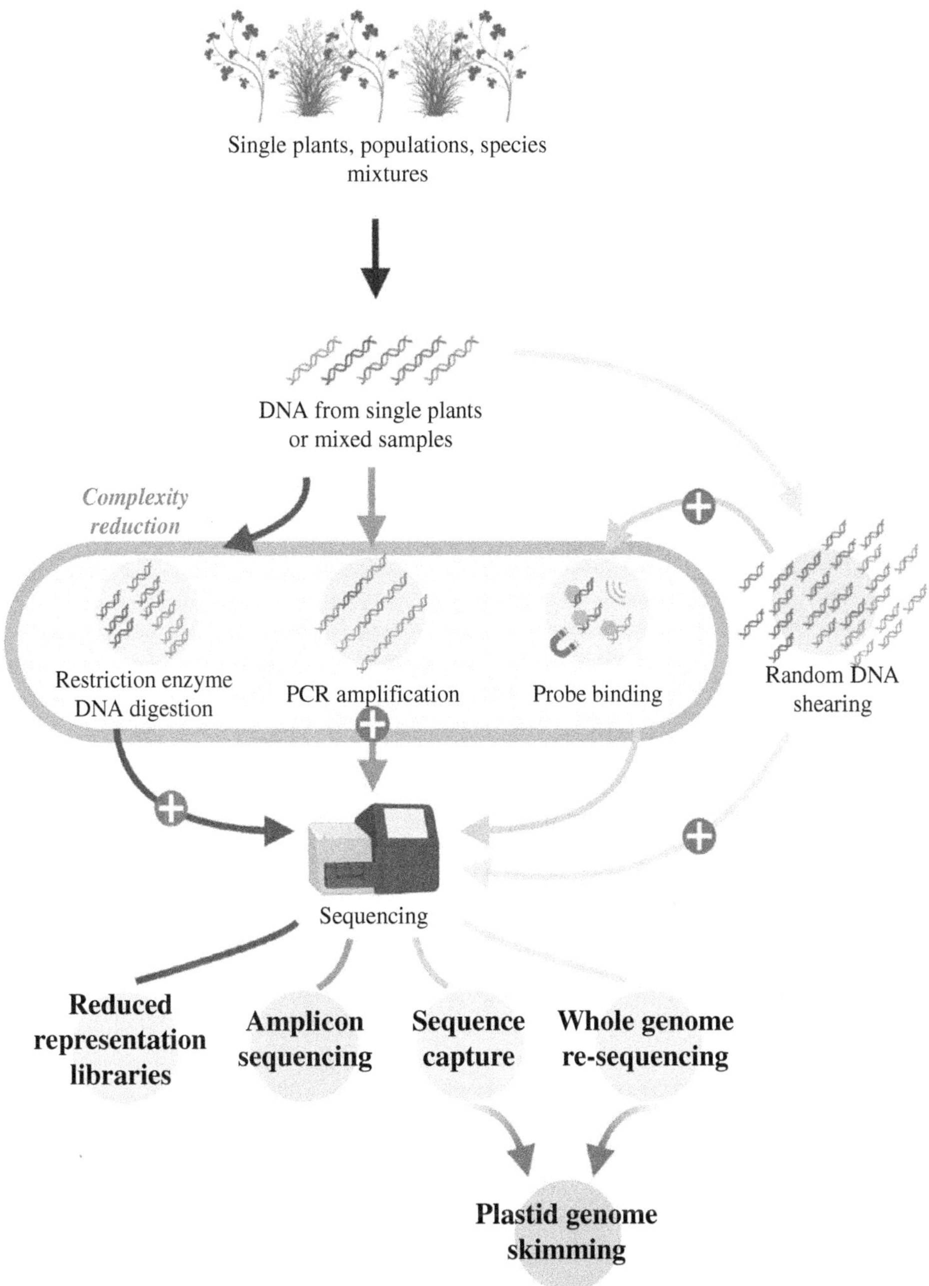


**Figure 3.1** Diagrammatic representation of various high-throughput-sequencing methods used in assessments of genetic diversity. Methods mentioned in grey oval represent the complexity reduction. Library preparation steps are indicated by plus sign. Red arrow indicates reduced representation libraries produces by starting DNA sample digestion with restriction enzyme. Orange arrow indicates amplicon sequencing while blue arrow represents the sequence capture. Whole-genome resequencing is indicated in purple arrow in which random fragments of digested DNA are sequenced. Green arrow represents plastid genome skimming responsible for plastid genome sequences recovery. The figure is reproduced from Loera-Sánchez et al. (2019) which is available under a Creative Commons Attribution 4.0 (CC-By 4.0) International License, which permits reproduction.

and accurate quantification of the target sequence. But the major drawbacks associated with this technique include it requires prior knowledge about the target sequences. With the advancements in technologies, transcriptomics came into existence which analyses all RNA molecules (tRNA, rRNA, mRNA, and other noncoding RNA) transcribed in an organism. Microarray is a widely employed technique to analyze the transcriptome. In a single experiment, this technique can measure the expression level of a thousand genes but there are several issues associated with its sensitivity and reproducibility.

The recent advancement in NGS includes next-generation RNA sequencing (RNA-seq) (Sharma et al. 2011; Sonah et al. 2016). This technique mainly focuses on the mRNA sequencing of only those genes that are expressed in the transcriptome (Chaudhary et al. 2019b). This technique helps in the identification of novel genes by de novo assembly without reference genome mapping. NGS also aids in the development of molecular markers. NGS has led to the exploration of thousands of markers on the entire genome resulting in an ease in the genome-wide association studies. These markers also aid in association mapping (Sonah et al. 2015; Zargar et al. 2015). These technologies have enabled us to understand the underlying process of gene expression and the development of resources for the analysis of marker-assisted breeding (MAS) and diversity analysis (Unamba et al. 2015).

## 3.2 Basic Steps Involved in Whole-Genome Sequencing and Resequencing

Whole-genome sequencing (WGS) can be divided into two groups, which include de novo WGS and whole-genome resequencing (WGR) (Bhat et al. 2020). WGS involves the genome sequence assembly for the first time while WGR compares genomic variability within individuals or populations (Patil et al. 2019). WGR requires the prior availability of reference genome for mapping and variant detection. Among WGS, de novo WGS involves the complete assembly of a species genome for the first time (Sevanthi et al. 2018). First, for the library preparation, high quality of genomic DNA is subjected to fragmentation followed by the addition of adaptors to the DNA fragments. For the detection of small structural variants such as INDELs or CNVs (copy number variations), short reads (350–550 bp insert size) from standard libraries are utilized while long-read data or mate-pair libraries with an insert size of around 2 to 20 kb will be required for the detection of large structural variants. For high-throughput sequencing, Illumina is often used. The sequences are mapped on the genome sequence based on similarity and local contigs are developed. While assembling the sequence, repetitive regions show difficulty in alignment with the short reads. In that case, mate pair-end sequencing reads aids in aligning large sequences which are also referred as scaffolds or supercontigs by linking and orienting contig. Unknown sequences gaps are denoted as Ns. The final result of a genome assembly corresponds to the contiguous scaffold sequences in a series separated by gaps.

In contrast to the WGS, WGR helps in the comparison of the variable sequences present between the genome of an individual or the population. In the case of WGR, the species genome sequence is a prerequirement for the read mapping. For example, in the case of an individual, genomic DNA of high quality is fragmented for library preparation in which adaptors are added to the fragments with an average insert size of 350–500 bp. With the help of high-throughput sequencing paired-end, short reads of about 100 bp are obtained. These short reads based on sequence similarity are mapped on the reference genome. When a particular nucleotide differs from the species-specific base single-nucleotide polymorphisms (SNPs) are detected. In some cases, SNPs might get lost as these are not present in the reference genome while some are heterozygous. And others may get lost due to low coverage. In the case of the population, the aim is to obtain genomic data from a wide range of individuals which are analyzed as a whole and are sequenced. These techniques have a wide range of applications in conservation and management (Fuentes-Pardo and Ruzzante 2017).

## 3.3 Whole-Genome Resequencing Mega Projects in Different Crops

### 3.3.1 1K Arabidopsis Genomes Resequencing Project

*Arabidopsis thaliana* belongs to the family Brassicaceae. It has 125–150 Mb diploid genome having around 30,000 protein-coding genes distributed over five chromosomes. Weigel and Mott (2009) initiated a project to report whole-genome sequence variation in 1001 accession. To sequence species-wide genome of the Arabidopsis, they proposed an approach with two different aspects. In the first aspect, by using the technologies such as Roche's 454 platform they generated a small number of sequences that approach the quality of *A. thaliana*'s original Col-o (Columbia) reference. For a large number of sequence, a relatively less expensive technology for example Applied Biosystem's SOLiD or Illumina's Genome Analyzer was used. Local haplotype similarity was exploited by using the information from the reference genome to draw a complete genome sequence. In a second aspect of this approach, the sampling was done for ten individuals from ten populations and geographical regions all around the Eurasia and at least one accession from North Africa ($10 \times 10 \times 10 + 1$). The aim of this 1K *Arabidopsis* genome project was to sequence a generalized genome that can encompass every *Arabidopsis* accession and every genome of the *A. thaliana* can be completely aligned against it (Weigel and Mott 2009).

### 3.3.2 3K Rice Genomes Resequencing Project

Rice is the principal staple food and is grown worldwide. It has high genetic diversity within genus and species and the wise use of rice diversity is a major factor for improved production. hence exploring the diversity at the genome level is required. A genome sequence program to sequence known diversity across the species was launched by the collaboration of CAAS (Chinese Academy of Agricultural Sciences), BGI (Beijing Genomics Institute), and IRRI (International Rice Research Institute) and named "The 3K rice genome project." The 3K rice genome project is the giga-dataset of the genome that is available publicly (averaging 14× depth of coverage), derived from 3K rice accession which are genetically and functionally diverse set collected from 89 countries. The aim of the 3K rice genome project was that the new population-specific genotyping arrays will be helpful for the genetic as well as breeding application and reveal population structure (Li et al. 2014a).

### 3.3.3 Soybean Whole-Genome Resequencing

The WGR approach has been used in soybean to identify the Quantitative trait loci (QTL) determining colonization of arbuscular mycorrhizal fungi (AFM). The microbial community like AFM can associate with 80% of the terrestrial plants and help host plants to uptake more nutrients, provide tolerance against stresses. The colonization and extent of benefits provided by AFM depend on the host genotypes. QTL is responsible for mycorrhizal responsiveness in different plants. Pawlowski and coworkers investigated the genetic components that are involved in the AFM association. The aim of the study was the genome-wide association analysis to identify the difference in AFM colonization in soybean genotypes and identification of genomic regions that are responsible for the colonization of AFM. They had used a genetically diverse set of 350 soybean genotypes inoculated with AFM (i.e. *Rhizophagus intraradices*). By using whole-genome resequencing-derived SNP dataset, they identified six QTL involved in the colonization of the AFM. The candidate genes identified in these QTL regions contain the homologs of the nodulin protein family and other genes responsible for symbiosis (Pawlowski et al. 2020).

### 3.3.4 Chickpea

*Cicer arietinum* is the best source for protein, β-carotene, and minerals such as iron, calcium, phosphorus, manganese, and zinc. Major abiotic stresses such as heat and drought can cause up to 70% loss in yield. Varshney and his coworkers utilize NGS technology to explore the germplasm wealth present in gene banks and provide information on genetic variation, domestication, and population structure of the 429 chickpea lines. They analyzed the 2.7 Tbp (terabase pair) raw data including 28.36 billion reads with around 6 Gbp (gigabase pairs) raw data per sample. By using mapped resequencing data, they reported a map of 4.97 million SNPs, 596 100 indels, 4931 CNVs, and 60 742 PAVs (presence–absence variation) in 429 reference genotypes set. Out of 4.97, million the most of the SNPs (i.e. 85%) were found in intergenic regions and around 4% were present in coding sequences. They identified 107 375 heterozygous and 20 544 homozygous SNPs by using SNP calling procedure. Genome-wide association provides all the resources and generates new avenues for chickpea breeding in the future (Varshney et al. 2019).

### 3.3.5 Pigeon pea

Pigeon pea is an important tropical grain legume fulfilling the nutritional security in most of the developing countries such as Africa, Asia, and tropical America. The completion of the draft genome using high-throughput techniques facilitated rapid detection of genomic variation among pigeon pea landraces, breeding lines, and wild species. Varshney and his coworkers generated the whole-genome resequencing data for 292 accessions that include breeding lines (117 accessions), landraces (166), 7 from wild species, and 2 others. They had generated 21.7 billion paired-end reads of 2.19 Tb sequence (101 bp in length) and mapped to the reference genome (Asha pigeon pea cultivar). These mapping results can be used for the identification of SNPs, INDELs, and larger-scale variants along with CNVs and PAVs, such findings can accelerate the sustainability and improve the yield of the pigeon pea (Varshney et al. 2017).

### 3.3.6 *Vitis*

*Vitis* is native to temperate regions and having 30 wild species indigenous to East Asia and 28 wild species from North America. Liang and coworkers resequenced the whole genome of 472 *Vitis* accessions that cover 48 out of the 60 extant species from different geographical distributions. They had generated around 4.1 Tb WGS data, 27.3 billion paired-end raw reads for 472 accessions. For the majority of accessions, the coverage was found more than 80%. By using different filter criteria, they identified the 77 726 929 SNPs, 10 278 017 indels, and 23K copy number of the variants. Next filtering yielded 27 859 960 SNPs, 3 854 659 indels with minor allele frequency (MAF) >0.005 and core set of about 12 549 273 SNPs and 9 lakh indels with MAF >0.005. Around 73.7% of the SNPs were present in the intergenic regions and around 4% of the SNPs were present in the coding region. All these outcomes from this study can play an important role in the improvement of the *Vitis* cultivars (Liang et al. 2019).

## 3.4 Whole-Genome Pooled Sequencing

Genotyping populations with NGS technologies are very convenient and appealing (Unamba et al. 2015). Whole-genome pooled sequencing popularly known as Pool-seq has been used to sequence multiple individuals instead of single individuals from a population separately (Fracassetti et al. 2015). The main advantage is that large number of samples can be assayed in a cost-effective manner. But sometimes, pooling is necessary when separating individuals is difficult and when sufficient DNA is not there to create individual libraries. Pooling is used for the estimation of SNP frequencies in population genetics. Pooling approach has been applied to study genetic loci governing a trait, to perform GWAS (Bastide et al. 2013), and to gather demographic history of population (Hellwege et al. 2017). During statistical analysis, random experimental errors in measuring allele frequencies from pooled DNA should be considered. Sophisticated pooling designs are being developed in order to take care of hidden population stratification, confounders, and interactions, allowing the analysis of haplotypes (Sham et al. 2002; Norton et al. 2004). Fracassetti et al. (2015) verified a pooled whole-genome resequencing (Pool-seq) technique on *Arabidopsis lyrata* in which only 1.6 ng DNA per individual was used for library preparation. They compared the SNP frequencies, acquired by pooling to those generated by individual-based genotyping-by-sequencing (GBS), and found, Pool-seq as a viable approach for obtaining population-level SNP frequency data with good accuracy in a larger number of samples. They also examined the effects of sample size, sequencing depth per individual, and variant caller on population SNP frequency estimations. Since differential amplification occurs for many SNPs, and this biasness should be rectified during the estimation of allele frequency from pooled DNA (Sham et al. 2002; Norton et al. 2004). Whole-genome sequencing of pooled F2 individuals has been used to identify mapping interval and candidate gene SNPs in maize (Klein et al. 2018).

NGS technique in combination with whole-genome shotgun sequencing can be used to genotype nonmutant Vs. mutant bulked segregant analysis (BSA) pools at markers across the genome (Klein et al. 2018). With the advent of NGS, even without precise mapping, BSA may be utilized to immediately identify a very narrow mapping interval, causative lesion, and so forth. The genomic interval containing the mutant gene of interest can be discovered using BSA-Seq data (Hill et al. 2013). In a pooled BSA sample, the unlinked marker will be genotypically heterozygous, unless F1 recombination has occurred between the mutant lesion and a specific marker, while the linked markers will be homozygous for the mutant parent genotype. BSA-Seq is a useful approach for genome reduction that allows to increase sequencing depth without raising costs.

## 3.5 Pinpointing Gene Through Whole-Genome Resequencing-based QTL Mapping

The NGS has been used for QTL mapping and pinpointing the genes of interest. QTL mapping approaches solve the issues caused by repetitive sequences and genome duplication in addition to combining SNP discovery, SNP validation, and genotyping (Xu et al. 2013). Huang et al. (2009) showed that QTL affecting plant height were linked to a region where the green revolution gene is situated (Huang et al. 2009; Xu et al. 2013). Rice recombinant inbred lines (RIL) populations were genotyped with the help of WGR (Huang et al. 2009). In rice, QTL mapping precision for grain width was evaluated from previously cloned QTL possessing the gene GW5 for grain width found to be localized in bin, its presumed genomic location (Xie et al. 2010). The complexity of the genome of some species for example soybean (paleopolyploid) hinders the applications of sequencing-based genotyping approaches. Since, due to two genome duplication events, 75% of the genes are present in multiple copies (Schmutz et al. 2010). These paralogs further pose issues due to the short sequence reads which cause undistinguished allelic variations. Besides, the abundant repetitive sequences, heterochromatic regions of soybean genome, provide technological challenges for sequence alignment. Using 246 RILs sequenced at an average depth of 0.19× (Xu et al. 2013). Using 246 RILs sequenced at an average depth of 0.19, Xu et al. (2013) employed NGS methods to find QTL for southern root-knot nematode (RKN) resistance in soybean. They identified and validated SNPs. RIL population was genotyped as well as the parental source of each SNP allele was deduced. Linkage map was subsequently established using 3509 bins and 3489 recombination intervals as molecular markers. Out of three QTLs identified, one major QTL was 29.7 kb in size and mapped to bin 10 of chromosome 10. This QTL possess three true and two pseudogenes. As a result of sequence differences and gene expression analyses, the RKN resistance candidate genes *Glyma10g02160* and *Glyma10g02150* were identified which encodes for pectin methylesterase inhibitor-pectin methylesterase and pectin methylesterase inhibitor, respectively. This method is widely used to enhance QTL mapping accuracy in crops with a reference genome.

## 3.6 Online Resources for Whole-Genome Resequencing Data

### 3.6.1 SNP Seek

SNP seek database is a user-friendly database that provides access to SNPs. Rice SNP seek database (www.snp-seek.irri.org) contains SNP genotyping data for 3K rice varieties obtained from 3000 Rice Genomes Projects (Locedie et al. 2016). About 20 million rice SNPs were identified from this Project for 3K rice variety sets against the reference Nipponbare genome (Alexandrov et al. 2015). We can use the interface to identify SNPs using gene ID (MSU locus name) or genomic region by specifying chromosome, start and end sites. This site also provides phenotype and variety information for 3K rice varieties from the International Rice Genebank Collection Information System (IRGCIS) (Locedie et al. 2016).

### 3.6.2 Rice Functional and Genomic Breeding

Rice Functional and Genomic Breeding (RFGB, www.rmbreeding.cn) is a user-friendly database developed for breeding applications based on SNPs and InDels from 3000 rice genome projects. RFGB database bridges gap between phenotypic and genotypic data sets of the sequenced genome (Chun-Chao et al. 2019). The RFGB database contains five major modules including Phenotype, Haplotype, SNP & InDel, Restore sequence, and Germplasm. Users can search variations (SNPs, InDel) using gene locus ID (ex. Os05g0187500, LOC_Os09g26999), chromosome region (ex. Chr1, 200001–500000), and keywords (ex. GW5, heading date). Further, haplotype analysis can be done using gene locus ID, chromosome region, and SNP sites (ex. Chr2-8120893) (Chun-Chao et al. 2019).

### 3.6.3 Genome Variation Map

The genome Variation Map (www.ngdc.cncb.ac.cn/gvm/home) is a data repository and retrieval system of genome variations in BIG Data Center (www.ngdc.cncb.ac.cn). It provides free open access to search for SNPs and short InDels from approximately 609 million genome variants for 25 plants, 13 animals, and 3 virus species. The current version of the Genome Var Map includes a total of 41 species, 202 projects, 64819 samples, 9 60 409 451 variants, 2 60 393 associations, and 95 submissions (Li et al. 2021). Users can retrieve variants by variation IDs, genomic coordinates, gene names, gene functions, and variant effects. The results can be downloaded directly in FASTA or VCF file formats.

## 3.7 Applications and Successful Examples of Whole-Genome Resequencing

One of the important applications of whole-genome resequencing is to explore the genetic diversity at the sequence level. Rice has known to be species that shows genetic diversity not at species level but within the genus as well. The 3000 rice genome sequencing has led to the capture of the sequence variation in the genome which aids to decipher population differentiation (local and global population), construct new reference genome, pan-genome of varieties, haplotyping, and SNP discovery (Li et al. 2014b). Further, previously characterized genes have also been analyzed for haplotype diversity which can be employed in haplotype-based breeding (Abbai et al. 2019). Another example includes the study of evolutionary history by genome resequencing in peach fruit (Yu et al. 2018).

For the identification of genetic traits associated with complex traits, genome wide association study (GWAS) has been employed. It helps in the recognition of many SNPs involved with the target trait either by correlation or comparative analysis. Abbai et al. (2019) selected 664 cultivated rice accessions for GWAS from the 3K rice genome. In the case of salt-tolerant traits, they identified the possible candidate gene and causal polymorphism by performing genetic variation, haplotyping, integrated gene information, and homology analysis. As the precise superior haplotype identification will have a great impact on the molecular breeding outcome. Similarly, Lin et al. (2021) explored weediness-related traits from the 3K rice genome project in the cultivated pool. They identified weediness-associated markers by performing GWAS within each subpopulation. One of the examples involved GWAS in rice for the salt-tolerant trait as shown in Figure 3.2 (Yuan et al. 2020). With the help of WGR, it has been possible to construct a high-quality HapMaps in maize and rice which help to perform GWAS and find unidentified genetic variations of agronomically important traits. The applications of WGR have led to a decrease in the cost of WGS as well as enabled the generation of large data of sequences.

The online resources available for rice, soybean, canola, maize, and wheat are widely being used for the marker development, allele mining, haplotypic evaluation as well as simple applications like gene family characterization (Deshmukh et al. 2016; André et al. 2017; Rasoolizadeh et al. 2018; Chaudhary et al. 2019c; de Ronne et al. 2020; Singh et al. 2019). The free access of resources more particularly in rice and soybean has accelerated basic as well as applied research in these crops.

Apart from the high-throughput applications, WGR is helpful for specific applications like characterization of varieties developed with marker-assisted breeding, characterization of introgression lines, and evaluation of transgenic events (Patil et al. 2018; Tayade et al. 2018; Shivaraj et al. 2019). Similarly, resequencing also helps for bulk evaluation-based approaches like Mut-Map and QTLseq (Bansal et al. 2019; Chaudhary et al. 2019a; Kumawat et al. 2019). Very recent advancements like genome editing also have implication of WGR (Mushtaq et al. 2019; Vats et al. 2019; Ansari et al. 2020). Mutations at target sites as well as at off-target can be easily verified with whole-genome resequencing.

## 3.8 Challenges for Whole-Genome Resequencing Studies

Although, various platforms such as Illumina, Roche 454, and SoLiD™ have significantly increased the throughput with a decreased error rate but many plant genomes represent unique challenges because of their repetitive nature which leads to a challenge for reliable genome assembly. This might be due to high copy number and transposable elements amplifying nature in a large number of plants leading to consequences in assembly. Another major constraint is that not all species of plant are homozygous inbred diploids.

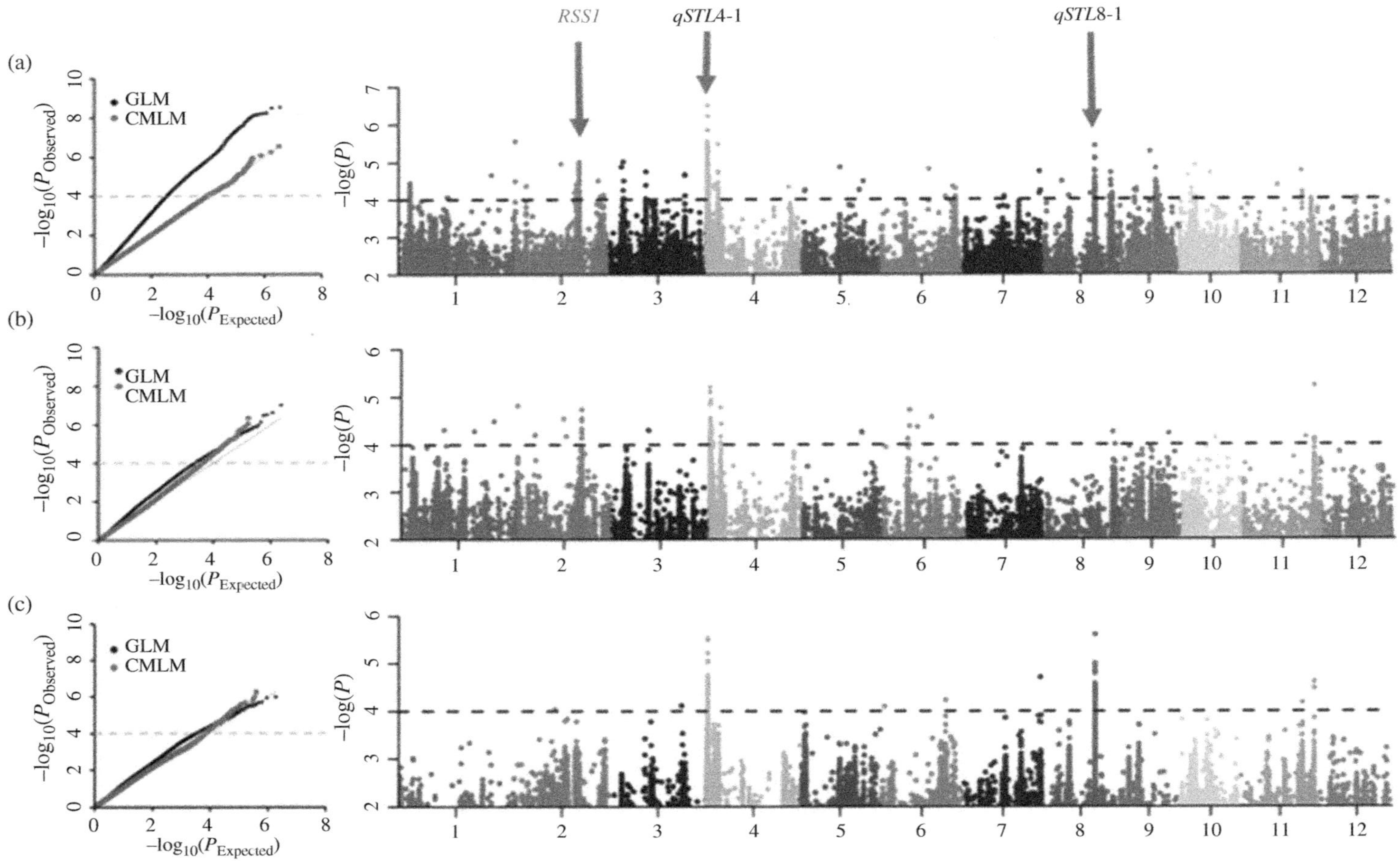


**Figure 3.2** Genome-wide association studies (GWAS) in rice seedling for salt-tolerant trait. For GWAS, quantile–quantile plots and Manhattan plots are represented for (a) *indica,* (b) *japonica* (c) represents population with the help of CMLM. In quantile–quantile plot, black and red points represent GLM and CMLM, respectively. In Manhattan plots, genes represented by red lines were previously cloned while QTLs were represented by black lines, which were considered to be important. The suggestive threshold ($P = 1.0 \times 10^{-4}$) for each population is indicated by dashed horizontal line. The figure is reproduced from Yuan et al. (2020) which is available under a Creative Commons Attribution 4.0 (CC-By 4.0) International License, which permits reproduction.

Currently, there is no WGR technique that meets all conservation geneticists' needs. However, with the advancement in next-generation sequencing technologies, these challenges can be overcome.

## 3.9 Summary

A new era has been brought in plant genetics with the rapid advancement in NGS technologies. With the help of this, a large amount of data is generated and used in the scientific community. These generated genomic sequences of diverse lines, especially the whole-genome resequencing aids in the identification of haplotypic/allelic variation. These techniques help in the identification of novel genes and alleles associated with the target genes or traits deployed in the improvement of crops. WGR-collected genomic data can aid with planning conservation and management of exploited species used commercially by assisting in the delimitation and monitoring of evolutionary units, as well as the prioritizing of endangered populations.

## References

Abbai, R., Singh, V.K., Nachimuthu, V.V. et al. (2019). *Plant Biotechnology Journal* 17: 1612–1622.

Alexandrov, N., Tai, S., Wang, W. et al. (2015). SNP-Seek database of SNPs derived from 3000 rice genomes. *Nucleic Acids Research* 43: D1023–D1027.

André, L.L., Sonah, H., Dias, W.P. et al. (2017). Genome-wide association study for resistance to the southern root-knot nematode (*Meloidogyne incognita*) in soybean. *Molecular Breeding* 37: 148.

Ansari, W.A., Chandanshive, S.U., Bhatt, V. et al. (2020). Genome editing in cereals: approaches, applications and challenges. *International Journal of Molecular Sciences* 21: 4040.

Bansal, R., Rana, N., Kumawat, S. et al. (2019). Advances in induced mutagenesis and mutation mapping approaches in rice. *Oryza* 59: 106–114.

Bastide, H., Betancourt, A., Nolte, V. et al. (2013). A genome-wide, fine-scale map of natural pigmentation variation in *Drosophila melanogaster*. *PLoS Genet* 9: e1003534.

Bhat, J.A., Deshmukh, R., Zhao, T. et al. (2020). Harnessing high-throughput phenotyping and genotyping for enhanced drought tolerance in crop plants. *Journal of Biotechnology* 324: 248–260.

Chaudhary, J., Alisha, A., Bhatt, V. et al. (2019a). Mutation breeding in tomato: advances, applicability and challenges. *Plants* 8: 128.

Chaudhary, J., Khatri, P., Singla, P. et al. (2019b). Advances in omics approaches for abiotic stress tolerance in tomato. *Biology* 8: 90.

Chaudhary, J., Shivaraj, S., Khatri, P. et al. (2019c). Approaches, applicability, and challenges for development of climate-smart soybean. In: *Genomic Designing of Climate-Smart Oilseed Crops*, 1–74. Cham: Springer.

Chun-Chao, W., Yu, H., Huang, J. et al. (2019). Towards a deeper haplotype mining of complex traits in rice with RFGB v2. 0. *Plant Biotechnology Journal* 18: 14.

de Ronne, M., Labbé, C., Lebreton, A. et al. (2020). Integrated QTL mapping, gene expression and nucleotide variation analyses to investigate complex quantitative traits: a case study with the soybean-*Phytophthora sojae* interaction. *Plant Biotechnology Journal* 18: 1492–1494.

Deshmukh, R.K., Sonah, H., and Singh, N.K. (2016). Intron gain, a dominant evolutionary process supporting high levels of gene expression in rice. *Journal of Plant Biochemistry and Biotechnology* 25: 142–146.

Fracassetti, M., Griffin, P.C., and Willi, Y. (2015). Validation of pooled whole-genome re-sequencing in *Arabidopsis lyrata*. *PLoS One* 10: e0140462.

Fuentes-Pardo, A.P. and Ruzzante, D.E. (2017). Whole-genome sequencing approaches for conservation biology: advantages, limitations and practical recommendations. *Molecular Ecology* 26: 5369–5406.

Goda, S.K. and Minton, N.P. (1995). A simple procedure for gel electrophoresis and northern blotting of RNA. *Nucleic Acids Research* 23: 3357.

Hellwege, J.N., Keaton, J.M., Giri, A. et al. (2017). Population stratification in genetic association studies. *Current Protocols in Human Genetics* 95: 1.22.21–21.22.23.

Hill, J.T., Demarest, B.L., Bisgrove, B.W. et al. (2013). MMAPPR: mutation mapping analysis pipeline for pooled RNA-seq. *Genome Research* 23: 687–697.

Huang, X., Feng, Q., Qian, Q. et al. (2009). High-throughput genotyping by whole-genome resequencing. *Genome Research* 19: 1068–1076.

Klein, H., Xiao, Y., Conklin, P.A. et al. (2018). Bulked-segregant analysis coupled to whole genome sequencing (BSA-Seq) for rapid gene cloning in maize. G3: Genes. *Genomes, Genetics* 8: 3583–3592.

Kumawat, S., Rana, N., Bansal, R. et al. (2019). Expanding avenue of fast neutron mediated mutagenesis for crop improvement. *Plants* 8: 164.

Li, C., Tian, D., Tang, B. et al. (2021). Genome variation map: a worldwide collection of genome variations across multiple species. *Nucleic Acids Research* 49: D1186–D1191.

Li, J.-Y., Wang, J., and Zeigler, R.S. (2014). The 3,000 rice genomes project: new opportunities and challenges for future rice research. *Gigascience* 3: 8.

Liang, Z., Duan, S., Sheng, J. et al. (2019). Whole-genome resequencing of 472 Vitis accessions for grapevine diversity and demographic history analyses. *Nature communications* 10: 1–12.

Lin, Y.-L., Wu, D.-H., Wu, C.-C. et al. (2021). Explore the genetics of weedy traits using rice 3K database. *Botanical Studies* 62: 2.

Locedie, M., Rommel, F.R., Nikki, B.F. et al. (2016). Rice SNP-seek database update: new SNPs, indels, and queries. *Nucleic Acids Research* 45: D1075–D1081.

Loera-Sánchez, M., Studer, B., and Kölliker, R. (2019). DNA-based assessment of genetic diversity in grassland plant species: challenges, approaches, and applications. *Agronomy* 9: 881.

Mushtaq, M., Mukhtar, S., Sakina, A. et al. (2019). Tweaking genome-editing approaches for virus interference in crop plants. *Plant Physiology and Biochemistry* 147: 242–250.

Norton, N., Williams, N., O'Donovan, M. et al. (2004). DNA pooling as a tool for large-scale association studies in complex traits. *Annals of Medicine* 36: 146–152.

Patil, G., Vuong, T.D., Kale, S. et al. (2018). Dissecting genomic hotspots underlying seed protein, oil, and sucrose content in an interspecific mapping population of soybean using high-density linkage mapping. *Plant Biotechnology Journal* 16: 1939–1953.

Patil, G.B., Lakhssassi, N., Wan, J. et al. (2019). Whole-genome re-sequencing reveals the impact of the interaction of copy number variants of the rhg1 and Rhg4 genes on broad-based resistance to soybean cyst nematode. *Plant Biotechnology Journal* 17: 1595–1611.

Pawlowski, M.L., Vuong, T.D., Valliyodan, B. et al. (2020). Whole-genome resequencing identifies quantitative trait loci associated with mycorrhizal colonization of soybean. *Theoretical and Applied Genetics* 133: 409–417.

Rana, N., Rahim, M.S., Kaur, G. et al. (2019). Applications and challenges for efficient exploration of omics interventions for the enhancement of nutritional quality in rice (*Oryza sativa* L.). *Critical Reviews in Food Science and Nutrition* 13: 1–17.

Rasoolizadeh, A., Labbé, C., Sonah, H. et al. (2018). Silicon protects soybean plants against *Phytophthora sojae* by interfering with effector-receptor expression. *BMC Plant Biology* 18: 97.

Ratnaparkhe, M.B., Marmat, N., Kumawat, G. et al. (2020). Whole genome re-sequencing of soybean accession EC241780 providing genomic landscape of candidate genes involved in rust resistance. *Current Genomics* 21: 504–511.

Schmutz, J., Cannon, S.B., Schlueter, J. et al. (2010). Genome sequence of the palaeopolyploid soybean. *Nature* 463: 178–183.

Sevanthi, A., Kandwal, P., Kale, P.B. et al. (2018). Whole genome characterization of a few EMS-induced mutants of upland rice variety Nagina 22 reveals a staggeringly high frequency of SNPs which show high phenotypic plasticity towards the wild-type. *Frontiers in Plant Science* 9: 1179.

Sham, P., Bader, J.S., Craig, I. et al. (2002). DNA pooling: a tool for large-scale association studies. *Nature Reviews Genetics* 3: 862–871.

Sharma, A., Kailasrao Deshmukh, R., Jain, N. et al. (2011). Combining QTL mapping and transcriptome profiling for an insight into genes for grain number in rice (*Oryza sativa* L.). *Indian Journal of Genetics and Plant Breeding* 71: 115.

Shivaraj, S., Dhakate, P., Sonah, H. et al. (2019). Progress toward development of climate-smart flax: a perspective on omics-assisted breeding. In: *Genomic Designing of Climate-Smart Oilseed Crops*, 239–274. Cham: Springer.

Singh, A., Singh, P.K., Sharma, A.K. et al. (2019). Understanding the role of the WRKY gene family under stress conditions in pigeonpea (*Cajanus Cajan* L.). *Plants* 8: 214.

Sonah, H., O'Donoughue, L., Cober, E. et al. (2015). Identification of loci governing eight agronomic traits using a GBS-GWAS approach and validation by QTL mapping in soya bean. *Plant Biotechnology Journal* 13: 211–221.

Sonah, H., Zhang, X., Deshmukh, R.K. et al. (2016). Comparative transcriptomic analysis of virulence factors in *Leptosphaeria maculans* during compatible and incompatible interactions with canola. *Frontiers in Plant Science* 7: 1784.

Tayade, R., Nguyen, T., Aeong Oh, S. et al. (2018). Effective strategies for enhancing tolerance to high-temperature stress in rice during the reproductive and ripening stages. *Plant Breeding and Biotechnology* 6: 1–18.

Unamba, C.I., Nag, A., and Sharma, R.K. (2015). Next generation sequencing technologies: the doorway to the unexplored genomics of non-model plants. *Frontiers in Plant Science* 6: 1074.

Varshney, R.K., Saxena, R.K., Upadhyaya, H.D. et al. (2017). Whole-genome resequencing of 292 pigeonpea accessions identifies genomic regions associated with domestication and agronomic traits. *Nature Genetics* 49: 1082.

Varshney, R.K., Thudi, M., Roorkiwal, M. et al. (2019). Resequencing of 429 chickpea accessions from 45 countries provides insights into genome diversity, domestication and agronomic traits. *Nature Genetics* 51: 857–864.

Vasupalli, N., Koramutla, M.K., Aminedi, R. et al. (2020). Omics approaches and biotechnological perspectives of arsenic stress and detoxification in plants. In: *Metalloids in Plants: Advances and Future Prospects*, 249–273. Wiley.

Vats, S., Kumawat, S., Kumar, V. et al. (2019). Genome editing in plants: exploration of technological advancements and challenges. *Cells* 8: 1386.

Weigel, D. and Mott, R. (2009). The 1001 genomes project for *Arabidopsis thaliana*. *Genome Biology* 10: 1–5.

Xie, W., Feng, Q., Yu, H. et al. (2010). Parent-independent genotyping for constructing an ultrahigh-density linkage map based on population sequencing. *Proceedings of the National Academy of Sciences of the United States of America* 107: 10578–10583.

Xu, X., Zeng, L., Tao, Y. et al. (2013). Pinpointing genes underlying the quantitative trait loci for root-knot nematode resistance in palaeopolyploid soybean by whole genome resequencing. *Proceedings of the National Academy of Sciences of the United States of America* 110: 13469–13474.

Yu, Y., Fu, J., Xu, Y. et al. (2018). Genome re-sequencing reveals the evolutionary history of peach fruit edibility. *Nature Communications* 9: 1–13.

Yuan, J., Wang, X., Zhao, Y. et al. (2020). Genetic basis and identification of candidate genes for salt tolerance in rice by GWAS. *Scientific Reports* 10: 9958.

Zargar, S.M., Raatz, B., Sonah, H. et al. (2015). Recent advances in molecular marker techniques: insight into QTL mapping, GWAS and genomic selection in plants. *Journal of Crop Science and Biotechnology* 18: 293–308.

# 4

# QTL Mapping Using Advanced Mapping Populations and High-throughput Genotyping

*Sweta Sinha[1], Brij Kishore Kushwaha[1], and Rupesh K. Deshmukh[2]*

[1] *Department of Molecular Biology and Genetic Engineering, Bihar Agricultural University, Sabour Bhagalpur, Bihar, India*
[2] *Agricultural Biotechnology, National Agri-Food Biotechnology Institute (NABI), Mohali, Punjab, India*

## 4.1 Introduction

The development of crop varieties for better adaptability and tolerance to environmental stresses has become the main challenge for plant breeders. It is, therefore, necessary for the breeder to identify and select superior genotypes to produce new varieties that can sustain agriculture under changing climatic conditions. Its success largely depends on the extent of genetic variation present in the germplasm. Since most of the observable phenotypic variations are quantitative and complex, the development and application of quantitative genetics approaches have improved our understanding about the genetic basis of complex traits. Variation in complex traits is controlled by segregation at multiple loci which are having small effects governed by many genes that may be sensitive to the environment are commonly known as quantitative trait loci (QTLs) (Mackay 2001). It is not possible to determine a QTL for such complex traits based on the segregation pattern of phenotypes in controlled crosses because the relationship between genotype and phenotype is not a simple ratio. With the advent of molecular genetics, understanding genomic information lead to achieving faster genetic gain than achieved based on phenotypic data only. The application of molecular markers has enabled the dissection of QTLs underlying complex traits. The molecular markers do exhibit Mendelian segregation patterns from one generation to the next. The increasing knowledge of QTLs for important agronomic traits in crops provides a vital tool for plant breeders for marker-assisted selection (MAS) (Ribaut and Hoisington, 1998; Uptmoor et al., 2008). Use of molecular breeding techniques has highly contributed in understanding the genetic basis of traits affecting the quality and yield in crop plants. The basic principle associated with the use of MAS in the context of QTL mapping is: If a QTL is linked to a marker, there will be a difference in mean values of the quantitative trait among individuals with different genotypes at the marker locus (Sax 1923). Most popular types of molecular markers used in QTL mapping are restriction fragment length polymorphisms (RFLP) (Beckmann and Soller 1986; Tanksley et al. 1989), amplified fragment length polymorphisms (AFLP) (Vos et al. 1995), simple-sequence repeats (SSRs)

*Genotyping by Sequencing for Crop Improvement*, First Edition.
Edited by Humira Sonah, Vinod Goyal, S.M. Shivaraj, and Rupesh K. Deshmukh.

(Powell et al. 1996), and single-nucleotide polymorphism (SNP) (Syvänen 2005). In recent years, the advent of SNPs and next-generation sequencing (NGS)-based genotyping for many crops has facilitated genome-wide genotyping at an affordable cost.

## 4.2 The Basic Objectives of QTL Mapping

- To identify the genomic regions that affect the trait of interest
- To estimate the genetic effect of the QTL on the trait
  - How much phenotypic variation is caused by the QTL?
  - What gene action is associated with the QTL?
  - Which allele is associated with a favorable effect?
- To detect markers closely linked to QTL so that it can be used for MAS in crop improvement programs

## 4.3 QTL Mapping Procedure

1) A mapping population: Bi-parental or multiparental
2) Phenotyping of mapping population for target traits
3) Genotyping of mapping population using markers polymorphic between parents
4) Statistical techniques to map QTLs

## 4.4 The General Steps for QTL Mapping

- Develop a suitable bi-parental or multiparental mapping population by crossing the selected parents. The selection of the mapping population is critical and depends on the genetic effect that we are looking for.
- The mapping population is evaluated for the target trait of interest in the target environment. The mapping population must be evaluated in multilocation/years under replicated trials to get precise phenotypic data.
- A parental polymorphism survey has to be performed to identify the markers polymorphic between the parents. These polymorphic markers will be used to genotype the mapping population. The polymorphic markers should be adequate in number and uniformly spaced throughout the chromosome covering the entire genome.
- The genotypic data generated on the mapping population will be used to construct a linkage map that depicts the position, order, and genetic distance between the markers.
- The genotypic and phenotypic data are analyzed to identify markers linked to the trait of interest using statistical programs.

## 4.5 Factors Influencing QTL Analysis

- Genetic properties of QTLs: The precision of QTLs depends upon the position of QTL, QTL effect, the interaction with other QTLs, genetic background, and QTL x environment interactions.

- Population size: The larger is the population size for QTL analysis, the greater is the chance of identifying QTLs with smaller effects, and more precisely is the position of QTL. Generally, the population size should be 200 or more, but it should not be less than 50.
- QTL × environment interaction: The expression of quantitative traits is affected by the environment. Generally, QTLs with relatively large effects are more stable over environments than those with small (<10% of the phenotypic variance) effects. Therefore, it is desirable to perform phenotypic evaluation over environments, locations, and years to assess the stability of QTL expression across environments.
- Experimental error: Phenotyping and genotyping of the mapping population are the major sources of experimental error. Therefore, phenotypic evaluation is conducted over locations and years/seasons in replication trials. During genotyping, the missing marker data can affect the order of markers in the linkage map and the distances between marker pairs.

## 4.6 QTL Mapping Approaches

QTLs are genomic regions and their genetic segregation within a population is statistically associated with a proportion of the phenotypic variation for a trait. The following approaches are used for QTL mapping.

1) Linkage mapping approach: It is the common approach for QTL mapping, wherein several individuals in a mapping population derived from bi-parental crosses are genotyped with a large number of markers and phenotyped for the target traits. Further, the individual segregants classified according to their marker genotypes and trait values of different classes are compared. It is labor-intensive, time-consuming, and costly approach (Salvi and Tuberosa 2005).
2) Bulked segregant analysis (BSA): It is simple, quick, and cheap. It includes genotyping of only a pair of bulked DNA samples from individuals with clear or extreme phenotypes of the target trait along with the parents (Michelmore et al. 1991). The principle behind BSA is that the polymorphic marker linked to the trait of interest will show an identical banding pattern between the parents and their respective bulks (Figure 4.1). For monogenic trait, the number of individuals per bulk can be 10–20, but in the case of QTL the number of individuals per bulk should be increased (Sun et al. 2010; Takagi et al. 2013b; Zou et al. 2016).
3) Selective genotyping: Selective genotyping is an extension of the BSA approach for QTL mapping using large mapping populations of over 500 individuals with a selective genotyping of individuals with extreme phenotypes in a population (Darvasi and Soller 1992). It reduces the amount of genotyping work.

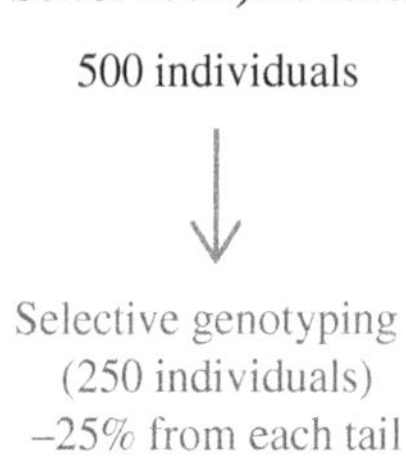

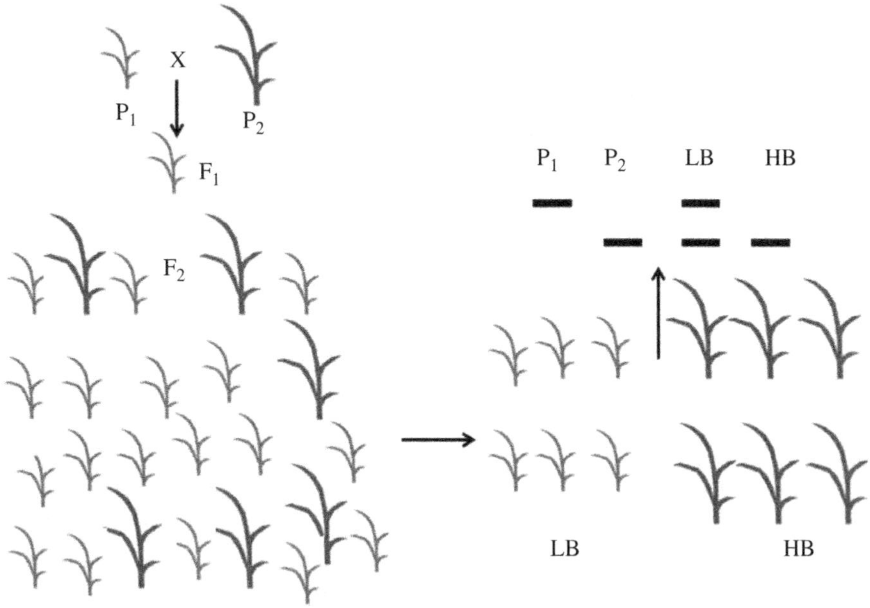


**Figure 4.1** Steps involved in bulked segregant analysis (BSA) used for QTL mapping.

4) Selective DNA pooling: This approach was proposed to reduce the cost of genotyping by selective DNA pooling and genotyping of individuals with extreme phenotypes in a population instead of genotyping the entire population in QTL mapping (Darvasi and Soller 1994).
5) X-QTL: Taking the advantage of high-throughput sequencing and power of BSA, Ehrenreich et al. (2010) proposed X-QTL for identifying QTLs in yeast using two populations with extreme phenotypes and genotype the bulked DNA by whole genome sequencing. This approach has been extensively used for QTLs analysis in yeast for utilization of xylose (Wenger et al. 2010), tolerance to heat stress (Parts et al. 2011), and ethanol stress (Swinnen et al. 2012).
6) SHOREmap: This approach was utilized for mapping QTLs in *Arabidopsis thaliana* to map a mutation (Schneeberger et al. 2009). SHOREmap uses genome-wide genotyping and candidate-gene sequencing in a single step through analysis of deep sequencing data from a large pool of recombinants to dramatically speed up the identification of causative point mutations and small deletions. It helps in de novo marker prediction mapping.
7) MutMap: MutMap is a rapid NGS-based method for the identification of candidate genes responsible for mutant phenotypes in EMS-induced mutants (Abe et al. 2012). This approach requires the crossing of selected mutant plants for a trait with the parental line and plants. The mutant phenotype is selected in the $F_2$ generation. The equal amounts of DNA from each selected plant are bulked and subjected to whole-genome sequencing (WGS) at sufficient depth. The short sequence of high-quality reads is aligned with the parental reference genome and the genomic positions with SNPs are identified (Figure 4.2). These SNPs would have been induced by the mutagen treatment (Singh and Singh 2015). Recently, the MutMap scheme was modified wherein it

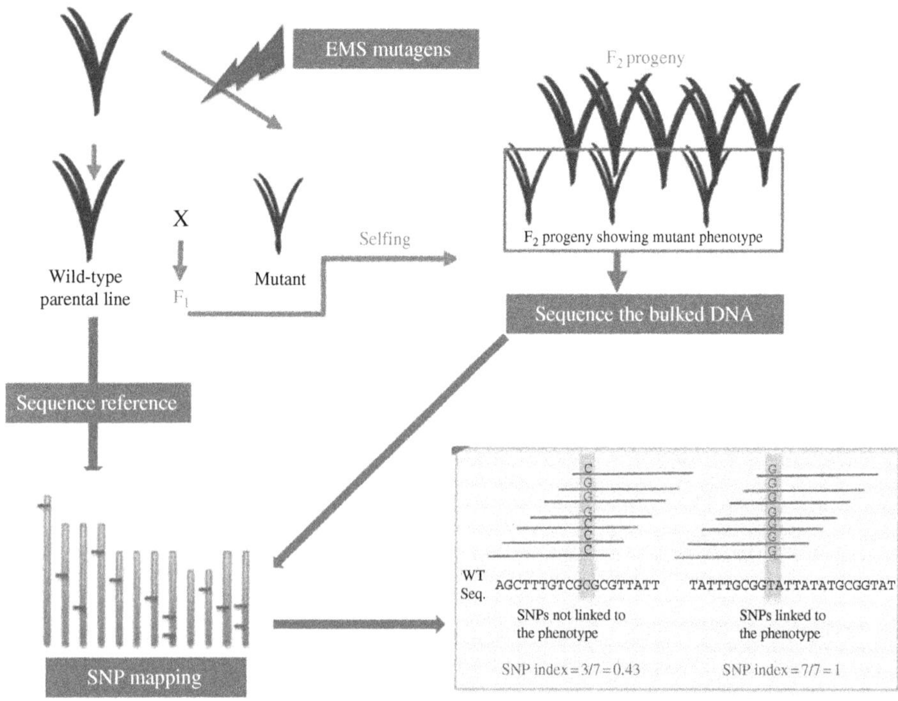


**Figure 4.2** Steps involved in MutMap approach used to map the QTLs of target trait.

avoids developing any population and directly uses mutant and parental lines for analysis which is referred to as MutMap+ (Fekih et al. 2013) while performing candidate gene discovery in the gap region which could not be sequenced during genome sequencing is known as MutMap-Gap (Takagi et al. 2013b).

8) QTL-seq: QTL-seq approach was proposed by Takagi et al. (2013b) for the rapid mapping of QTLs in rice by whole-genome resequencing of DNA bulks of extremely high phenotype and the extreme low phenotype. The short sequence high quality reads obtained from NGS is aligned with the reference genome sequence of one of the parents of the mapping population. Then, SNP-index plots of high and low bulks are compared. The genomic region having a QTL affecting the target trait would display contrasting patterns of SNP indices in the plots for the two bulks (Figure 4.3).
9) Bulked segregant RNA sequencing (BSR-seq): It uses RNA in place of DNA from the two phenotypic extreme bulks to identify markers tightly linked to the gene responsible for the target trait (Liu et al. 2012) (Figure 4.4). BSR-seq efficiently maps the genes even in populations for which no polymorphic markers have been previously identified. It also provides information on the effects of the mutant on global patterns of gene expression at no extra cost, which greatly simplifies gene cloning experiments. BSR-seq was used to demonstrate the cloning of the *glossy3* (*gl3*) gene of maize, which affects epicuticular wax deposition on juvenile leaves to generate the glossy phenotype. Moreover, bulked segregant RNA sequencing on the $F_2$ population was used to study the genetic architecture of the number of kernel rows in maize (Liu et al. 2016) as well as for leaf traits of pea (Zheng et al. 2018).

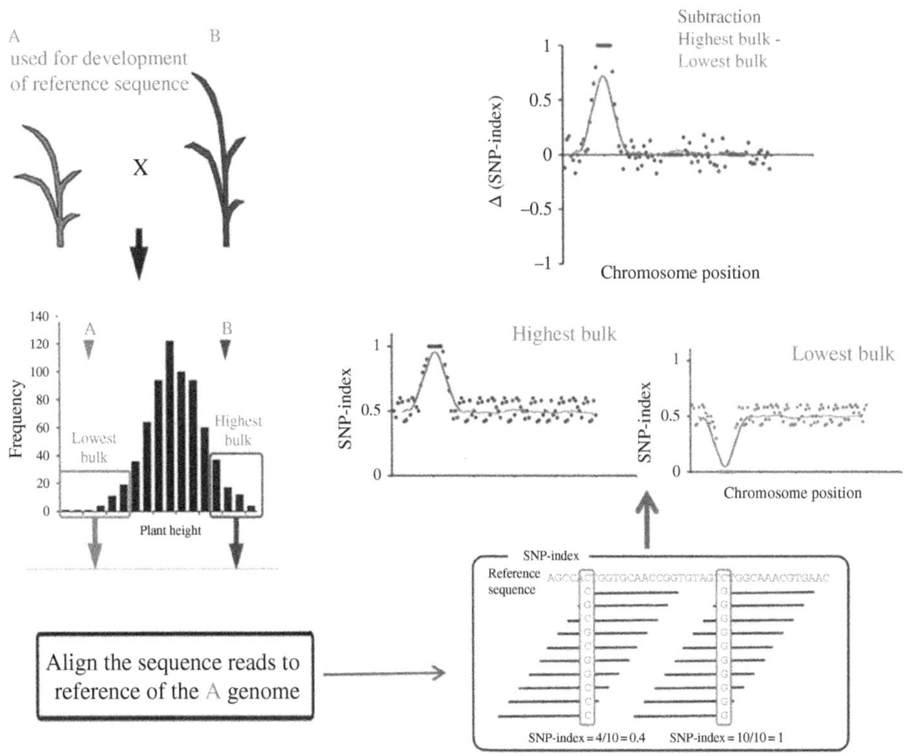


**Figure 4.3** Steps involved in the QTL-seq approach used to map the QTLs of the target trait.

10) Indel-seq: A mapping approach that mainly focuses on variations identified based on insertions and deletions (Singh et al. 2017). To date, the presence of insertions and deletions in the candidate genomic region has been largely ignored and not targeted for trait mapping in any of the approaches. This approach has larger practical utility with the fact that most of the cloned genes in rice and other crop possess Indels in the reported candidate genes (Figure 4.5).

## 4.7 Statistical Methods for QTL Mapping

Understanding the genetic basis of quantitative traits has been a major goal for plant breeders in breeding programs. However, the combined efforts of molecular biologists, plant breeders, and statistical geneticists lead to the development of the theory and methods for the detection of QTLs in different types of mapping populations (Doerge 2002; Sneller et al. 2009). The following are the methods often performed to determine QTLs:

1) Single-marker analysis (SMA): It is the simplest method of QTL mapping which does not require the construction of a linkage map and can be performed with basic statistical software programs. The phenotypic means between each of the marker genotype classes in the mapping population are subjected to the test of significance. A significant difference indicates the marker to be linked to a QTL governing the trait. The procedure is repeated with

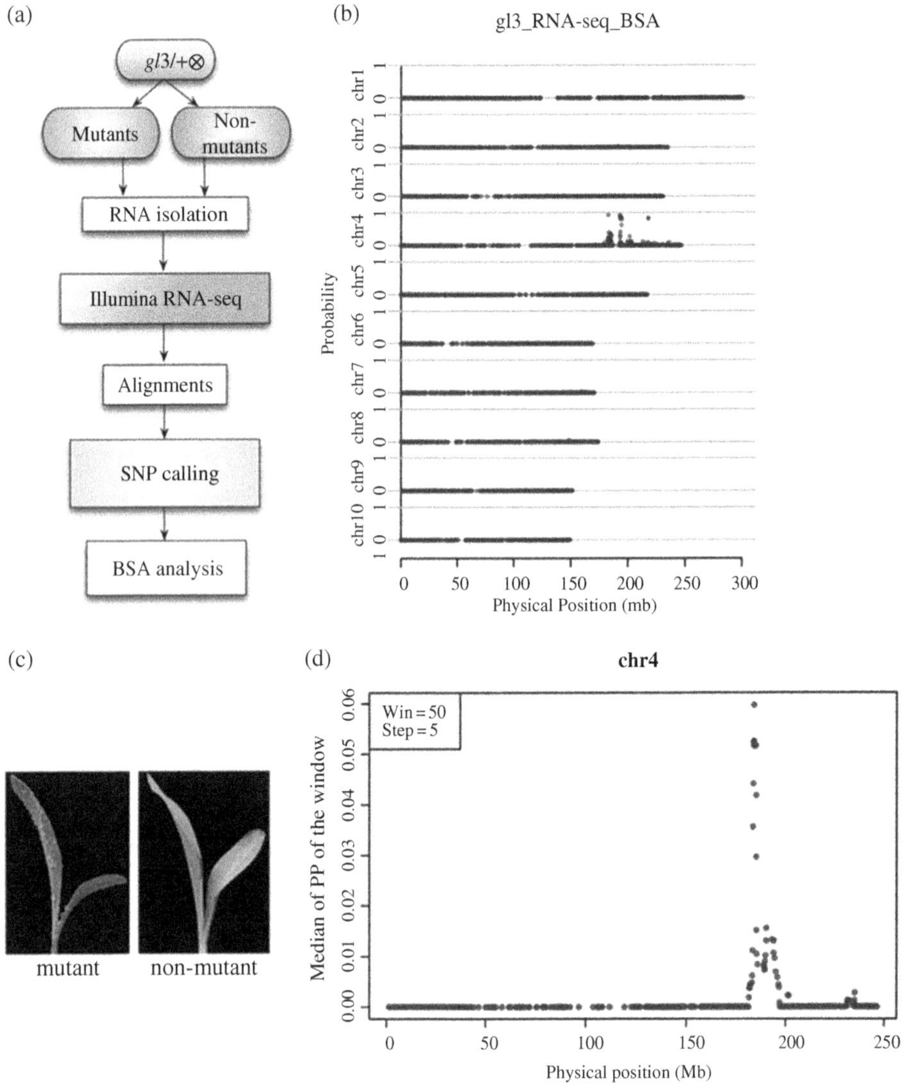


**Figure 4.4** Steps involved in bulked segregant RNA sequencing (BSR-Seq) used to map the QTLs of the target trait. *Source:* Liu et al., 2012, figure 01(p.03), CC BY 4.0

all the polymorphic markers data for the identification of the QTLs. The magnitude of difference between the phenotypic means of marker classes provides an estimate of the QTL effect. To test the significance of phenotypic means between marker classes, statistical tests used are *t*-tests, analysis of variance (ANOVA), linear regression, maximum likelihood estimation, etc. The major disadvantages of SMA are (i) the probability of detecting QTL decreases as the distance between the marker and QTL increases; (ii) the recombination rate cannot be estimated, as a result, the position of QTL in the genome remains unknown.

2) Simple interval mapping (SIM): Lander and Botstein (1989) developed the concept of SIM. This approach requires a linkage map as it uses adjacent marker pairs to define

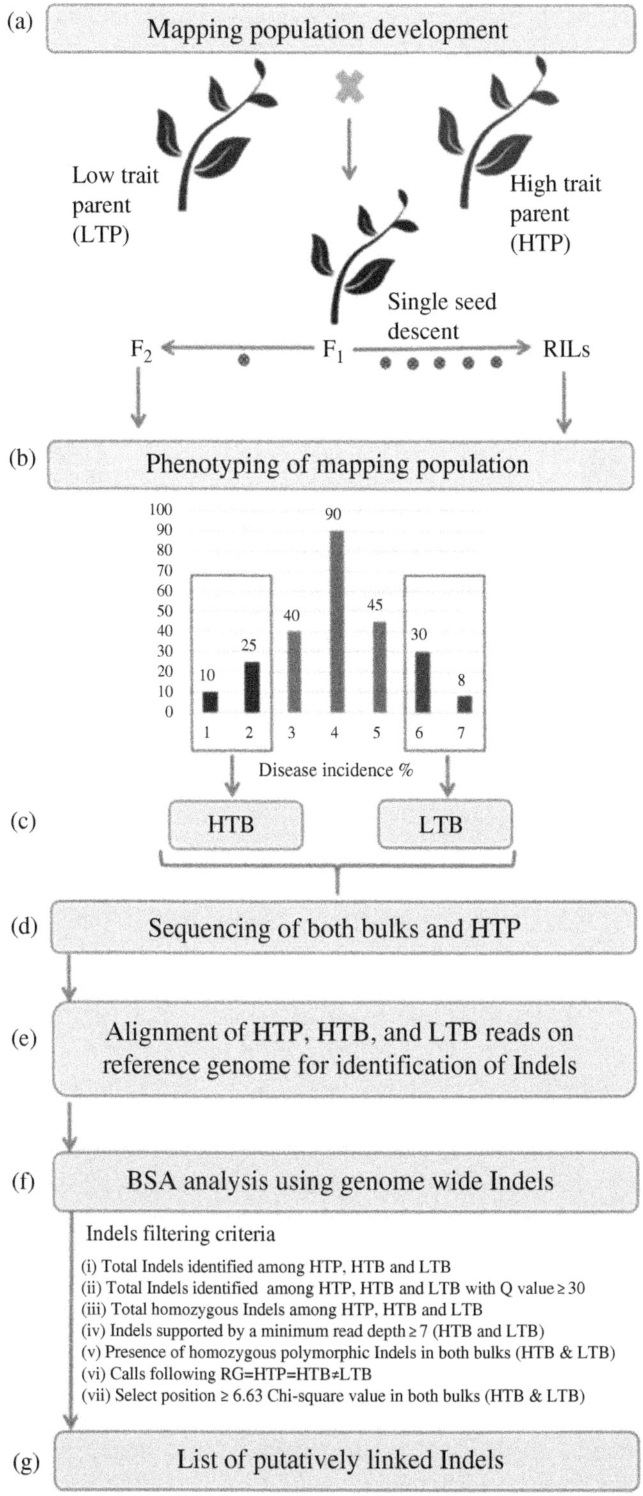


**Figure 4.5** Steps involved in Indel sequencing used to map the QTLs of the target trait.

marker intervals and detects QTLs within each of the intervals. It systematically searches for the QTL at every 1 or 2 cM within each marker interval. The maximum likelihood estimation method is used to estimate the LOD score, which indicates the likelihood of a QTL being present in the marker interval. The LOD score is plotted against the marker positions and a position where LOD score peaks as well as exceeds the threshold value is declared as the QTL. The QTL position is described by a confidence interval called a confidence or support interval. The estimate of the QTL effect is reliable than that of SMA. However, the major drawback in the SIM method is that it hypotheses presence of a single QTL for a trait, which is practically not possible. This leads to the detection of ghost QTLs and fails to detect the two linked QTLs with opposite effects.

3) Composite interval mapping (CIM): The method of CIM was developed by Jansen (1994) and Zeng (1994). It combines interval mapping with regression analysis. CIM first carries out SMA and identifies the significant markers and then builds up the multi-QTL model using a stepwise or forward regression method. Each of the markers is added sequentially and re-evaluated for significance. All the markers that remain significant when brought together are fitted into the model as cofactors and interval mapping is performed. This method significantly increases the QTL detection power and also resolves the linked QTLs into two separate QTLs, thereby estimating the QTL effects with better precision.
4) Inclusive composite interval mapping (ICIM): The ICIM is a modification of the CIM method which uses all the marker information to generate the linear regression model of CIM (Li et al. 2007a). It uses stepwise regression analysis to search the significant markers for the QTL analysis and ultimately identifies the significant QTLs affecting the trait. ICIM is simpler, less biased, and detects precise QTLs that have lower false positives. It also has the power of detecting additive, dominant, and epistatic QTLs using biparental populations.
5) Joint inclusive composite interval mapping (JICIM): The JICIM method is an extended form of ICIM (Li et al. 2011). In this method, one performs QTL mapping using multiple cross populations that have one common parent, e.g. nested association mapping (NAM) population data (Li et al. 2011). It includes stepwise regression analysis and then scans for marker intervals which is similar to the ICIM method. It also estimates the additive effect of each detected QTL in every family and detects rare QTL.
6) Multiple interval mapping (MIM): The MIM method was developed by Kao et al. (1999) and it uses multiple marker intervals for QTL mapping. MIM searches for the best genetic model by applying different model selection methods like the forward search method and forward and backward selection methods. However, MIM provides different results on the different model selection methods. It also estimates epistasis and interaction between QTL and the environment. This method is computationally intensive and detection of QTL depends on the selection of suitable models.
7) Bayesian multiple QTL mapping: Yi and Xu (2001) developed Bayesian multiple QTL mapping methods for complex multiple crosses. Using the Bayesian multiple QTL mapping methods, multiple QTLs can be detected with QTL by environment interaction using the Markov Chain Monte Carlo algorithm. It is particularly suitable when high-density maps are available and the genotype data without missing values. In this method, first, a prior distribution is selected then the posterior distribution is derived from this and finally, inferences are drawn from the posterior distribution. It is not widely used for QTL mapping because of the complexities of computation, difficulty in selecting suitable prior distribution, and lack of user-friendly software.

## 4.8 Software for QTL Mapping

Identification of QTL that influences quantitative traits plays a key role in uncovering the genetic basis of complex traits. The development of statistical methods and QTL mapping software enables the identification of genomic regions associated with phenotypic variation of traits. Each software has unique strengths in which they are particularly useful. Different software is available for QTL mapping.

## 4.9 Bi-parental Mapping Populations

A population used for the identification of genomic regions controlling phenotypic traits is known as a mapping population. Mapping populations developed by the crossing of two genetically diverse parents are known as bi-parental mapping populations. The parents used for crossing may be from the same species or related species. But it is essential to develop a suitable mapping population using parents that are highly diverse phenotypically and genotypically for the target trait of interest. Hence, it is important to perform the polymorphism survey to ascertain variation at the phenotypic and genotypic levels between the two parents before crossing. The variation at the genotypic level, i.e., at the DNA level is essential to find the recombination events. The more variation exists at the DNA level, the easier it is to find polymorphic informative markers. If the parents are diverse with respect to phenotype for a given trait, there is a great chance that genetic variation is present between the parents. However, environmental factors might affect the phenotype, which may have no genetic basis. Similarly, the absence of phenotypic variation between the selected parents does not necessarily mean that there is no genetic variation for the trait of interest, as distinct sets of genes could cause the same phenotype (Mackay, 2001; Ribaut and Hoisington, 1998).

There are several types of bi-parental mapping population that may be produced from a cross between two parental lines for the target trait which can be used for QTL mapping (Figure 4.6). The following are different types of mapping populations:

1) **$F_2$ population:** A $F_2$ mapping population produced by selfing or intermating of the $F_1$ individuals obtained from a cross between the selected parents. $F_2$ individuals are the product of a single meiotic cycle. In an $F_2$ population, the ratio expected for the dominant marker is 3 : 1 and for the codominant marker is 1 : 2 : 1. $F_2$ populations are relatively easy to produce with minimum effort and are considered the best population for preliminary mapping. Therefore, the target traits are scored on single plants. These populations cannot be used for fine mapping and have limited use in QTL mapping. Each individual is genetically different in $F_2$ populations and cannot be evaluated in replicated trials over locations and years. Hence, the effect of genotype x environment interaction on the expression of quantitative traits cannot be precisely estimated. Similarly, the QTL×QTL interaction cannot be detected but QTLs having additive effects can be detected. It is not an immortal population and hence impossible to develop an exact replica or increase seed amount.
2) **$F_2$-derived $F_3$ population ($F_{2:3}$):** $F_{2:3}$ population is obtained by selfing the $F_2$ individuals for a single generation. The seeds from each $F_2$ plant are harvested separately

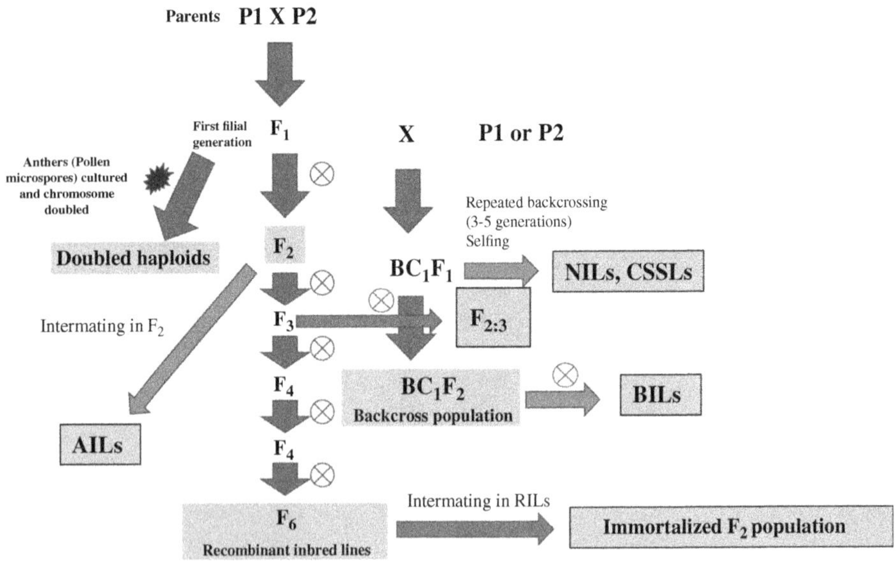


**Figure 4.6** Schematic representation of different types of biparental mapping populations used for mapping QTLs of target traits.

and each $F_2$ plant is represented as an individual plant progeny. For genotyping, DNA is obtained from individual $F_2$ plants or it can be from a bulk of at least 20 plants of each F3 family and this bulked DNA may represent the genotype of the parental $F_2$ plant (Yu et al. 1997). It is also not immortal like the $F_2$ population. It is suitable for mapping quantitative traits and recessive genes.

3) **Backcross population:** Backcross populations are derived by crossing $F_1$ plants to one of the two parents used in the initial cross. Generally, for genetic analysis, backcross with a recessive parent is used which is known as testcross. It also requires less time to develop and recombination information is based on only one parent. Backcross populations can be utilized for marker-assisted backcrossing breeding to introgress the target traits. Like $F_2$ populations, the backcross populations are not perpetual and cannot be evaluated in replicated trials, so unsuitable for QTL mapping. However, it is suitable for oligogene mapping. Using this population, the negative alleles from the recurrent parent cannot be detected and QTL x QTL interaction cannot be estimated.
4) **Doubled haploids (DHs):** DH plants are developed by chromosome doubling of haploid plants obtained by another culture or pollen culture from $F_1$ plants. DH populations are permanent mapping populations so they can be evaluated in replicated trials over location and years and multiplied without any genotypic change. They are suitable for mapping both quantitative and qualitative traits. Homozygous lines are produced in a short period, thus saving time. DHs are comparable to $F_2$ as both are products of one meiotic cycle. In a DH population, the expected ratio for dominant or codominant markers is 1 : 1. Its development involves tissue culture techniques so technical skill is required in comparison to other mapping populations. This population is not useful for mapping heterosis QTLs.

5) **Recombinant inbred lines (RILs):** RILs are produced by continuous selfing or sib mating of individual $F_2$ plants until complete homozygosity is achieved by following the single-seed descent (SSD) method. In the SSD method, one seed is harvested from each plant of the $F_2$, and in later generations seeds from all the plants are bulked and planted to raise the next generation and this procedure is followed for five or more generations. In the end, seeds from each plant are harvested separately to obtain as many RILs as the lines fixed for many recombination events. The genetic segregation ratio for both dominant and co-dominant markers is 1 : 1. RILs are immortal populations so they can be evaluated in replicated trials over location and years and multiplied without any further segregation since achieved homozygosity. RILs have been widely used for the mapping of QTLs. RILs are obtained after several cycles of meiosis, therefore very useful for the identification of tightly linked markers. However, it requires many seasons or generations to develop, and developing RILs in crops with high inbreeding depression is difficult.
6) **Immortalized $F_2$ population:** Immortalized $F_2$ populations are produced by intercrossing a set of RILs. It is an immortal population so it can be evaluated in replicated trials over location and years. It can be used for heterotic QTL detection. It requires the making of a large number of crosses for development.
7) **Near-isogenic lines (NILs):** NILs are developed by either repeated selfing or backcrossing the $F_1$ plants to the recurrent parents. NILs are similar to recurrent parents except for the gene of interest when developed through backcrossing. However, NILs developed through selfing are similar in pair except for the gene of interest but differ a lot from the recurrent parent. The genetic segregation ratio for both dominant and co-dominant markers is 1 : 1. Like DHs and RILs, NILs are immortal populations. They are useful for positional cloning. It requires many generations to develop and the linkage drag problem is associated with NILs.
8) **Chromosomal segment substitution lines (CSSLs):** They are a homozygous line with distinct chromosome segments from a donor parent in the chromosome background of the recurrent parent. It is useful for the mapping of oligogenes as well as QTLs. Linkage drag is a major problem in CSSLs as undesirable traits might have been linked to the target gene.
9) **Backcross inbred lines (BILs):** BILs are obtained by backcrossing the $F_1$ from a cross between one of the two parents used in the initial cross and then continue selfing of the $BC_1F_1$ progeny to get homozygous lines.
10) **Advanced intercross lines (AILs):** AIL population is produced by intermating the $F_2$ individuals and subsequent generations from a suitable cross. Heterozygosity is maintained by intermating in the segregating generations and thus it allows recombination between the QTLs and the markers linked to them in every generation and leads to more precise QTL identification.

## 4.10 QTL Mapping Using Bi-parental Populations

QTL mapping depends on well-defined populations developed by crossing two parents. Bi-parental mapping populations are simple to develop and therefore most widely used for identifying regions of genome controlling phenotypic variation. For QTL mapping, it is essential to develop a suitable experimental bi-parental mapping population using parental lines that are phenotypically diverse for the target trait (highly resistant/tolerant and susceptible lines) as well as genetically diverse. The choice of a bi-parental mapping population depends upon the objectives of the experiment, the timeline as well as the availability of markers for carrying QTL analysis. The development of biparental mapping populations requires time and effort. However, this may not be feasible for some species like tree species also each population type has its advantages and disadvantages. $F_2$ and $F_2$-derived $F_3$ populations can measure the effects of additive and dominance gene actions for a specific locus while RIL and DH populations are homozygous, therefore only additive gene action can be measured. In the backcross population, negative alleles from the recurrent parent cannot be detected and biased in an estimate of gene effects when dominance gene action is present.

QTL mapping using biparental populations tells only a small portion of the genetic architecture for a trait of interest because only two alleles that differ between the two parental lines will be detected. Thus, it has low-resolution power as only a few meiotic events occur during the hybridization between the parents. A QTL detected in a biparental population need to be validated in different genetic backgrounds to test their applicability which requires additional time and effort. However, biparental mapping is a powerful tool for detecting complex traits in plants.

## 4.11 Multiparental Mapping Populations

QTL mapping is used to identify the genomic location controlling phenotypic variation. Generally, bi-parental populations are routinely used for QTL mapping which includes $F_2$, BILs or RILs, NILs, AILs, double haploid populations which derived from only two parents. Therefore, only two alleles are analyzed and genetic recombination in bi-parental populations is limited especially in $F_2$ or backcross populations which limits the resolution for QTL detection. Recently, a multiparental population strategy has been proposed to interrogate multiple alleles and to provide an increased level of recombination and mapping resolution. The main objective of developing multiparental populations from more than two founder or parental inbred lines is to promote intercrossing and shuffling of the genome. Therefore, it possesses far greater genetic variation and phenotypic variability than bi-parental populations.

Multiparental populations have been developed in numerous crops and have been successfully utilized in QTL mapping studies. The two most commonly developed forms of the multiparental population in crop breeding are NAM and multifounder advanced generation inter-cross (MAGIC) populations.

### 4.11.1 Nested Association Mapping (NAM)

NAM populations are obtained by crossing a single inbred parent to a successive set of diverse inbred lines. It will provide increased power and precision of QTL mapping by combining the advantages of both bi-parental populations and association mapping. Yu et al. (2008) developed the first NAM population in maize which is derived from crosses between the reference line B73 and 25 diverse inbred lines to generate 5000 RILs. The NAM population harbors thousands of recombination events however recombination and segregation distortion differ among different families which can reduce the precision of genetic dissection of quantitative traits (McMullen et al. 2009). In barley, a NAM-derived advanced backcross population has been developed which combines wild landraces with the exotic background Rasmusson (Nice et al. 2016).

### 4.11.2 Multi-advanced Generation Inter-cross Populations (MAGIC)

MAGIC population is obtained by inter-crossing multiple parental lines (usually 4, 8, or 16) in a balanced funnel crossing system and subsequently the population advanced through multiple generations of intermating or SSD. Thus obtained RILs are highly recombined assortments of the founder genomes. Multicross populations were first proposed for mice known as heterogeneous stock and collaborative cross populations (Mott et al. 2000; Valdar et al. 2006; Threadgill and Churchill 2012) later the concept was adopted in plants by Mackay and Powell (2007). Huang et al. (2011) developed the Arabidopsis multiparent recombinant inbred line (AMPRIL) population from diallel crossing of eight accessions of Arabidopsis from diverse geographical origins. The MAGIC population gives high mapping resolution due to high level of recombination and low LD. For this reason, the MAGIC approach has been widely applied to many crop and plant species for genetic research and breeding (Huang et al. 2015)

The MAGIC populations have been developed in Arabidopsis (Kover et al. 2009), wheat (Huang et al. 2012; Mackay et al. 2014), rice (Bandillo et al. 2013), tomato (Pascual et al. 2013), sorghum (Higgins et al. 2014), barley (Sannemann et al. 2015), and maize (Dell'Acqua et al. 2015). A high-density MAGIC linkage map has been developed in wheat (Gardner et al. 2016).

## 4.12 QTL Mapping Using Multiparental Populations

In recent years, the novel multiparental populations have greatly transformed the way of detecting QTLs in various plant species. The two multiparent populations namely, MAGIC and NAM populations are very popular and are used for genome-wide association studies, QTL mapping as well as used for variety development. Multiparental populations have a large genetic variation among parental lines and a large population size; therefore, they are powerful multiple-cross to study the complex traits. Multiparental populations can be used for studying the interactions of genome segment introgressions and chromosomal recombinations. However, efficient and advanced statistical methods are required for QTL linkage mapping of genetic populations derived from multiparental populations.

NAM population generated from a multiple-cross mating design sharing one common parent facilitates high power and high resolution for quantitative trait mapping. The first NAM population was reported in maize. This maize NAM population has 25 biparental families sharing the common parent B73. A total of 5000 lines were produced from 200 RILs of each cross. The 4699 RILs and 1106 markers were used for QTL mapping (McMullen et al. 2009). Similarly, Hayashi and Iwata (2009) used a NAM population consisting of 2000 individuals and 84 evenly distributed markers in the whole genome to identify QTL using a Bayesian method of QTL mapping. Li et al. (2011) developed an algorithm for JICIM method to detect QTL in maize and an Arabidopsis NAM population. The maize

NAM population has been broadly used to study the genetic architecture of several morphological and disease resistance traits (Buckler et al. 2009; Tian et al. 2011; Cook et al. 2012). Bajgain et al. (2016) used NAM in wheat for stem rust resistance using genotyping by sequencing approach.

Two wild barley (*Hordeum vulgare* spp. spontaneum) NAM populations, the AB-NAM and HEB-25, were used to identify QTL for complex traits, including days to flowering (DTF) (Maurer et al. 2015; Nice et al. 2016, 2017). Hemshrot et al. (2019) developed a NAM population of barley with 88 cultivated accessions crossed to a common parent (Rasmusson). They generated 6160 $F_5$ RILs with an average of 69 and a range of 37–168 RILs per family and performed genotyping with 7773 SNPs with an average of 3889 SNPs segregating per family. They detected 23 QTL associated with flowering time and found that a sample size of 50 families or 3000 individuals results in the highest power for QTL detection for flowering time in barley.

MAGIC populations are derived from a complex cross/outcross population involving several parental lines resulting in RILs. The parental lines may include breeding lines and germplasm lines of interest to breeders. The progeny from the complex cross or outcross may be intermated for one or more generations. Since these populations are derived from several parents, which depicts segregation for multiple traits and multiple QTLs for each trait. As well as show more than two alleles for individual QTLs. Thus, it complicates recombination frequency estimation and linkage map construction. However, precise linkage analysis and mapping of QTL are essential for MAGIC populations. Several softwares have been developed for linkage map construction and QTL mapping using MAGIC populations which include R/happy (Mott et al. 2000), R/qtl (Broman et al. 2003), R/mpMap (Huang and George 2011), and GAPL (Zhang et al. 2017, 2019). Among these packages, GAPL is a stand-alone package which is used to construct accurate linkage maps in a short time (Zhang et al. 2019). ICIM operates in GAPL is used for accurate estimation of QTL positions and effects in both four and eight parent pure-line populations with high detection power and less false discovery rate (Shi et al. 2019).

A large wheat MAGIC population was used for mapping QTLs governing complex traits like plant height and hectolitre weight (Huang et al. 2012). Bandillo et al. (2013) developed MAGIC rice populations using eight founder parents and 1328 lines produced by SSD. The MAGIC population was genotyped using a genotyping-by-sequencing (GBS) method and phenotyped for different traits such as salinity, submergence tolerance, resistance to blast and bacterial blight, and grain quality. At IRRI, rice MAGIC lines were developed for dissection of complex traits such as yield, drought tolerance, and quantitative disease resistance.

Stadlmeier et al. (2018) developed an eight-founder MAGIC population in common wheat (*Triticum aestivum* L.) for QTL mapping for seedling resistance to powdery mildew. The population was composed of 394 $F_{6:8}$ RILs and genotyped with 5435 SNP markers. Three major and two minor QTL was detected with a total explained phenotypic variance of 72.5% using the mpMap package.

## 4.13 Use of High-throughput Genotyping for QTL Mapping

Conventional genotyping methods for QTL mapping are often hindered by the expense and time required to screen the hundreds to thousands of individuals of a segregating mapping population with hundreds to thousands of polymorphic markers. Also, the low marker coverage genotyping results in poor mapping resolution of QTLs. Nevertheless, the rapid advances in high-throughput sequencing and genotyping technologies have recently facilitated the mapping of QTL more accurately. In recent years, SNPs have attracted substantial attention and have emerged as the marker of choice in crop breeding due to their abundance and amenability to high-throughput genotyping. Many SNP genotyping methods available are capable of efficient, fast, and high-throughput sample processing for QTL mapping. The available SNP genotyping methods are either based on sequence information to generate genotyping assays or directly sequencing a large number of individuals.

### 4.13.1 PCR-based SNP Genotyping

One of the earliest methods for PCR-based SNP genotyping was the TaqMan assay which is based on allele-specific fluorescently tagged probes and detected by using a real-time polymerase chain reaction (PCR) machine (De La Vega et al. 2005). Another popular and leading SNP genotyping technology is Kompetitive allele-specific PCR (KASP) (Semagn et al. 2014), which discriminates tagged alleles based on endpoint fluorescence detection method by using a fluorescence microplate reader or real-time PCR machine. KASP genotyping assay includes two forward primers, each specific to an allele and tagged with a particular fluorophore, and one common reverse primer to discriminate between two alleles. The software is required to analyze the data and it is included with the plate reader. The TaqMan and the KASP assays have been widely used for SNP genotyping due to their high throughput, low cost, fast, and tolerance of variation in the quality and quantity of input DNA (Figure 4.7). However, the KASP assays can detect only two alleles.

Another method for SNP genotyping is mass spectrometry (MS), which uses the Sequenom MassARRAY platform and detects multiple alleles at a given locus. The assay comprises of a preliminary locus-specific PCR followed by a single base extension by employing mass-modified dideoxy nucleotide terminators of an oligonucleotide primer

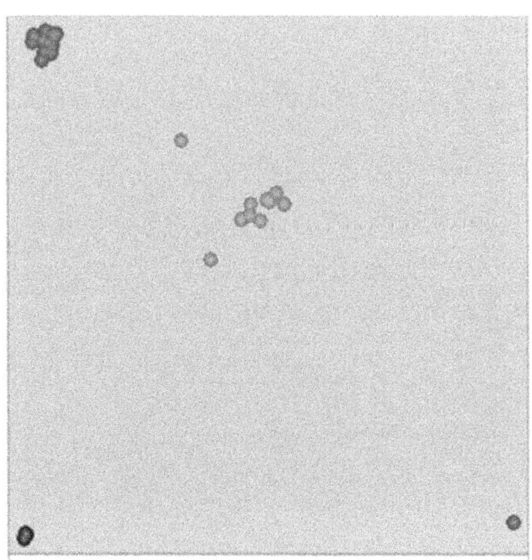

**Figure 4.7** Genotyping of segregating population using KASPar assay.

which anneals immediately upstream of the polymorphic site of interest. The distinct mass of the extended primer identifies the SNP allele using MALDI-TOF MS and data analyzed with software included with the system (Figure 4.8).

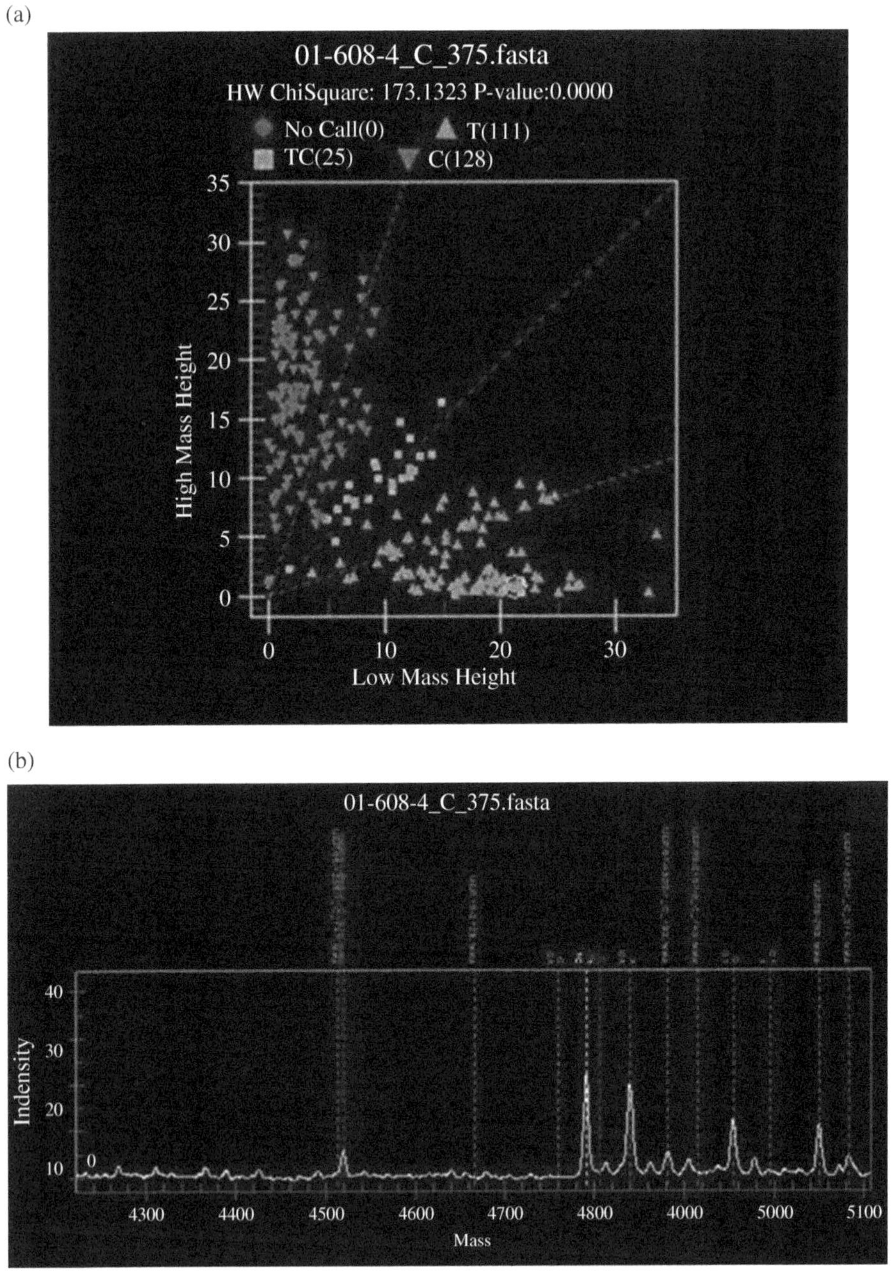


**Figure 4.8** Genotyping of segregating population using Sequenom MassARRAY system for SNP marker.

## 4.14 Next-Generation Sequencing-based Genotyping

In recent years, NGS technologies are being increasingly used in QTL mapping studies due to the rapid decrease in the cost of sequencing. Thus, enabling the discovery of the cost-effective marker and genotyping to map genomic regions controlling traits of interest more accurately for which no reference genome or genotypic information is available. NGS-based genotyping approaches such as restriction-site-associated DNA sequencing (Baird et al. 2008; Zhou et al. 2014), GBS (Elshire et al. 2011; Sonah et al. 2013), and whole-genome resequencing (Xu et al. 2013; Qi et al. 2014) has been used in crops for QTL mapping. These newer methods are based on sequencing of the whole genome or targeted genome and avoid the limitation of marker availability. The emergence of NGS-based high-throughput genotyping platforms has sufficient marker density and is used successfully in complex crop species (Table 4.1).

### 4.14.1 Restriction-Site-Associated DNA Sequencing (RAD-seq)

The restriction site-associated genomic DNA (RAD) is used for high-density SNP discovery and genotyping and to increase the resolution of QTL mapping (Spindel et al., 2013). RAD-seq uses methylation-sensitive enzymes for cutting genome sequences which reduces redundancy. RAD-seq method has been reported in many crops, such as barley, grape, pear, eggplant, alfalfa, soybean, sorghum, etc. for QTL identification.

### 4.14.2 Genotyping-by-Sequencing

The resolution of QTL mapping has been improved significantly by the increasing number of markers with the development of GBS. GBS can be used for simultaneous marker discovery and genotyping. In GBS, the cost per sample can be reduced by multiplexing many samples (He et al. 2014). Restriction enzyme digestion is included in GBS to reduce the genome complexity followed by the addition of DNA barcoded adapters to produce multiplex libraries of samples ready for NGS sequencing, which makes it possible to map traits of interest within mapping populations of plant species with large and complex genomes such as wheat, cotton, etc. An improved GBS protocol has been developed (Poland et al. 2012) which makes use of two-enzyme (PstI/MspI) digestion thereby reducing genome complexity by a greater degree and also results in more uniform libraries for sequencing than the original single enzyme protocol (Elshire et al. 2011). Sonah et al. (2013) also modified the library

**Table 4.1** The different software used for quantitative loci (QTL) mapping.

| S. No | Software | Description |
|---|---|---|
| 1) | Map Manager QT, Mapmaker/QTL, QTL Cartographer, R/qtl, PLABQTL, QTLNetwork | Mapping quantitative trait loci in populations derived from backcrosses, intercrosses, and recombinant inbred lines |
| 2) | R/qtl2, HAPPY, DOQTL, IciMapping | QTL analysis in multiparent populations |

preparation procedure to enhance both the number of SNPs called and their coverage depth by selective amplification which resulted in high efficiency and reduction in per-sample cost. BSAs combined with the GBS method was used to map QTLs for broomrape resistance in four sunflower lines (Imerovski et al. 2019)

### 4.14.3 Whole-Genome Resequencing

Whole-genome resequencing has been extensively used in genotyping for QTLs identification in recent years. BSAs combined with WGS have been applied for the rapid identification of QTLs for important agronomic traits of crops. Yang et al. (2017) used pools of low and high bulk samples from 280 $F_2$ individuals to detect the genomic regions in the rice for nitrogen use efficiency by applying BSA, WGS, and single-nucleotide polymorphism index (SNP-index) methods.

BSA-seq has been successfully applied for mapping traits in cucumber such as early flowering (Lu et al. 2014), flesh thickness (Xu et al. 2015), and downy mildew resistance (Win et al. 2019). Pujol et al. (2019) mapped a monogenic locus for resistance to Cucumber vein yellowing virus in cucumber by using a BSA approach coupled with whole-genome resequencing. They selected 15 resistant and 15 susceptible homozygous $F_2$ individuals for mapping using whole-genome resequencing. The resequencing data resulted in a large number of SNP markers and lead to the identification of precise locus for resistance to cucumber vein yellowing virus (Table 4.2).

**Table 4.2** Studies that utilized high-throughput genotyping for QTL mapping.

| Crop | Trait | Mapping population | References |
|---|---|---|---|
| Maize | Dozen recessive mutants | $F_2$ | Liu et al. (2010) |
| Arabidopsis | Sulfur and selenium content | $F_2$ | Becker et al. (2011) |
| Cotton | Salt resistance | $BC_2F_1$ | Rodriguez-Uribe et al. (2011) |
| Rice | Pale green leaves and semi dwarfism | $F_2$ | Abe et al. (2012) |
| Maize | Root-lodging | $F_2$ | Farkhari et al. (2013) |
| Rice | Male sterility | $F_2$ | Frouin et al. (2013) |
| Rice | Blast disease and seedling vigor | RILs | Takagi et al. (2013a) |
| Rice | Cold tolerance | $F_3$ | Yang et al. (2013) |
| Wheat | Leaf rust | $F_{3:4}$ | Forrest et al. (2014) |
| Common bean | Bean common mosaic virus | Natural population | Bello et al. (2014) |
| Pepper | Phytophthora root rot | $F_2$ | Liu et al. (2014) |
| Poplar | Photosynthetic traits | $F_2$ | Wang et al. (2014a) |

**Table 4.2** (Continued)

| Crop | Trait | Mapping population | References |
|---|---|---|---|
| Maize | gl3 gene | $F_2$ | Liu et al. (2012) |
| Wheat | Grain protein content | RSLs | Trick et al. (2012) |
| Sunflower | Downy mildew | $F_2$ | Livaja et al. (2013) |
| Cotton | Short-fiber mutant | $F_2$ | Thyssen et al. (2014) |
| Cucumber | Early flowering | $F_2$ | Lu et al. (2014) |
| Wheat | Yellow rust | $F_2$ | Ramirez-Gonzalez et al. (2014) |
| Radish | Cytoplasmic male sterility | $F_2$ | Lee et al. (2014b) |
| Sand pear | Pericarp russet pigmentation | $F_2$ | Wang et al. (2014b) |
| Cotton | Nulliplex-branch | $F_2$ | Chen et al. (2015) |
| Tomato | Fruit weight and locule number | $F_2$ | Illa-Berenguer et al. (2015) |
| Chickpea | 100-seed weight | $F_5$ | Das et al. (2015) |
| Rice | Salt tolerance | $F_2$ | Takagi et al. (2015) |
| Maize | Multiple traits | Natural population | Yang et al. (2015) |
| Onion | Restorer-of-fertility | $F_{2:5}$ | Kim et al. (2015) |
| Pigeonpea | Fusarium Wilt and SMD | RILs | Singh et al. (2016b) |
| Chickpea | 100-seed weight, root/plant dry weight | RILs | Singh et al. (2016a) |
| Ground nut | Rust and late leaf spot | RILs | Pandey et al. (2017) |
| Wheat | Flag leaf traits | RILs | Hussain et al. (2017) |
| Maize | Maize kernel row number | $F_2$ | Liu et al. (2016) |
| Chickpea | Flowering time | $F_9$ | Srivastava et al. (2017) |
| Rice | Grain length and weight | NIL, $F_2$ | Yaobin et al. (2018) |
| Soybean | Plant height | $F_2$ | Zhang et al. (2018) |
| Tomato | Early flowering | $F_2$ | Ruangrak et al. (2018) |
| Broccoli | Flowering time | $F_2$, $F_{2:3}$ | Shu et al. (2018) |
| Pea | Leaf shape | $F_2$ | Zheng et al. (2018) |
| Tomato | Heat tolerance | $F_2$ | Wen et al. (2019) |
| Groundnut | Fresh seed dormancy | RILs | Kumar et al. (2020) |
| Rice | Spikelet fertility | $F_2$ | Nubankoh et al. (2020) |

## 4.15 Challenges with QTL Mapping Using Multiparental Populations and High-throughput Genotyping

Presently, the use of multiparent populations is an attractive approach for QTL mapping because the availability of high-throughput genotyping platforms and advances in statistical methods can aid in analyzing data derived from such populations. However, the development of multiparent populations needs a greater investment of time and effort compared to bi-parental populations. It requires more founders and large populations for the precision of QTL detection. The power of QTL detection is also hampered by genotyping and phenotyping resources. Compare to MAGIC, NAM is more highly influenced by the choice of the recurrent (common) parent. Thus, NAM should be extended to include more than one recurrent parent to increase diversity for trait mapping (Gao et al. 2010). Statistical methods and software for QTL analysis are required to incorporate multiple alleles from multiple-parent populations.

The disadvantage of NGS-based genotyping is that it yields large amount of missing data due to the low depth of sequencing. Hence imputation of the data is necessary to assign values before performing any genetic analysis. Accurate imputation methods depend on certain level of LD between adjacent markers or complete sequence of the reference genome of germplasm that is believed to include most of the haplotypes found in the panel to be imputed. Many algorithms have been developed to impute genotypes. Imputation of the data improves the power of detection of marker-trait associations for the trait of interest. However, for such crop species which are not having the reference genome information and complete reference linkage map, the imputation becomes more challenging.

## References

Abe, A., Kosugi, S., Yoshida, K. et al. (2012). Genome sequencing reveals agronomically important loci in rice using MutMap. *Nature Biotechnology* 30: 174–178.

Baird, N.A., Etter, P.D., Atwood, T.S. et al. (2008). Rapid SNP discovery and genetic mapping using sequenced RAD markers. *PLoS One* 3: e3376.

Bajgain, P., Rouse, M.N., Tsilo, T.J. et al. (2016). Nested association mapping of stem rust resistance in wheat using genotyping by sequencing. *PLoS One* 11: 1–22.

Bandillo, N., Raghavan, C., Muyco, P. et al. (2013). Multi-parent advanced generation inter-cross (magic) populations in rice: progress and potential for genetics research and breeding. *Rice* 6: 11.

Becker, A., Chao, D.-Y., Zhang, X. et al. (2011). Bulk segregant analysis using single nucleotide polymorphism microarrays. *PLoS One* 6: e15993.

Beckmann, J. and Soller, M. (1986). Restriction fragment length polymorphisms and genetic improvement of agricultural species. *Euphytica* 35: 111–124.

Bello, M.H., Moghaddam, S.M., Massoudi, M. et al. (2014). Application of in silico bulked segregant analysis for rapid development of markers linked to bean common mosaic virus resistance in common bean. *BMC Genomics* 15: 903.

Broman, K.W., Wu, H., Sen, S. et al. (2003). R/qtl: QTL mapping in experimental crosses. *Bioinformatics* 19 (7): 889–890.

Buckler, E.S., Holland, J.B., Bradbury, P.J. et al. (2009). The genetic architecture of maize flowering time. *Science* 7 (325): 714–718.

Chen, W., Yao, J., Chu, L. et al. (2015). Genetic mapping of the nulliplex-branch gene (*gb_nb1*) in cotton using nextgeneration sequencing. *Theoretical and Applied Genetics* 128: 539–547.

Cook, J.P., McMullen, M.D., Holland, J.B. et al. (2012). Genetic architecture of maize kernel composition in the nested association mapping and inbred association panels. *Plant Physiology* 158: 824–834.

Darvasi, A. and Soller, M. (1992). Selective genotyping for determination of linkage between a marker locus and a quantitative trait locus. *Theoretical and Applied Genetics* 85: 353–359.

Darvasi, A. and Soller, M. (1994). Selective DNA pooling for determination of linkage between a molecular marker and a quantitative trait locus. *Genetics* 138 (4): 1365–1373.

Das, S., Upadhayaya, H.D., Srivastava, R. et al. (2015). Genome-wide insertion-deletion (InDel) marker discovery and genotyping for genomics-assisted breeding applications in chickpea. *DNA Research* 22: 377–386.

De La Vega, F.M., Lazaruk, K.D., Rhodes, M.D. et al. (2005). Assessment of two fexible and compatible SNP genotyping platforms: TaqMan® SNP genotyping assays and the SNPlexTM genotyping system. *Mutation Research* 573: 111–135.

Dell'Acqua, M., Gatti, D.M., Pea, G. et al. (2015). Genetic properties of the magic maize population: a new platform for high definition QTL mapping in Zea mays. *Genome Biology* 16: 167.

Doerge, R.W. (2002). Mapping and analysis of quantitative trait loci in experimental populations. *Nature Reviews. Genetics* 3: 43–52.

Ehrenreich, I., Torabi, N., Jia, Y. et al. (2010). Dissection of genetically complex traits with extremely large pools of yeast segregants. *Nature* 464: 1039–1042.

Elshire, R.J., Glaubitz, J.C., Sun, Q. et al. (2011). A robust, simple genotyping-by-sequencing (GBS) approach for high diversity species. *PLoS One* 6: e19379.

Farkhari, M., Krivanek, A., Xu, Y. et al. (2013). Root-lodging resistance in maize as an example for high-throughput genetic mapping via single nucleotide polymorphism-based selective genotyping. *Plant Breeding* 132: 90–98.

Fekih, R., Takagi, H., Tamiru, M. et al. (2013). MutMap+: genetic mapping and mutant identification without crossing in rice. *PLoS One* 8: e68529.

Forrest, K., Pujol, V., Bulli, P. et al. (2014). Development of a SNP marker assay for the Lr67 gene of wheat using a genotyping by sequencing approach. *Molecular Breeding* 34: 2109–2118.

Frouin, J., Filloux, D., Taillebois, J. et al. (2013). Positional cloning of the rice male sterility gene msIR36 widely used in the inter-crossing phase of recurrent selection schemes. *Molecular Breeding* 33: 555–567.

Gao, X., Becker, L.C., Becker, D.M. et al. (2010). Avoiding the high Bonferroni penalty in genome-wide association studies. *Genetic Epidemiology* 34: 100–105.

Gardner, K.A., Wittern, L.M., and Mackay, I.J. (2016). A highly recombined, high-density, eight-founder wheat magic map reveals extensive segregation distortion and genomic locations of introgression segments. *Plant Biotechnology Journal* 14: 1406–1417.

Hayashi, T. and Iwata, H. (2009). Bayesian QTL mapping for multiple families derived from crossing a set of inbred lines to a reference line. *Heredity* 102: 497–505.

He, J., Zhao, X., Laroche, A. et al. (2014). Genotyping-by-sequencing (GBS), an ultimate marker-assisted selection (MAS) tool to accelerate plant breeding. *Frontiers in Plant Science* 5: 484.

Hemshrot, A., Poets, A.M., Tyagi, P. et al. (2019). Development of a multiparent population for genetic mapping and allele discovery in six-row barley. *Genetics* 213: 595–613.

Higgins, R.H., Thurber, C.S., Assaranurak, I. et al. (2014). Multiparental mapping of plant height and flowering time QTL in partially isogenic sorghum families. *G3* 4: 1593–1602.

Huang, B.E. and George, A.W. (2011). R/mpMap: a computational platform for the genetic analysis of multi-parent recombinant inbred lines. *Bioinformatics* 27: 727–729.

Huang, X., Paulo, M.J., Boer, M. et al. (2011). Analysis of natural allelic variation in Arabidopsis using a multiparent recombinant inbred line population Proc. *Proceedings of the Nationall Academy Sciences of the United States of America* 108: 4488–4493.

Huang, B.E., George, A.W., Forrest, K.L. et al. (2012). A multiparent advanced generation inter-cross population for genetic analysis in wheat. *Plant Biotechnology Journal* 10: 826–839.

Huang, B.E., Verbyla, K.L., Verbyla, A.P. et al. (2015). MAGIC populations in crops: current status and future prospects. *TheorAppl Genetics* 128 (6): 999–1017.

Hussain, W., Baenziger, P.S., Belamkar, V. et al. (2017). Genotyping-by-sequencing derived high-density linkage map and its application to QTL mapping of flag leaf traits in bread wheat. *Scientific Reports* 7: 16394.

Illa-Berenguer, E., Van Houten, J., Huang, Z., and van der Knaap, E. (2015). Rapid and reliable identification of tomato fruit weight and locule number loci by QTL-seq. *Theoretical and Applied Genetics* 128: 1329–1342.

Imerovski, I., Dedić, B., Cvejić, S. et al. (2019). BSA-seq mapping reveals major QTL for broomrape resistance in four sunflower lines. *Molecular Breeding* 39: 41.

Jansen, R.C. (1994). High resolution of quantitative traits into multiple loci via interval mapping. *Genetics* 136: 1447–1455.

Kao, C.H., Zeng, Z.B., and Teasdale, R.D. (1999). Multiple interval mapping. *Genetics* 152: 1203–1216.

Kim, S., Kim, C.-W., Park, M. et al. (2015). Identification of candidate genes associated with fertility restoration of cytoplasmic male-sterility in onion (*Allium cepa* L.) using a combination of bulked segregant analysis and RNA-seq. *Theoretical and Applied Genetics* 128: 2289–2299.

Kover, P.X., Valdar, W., and Trakalo, J. (2009). A multiparent advanced generation inter-cross to fine-map quantitative traits in *Arabidopsis thaliana*. *PLoS Genetics* 5: 7.

Kumar, R., Janila, P., Vishwakarma, M.K. et al. (2020). Whole-genome resequencing-based QTL-seq identified candidate genes and molecular markers for fresh seed dormancy in groundnut. *Plant Biotechnology Journal* 18 (4): 992–1003.

Lander, F.S. and Botstein, D. (1989). Mapping Mendelian factors underlying quantitative traits using RFLP linkage maps. *Genetics* 121: 185–199.

Lee, Y.-P., Cho, Y., and Kim, S. (2014b). A high-resolution linkage map of the Rfd1 a restorer-of-fertility locus for cytoplasmic male sterility in radish (*Raphanus sativus* L.) produced by a combination of bulked segregant analysis and RNA-Seq. *Theoretical and Applied Genetics* 127: 2243–2252.

Li, H., Ye, G., and Wang, J. (2007a). A modified algorithm for the improvement of composite interval mapping. *Genetics* 175: 361–374.

Li, H., Bradbury, P., Ersoz, E. et al. (2011). Joint QTL linkage mapping for multiple-cross mating design sharing one common parent. *PLoS One* 6 (3): e17573.

Liu, S., Chen, H.D., Makarevitch, I. et al. (2010). High-throughput genetic mapping of mutants via quantitative single nucleotide polymorphism typing. *Genetics* 184: 19–26.

Liu, S., Yeh, C.T., Tang, H.M. et al. (2012). Gene mapping via bulked segregant RNA-Seq (BSR-Seq). *PLoS One* 7 (5): e36406.

Liu, W.-Y., Kang, J.-H., Jeong, H.-S. et al. (2014). Combined use of bulked segregant analysis and microarrays reveals SNP markers pinpointing a major QTL for resistance to *Phytophthora capsici* in pepper. *Theoretical and Applied Genetics* 127: 2503–2513.

Liu, C., Zhou, Q., Dong, L. et al. (2016). Genetic architecture of the maize kernel row number revealed by combining QTL mapping using a high-density genetic map and bulked segregant RNA sequencing. *BMC Genomics* 17: 915.

Livaja, M., Wang, Y., Wieckhorst, S. et al. (2013). BSTA: a targeted approach combines bulked segregant analysis with next- generation sequencing and de novo transcriptome assembly for SNP discovery in sunflower. *BMC Genomics* 14: 628.

Lu, H., Lin, T., Klein, J. et al. (2014). QTL-seq identifies an early flowering QTL located near flowering locus T in cucumber. *Theoretical and Applied Genetics* 127: 1491–1499.

Mackay, T.F.C. (2001). The genetic architecture of quantitative traits. *Annual Review of Genetics* 35: 303–339.

Mackay, I.J. and Powell, W. (2007). The significance and relevance of linkage disequilibrium and association mapping in crops. *Trends in Plant Science* 12: 53.

Mackay, I.J., Bansept-Basler, P., Barber, T. et al. (2014). An eight-parent multiparent advanced generation inter-cross population for winter-sown wheat: creation, properties and validation. *G3* 4: 1603–1610.

Maurer, A., Draba, V., Jiang, Y. et al. (2015). Modelling the genetic architecture of flowering time control in barley through nested association mapping. *BMC Genomics* 16: 290.

McMullen, M.D., Kresovich, S., Villeda, H.S. et al. (2009). Genetic properties of the maize nested association mapping population. *Science* 325: 737–740.

Michelmore, R.W., Paran, I., and Kesseli, R.V. (1991). Identification of markers linked to disease-resistance genes by bulked segregant analysis: a rapid method to detect markers in specific genomic regions by using segregating populations. *Proceedings of the Nationa Academy Sciences of the United States of America* 88: 9828–9832.

Mott, R., Talbot, C.J., Turri, M.G. et al. (2000). A new method for fine-mapping quantitative trait loci in outbred animal stocks. *Proceedings of the Nationall Academy Sciences of the United States of America* 97: 12649–12654.

Nice, L.M., Steffenson, B.J., Brown-Guedira, G.L. et al. (2016). Development and genetic characterization of an advanced backcross–nested association mapping (AB-NAM) population of wild3 cultivated barley. *Genetics* 203: 1453–1467.

Nice, L.M., Steffenson, B.J., Blake, T.K. et al. (2017). Mapping agronomic traits in a wild barley advanced backcross–nested association mapping population. *Crop Science* 57: 1199–1210.

Nubankoh, P., Wanchana, S., Saensuk, C. et al. (2020). QTL-seq reveals genomic regions associated with spikelet fertility in response to a high temperature in rice (*Oryza sativa* L.). *Plant Cell Reports* 39 (1): 149–162.

Pandey, M.K., Khan, A.W., Singh, V.K. et al. (2017). QTL-seq approach identified genomic regions and diagnostic markers for rust and late leaf spot resistance in groundnut (*Arachis hypogaea* L.). Plant Biotech. *J.* 15: 927–941.

Parts, L., Cubillos, F.A., Warringer, J. et al. (2011). Revealing the genetic structure of a trait by sequencing a population under selection. *Genome Research* 21: 1131–1138.

Pascual, L., Xu, J., Biais, B. et al. (2013). Deciphering genetic diversity and inheritance of tomato fruit weight and composition through a systems biology approach. *Journal of Experimental Botany* 64: 5737–5752.

Poland, J.A., Brown, P.J., Sorrells, M.E. et al. (2012). Development of high-density genetic maps for barley and wheat using a novel two-enzyme genotyping-by-sequencing approach. *PLoS One* 7 (2): e32253.

Powell, W., Machray, G.C., and Provan, J. (1996). Polymorphism revealed by simple sequence repeats. *Trends in Plant Science* 1: 215–222.

Pujol, M., Alexiou, K.G., Fontaine, A.S. et al. (2019). Mapping cucumber vein yellowing virus resistance in cucumber (*Cucumis sativus* L.) by using BSA-seq analysis. *Frontiers in Plant Science* 10: 1583.

Qi, X., Li, M.-W., Xie, M. et al. (2014). Identification of a novel salt tolerance gene in wild soybean by whole-genome sequencing. *Nature Communications* 5: 4340.

Ramirez-Gonzalez, R.H., Segovia, V., Bird, N. et al. (2014). RNA-Seq bulked segregant analysis enables the identification of high-resolution genetic markers for breeding in hexaploid wheat. *Plant Biotechnology Journal* 13: 613–624.

Ribaut, J.M. and Hoisington, D. (1998). Marker-assisted selection: new tools and strategies. *Trends in Plant Science* 3: 236–239.

Rodriguez-Uribe, L., Higbie, S.M., Stewart, J.M. et al. (2011). Identification of salt responsive genes using comparative microarray analysis in Upland cotton (*Gossypium hirsutum* L.). *Plant Science* 180: 461–469.

Ruangrak, E., Su, X., Huang, Z. et al. (2018). Fine mapping of a major QTL controlling early flowering in tomato using QTLseq. *Canadian Journal of Plant Science* 98: 1.

Salvi, S. and Tuberosa, R. (2005). To clone or not to clone plant QTLs: present and future challenges. *Trends in Plant Science* 10: 297–304.

Sannemann, W., Huang, B.E., Mathew, B. et al. (2015). Multi-parent advanced generation inter-cross in barley: high-resolution quantitative trait locus mapping for flowering time as a proof of concept. *Molecular Breeding* 35: 86.

Sax, K. (1923). The association of size differences with seed-coat pattern and pigmentation in *Phaseolus vulgaris*. *Genetics* 8: 552.

Schneeberger, K., Ossowski, S., Lanz, C. et al. (2009). SHOREmap: simultaneous mapping and mutation identification by deep sequencing. *Nature Methods* 6: 550–551.

Semagn, K., Babu, R., Hearne, S. et al. (2014). Single nucleotide polymorphism genotyping using Kompetitive Allele Specifc PCR (KASP): overview of the technology and its application in crop improvement. *Molecular Breeding* 33: 1–14.

Shi, L., Song, J., Guo, C. et al. (2019). A CACTA-like transposable element in the upstream region of BnaA9.CYP78A9 acts as an enhancer to increase silique length and seed weight in rapeseed. *The Plant Journal* 98 (3): 524–539.

Shu, J., Liu, Y., Zhang, L. et al. (2018). QTL-seq for rapid identification of candidate genes for flowering time in broccoli × cabbage. *Theoretical and Applied Genetics* 131: 917–928.

Singh, B.D. and Singh, A.K. (2015). *Marker-Assisted Plant Breeding: Principles and Practices*. New Delhi, Heidelberg, New York, Dordrecht, London: Springer.

Singh, V.K., Khan, A.W., Jaganathan, D. et al. (2016a). QTL-seq for rapid identification of candidate genes for 100-seed weight and root/total plant dry weight ratio under rainfed conditions in chickpea. *Plant Biotechnology Journal* 14: 2110–2119.

Singh, V.K., Khan, A.W., Saxena, R.K. et al. (2016b). Next generation sequencing for identification of candidate genes for *Fusarium wilt* and sterility mosaic disease in pigeonpea (*Cajanus cajan*). *Plant Biotechnology Journal* 4: 1183–1194.

Singh, V.K., Khan, A.W., Saxena, R.K. et al. (2017). Indel-seq: a fast-forward genetics approach for identification of trait-associated putative candidate genomic regions and its application in pigeonpea (*Cajanus cajan*). *Plant Biotechnology Journal* 15 (7): 906–914.

Sneller, C.H., Mather, D.E., and Crepieux, S. (2009). Analytical approaches and population types for finding and utilizing QTL in complex plant populations. *Crop Science* 49: 363–380.

Sonah, H., Bastien, M., Iquira, E. et al. (2013). An improved genotyping by sequencing (GBS) approach offering increased versatility and efficiency of SNP discovery and genotyping. *PLoS One* 8: e54603.

Spindel, J., Wright, M., Chen, C. et al. (2013). Bridging the genotyping gap: using genotyping by sequencing (GBS) to add high-density SNP markers and new value to traditional bi-parental mapping and breeding populations. *Theoretical and Applied Genetics* 126 (11): 2699–2716.

Srivastava, R., Upadhyaya, H.D., Rajendra, K. et al. (2017). A multiple QTL-seq strategy delineates potential genomic loci governing flowering time in chickpea. *Frontiers in Plant Science* 8: 1105.

Stadlmeier, M., Hartl, L., and Mohler, V. (2018). Usefulness of a multiparent advanced generation intercross population with a greatly reduced mating design for genetic studies in winter wheat. *Frontiers in Plant Science* 9: 1825.

Sun, X., Marza, F., Ma, H. et al. (2010). Mapping quantitative trait loci for quality factors in an inter-class cross of US and Chinese wheat. *Theoretical and Applied Genetics* 120: 1041–1051.

Swinnen, S., Schaerlaekens, K., Pais, T. et al. (2012). Identification of novel causative genes determining the complex trait of high ethanol tolerance in yeast using pooled-segregant whole-genome sequence analysis. *Genome Research* 22: 975–984.

Syvänen, A.C. (2005). Toward genome-wide SNP genotyping. *Nature Genetics* 37: S5–S10.

Takagi, H., Uemura, A., Yaegashi, H. et al. (2013a). MutMap-Gap: whole-genome resequencing of mutant F2 progeny bulk combined with de novo assembly of gap regions identifies the rice blast resistance gene Pii. *The New Phytologist* 200: 276–283.

Takagi, H., Abe, A., Yoshida, K. et al. (2013b). QTL-seq: rapid mapping of quantitative trait loci in rice by whole genome resequencing of DNA from two bulked populations. *The Plant Journal* 74: 174–183.

Takagi, H., Tamiru, M., Abe, A. et al. (2015). MutMap accelerates breeding of a salt-tolerant rice cultivar. *Nature Biotechnology* 33: 445–449.

Tanksley, S., Young, N., Paterson, A. et al. (1989). RFLP mapping in plant breeding: new tools for an old science. *Nature Biotechnology* 7: 257–264.

Threadgill, D.W. and Churchill, G.A. (2012). Ten years of the collaborative cross. *Genetics* 190: 291–294.

Thyssen, G.N., Fang, D.D., Turley, R.B. et al. (2014). Next generation genetic mapping of the Ligon-lintless-2 (Li 2) locus in upland cotton (*Gossypium hirsutum* L.). *Theoretical and Applied Genetics* 127: 2183–2192.

Tian, F., Bradbury, P.J., Brown, P.J. et al. (2011). Genome-wide association study of leaf architecture in the maize nested association mapping population. *Nature Genetics* 43: 159–162.

Trick, M., Adamski, N., Mugford, S.G. et al. (2012). Combining SNP discovery from next-generation sequencing data with bulked segregant analysis (BSA) to fine-map genes in polyploid wheat. *BMC Plant Biology* 12: 14.

Uptmoor, R., Schrag, T., Stützel, H. et al. (2008). Crop model based QTL analysis across environments and QTL based estimation of time to floral induction and flowering in *Brassica oleracea*. *Molecular Breeding* 21: 205–216.

Valdar, W., Flint, J., and Mott, R. (2006). Simulating the collaborative cross: power of quantitative trait loci detection and mapping resolution in large sets of recombinant inbred strains of mice. *Genetics* 172: 1783–1797.

Vos, P., Hogers, R., Bleeker, M. et al. (1995). AFLP: a new technique for DNA fingerprinting. *Nucleic Acids Research* 23: 4407–4414.

Wang, B., Du, Q., Yang, X. et al. (2014a). Identification and characterization of nuclear genes involved in photosynthesis in Populus. *BMC Plant Biology* 14: 81.

Wang, Y., Dai, M., Zhang, S. et al. (2014b). Exploring candidate genes for pericarp russet pigmentation of sand pear (*Pyrus pyrifolia*) via RNA-Seq data in two genotypes contrasting for pericarp color. *PLoS One* 9: e83675.

Wen, J., Jiang, F., Weng, Y. et al. (2019). Identification of heat-tolerance QTLs and high-temperature stress-responsive genes through conventional QTL mapping, QTL-seq and RNA-seq in tomato. *BMC Plant Biology* 19: 398.

Wenger, J.W., Schwartz, K., and Sherlock, G. (2010). Bulk segregant analysis by high-throughput sequencing reveals a novel xylose utilization gene from *Saccharomyces cerevisiae*. *PLoS Genetics* 6: e1000942.

Win, K.T., Zhang, C., Silva, R.R. et al. (2019). Identification of quantitative trait loci governing sub gynoecy in cucumber. *Theoretical and Applied Genetics* 132: 1505–1521.

Xu, X., Zeng, L., Tao, Y. et al. (2013). Pinpointing genes underlying the quantitative trait loci for root-knot nematode resistance in palaeopolyploid soybean by whole genome resequencing. *Proceedings of the Nationall Academy Sciences of the United States of America* 110: 13469–13474.

Xu, X., Lu, L., Zhu, B. et al. (2015). QTL mapping of cucumber fruit flesh thickness by SLAF-seq. *Scientific Reports* 5: 1–9.

Yang, Z., Huang, D., Tang, W. et al. (2013). Mapping of quantitative trait loci underlying cold tolerance in rice seedlings via high-throughput sequencing of pooled extremes. *PLoS One* 8: e68433.

Yang, J., Jiang, H., Yeh, C. et al. (2015). Extreme-phenotype genome-wide association study (XP-GWAS): a method for identifying trait-associated variants by sequencing pools of individuals selected from a diversity panel. *The Plant Journal* 84: 587–596.

Yang, X., Xia, X., Zhang, Z. et al. (2017). QTL mapping by whole genome re-sequencing and analysis of candidate genes for nitrogen use efficiency in rice. *Frontiers in Plant Science* 8: 1634.

Yaobin, Q.I., Peng, C.H., Yichen, C.H. et al. (2018). QTL-Seq identified a major QTL for grain length and weight in rice using near isogenic F2 population. *Rice Science* 25: 121–131.

Yi, N. and Xu, S. (2001). Bayesian mapping of quantitative trait loci under complicated mating designs. *Genetics* 157: 1759–1771.

Yu, S.B., Li, J.X., Xu, G.C. et al. (1997). Importance of epistasis as the genetic basis of heterosis in an elite rice hybrid. *Proceedings of the Nationall Academy Sciences of the United States of America* 94: 9226–9231.

Yu, J., Holland, J.B., McMullen, M.D. et al. (2008). Genetic design and statistical power of nested association. *Genetics* 178: 539–551.

Zeng, Z.B. (1994). Precision mapping of quantitative trait loci. *Genetics* 136: 1457–1468.

Zhang, S., Meng, L., Wang, J. et al. (2017). Background controlled QTL mapping in pure-line genetic populations derived from four-way crosses. *Heredity* 119 (4): 256.

Zhang, X., Wang, W., Guo, N. et al. (2018). Combining QTL-seq and linkage mapping to fine map a wild soybean allele characteristic of greater plant height. *BMC Genomics* 19: 226.

Zhang, L., Meng, L., and Wang, J. (2019). Linkage analysis and integrated software GAPL for pure-line populations derived from four-way and eight-way crosses. *Crop Journal* 7 (3): 283–293.

Zheng, Y., Xu, F., Li, Q. et al. (2018). QTL mapping combined with bulked segregant analysis identify SNP markers linked to leaf shape traits in *Pisum sativum* using SLAF sequencing. *Frontiers in Genetics* 9: 615.

Zhou, X., Xia, Y., Ren, X. et al. (2014). Construction of a SNP-based genetic linkage map in cultivated peanut based on large scale marker development using next-generation double-digest restriction-site-associated DNA sequencing (ddRADseq). *BMC Genomics* 15: 351.

Zou, J., Zhao, Y.S., Liu, P.F. et al. (2016). Seed quality traits can be predicted with high accuracy in brassica napus using genomic data. *PLoS One* 11 (11): e0166624.

# 5

# Genome-Wide Association Study: Approaches, Applicability, and Challenges

*Akshay S. Sakhare*[1], *Suneetha Kota*[2], *Santosh Rathod*[2], *Brajendra Parmar*[2], *and Viswanathan Chinnusamy*[1]

[1] *Division of Plant Physiology, ICAR-IARI, New Delhi, India*
[2] *ICAR-IIRR, Hyderabad, Telangana, India*

## 5.1 Introduction

Understanding the contribution of genetic variation to phenotypic variation forms the basis the genetic studies in any crop species. The phenotypic variation explained for different traits is either governed by discrete variables or continuous variables. The inheritance pattern of monogenic and polygenic inheritance has been delineated. Several methods have been developed and utilized such as traditional linkage mapping, genome-wide association studies, nested association mapping (NAM), and genomic selection in dissecting the underlying genetic basis of polygenic traits. Association mapping, also known as linkage disequilibrium (LD) mapping, detects the genetic associations among various quantitative traits. There are two approaches to association mapping: genome-wide association studies (GWAS) and candidate gene-based association mapping. With the advent of high-throughput genome sequencing technology, new computational biological approaches, and release of crops reference genomes, fast and precise resequencing (genotyping) of a large number of different germplasm of crops is possible. There are various types of genetic sequence variants in the genome. This sequence information is used to study the natural variations in the genome sequences among the diverse germplasm within a given crop. The recent next-generation sequencing tools and the advent of GWAS have facilitated understanding of the genetic basis of complex traits to a greater extent. In evolutionary ecology, the genetic variation created through various means like selection, mutation, migration, and genetic drift forms the basis of new adaptive variation. The variation brought forward might be fitness-related traits or productivity traits related to yield and quality

*Genotyping by Sequencing for Crop Improvement*, First Edition.
Edited by Humira Sonah, Vinod Goyal, S.M. Shivaraj, and Rupesh K. Deshmukh.

encountering diverse ecological conditions. GWAS has emerged as a powerful approach for genetic dissection of underlying natural phenotypic variation governing several traits that are exposed to natural selection or phenotypes in realistic environments. GWAS utilizes historical recombination events that have occurred in many generations spanning over the period of thousands of years. GWAS is based on the fact that there were common ancestors in ancient times for the diverse germplasm of modern times and therefore some of the chromosomal segments in these germplasms are common. These common chromosome segments, which originate from the common ancestor without recombination event, will have the same genetic sequence. If there is a marker or an allele or a haplotype or a quantitative trait loci (QTL) located in this common chromosomal segment, all the individuals carrying these common chromosomal segments will also carry the same marker or an allele or a haplotype or a QTL. GWAS utilizes the existing LD to detect genetic variation associated with the phenotypic variation for the traits. Therefore, GWAS hypothesis states that a particular phenotypic trait shown by a group of individuals of a population might be associated with genetic changes in the common chromosomal segment in their recent ancestor, which led to the development of a particular phenotypic trait (Hayes 2013; Liu and Yan 2019). GWAS usually considers the coverage of the whole genome with markers distributed densely covering all the chromosomes. In candidate gene association analysis, markers based on the specific genomic regions that are associated with the trait under study, usually based on previous QTL studies or gene function relevant to the trait variation. GWAS uses LD to detect the associations among various common chromosomal segments and phenotypic traits hence utilize historical recombination events that have occurred in many generations spanning over the period of thousands of years in the determination of genetic architecture of the traits. LD shows an association between markers and changed or mutated chromosomal segments at the population level. In the traditional QTL mapping method, the diverse parents for the trait of interest were utilized in hybridization to create segregating mapping populations. Allele(s) governing the specific trait was detected and an association between the genotypic and phenotype is established through linkage analysis. Different kinds of mapping populations were developed and utilized in linkage mapping studies, while in GWAS natural unstructured populations form the basis of marker–trait association study.

## 5.2 Methodology to Conduct GWAS in Crops

The methodology to conduct GWAS in crops has been described in the following points and illustrated graphically in Figure 5.1

1) Selection of population under study with due consideration of the population size.
2) Accurate phenotyping of all the individuals of the population for the trait (s) under study, preferably repeated replications over locations and years.
3) Estimation of broad-sense heritability and the phenotypic data for the estimation of BLUE (fixed effects) or BLUP (random effects) for the genotypic values.

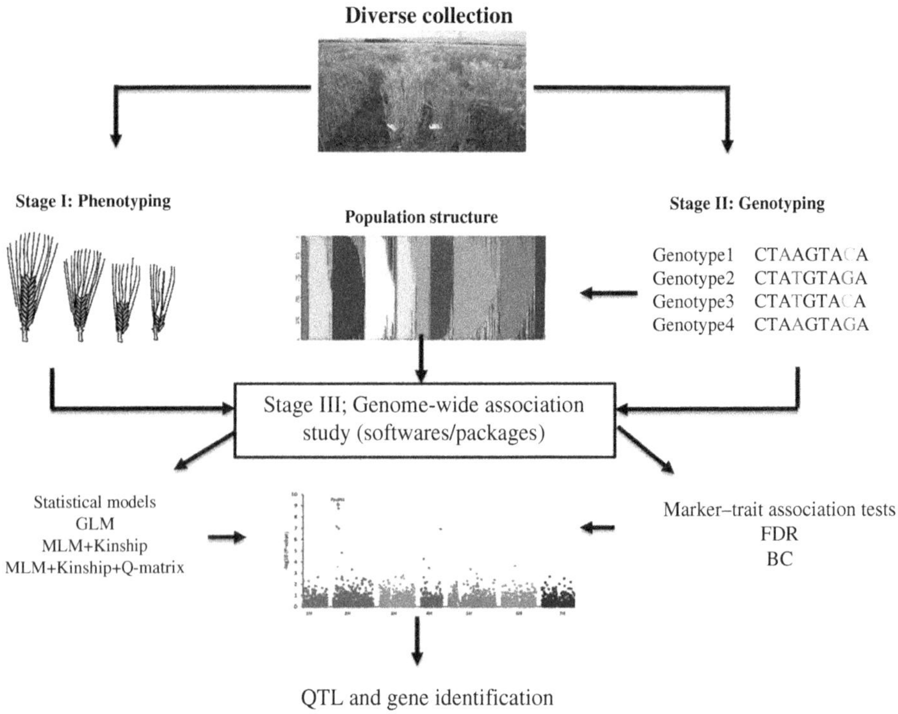


**Figure 5.1** Methodology to conduct GWAS in crops. It can be divided into three stages: stage I: phenotyping, stage II: genotyping, and stage III: genome-wide association study using statistical tools to identify QTL and gene. *Source:* Alqudah 2020, p. 51, with permission of Elsevier.

4) Genotyping of the individuals of the population using molecular markers. Genotyping-by-sequencing (GBS) is a widely used method. The sequence data are to be filtered for missing data, minor allele frequency (MAF), and heterozygosity.
5) Population structure to be determined for selection of appropriate GWAS model.
6) The general linear model (GLM) which does not consider population structure and mixed linear model (MLM) considering population structure through Kinship + Q matrix + PCA are the statistical models often proposed for performing GWAS.
7) The phenotypic and genotypic data are analyzed together using appropriate software as described in Table 5.1.
8) Significant marker–trait associations were determined by false discovery rate (FDR) or Bonferroni correction (BC) through multiple comparisons to test the significant markers fitting the threshold. BC is used to define the threshold of significant markers for more than one trait simultaneously. The FDR is the test that gives an actual estimate of significant results of marker–trait associations. FDR is calculated for each trait independently and is more flexible, unlike BC. The marker–trait association can be determined at the significance level of 0.01 and 0.05. High FDR is utilized to determine the genetic basis of the trait and low FDR is used to identify candidate genes governing the trait under study for dissecting the genetic and molecular levels.

**Table 5.1** Popular bioinformatics software and tools available for GWAS analysis.

| S. No. | Software's/Tools/R Package | Features | Link |
|---|---|---|---|
| 1 | GAPIT (Genome Association and Prediction Integrated Tool)<br>Version: 3.0 | GAPIT is a commonly used GWAS tool which is having the advanced procedures of GWAS namely. Q + K mixed linear model, Compressed MLM, and SUPER | http://www.zzlab.net/GAPIT |
| 2 | TASSEL (Trait Analysis by association Evolution and Linkage)<br>Version: 5.0 | TASSEL uses GLM and MLM statistical approaches for association mapping. The newer version of MLM reduces Type I error (False Positive) in association mapping with complex pedigrees, families, founding effects, and population structure | https://www.maizegenetics.net/tassel |
| 3 | GWASpoly | It helps to perform GWAS with biallelic single nucleotide polymorphisms (SNPs) in autopolyploids and diploids, it uses mixed modeling of GWAS | https://potatobreeding.cals.wisc.edu/software |
| 4 | Netgwas: An R Package for Network-Based Genome-Wide Association Studies<br>Version 1.11 | Netgaws is designed to construct linkage map, reconstructing linkage disequilibrium (LD) networks from multi–loci genotype data, and for developing high dimensional Genotype*Phenotype networks | https://cran.r-project.org/web/packages/netgwas/index.html |
| 5 | GWASbyCluster<br>Version 0.1.7 | An R package that helps in identifying significant SNPs in genome-wide association studies (GWAS) using Bayesian hierarchical clustering | https://cran.rstudio.com/web/packages/GWASbyCluster/index.html |
| 6 | SNPassoc<br>Version 1.9-2 | This package is helpful in carrying out descriptive statistics, calculation of Hardy–Weinberg equilibrium, analysis of association based on GLM, and analysis of haplotype and epistasis interaction | https://cran.r-project.org/web/packages/SNPassoc/index.html |
| 7 | GWASTools (Tools for Genome-Wide Association Studies)<br>Version:1.34.0 | GWASTools is a R/Bioconductor package for quality control and analysis of GWAS. This tool uses data stored in several formats and is even capable of handling large-sized SNP files | https://bioconductor.org/packages/release/bioc/html/GWASTools.html |
| 8 | GWASinspector<br>Version 1.4.7.1 | This package performs quality control analysis of GWAS results and it also provides the user with a comprehensive report | https://cran.r-project.org/web/packages/GWASinspector/index.html |

## 5.3 Statistical Modeling in GWAS

As GWAS is a cause and effect problem, regression analysis is the most prominent statistical technique applied to understand the relationship between marker effect and phenotypes of interest. Single-marker regression analysis is the simple method that tests the association between marker and trait. In a multiple linear regression model, we assume that error terms are assumed to be normally distributed. In real-life situation, the data set may not follow the assumption of normality so a GLM is an alternative to linear models in case residual terms follow non-Gaussian assumptions. As the number of markers increases, the situation leads to the most prominent $p > n$ statistical problem, methods like stepwise regression analysis, principal component analysis (PCA) are applied in the past to overcome the $p > n$ problem. The GLM model reduces the spurious association caused by differences in relatedness between pairs of individuals. The population structure and unequal relatedness are simultaneously modeled using MLM, in which group of related individuals or principle components are considered as fixed effect and kinship (covariance matrix) as a random effect, the MLM model outperforms the other model in terms of statistical power. In another alternative method in which the individuals are clustered using clustering methods by optimizing the number of groups and then MLM model is applied on the grouped individuals termed as compressed MLM which improves statistical power of GWAS. Another improved method over compressed mixed linear model (CMLM) is enriched CMLM which utilizes different clustering algorithms UPGMA, complete linkage, WPGMA, etc., and optimizes the most effective algorithm for CMLM (Li et al. 2014). As stated earlier GWAS is a $p > n$ problem, another method to address this problem was given by Wang et al. (2014) called as SUPER (Settlement of MLM Under Progressively Exclusive relationship) method in which the whole genome is divided into small groups or bins which is represented by significant markers and influential bins were selected based on maximum likelihood estimation (MLE) method.

Alternative to statistical models of GWAS, there is several machine learning or artificial intelligence algorithms are available, which can be applied to GWAS to study the underlying complex relationship between marker effects and phenotypes. One of the major advantages of machine learning algorithms is they are free of initial population structure distribution and they consider the most complex chaotic relationships between marker effects and phenotypes which is failed by statistical models. Some of the prominent machine learning models applied in GWAS are ANN (artificial neural network), decision trees, Bayesian ANN, K nearest neighbor (KNN), etc. (Szymczak et al. 2009; Romagnoni et al. 2019). Supervised machine learning methods viz. Random Forest, Support vector regression can also be explored in GWAS studies. Penalized methods such as penalized regression model, Ridge regression, LASSO (least absolute shrinkage and selection operator) regression, Elastic Net, etc. are alternative methods to stepwise regression analysis and PCA for significant marker selection before proceeding to model part of GWAS. Based on our understanding of statistical models in GWAS, one can state that no models are universal to all the data set, they have their own merits and demerits, which should be chosen carefully based on the underlying relationship existing among the individuals.

The basic regression equation used in GWAS is expressed as follows;

$$Y = X\beta + Zu + e \tag{5.1}$$

where

$\boldsymbol{Y}$ = Vector of a phenotype or trait of interest or dependent variable

$\beta$ = Vector of regression coefficient depicts genetic marker effects and intercept

$\boldsymbol{u}$ = Vector of size s (the number of groups) for unknown random additive genetic effects from multiple background QTL for individuals/lines following distribution with a mean of zero and covariance matrix of $\boldsymbol{G}$

$\boldsymbol{X}$ = Design matrices for $\beta$

$\boldsymbol{Z}$ = Design matrices for $\boldsymbol{u}$

$\boldsymbol{e}$ = Residual effect normally distributed with zero mean & covariance ($R$)

$\boldsymbol{u}$ and $\boldsymbol{e}$ vectors are assumed to be normally distributed with a mean of zero and a variance of $u$ and $e$.

Hence,

$$\text{Var}\begin{pmatrix} u \\ e \end{pmatrix} = \begin{pmatrix} G & \mathbf{0} \\ \mathbf{0} & R \end{pmatrix} \tag{5.2}$$

Covariance matrix ($\boldsymbol{G}$) expressed as $G = \mathbf{2}K\sigma_{\mathrm{a}}^2$

where

$\boldsymbol{K}$ = Kinship matrix with element $k_{ij}$ $(i, j = 1,2,\ldots s)$

$\sigma_a^2$ = Additive genetic variance

Homogeneous variance assumed for the residual effect; i.e., $R = I\sigma_{\mathrm{e}}^2$

where

$\boldsymbol{I}$ = Identity matrix

$\sigma_e^2$ = Residual variance

## 5.4 Efficiency of GWAS with Different Marker Types

The selection of molecular markers for genotyping in association studies is an important factor in detecting the heritable differences within a population that can identify associated genetic loci and traits. Markers selected are high throughput and reproducible. Different types of markers such as simple sequence repeats (SSR), expressed sequence tag (EST), restriction fragment length polymorphism (RFLP), random amplified polymorphic DNA (RAPD), amplified fragment length polymorphism (AFLP), diversity arrays technology (DArT), and single nucleotide polymorphisms (SNPs) have contributed to GWAS. Earlier studies on GWAS were conducted with small populations and a limited number of markers namely SSR, sequence-specific amplification polymorphisms (SSAP), AFLP, and DArT. The information generated with SSR markers in complex genomes like sugarcane could be able to distinguish between ancestral and improved accessions within the crop. Using the GWAS approach with SSR markers, significant marker–traits association for soluble solid content, stalk height, stalk number, stalk weight, and cane yield was detected in sugarcane (Barreto et al. 2019). Similarly, in Pearl Millet, another outcrossing species with residual heterozygosity, the initial GWAS studies were carried out using SSRs to detect genetic determinants governing flowering and crop adaptation-related traits (Saïdou et al. 2009; Mariac et al. 2011; Anuradha et al. 2017). Using 250 full-sib progenies and 34 SSR markers in GWAS identified the associations linked to flowering time, plant height (PH), panicle length, and grain yield (GY) (Kannan et al. 2014). Using 285 DArT markers and phenotypic data from 151 Pearl Millet inbred Germplasm Association Panel (PMiGAP) lines from West Africa conducted GWAS for flowering time, and GY under low P conditions (Gemenet et al. 2015). Association mapping involves whole-genome scan and hence requires more number of makers to cover the genome. Molecular markers such as SNPs have become handy to dissect the functional impacts of individual trait variants. Markers such as SNPs and SSR are profoundly applied in determining the relative kinship matrix and population structure (Zhu et al. 2008) each with its own merits and demerits. GWAS with SNPs is of particular importance to study various genetic variations because SNPs are the most abundant throughout the genome and easily identifiable genetic variations. SNPs are mostly biallelic in nature as compared to SSRs which are multiallelic and hence make them less informative than SSRs. However, the expected heterozygosity is lower in SNPs than SSR and hence assessment of population structure and family relatedness is achieved successfully in most crops. The effectiveness of SNPs in detecting the phenotypic variation within a population depends on several factors such as allele frequency, population size, sample effects, etc. (Wessinger et al. 2018). Despite these factors, SNPs are the most preferred makers for association studies even in complex polyploid species. In recent years, large association panels with a large number of SNP markers are being utilized for efficient analysis of complex traits through GWAS.

## 5.5 Computational Tools for GWAS

There are several software tools and R packages are available for GWAS analysis, some of them are given in Table 5.1 with their important features.

## 5.6 GWAS Challenges for Complex Traits

The vast diversity displayed by plants in various traits such as morphology, physiology, stress responses, and disease susceptibility is the result of the genetic complement of the complex traits which emanate from the interaction of multiple loci and the allelic effects regulated by environmental cues. Therefore, it is crucial to understand the association between the diverse phenotypes of complex traits and the underlying genomic sequence variation. This knowledge will be helpful in enhancing the pace of breeding to develop climate-resilient crop plants (Mackay et al. 2009). Complex traits are governed by many genes and every gene has a small additive effect on the trait. These genes are widely distributed in the genome. Complex traits or quantitative traits do not follow Mendelian laws. Examples of a complex trait are yield, PH, stress tolerance, disease susceptibility, etc. As complex traits are polygenic in nature, every gene governing a complex trait will have two or more alleles. Therefore, the expressed phenotype of a complex trait will be the result of the interaction of several loci with various allelic effects. Therefore, the phenotype governed by a particular genotype in given environmental conditions is unique to that environmental conditions and it may vary with the subsequent change in the environment. This genetic variation for complex traits is critical for survival under diverse natural environments and crop productivity (Mitchell-Olds 2010). To differentiate between the phenotypic effects of adjacent genes; GWAS utilizes historical recombination events that have occurred in many generations spanning over the period of thousands of years. As mentioned earlier, GWAS detects association between a phenotype and SNP markers on the basis of LD. However, some of the SNPs differ with population, therefore, it is difficult for GWAS to identify QTLs or markers or SNPs that affect phenotype that is associated with population structure. Complex traits are polygenic and every gene has a small additive effect on the trait, therefore all or some of the genes may not show any association with a marker or loci or QTL. Therefore, in this case, it is challenging for GWAS to identify an entire set of genes governing a particular phenotype. Moreover, epistatic effects, i.e. one gene suppresses the expression of other genes, further complicates the GWAS analysis of complex traits. Although the GWAS approach has been utilized in detecting several loci governing traits, the limitations of GWAS include population structure, genetic relatedness, computational challenges, FDRs, rare alleles, and reproducibility of detected loci and their marker associations.

## 5.7 Factors Challenging the GWAS for Complex Traits

1) **Polygenic nature of complex traits**: Polygenic nature of complex traits is the root cause of the problems faced by GWAS analysis of complex traits. Every gene or loci governing a complex trait may have two or more alleles which increases the number of different combinations of the alleles may have.
2) **Epistatic effects:** Epistatic effect is when a gene or allele suppresses/masks the expression of the other gene or allele. Epistasis masks the effects of many QTLs and therefore it makes mapping difficult and remains a major obstacle in understanding the genetic basis of complex traits (Phillips 2008). The presence of such an effect makes it difficult to associate a phenotype with a genotype.
3) **Pleiotropic effects**: Polygenic nature of complex traits is inevitably affected by the phenomenon of pleiotropy. Pleiotropy is when a gene or allele controls the expression of many genes. Under pleiotropic effects, the individual gene or allele effects may not be demarcated and hence it remains obscure. Under such a situation deciphering the entire genetic compliment of a complex trait becomes challenging.
4) **Heritable phenotypic variation:** The proportion of heritable variation is important for any genetic study. The power of detection of marker–trait association is often limited by the low heritability of polygenic traits. Genotype x environment interaction is another factor for reduced heritability.
5) **Population size:** GWAS carried out with a small population size reduces the power of detection of marker–trait association. Large populations representing global diversity with individuals representing different geographical regions, ecosystems will maximize the genetic variance and in the detection of specific alleles governing the trait under study.

6) **Population structure:** Structured population is another limitation for GWAS that can lead to a spurious association between genotype and phenotype. Often researchers exercise a selection of population-based on geographic region or habitat wherein all the individuals in a population are not distantly related to each other at the genetic level. Population structure can be controlled by statistical approaches like STRUCTURE analysis and PCA and mixed-model approach.
7) **Allele frequency:** GWAS is mainly focused on common variants and major allele frequency within the population. Low allele frequency and rare alleles with MAF at less than 5% in few individuals across the population reduces the resolution power of GWAS.
8) **Linkage disequilibrium (LD):** LD is another indicator in GWAS of nonrandom association of alleles at different loci. A high LD value means a low number of required markers for the whole genome. The coefficient of LD measures the likelihood of association between two loci sharing a common history of mutation and recombination. LD decay is the rate at which LD declines with genetic or physical distance. Rapid LD decay means large numbers of markers are needed for whole-genome association.

## 5.8 GWAS Applications in Major Crops

GWAS has been applied in the majority of crops including diploid as well as in polyploid crops such as rice, wheat (Sun et al. 2017) maize (Samayoa et al. 2019), millets, and cotton (Dilnur et al. 2019) for genetic dissection of important agronomic, physiological and biochemical traits (Table 5.2).

### 5.8.1 Maize

Maize is the first among the crop species where association mapping was applied (Thornsberry et al. 2001). Significant progress has been made in maize since the release of the reference genome of B73 and the application of GWAS in dissecting genotypic variations corresponding to the phenotype of target traits. Analysis of data generated through GWAS using diverse populations for genomic and metabolomics paved the way to understand the genetic basis of quantitative traits and their interaction with the environmental factors. The first association study as whole genome-wide scan association mapping applied to detect genetic loci governing oleic acid content in maize kernels among 553 maize inbred lines (Belo et al. 2008). A fatty acid desaturase, fad2, was identified as a candidate gene responsible for the allelic differences in oleic acid content in maize. Later on, GWAS for several agriculturally important traits at an agronomic, genetic, molecular, and metabolic level have been comprehensively studied and several candidate genes were delineated.

GWAS-based identification of rare and favorable alleles of two genes namely *LcyE* for lycopene epsilon cyclase and *crtRB1* for β-carotene hydroxylase 1 governing high provitamin A has been carried out in Maize (Harjes et al. 2008; Yan et al. 2010). The genetic basis of photoperiod sensitivity has been dissected through GWAS with the identification of major gene *ZmCCT* governing photoperiod sensitivity (Hung et al. 2012; Yang et al. 2013), where the allelic differences govern the gene expression. Reduced expression of *ZmCCT*

**Table 5.2** Genome-wide association studies (GWAS) conducted for dissection of several important traits in major crops.

| Crop | Traits studied | Population type and size | Markers | References |
|---|---|---|---|---|
| Maize | Flowering time | 5000 NAM RILs +281 Inbred line panel (ILP) | 1.1K SNP | Buckler et al. (2009) |
| | Flowering time, Northern corn leaf blight resistance | 1487 European ILs | 359 SSR and 8.2K SNP | Van Inghelandt et al. (2012) |
| | Day-length adaptation | 5000 NAM RILs + 481 ILP | 1.1K SNP | Hung et al. (2012) |
| | Flowering related traits | 508 ILP | 557K SNP | Yang et al. (2013) |
| | Oil biosynthesis | 368 ILP | 557K SNP | Li et al. (2013) |
| | Provitamin A carotenoids | 281 ILP | 284K SNP | Owens et al. (2014) |
| | 17 Agronomic traits | 513 ILP | 557K SNP | Yang et al. (2014) |
| | Yield under drought, aflatoxin resistance, flowering time traits, and other agronomic traits | 346 ILP | 60K SNP | Farfan et al. (2015) |
| | Grain yield, days to pollen shed (PS), plant height (PH), and ear height (EH) | 529 MAGIC lines | 54K SNP | Dell'Acqua et al. (2015) |
| | Rough dwarf disease resistance | 527 ILP | 557K SNP | Chen et al. (2015) |
| | Mediterranean corn borer resistance | 302 ILP | 246K SNP | Samayoa et al. (2015) |
| | Iron homeostasis | 267 ILP | 438K SNP | Benke et al. (2015) |
| | Aflatoxin accumulation resistance | 300 ILP | 261K SNP | Warburton et al. (2015) |
| | Flowering time | 5000 + 2000 NAM RILs +1745 Ames ILP | 950K SNP | Li et al. (2016) |
| | Drought related metabolic and physiological traits | 318 ILP | 157K SNP | Zhang et al. (2016) |
| | Forage quality, seedling stage drought tolerance | 368 ILP | 557K SNP | Wang et al. (2016) |
| | Cold tolerance associated traits: days to emergence, relative leaf chlorophyll content/quantum efficiency of photosystem II | 306 dent and 292 European flint ILP | 50K SNP | Revilla et al. (2016) |

*(Continued)*

**Table 5.2** (Continued)

| Crop | Traits studied | Population type and size | Markers | References |
|---|---|---|---|---|
| | Tar spot complex resistance | 890 ILP | 56K SNP | Mahuku et al. (2016) |
| | Fusarium ear rot resistance | 818 tropical ILP | 43.4K SNP | Chen et al. (2016) |
| | Husk traits | 508 ILP | 557K SNP | Cui et al. (2016) |
| | Starch content regulation | 263 ILP | 56K SNP | Liu et al. (2016) |
| | Kernel and ear traits related to yield | 292 ILP | 50K SNP | Zhu et al. (2018) |
| | Stalk biomass and anatomical traits | 942 ILP | 899K SNP | Mazaheri et al. (2019) |
| Rice | Agronomic, drought, quality traits | 517 landraces | ~3.6 million SNP | Huang et al. (2010) |
| | Root cone angle | 162 *indica* and 169 *japonica* rice accessions | SNP | Bettembourg et al. (2017) |
| | Physiological, developmental, and morphological traits | 413 accessions | 44K SNP | Zhao et al. (2011) |
| | Flowering time and grain yield traits | 950 accessions | SNP | Huang et al. (2011) |
| | Root system architecture and biomass traits | 167 *japonica* | SNP | Courtois et al. (2013) |
| | Root-knot nematode resistance | 332 accessions | 44K SNP | Dimkpa et al. (2016) |
| | Agronomic traits | 176 *japonica* | SNP | Yano et al. (2016) |
| | 21 root traits under normal and drought conditions | 529 accessions | SNP | Li et al. (2017) |
| | Salinity tolerance | 306 accessions | 700K SNP | Patishtan et al. (2018) |
| | Plant morphology, grain quality, plant development, and agronomic traits | 413 *indica* | SNP | Zhao et al. (2011) |
| | Grain shape | 126 MAGIC lines | SNP | Ogawa et al. (2018) |
| | Early root vigor for direct-seeded rice | 307 accessions | SNP | Wang et al. (2018) |
| Wheat | Grain yield | 166 elite wheat varieties | 90K and 660K SNP array | Li et al. (2019c) |
| | Yield and yield contributing traits | 320 diverse spring wheat | 9K SNPs | Gahlaut et al. (2019) |

**Table 5.2** (Continued)

| Crop | Traits studied | Population type and size | Markers | References |
|---|---|---|---|---|
| | Agro-morphological traits | 404 wheat accessions | 35K Axiom array | Sheoran et al. (2019) |
| | Agronomic traits | 493 durum wheat accessions | 9K iSelect assay | Wang et al. (2019) |
| | Yield related traits | 322 soft red winter wheat | Illumina HiSeq 2500 | Ward et al. (2019) |
| | Drought tolerance and biomass allocation | 100 wheat genotypes | 28,356 DArT markers | Mathew et al. (2019) |
| | Root and shoot traits | 323 wheat accessions | 660K SNP array | Li et al. (2019a) |
| | Drought and heat-responsive agronomic traits | 277 wheat accessions | 660K SNP array | Li et al. (2019b) |
| | Stripe rust resistance | 120 winter wheat | 90K SNP array | Cheng et al. (2020) |
| | Agronomic traits and Fusarium head blight resistance | 558 wheat accessions | 90K SNP + 26 KASP | Mellers et al. (2020) |
| | Plant architectural traits | 319 wheat germplasm | 90K SNP | Wang et al. (2020) |
| | Yield, quality, and disease-related traits | 1,325 winter wheat | 15K wheat SNP | Tsai et al. (2020) |
| Barley | Heading date, plant height, Thousand grain weight (TGW), starch content, crude protein | 224 | 1.5K SNP | Pasam et al. (2012) |
| | TGW, glume fineness, extract, and friability | 183 European cultivars | 253 DArT and 22 SSR | Matthies et al. (2012) |
| | Grain yield, TGW, agronomic, and quality traits | 106 German winter lines | 1169 DArT | Lex et al. (2014) |
| | Drought tolerance related traits: grain yield, TGW, peduncle, leaf, and spike | 185 germplasm | 710 DArT, 61 SNP, and 45 SSR | Varshney et al. (2012) |
| | Grain yield-associated traits | 379 Pan-European Cultivars | 9K SNP | Xu et al. (2018) |
| | Water use efficiency, water content, and relative water content | 109 drought tolerant collection | 5153 DArT | Wojcik-Jagla et al. (2018) |
| | Spike length, plant height, and grain number | 148 European cultivars | 407 SSR | Jabbari et al. (2018) |
| Brassica | Branch number | 327 accessions | 33K SNP | He et al. (2017) |
| Sugarcane | Yield component traits | 134 accessions | 100 SSR | Barreto et al. (2019) |

*(Continued)*

**Table 5.2** (Continued)

| Crop | Traits studied | Population type and size | Markers | References |
|---|---|---|---|---|
| Cotton | Drought tolerance | 319 germplasm | 55K SNP | Hou et al. (2018) |
| | Fiber quality traits and yield components | 231 RILs | 122 SSR and 4.7K SNP | Liu et al. (2018) |
| | Plant architecture | 355 accessions | 93K SNPs | Su et al. (2018) |
| | Salinity tolerance | 713 germplasm | 10.5K SNPs | Sun et al. (2018) |
| | Protein, oil, and five fatty acids | 316 accessions | 390K SNPs | Du et al. (2018) |

gene promotes early flowering and thus allows adaptation of maize to higher latitudes. In another study, a major PH QTL, qph1, was mapped in *Brachytic2* (*Br2*) coding sequence on maize chromosome 1. A rare and occurring SNP in qph1 detected in temperate maize germplasm resulted in an amino acid substitution and optimized PH. The reduced PH in turn allowed higher plant density and hence higher yields (Xing et al. 2015). In a different study, plant architecture-related traits such as leaf angle which determines solar radiation absorption governed by lg1 and lg2 loci were identified (Tian et al. 2011).

### 5.8.2 Rice

In rice, GWAS has been successfully applied to dissect the genetic architecture of complex traits *viz.*, agronomic, yield, quality, abiotic, and biotic stress related traits (Huang et al. 2010, 2011; Kumar et al. 2015; Ma et al. 2016; Dimkpa et al. 2016; Zhao et al. 2011; Wang et al. 2018; Han et al. 2018; Yu et al. 2018). GWAS is being applied with more genetically diverse populations than biparental populations (Han and Huang 2013). However, complicated population structures hinder the accurate detection of genetic effects governing the traits (Ogawa et al. 2018). GWAS method was applied with selected populations that are not structured genetically but possessing phenotypic diversity for agronomic traits (Yano et al. 2016) and grain length (Ogawa et al. 2018). GWAS was employed in 517 rice landraces to uncover the genetic basis of 14 agronomic traits and detected 6 loci which explained ~36% of the phenotypic variance associated with previously identified genes (Huang et al. 2010). GWAS based on 413 diverse accessions of *O. sativa* collected from 82 countries and SNP genotyping for 34 traits in rice revealed genetic architecture associated complex traits related to plant morphology, grain quality, plant development, and agronomic traits with subpopulation structure and response to environment (Zhao et al. 2011). Genetic loci associated with flowering time and grain related traits were dissected using GWAS in 950 worldwide rice varieties (Huang et al. 2011). Courtois et al. (2013) performed an association mapping for traits related to root system architecture using a panel of 167 japonica accessions and reported 19 associations significant at $P < 1e{-}05$, 78 were significant at $P < 1e{-}04$ with significant associations were detected for deep root mass and the number of deep roots. Among the significant loci identified, 51 were unique and others were co-localized for several different traits. Among the 51 loci, 19 were in predicted genes, and ten of these encoded expressed or hypothetical proteins without known functions. About 37 loci were associated with a root trait and 20 of them were co-localized with root QTLs on chromosomes 1, 2, 3, 4, 5, 7, 8, 9, 10, and 11. Several likely candidate genes were found in close proximity to the identified loci. GWAS was conducted for a ratio of deep rooting (RDR) in the whole collection of 237 genotypes, the *indica* subpopulation, and in the *japonica,* subpopulation detected a significant association on chromosome 1 in all the three groups. In the collection of 237 genotypes, 48 associated SNPs were identified that clustered into seven regions located on chromosomes 1, 3, 4, 6, and 7. In the *indica* subpopulation, unlike the *japonica* subpopulation or the whole collection, a significant association on chromosome 2, overlapped with that of major QTL *qRDR-2* indicated that the deep-rooting group had selective sweep regions on chromosomes 1 and 2, specifically in the major QTL region on chromosome 2. Seven of the nine candidate SNPs identified by association mapping were also validated in RDR extreme groups that possessed different alleles of the trait (Lou et al. 2015). GWAS with SNP panel of exclusively 180 Vietnam rice accessions of *indica* and *japonica* groups for root traits which include root thickness, root length (RL), root mass in different layers, the number of crown roots, as well as many derived roots and shoot traits, detected several QTLs in the *indica* and *japonica* subpanels or in the whole panel, of which association each for root thickness on chromosome 2 and for crown root number (RNo) on chromosome 11 found to be highly significant (Phung et al. 2016). GWAS on 162 *indica* and 169 *japonica* rice accession panels for root cone angle reported 15 and 40 QTLs for the *indica* and *japonica* panels, respectively. Candidate gene analysis focused mainly on auxin-related genes, kinases, and genes involved in root developmental processes within the reported QTLs led to the identification of eight genes *viz., OsSAUR12, OsCKX4, OsSPR1, OsCSLD1, STRUBBELIG-RECEPTOR FAMILY 8, BRASSINOTEROID INSENSITIVE 1, Hw1*, and *PAL* close to significant markers (Bettembourg et al. 2017). Using 529 rice accessions as natural population adopted GWAS for genetic dissection of root traits, identified 110 significant association loci containing many reported and unknown candidate genes related to 21 root development traits including maximum root length (MRL), root volume, and root dry weight (RDW) under normal and drought stress conditions at maturity stage. Candidate gene analysis of the associated loci led to the co-localization of 11 root-related genes including previously reported *DRO1, WOX11*, and *OsPID*. The usefulness of GWAS results in the identification of causal genes governing root traits with *Nal1* and *OsJAZ* genes was demonstrated (Li et al. 2017). The genetic control of rooting behavior for 35 root traits among 274 *indica* panel elucidated in GWAS. Association analysis detected 104 and 106 significant loci under control and water-deficit stress conditions, respectively. Twenty-two out of 104 associations and 10 out of 106 associations were found to have tight linkages or pleiotropic effects linked with more than one loci or genes. QTLs for total root length (TRL) identified on chromosome 4 were found to be associated with root weight (RW) and root-shoot ratio. Candidate genes governing root growth and development in the associated regions were also delineated (Kadam et al. 2017). GWAS method was applied in *indica* predominant diversity panel in rice for seedling vigor traits including crown RNo, MRL, TRL, root surface area (RSA), and root tip number (TIPS), shoot length (SL), RDW and shoot dry weight (SDW), identified two QTLs for MRL on chromosomes 5 and 6, two for RNo on chromosomes 1 and 8, one for TRL on chromosome 11, and one for TIPS on chromosome 11 (Wang et al. 2018).

### 5.8.3 Wheat

Wheat (*Triticum aestivum* L) is one of the major cereal crops cultivated worldwide. Compared to other cereals, wheat being hexaploid possesses the largest genome among cereals (William et al. 2007). The large size of the *T. aestivum* genome is a major constraint to sequencing as physical map construction remains a challenge. The availability of high-density wheat iSelect 90K SNP array facilitated GWAS for complex traits. Several studies have established the effectiveness of GWAS in the identification of genomic regions associated with different traits. Marker–trait associations for plant architecture-related traits, GY, yield components, agronomic traits, adaptation traits, several morphological traits, biotic, and abiotic stresses were identified through GWAS. Awns play a major role in seed dispersal and photosynthesis of spikes as they contribute about 40–80% of spike to the carbon exchange rate depending upon the surface area of awns and are also involved in water use efficiency. Hence, long awns as selection criteria for high yield in wheat has been reported (Rebetzke et al. 2016). In wheat, three major awn inhibitors namely Hd (Hooded, chromosome 4A), B1 (Tipped 1, 5A), and B2 (Tipped 2, 6B) were identified. A major genetic locus for awning was detected on the long arm of chromosome 5A using GWAS. A significant SNP associated with awning was identified in MAGIC population (Mackay et al. 2014). Recent studies identified a C2H2 zinc finger transcription factor as a candidate gene for awning in GWAS and biparental mapping studies (DeWitt et al. 2019; Wang et al. 2019; Niu et al. 2020). GWAS identified genetic loci governing Fusarium head blight resistance (Wang et al. 2017). GWAS has been utilized for the identification of disease resistance in wheat. SNPs associated with seedling and adult stage resistance were detected for stripe rust, caused by *Puccinia striiformis* f. sp. *tritici* (Pst) in GWAS of 181 synthetic hexaploid wheat (Zegeye et al. 2014). Analysis of a worldwide collection of 1000 spring wheat accessions in GWAS for yellow rust identified 97 significant loci associated in three environments (Maccaferri et al. 2015). In GWAS of 182 durum wheat landraces, Liu et al. (2017) identified 68 SNP loci that are associated with seedling resistance to one or more races. In Ethiopian germplasm, 15 SNP loci that are significantly associated with seedling and adult plant resistance (APR) to stripe rust were identified (Muleta et al. 2017). Genomic regions associated with stripe rust in wheat were identified through GWAS in 120 China winter wheat accessions using a 90K wheat SNP array (Cheng et al. 2020). Single and multilocus GWAS was applied for plant stature-related traits including PH, flag leaf characteristics, and tiller number per plant, and detected significant associations across multiple test environments (Wang et al. 2020). Using a panel of 470 accessions with an Illumina iSelect 90K SNP array and Kompetitive allele-specific PCR (KASP) markers identified allelic combination at genes governing agronomic traits and Fusarium head blight resistance via the Fhb1 locus (Mellers et al. 2020). Neumann et al. (2011) conducted GWAS for 20 different traits in a winter wheat core set using DArT markers where significant associations were detected for GY, PH, and disease resistance. In another study, a diverse set of 94 wheat lines were used to detect the marker–trait associations for kernel weight, kernel hardness, grain volume, grain protein concentration, and PH (Wang et al. 2014). Genomic regions associated with GY and yield components were identified in a 287 elite wheat association mapping initiative (WAMI) panel (Sukumaran et al. 2014; Lopes et al. 2015). In a panel of 404 diverse wheat genotypes and 35K Axiom array, GWAS identified SNP loci and their underlying candidate genes associated with the 36 agro-morphological traits (Sheoran et al. 2019). The genetic basis of 17 agronomically important traits and a drought wilting score was uncovered using GWAS in 493 worldwide durum wheat collection and identified 90 marker–trait associations and 14 putative candidate genes (Wang et al. 2019). GWAS on yield and yield components among 322 soft red winter wheat lines across four environments using a GBS identified significant marker–trait associations. The suggestive candidate genes contributing to GY include aberrant panicle organization (APO1) protein and a gibberellin oxidase protein (GA2ox-A1) affecting grain number and B2 heat stress response protein affecting the trait seeds per head (Ward et al. 2019). GWAS was conducted for drought tolerance and biomass traits including days to heading (DTH), days to maturity (DTM), shoot biomass (SB), root biomass (RB), root to shoot ratio (RS), and GY among 100 wheat genotypes identified significant marker–trait associations (Mathew et al. 2019). About 323 wheat accessions were investigated for root and shoot traits including PH, canopy temperature (CT), and grain yield per plant (YPP) indicated that deep-rooted accessions had lower CT and higher YPP than those with shallow roots where significantly associated loci were detected (Li et al. 2019a). In a study involving 277 wheat accessions tested over 30 environments for drought, heat, and combined heat and drought stress conditions, GWAS revealed significant associations loci for agronomic traits and multiple abiotic stress tolerance across different treatments (Li et al. 2019b). GWAS also identified SNPs associated with starch content in wheat (Tsai et al. 2020).

### 5.8.4 Barley

Barley is often considered as a basic crop for human civilization with about 70% of its production is used as animal feed and the rest for malting and food purpose. Barley is mostly grown in temperate regions of the world. The natural variation maintained through domestication of barely over 10 000 years conserved several developmental and adaptive traits. GWAS has elucidated the genetic basis of phenotypic variation for barley adaption and yield improvement. Improvement of barley for several traits using GWAS has been attempted at ICRISAT (Figure 5.2)

Several useful QTLs and genes were identified using GWAS that facilitated their use in marker-assisted selection. Most of the populations are genotyped using 9K iSelect Illumina Infinium array or 50K Illumina Infinium iSelect genotyping array with a high density of quality markers and better genome coverage (Bayer et al. 2017). The effectiveness of GWAS combined with and high-throughput sequencing technology in dissecting the complex traits in barley has been demonstrated to be a more efficient approach in utilizing the available genetic resources. The genetic loci underlying the phenotypic variation for different developmental and agronomic traits such as heading date, PH, tiller number, grain number, biotic and abiotic stress tolerance, yield, and quality have been elucidated using the GWAS approach (Wang et al. 2012; Pasam et al. 2012; Digel et al. 2016; Bykova et al. 2017; Alqudah et al. 2016, 2018). Several studies have demonstrated the efficiency of GWAS with high-throughput genotyping platform in detecting QTL and allelic variation as well as identification of candidate genes controlling target traits in barley. Using GWAS, *ANTHOCYANINLESS 2* (*ANT2*) is the first gene identified and cloned (Cockram et al. 2010). The heading time gene related to responses to long-day photoperiod governed by the allelic

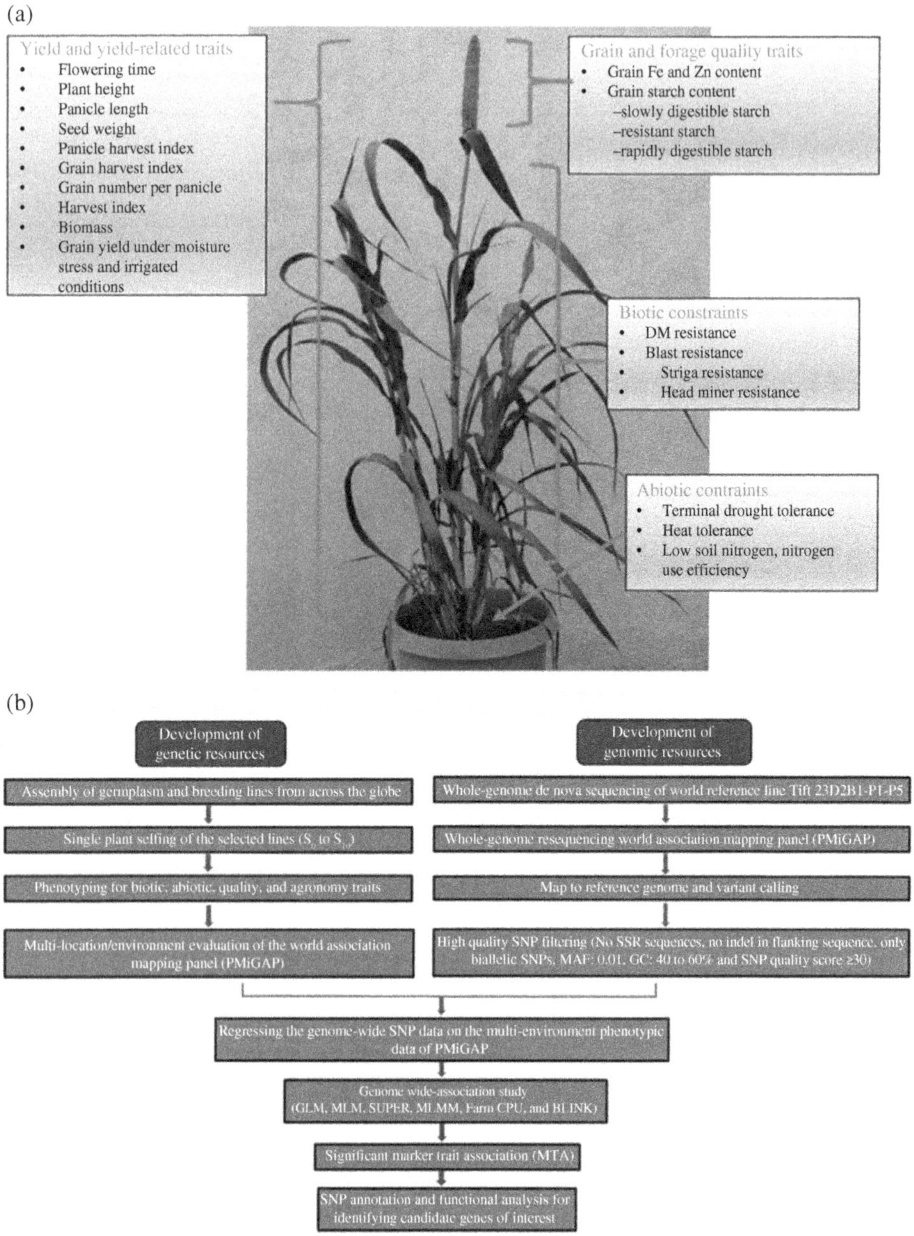


**Figure 5.2** (a) The figure illustrating the traits for which genome-wide association studies (GWAS) are being attempted at ICRISAT, India. (b) Workflow for genome-wide association studies (GWAS) pipeline. *Source:* Srivastava et al. (2020)/figure 01(p.06)/CC BY 4.0.

diversity *PSEUDO-RESPONSE REGULATOR* (*HvPRR37*)/*PHOTOPERIOD RESPONSE LOCUS1* (*Ppd-H1*) gene in barley. *Ppd-H1* regulates the photoperiod-sensitive accessions in differentiating early and late heading accessions (Turner et al. 2005; Alqudah et al. 2014).

Allelic variations at *Ppd-H1* controls cell proliferation period and leaf maturation contributing to leaf size in winter barley (Digel et al. 2016). A multilocus genome-wide association study was applied for genetic dissection of genes governing 14 agronomic traits in double haploid (DH) population and 10 candidate genes were identified from gene annotation. The *Vrs1* gene, governing the number of rows in the spike, was found to be associated with a number of spikelets in the main spike, a number of spikelet per plant, grain number per plant, grain number per spike, and 1000-grain weight tested over multiple environments and detected by multilocus GWAS methods (Thirulogachandar et al. 2017; Hu et al. 2018). GWAS was applied to dissect alleles underlying the six-rowed spike 2 (*VRS2*) gene in a worldwide barely collection. The functional VRS2 protein was found to be in significant association leaf area and leaf and tiller number (Youssef et al. 2017). GWAS also elucidated the natural selection of adaptive variation for late heading in European accessions in turn improves the adaptation of other developmental traits related to increased GY (Digel et al. 2016; Alqudah et al. 2016, 2018). GWAS has been used in delineating the genetic basis of a wide range of traits and detected novel QTLs for germination, seedling architecture for PH, tillering, leaf area, phase transition, and developmental stages (Alqudah et al. 2014, 2016, 2018). GWAS approach was also applied for understanding genetic mechanisms of biotic and abiotic stress tolerance using diverse populations of barley (Thabet et al. 2018). GWAS was applied to explore the underlying associated drought tolerance traits such as GY and physiological traits like water use efficiency and water content (Varshney et al. 2012; Wojcik-Jagla et al. 2018); stem rust resistance (Belcher et al. 2018); genetic architecture of yield component and quality traits (Xu et al. 2018; Tsai et al. 2020).

### 5.8.5 Pearl Millet

Pearl millet is one of the staple food crops grown in arid and semi-arid regions, often referred to as a climate-resilient, drought-tolerant, nutrient-rich crop contributing 30–40% of inorganic nutrients with an adequate level of iron and zinc to combat micronutrient deficiency. Pearl Millet with the high outcrossing rate with inbreeding depression, heterozygous nature, and residual heterozygosity imposes a major limitation in the crop improvement program including the application of GWAS which warrants relatively a greater number of markers. The crop remained neglected until the twenty-first century, and with the development of various genetic and genomic resources such as genome-wide association studies feasible with the developments in high-throughput GBS, RAD-sequencing, and whole-genome resequencing (WGRS) it gained importance. A panel for GWAS called "Pearl Millet inbred Germplasm Association Panel (PMiGAP)" was developed from germplasm lines from several countries through WGRS approach. In Pearl millet, GWAS has been conducted to map genetic loci governing drought tolerance, grain Fe and Zn content, agronomic traits, nitrogen use efficiency, abiotic and biotic stresses, quality, GY (Hu et al. 2015). Association studies for a flowering time at the phytochrome C (PHYC) (866 bp) locus were carried out using 90 inbred and 598 pearl millet varieties from India, East, and West Africa following genotyping with 27 SSR and 6 AFLP markers (Saïdou et al. 2009, 2014). Association studies involving 90 inbred for target traits such as flowering time, stem diameter, PH, spikelet length, and spikelet density conducted and identified the MADS-box gene family member, PgMADS11 gene, which is important in somatic and reproductive phase of development in the plant during its adaptation toward climate change (Mariac et al. 2011). GWAS with selective SSR markers with the PH, flowering time, panicle length, stover, and GY were genetically dissected (Kannan et al. 2014). Using 151 PMiGAP lines from West Africa tested across six environments under high and low P conditions, GWAS was conducted with 285 DArT markers and phenotypic data, identified significant marker–trait associations for flowering time, and GY under low P conditions (Gemenet et al. 2015). Significant marker–trait association was detected for grain iron and zinc through GWAS among 130 diversified lines and 250 SSR and 17 genic markers (Anuradha et al. 2017).

### 5.8.6 Sugarcane

Sugarcane (*Saccharum* spp.) is an important industrial crop, which has a complex large genome with varying ploidy levels and frequent aneuploidy. These characters render more challenges to geneticists and breeders to implement strategies for crop improvement. The sugarcane breeding programs are mainly focused on the development of cultivars with high biomass production, high sucrose content, and stress resistance. Molecular approaches for trait locus (QTL) mapping and genome-wide association studies (GWAS) enabled a better understanding of the genetic architecture of complex traits. The highly heterozygous outcrossing nature of sugarcane crop where different segregating alleles at each segregating loci limits the application of the traditional QTL mapping approach due to a limited number of available markers and low resolution of genomic regions identified. Alternatively, GWAS has been widely conducted to identify marker–trait associations among genetically diverse populations (Racedo et al. 2016). GWAS was performed using 134 accessions and SSR markers for traits such as soluble solid content, stalk height, stalk number, stalk weight, and cane yield (Barreto et al. 2019).

## 5.9 Candidate Gene Identification at GWAS Loci

Apart from GWAS, candidate genes at GWAS loci affecting target traits were identified through candidate-gene-based association analysis in different crop plants (Table 5.3). SNPs within specific genes are important for candidate-gene-based association analysis. The method involves resequencing of amplicons for their associated SNP identified from a large population of numerous genetically diverse lines and/or within the specific genes. In general, to identify a candidate gene, promoter SNPs, exon, intron, and untranslated 5′/3′ regions are targeted along with coding regions that have less level of nucleotide diversity than the noncoding part (Zhu et al. 2008) while taking into account the rate of LD decay and SNP distribution. Several genes associated with complex traits have been identified through integrating GWAS and gene expression analyses (Table 5.3). In Arabidopsis, a novel gene F-box *KUK* is associated with root cell length (Meijón et al. 2014) and a brassinosteroid signaling kinase encoding *BSK3* for root elongation under mild nitrogen deficiency (Jia et al. 2019) were identified through GWAS. Among the 143 significant associations for various root traits, 11 reported root-related gens were identified in GWAS among 529 rice accessions and genes *Nal1* and *OsJAZ1* were validated through expression

**Table 5.3** Candidate genes identified through genome-wide association studies (GWAS).

| Crop | Trait | Candidate gene | Gene key functions | References |
|---|---|---|---|---|
| Pearl millet | Flowering time | *PgMADS11* | MADS-box, somatic, and reproductive phase development | Mariac et al. (2011) |
| Arabidopsis | Root cell length | F-box *KUK* | F-box family | Meijón et al. (2014) |
| Arabidopsis | Root elongation | *BSK3* | Brassinosteroid signaling kinase (BSK) | Jia et al. (2019) |
| Rice | Root system architecture | *DRO1* | IGT family, root system architecture, and drought avoidance | Li et al. (2017) |
| Rice | Crown root formation and development | *WOX11* | WUSCHEL-related homeobox gene family | Li et al. (2017) |
| Rice | Root hair formation | *OsAPY* | Aspartic proteases | Li et al. (2017) |
| Rice | Crown root development | *RR2* | Cytokinin-signalling | Li et al. (2017) |
| Rice | Crown root development | *WOX4* | WUSCHEL-related homeobox family, and cytokinin signaling | Li et al. (2017) |
| Rice | Root gravitropism, auxin transport | *OsPID* | PINOID, root development | Li et al. (2017) |
| Rice | Crown root development | *DGL1* | GATA family, microtubule organization, gibberellin signaling | Li et al. (2017) |
| Rice | Root length, lateral root development | *OsCKI1* | Casein kinase I-like gene family | Li et al. (2017) |
| Rice | Root hair development | *OsRHL1* | RSL class II subfamily | Li et al. (2017) |
| Rice | Leaf width regulation, adventitious root, and spikelet development | *Nal1* | YABBY transcription factor | Li et al. (2017) |
| Rice | Maximum root length | *OsJAZ1* | Jasmonate ZIM-domain protein, upregulated under drought stress | Zhao et al. (2018) |
| Rice | Maximum root length | *OsRL1.1* | RING-H2 finger protein | Zhao et al. (2018) |
| Rice | Maximum root length | *OsGH3.1* | Probable indole-3-acetic acid amido synthetase | Zhao et al. (2018) |

(*Continued*)

**Table 5.3** (Continued)

| Crop | Trait | Candidate gene | Gene key functions | References |
|---|---|---|---|---|
| Rice | Maximum root length | *OsRL3.1* | RPP13-like protein 1 | Zhao et al. (2018) |
| Rice | Maximum root length | *OsRL3.3* | Coatomer alpha subunit, putative | Zhao et al. (2018) |
| Rice | Maximum root length | *OsRL7.1* | WD domain, G-beta repeat domain-containing protein | Zhao et al. (2018) |
| Rice | Maximum root length | *OsRL8.2* | Transferase family protein | Zhao et al. (2018) |
| Rice | Maximum root length | *OsRL11.1* | NBS-LRR type disease resistance protein, putative, expressed | Zhao et al. (2018) |
| Rice | Basal root thickness | *OsRT4.1* | Amidase family protein, putative, expressed | Zhao et al. (2018) |
| Rice | Basal root thickness | *OsRT4.2* | MATE efflux family protein | Zhao et al. (2018) |
| Rice | Basal root thickness | *OsRT4.4* | Methyltransferase domain-containing protein | Zhao et al. (2018) |
| Rice | Basal root thickness | *OsRT4.7* | OsFBK14–Fbox domain and kelch repeat-containing protein | Zhao et al. (2018) |
| Rice | Basal root thickness | *OsRT8.1* | SET domain-containing protein | Zhao et al. (2018) |
| Rice | Basal root thickness | *OsRT8.2* | CCT/B-box zinc finger protein | Zhao et al. (2018) |
| Maize | Average root diameter | *GRMZM2G138338* | Leucine-rich receptor-like protein kinase family protein | Wang et al. (2019) |
| Maize | Average seminal root length | *GRMZM2G174797* | ELMO/CED-12 family protein | Wang et al. (2019) |
| Maize | Lateral root length, root surface area | *GRMZM2G476902* | Armadillo/beta-catenin repeat family protein | Wang et al. (2019) |
| Maize | Total root length | *GRMZM2G397965* | Vignain precursor/(SAG12) senescence-associated gene 12 | Wang et al. (2019) |
| Maize | Root dry weight | *GRMZM2G031528* | Heavy metal transport/detoxification superfamily protein | Wang et al. (2019) |

analysis (Li et al. 2017). Among the 44 and 97 QTL candidate genes for RL and root thickness identified by GWAS among 795 rice accessions and other integrative methods, five RL QTL candidates were validated by T-DNA insertional mutation and transcriptomic analyses (Zhao et al. 2018). In maize using a panel of 297 inbred lines for GWAS for root architecture at the seedling stage, RNA-seq was conducted with three long-root genotypes and three short-root genotypes to obtain the expression levels of candidate genes which led to the identification of potential candidate genes for root traits (Wang et al. 2019)

## 5.10 Meta-GWAS

Identification of genomic region(s) associated with complex traits has been a major focus of QTL mapping and genome-wide association studies. Several QTLs governing major traits in different crops have been detected. However, the utility of these QTLs for introgression in breeding has been limited for various reasons. Very few QTLs identified were fine mapped, cloned, and hence introgressed via marker-assisted breeding. Comparison, identification of common genomic region(s), and precise utilization of QTLs detected across different studies would be highly informative in targeted breeding for the most important traits but it is tedious. Meta-QTL analysis, a statistical approach that enables to combine the QTL results from independent studies into a single consensus QTL. Meta-analysis is a robust technique in the identification of genomic regions associated with traits of interest. Meta-QTL refers to a consensus QTL at a confidence interval (CI) of 95% identified from a set of QTLs from various studies. Previously, this approach has been reported for various traits identified through linkage analysis in rice and other crops including GY (Swamy et al. 2011), root traits (Courtois et al. 2009), protein concentration (Zhao-Ming et al. 2011), panicle related traits (Wu et al. 2016), grain size/weight (Daware et al. 2017), disease resistance traits (Lu et al. 2018), and seedling stage salinity (Islam et al. 2019). In meta-GWAS analysis, the QTLs identified through GWAS from different studies were optimized and results are combined into a single output. GWAS-based genetic studies on natural or unstructured populations involve the unbalanced datasets where individuals are measured at different time and environments. GWAS conducted in different studies for the same trait is expected to exhibit a certain level of heterogeneity for identified QTLs due to different genotyping markers and methods, experimental conditions, trait measurements, gene and environment interaction, population type, and size as well as the analysis method through which QTLs detected. Meta-GWAS analysis is mostly carried out either by combining the *P*-values using Fisher's method in which all the studies are weighted equally. An improved method based on weighted *Z*-score method, in which *P*-values are transformed to *Z*-scores in a one-to-one transformation and allow different weights for different studies (Whitlock 2005), implemented using software METAL (Willer et al. 2010). Another alternative to combining *P*-values and/or *Z*-scores is to combine estimates usually done either with fixed-effects or random-effects models. In fixed-effects meta-analysis, genetic effects are assumed as similar across different studies, thus provides narrow CIs significantly lower *P*-values for the variants than random-effects models (Nakaoka and Inoue 2009). Meta-analysis of genome-wide associations has been utilized in the genetic dissection of complex human diseases. However, very few reports are available in crops despite several GWAS conducted in major crops for various traits. Using 4095 advanced breeding lines from CIMMYT bread wheat breeding program over the period of 2009 to 2014, a meta-genome-wide association study (GWAS) identified marker–trait associations, allele effects, candidate genes for processing, and end-use quality traits (Battenfield et al. 2018). The meta-analysis of genome-wide association studies (GWAS) using 775 tomato accessions and 2 316 117 SNPs from three GWAS panels and identified 305 significant associations for sugars, acids, amino acids, and flavor-related volatiles (Zhao et al. 2019).

## 5.11 GWAS vs. QTL Mapping

QTL mapping and GWAS exploit chromosomal recombination or crossing over to detect associations between traits and the genomic regions associated with the traits. QTL mapping uses recent recombination whereas GWAS uses historical recombinations. QTL mapping or Bi-parental association mapping or linkage mapping is based on a cross between contrasting parents for the trait of interest. Therefore, only one recombination event is recorded in the population. Whereas in GWAS, there are several recombination events in a diverse population. Therefore, it can test several alleles of a trait in the same experiment. In fact, GWAS takes advantage of historical recombination events that occurred over several generations by way of LD. QTL mapping uses linkage to make association between a trait and a chromosomal region. GWAS uses LD to make association between a trait and a chromosomal region at population level. Therefore, in GWAS the mapping resolution is high, but there is a chance of spurious associations. In QTL mapping, only two alleles are found whereas in GWAS a set of several alleles per loci are found. In QTL, mapping a linkage map with markers is created for a particular trait of interest, which is specific for that trait and cannot be used for any other trait. In GWAS, if a population is genotyped once, it can be used for any other trait of that organism. QTL mapping typically provides a resolution of a linkage map of the magnitude of 5–10 cM. This magnitude can accommodate several hundred genes (Buckler and Thornsberry 2002) whereas GWAS can identify genes governing a trait of interest (Hung et al. 2012; Yang et al. 2013). QTL mapping requires fewer markers whereas GWAS requires a very high number of markers. In GWAS, for a strong association between marker and phenotype diverse population is needed whereas in QTL mapping a contrasting parent for a trait of interest is required. QTL mapping is laborious and time consuming whereas GWAS is comparatively less laborious and time savvy.

## References


Alqudah, A.M., Sharma, R., Pasam, R.K. et al. (2014). Genetic dissection of photoperiod response based on GWAS of pre-anthesis phase duration in spring barley. *PLoS One* 9 (11): e113120.

Alqudah, A.M., Koppolu, R., Wolde, G.M. et al. (2016). The genetic architecture of barley plant stature. *Frontiers in Genetics* 7: 117.

Alqudah, A.M., Youssef, H.M., Graner, A. et al. (2018). Natural variation and genetic make-up of leaf blade area in spring barley. *TAG. Theoretical and Applied Genetics. Theoretische Und Angewandte Genetik* 131 (4): 873–886.

Alqudah, A.M., Sallam, A., Baenziger, P.S. et al. (2020). GWAS: Fast-forwarding gene identification and characterization in temperate Cereals: lessons from Barley–A review. *Journal of Advanced Research* 22: 119–135.

Anuradha, N., Satyavathi, C.T., Bharadwaj, C. et al. (2017). Deciphering genomic regions for high grain iron and zinc content using association mapping in pearl millet. *Frontiers in Plant Science* 8: 412. https://doi.org/10.3389/fpls.2017.00412.

Barreto, F.Z., Rosa, J.R.B.F., Balsalobre, T.W.A. et al. (2019). A genome-wide association study identified loci for yield component traits in sugarcane (*Saccharum* spp.). *PLoS One* 14 (7): e0219843. https://doi.org/10.1371/journal.pone.0219843.

Battenfield, S.D., Sheridan, J.L., Silva, L.D.C.E. et al. (2018). Breeding-assisted genomics: applying meta-GWAS for milling and baking quality in CIMMYT wheat breeding program. *PLoS One* 13 (11): e0204757. https://doi.org/10.1371/journal.pone.0204757.

Bayer, M.M., Rapazote-Flores, P., Ganal, M. et al. (2017). Development and evaluation of a barley 50k iSelect SNP array. *Frontiers in Plant Science* 8: 1792.

Belcher, A.R., Cuesta-Marcos, A., Smith, K.P. et al. (2018). TCAP FAC-WIN6 elite barley GWAS panel QTL. I. Barley stripe rust resistance QTL in facultative and winter six-rowed malt barley breeding programs identified via GWAS. *Crop Science* 58 (1): 103–119.

Belo, A., Zheng, P., Luck, S. et al. (2008). Whole genome scan detects an allelic variant of fad2 associated with increased oleic acid levels in maize. *Molecular Genetics and Genomics* 279: 1–10.

Benke, A., Urbany, C., and Stich, B. (2015). Genome-wide association mapping of iron homeostasis in the maize association population. *BMC Genetics* 16: 1.

Bettembourg, M., Dardou, A., Audebert, A. et al. (2017). Genome-wide association mapping for root cone angle in rice. *Rice* https://doi.org/10.1186/s12284-017-0184-z.

Buckler, E.S. and Thornsberry, J.M. (2002). Plant molecular diversity and applications to genomics. *Current Opinion in Plant Biology* 5: 107–111.

Buckler, E.S., Holland, J.B., Bradbury, P.J. et al. (2009). The genetic architecture of maize flowering time. *Science* 325 (5941): 714–718. https://doi.org/10.1126/science.1174276.

Bykova, I.V., Lashina, N.M., Efimov, V.M. et al. (2017). Identification of 50 K Illumina-chip SNPs associated with resistance to spot blotch in barley. *BMC Plant Biology* 17 (Suppl 2): 250.

Chen, G., Wang, X., Hao, J. et al. (2015). Genome-wide association implicates candidate genes conferring resistance to maize rough dwarf disease in maize. *PLoS One* 10: e0142001.

Chen, J., Shrestha, R., Ding, J. et al. (2016). Genome-wide association study and QTL mapping reveal genomic loci associated with fusarium ear rot resistance in tropical maize germplasm. *G3 (Bethesda)* 6: 3803–3815.

Cheng, B., Gao, X., Cao, N. et al. (2020). Genome-wide association analysis of stripe rust resistance loci in wheat accessions from southwestern China. *Journal of Applied Genetics* 61: 37–50. https://doi.org/10.1007/s13353-019-00533-8.

Cockram, J., White, J., Zuluaga, D.L. et al. (2010). Genome-wide association mapping to candidate polymorphism resolution in the unsequenced barley genome. *Proceedings of the National Academy of Sciences* 107 (50): 21611–21616.

Courtois, B., Ahmadi, N., Khowaja, F. et al. (2009). Rice drought tolerance QTLs: meta-analysis improves resolution to a few candidate gene level. 6th International Rice Genetics Symposium, Manila, Philippines, 06–19 November 2009.

Courtois, B., Audebert, A., Dardou, A. et al. (2013). Genome-wide association mapping of root traits in a Japonica rice panel. *PLoS One* 8: 1–18. https://doi.org/10.1371/journal.pone.0078037.

Cui, Z., Luo, J., Qi, C. et al. (2016). Genome-wide association study (GWAS) reveals the genetic architecture of four husk traits in maize. *BMC Genomics* 17: 946.

Daware, A.V., Srivastava, R., Singh, A.K. et al. (2017). Regional association analysis of MetaQTLs delineates candidate grain size genes in rice. *Frontiers in Plant Science* 8: 807. https://doi.org/10.3389/fpls.2017.00807.

Dell'Acqua, M., Gatti, D., Pea, G. et al. (2015). Genetic properties of the MAGIC maize population: a new platform for high definition QTL mapping in Zea mays. *Genome Biology* 16: 167.

DeWitt, N., Guedira, M., Lauer, E. et al. (2019). Sequence based mapping identifies AWNS1, a candidate transcription repressor underlying awn suppression at the B1 locus in wheat. *bioRxiv* 519264. https://doi.org/10.1101/519264.

Digel, B., Tavakol, E., Verderio, G. et al. (2016). Photoperiod-H1 (Ppd-H1) controls leaf size. *Plant Physiology* 172 (1): 405–415.

Dilnur, T., Peng, Z., Pan, Z. et al. (2019). Association analysis of salt tolerance in Asiatic cotton (*Gossypium arboretum*) with SNP markers. *International Journal of Molecular Sciences* 20 (9): 2168. https://doi.org/10.3390/ijms20092168.

Dimkpa, S.O., Lahari, M.Z., Shrestha, R. et al. (2016). A genome-wide association study of a global rice panel reveals resistance in *Oryza sativa* to root-knot nematodes. *Journal of Experimental Botany* 67 (4): 1191–1200. http://hdl.handle.net/1854/LU-8203288.

Du, X., Liu, S., Sun, J. et al. (2018). Dissection of complicate genetic architecture and breeding perspective of cottonseed traits by genome-wide association study. *BMC Genomics* 19: 451.

Farfan, I., De La Fuente, G., Murray, S. et al. (2015). Genome wide association study for drought, aflatoxin resistance, and important agronomic traits of maize hybrids in the sub-tropics. *PLoS One* 10: e0117737.

Gahlaut, V., Jaiswal, V., Singh, S. et al. (2019). Multi-locus genome wide association mapping for yield and its contributing traits in hexaploid wheat under different water regimes. *Scientific Reports* **9**: 19486. https://doi.org/10.1038/s41598-019-55520-0.

Gemenet, D.C., Hash, C.T., Sanogo, M.D. et al. (2015). Phosphorus uptake and utilization efficiency in West African pearl millet inbred lines. *Field Crops Research* 171: 54–66. https://doi.org/10.1016/j.fcr.2014.11.001.

Han, B. and Huang, X. (2013). Sequencing-based genome-wide association study in rice. *Current Opinion in Plant Biology* 16: 133–138. https://doi.org/10.1016/j.pbi.2013.03.006.

Han, J., Shin, N., Jang, S. et al. (2018). Identification of quantitative trait loci for vigorous root development under water-deficiency conditions in rice. *Plant Breeding and Biotechnology* 2018: 147–158.

Harjes, C.E., Rocheford, T.R., Bai, L. et al. (2008). Natural genetic variation in lycopene epsilon cyclase tapped for maize biofortification. *Science* 319: 330–333.

Hayes, B. (2013). Overview of statistical methods for genome-wide association studies (GWAS). In: *Genome-Wide Association Studies and Genomic Prediction. Methods in Molecular Biology*

*(Methods and Protocols)*, vol. 1019 (ed. C. Gondro, J. van der Werf and B. Hayes). Totowa, NJ: Humana Press.

He, Y., Wu, D., Wei, D. et al. (2017). GWAS, QTL mapping and gene expression analyses in *Brassica napus* reveal genetic control of branching morphogenesis. *Scientific Reports* 7: 15971.

Hou, S., Zhu, G., Li, Y. et al. (2018). Genome-wide association studies reveal genetic variation and candidate genes of drought stress-related traits in cotton (*Gossypium hirsutum* L). *Frontiers in Plant Science* 9: 1276.

Hu, Z., Mbacké, B., Perumal, R. et al. (2015). Population genomics of pearl millet (*Pennisetum glaucum* (L.) R. Br.): comparative analysis of global accessions and Senegalese landraces. *BMC Genomics* 16: 1048. https://doi.org/10.1186/s12864-015-2255-0.

Hu, X., Zuo, J., Wang, J. et al. (2018). Multi-locus genome-wide association studies for 14 main agronomic traits in barley. *Frontiers in Plant Science* 9: 1683. https://doi.org/10.3389/fpls.2018.01683.

Huang, X., Wei, X., Sang, T. et al. (2010). Genome-wide association studies of 14 agronomic traits in rice landraces. *Nature Genetics* 42: 961.

Huang, X., Zhao, Y., Wei, X. et al. (2011). Genome-wide association study of flowering time and grain yield traits in a worldwide collection of rice germplasm. *Nature Genetics* 44: 32.

Hung, H., Shannon, L., Tian, F. et al. (2012). ZmCCT and the genetic basis of day-length adaptation underlying the post domestication spread of maize. *Proceedings of the National Academy of Sciences of the United States of America* 109: 1913–1921.

Islam, M.S., Ontoy, J., and Subudhi, P.K. (2019). Meta-analysis of quantitative trait loci associated with seedling-stage salt tolerance in rice (*Oryza sativa* L.). *Plants (Basel, Switzerland)* 8 (2): 33. https://doi.org/10.3390/plants8020033.

Jabbari, M., Fakheri, B.A., Aghnoum, R. et al. (2018). GWAS analysis in spring barley (*Hordeum vulgare* L.) for morphological traits exposed to drought. *PLoS One* 13 (9): e0204952.

Jia, Z., Giehl, R.F.H., Meyer, R.C. et al. (2019). Natural variation of BSK3 tunes brassinosteroid signaling to regulate root foraging under low nitrogen. *Nature Communications* 10: 2378.

Kadam, N.N., Tamilselvan, A., Lawas, L.M.F. et al. (2017). Genetic control of plasticity in root morphology and anatomy of rice in response to water deficit. *Plant Physiology* 174: 2302–2315. https://doi.org/10.1104/pp.17.00500.

Kannan, B., Senapathy, S., Raj, B. et al. (2014). Association analysis of SSR markers with phenology, grain, and stover yield related traits in Pearl Millet (*Pennisetum glaucum* (L.) R. Br.). *Scientific World Journal* 2014: 1–14. https://doi.org/10.1155/2014/562327.

Kumar, V., Singh, A., Mithra, S.V.A. et al. (2015). Genome-wide association mapping of salinity tolerance in rice (*Oryza sativa*). *DNA Research* 22: 133–145. https://doi.org/10.1093/dnares/dsu046.

Lex, J., Ahlemeyer, J., Friedt, W. et al. (2014). Genome-wide association studies of agronomic and quality traits in a set of German winter barley (*Hordeum vulgare* L.) cultivars using Diversity Arrays Technology (DArT). *Journal of Applied Genetics* 55 (3): 295–305.

Li, H., Peng, Z., Yang, X. et al. (2013). Genome-wide association study dissects the genetic architecture of oil biosynthesis in maize kernels. *Nature Genetics* 45: 43–50.

Li, M., Liu, X., Bradbury, P. et al. (2014). Enrichment of statistical power for genome-wide association studies. *BMC Biology* **12**: 73. https://doi.org/10.1186/s12915-014-0073-5.

Li, Y., Li, C., Bradbury, P. et al. (2016). Identification of genetic variants associated with maize flowering time using an extremely large multigenetic background population. *The Plant Journal* 86: 391–402.

Li, X., Guo, Z., Lv, Y. et al. (2017). Genetic control of the root system in rice under normal and drought stress conditions by genome-wide association study. *PLoS Genetics* 13: e1006889. https://doi.org/10.1371/journal.pgen.1006889.

Li, L., Peng, Z., Mao, X. et al. (2019a). Genome-wide association study reveals genomic regions controlling root and shoot traits at late growth stages in wheat. *Annals of Botany* 124 (6): 993–1006. https://doi.org/10.1093/aob/mcz041.

Li, L., Mao, X., Wang, J. et al. (2019b). Genetic dissection of drought and heat-responsive agronomic traits in wheat. *Plant, Cell & Environment* 42 (9): 2540–2553. https://doi.org/10.1111/pce.13577.

Li, F., Wen, W., Liu, J. et al. (2019c). Genetic architecture of grain yield in bread wheat based on genome-wide association studies. *BMC Plant Biology* 19 (1): 1–19.

Liu, H.J. and Yan, J. (2019). Crop genome-wide association study: a harvest of biological relevance. *The Plant Journal* 97 (1): 8–18. https://doi.org/10.1111/tpj.14139.

Liu, N., Xue, Y., Guo, Z. et al. (2016). Genome-wide association study identifies candidate genes for starch content regulation in maize kernels. *Frontiers in Plant Science* 7: 1046.

Liu, W., Maccaferri, M., Rynearson, S. et al. (2017). Novel sources of stripe rust resistance identified by genome-wide association mapping in Ethiopian durum wheat (*Triticum turgidum* ssp. *durum*). *Frontiers in Plant Science* 8: 774.

Liu, R., Gong, J., Xiao, X. et al. (2018). GWAS analysis and QTL identification of fiber quality traits and yield components in upland cotton using enriched high density SNP markers. *Frontiers in Plant Science* 9: 1067.

Lopes, M.S., Dreisigacker, S., Peña, R.J. et al. (2015). Genetic characterization of the wheat association mapping initiative (WAMI) panel for dissection of complex traits in spring wheat. *Theoretical and Applied Genetics* 128: 453–464. https://doi.org/10.1007/s00122-014-2444-2.

Lou, Q., Chen, L., Mei, H. et al. (2015). Quantitative trait locus mapping of deep rooting by linkage and association analysis in rice. *Journal of Experimental Botany* 66: 4749–4757. https://doi.org/10.1093/jxb/erv246.

Lu, Q., Liu, H., Hong, Y. et al. (2018). Consensus map integration and QTL meta-analysis narrowed a locus for yield traits to 0.7 cM and refined a region for late leaf spot resistance traits to 0.38 cM on linkage group A05 in peanut (*Arachis hypogaea* L.). *BMC Genomics* 19 (1): 887. https://doi.org/10.1186/s12864-018-5288-3.

Ma, X., Feng, F., Wei, H. et al. (2016). Genome-wide association study for plant height and grain yield in rice under contrasting moisture regimes. *Frontiers in Plant Science* 7: 1801. https://doi.org/10.3389/fpls.2016.01801.

Maccaferri, M., Zhang, J., Bulli, P. et al. (2015). A genome-wide association study of resistance to stripe rust (*Puccinia striiformis* f. sp. tritici) in a worldwide collection of hexaploid spring wheat (*Triticum aestivum* L.). *G3 (Bethesda, Md.)* 5 (3): 449–465. https://doi.org/10.1534/g3.114.014563.

Mackay, T.F.C., Stone, E.A., and Ayroles, J.F. (2009). The genetics of quantitative traits: challenges and prospects. *Nature Reviews. Genetics* 10: 565–577.

Mackay, I.J., Bansept-Basler, P., Barber, T. et al. (2014). An eight-parent multiparent advanced generation inter-cross population for winter-sown wheat: creation, properties, and validation. *G3 (Bethesda, Md.)* 4 (9): 1603–1610. https://doi.org/10.1534/g3.114.012963.

Mahuku, G., Chen, J., Shrestha, R. et al. (2016). Combined linkage and association mapping identifies a major QTL (qRtsc8-1), conferring tar spot complex resistance in maize. *TAG. Theoretical and Applied Genetics. Theoretische Und Angewandte Genetik 129* (6): 1217–1229. https://doi.org/10.1007/s00122-016-2698-y.

Mariac, C., Jehin, L., Saïdou, A.A. et al. (2011). Genetic basis of pearl millet adaptation along an environmental gradient investigated by a combination of genome scan and association mapping. *Molecular Ecology* 20: 80–91. https://doi.org/10.1016/j.jas.2010.09.007.

Mathew, I., Shimelis, H., Shayanowako, A. et al. (2019). Genome-wide association study of drought tolerance and biomass allocation in wheat. *PLoS One* 14 (12): e0225383. https://doi.org/10.1371/journal.pone.0225383.

Matthies, I.E., van Hintum, T., Weise, S. et al. (2012). Population structure revealed by different marker types (SSR or DArT) has an impact on the results of genome-wide association mapping in European barley cultivars. *Molecular Breeding* 30 (2): 951–966.

Mazaheri, M., Heckwolf, M., Vaillancourt, B. et al. (2019). Genome-wide association analysis of stalk biomass and anatomical traits in maize. *BMC Plant Biology* 19: 45. https://doi.org/10.1186/s12870-019-1653-x.

Meijón, M., Satbhai, S.B., Tsuchimatsu, T. et al. (2014). Genome-wide association study using cellular traits identifies a new regulator of root development in *Arabidopsis*. *Nature Genetics* 46: 77.

Mellers, G., Aguilera, J.G., Bird, N. et al. (2020). Genetic characterization of a wheat association mapping panel relevant to Brazilian breeding using a high-density single nucleotide polymorphism array. *G3 (Bethesda, Md.)* 10 (7): 2229–2239. https://doi.org/10.1534/g3.120.401234.

Mitchell-Olds, T. (2010). Complex-trait analysis in plants. *Genome Biology* 11: 113.

Muleta, K.T., Bulli, P., Rynearson, S. et al. (2017). Loci associated with resistance to stripe rust (*Puccinia striiformis* f. sp. tritici) in a core collection of spring wheat (*Triticum aestivum*). *PLoS One* 12 (6): e0179087. https://doi.org/10.1371/journal.pone.0179087.

Nakaoka, H. and Inoue, I. (2009). Meta-analysis of genetic association studies: methodologies, between-study heterogeneity and winner's curse. *Journal of Human Genetics* 54: 615–623.

Neumann, K., Kobiljski, B., Denčić, S. et al. (2011). Genome-wide association mapping: a case study in bread wheat (*Triticum aestivum* L.). *Molecular Breeding* 27 (1): 37–58. https://doi.org/10.1007/s11032-010-9411-7.

Niu, J., Zheng, S., Shi, X. et al. (2020). Fine mapping and characterization of the awn inhibitor B1 locus in common wheat (*Triticum aestivum* L.). *Crop Journal* https://doi.org/10.1016/j.cj.2019.12.005.

Ogawa, D., Nonoue, Y., Tsunematsu, H. et al. (2018). Discovery of QTL alleles for grain shape in the Japan-MAGIC rice population using haplotype information. *G3 (Bethesda)* 8: g3.200558.2018. https://doi.org/10.1534/g3.118.200558.

Owens, B., Lipka, A., Magallanes-Lundback, M. et al. (2014). A foundation for provitamin A biofortification of maize: genome-wide association and genomic prediction models of carotenoid levels. *Genetics* 198: 1699–1716.

Pasam, R.K., Sharma, R., Malosetti, M. et al. (2012). Genome-wide association studies for agronomical traits in a world wide spring barley collection. *BMC Plant Biology* 12: 16.

Patishtan, J., Hartley, T.N., Fonseca de Carvalho, R. et al. (2018). Genome-wide association studies to identify rice salt-tolerance markers. *Plant, Cell & Environment* 41: 970–982.

Phillips, P.C. (2008). Epistasis - the essential role of gene interactions in the structure and evolution of genetic systems. *Nature Reviews. Genetics* **9**: 855–867.

Phung, N.T.P., Mai, C.D., Hoang, G.T. et al. (2016). Genome-wide association mapping for root traits in a panel of rice accessions from Vietnam. *BMC Plant Biology* 16: https://doi.org/10.1186/s12870-016-0747-y.

Racedo, J., Gutierrez, L., Perera, M.F. et al. (2016). Genome-wide association mapping of quantitative traits in a breeding population of sugarcane. *BMC Plant Biology* 16: 142. https://doi.org/10.1186/s12870-016-0829-x PMID: 27342657.

Rebetzke, G.J., Bonnett, D.G., and Reynolds, M.P. (2016). Awns reduce grain number to increase grain size and harvestable yield in irrigated and rainfed spring wheat. *Journal of Experimental Botany* 67 (9): 2573–2586. https://doi.org/10.1093/jxb/erw081.

Revilla, P., Rodriguez, V., Ordas, A. et al. (2016). Association mapping for cold tolerance in two large maize inbred panels. *BMC Plant Biology* 16: 127.

Romagnoni, A., Jégou, S., Steen, K.V. et al. (2019). Comparative performances of machine learning methods for classifying crohn disease patients using genome-wide genotyping data. *Scientific Reports* 9 (1): 1–18. https://www.nature.com/articles/s41598-019-46649-z.

Saïdou, A.A., Mariac, C., Luong, V. et al. (2009). Association studies identify natural variation at PHYC linked to flowering time and morphological variation in pearl millet. *Genetics* 182: 899–910. https://doi.org/10.1186/1471-2164-14-159.

Saïdou, A.A., Clotault, J., Couderc, M. et al. (2014). Association mapping, patterns of linkage disequilibrium and selection in the vicinity of the PHYTOCHROME C gene in pearl millet. *Theoretical and Applied Genetics* 127: 19–32. https://doi.org/10.1534/genetics.109.102756.

Samayoa, L.F., Malvar, R.A., Olukolu, B.A. et al. (2015). Genome-wide association study reveals a set of genes associated with resistance to the Mediterranean corn borer (*Sesamia nonagrioides* L.) in a maize diversity panel. *BMC Plant Biology* **15**: 35. https://doi.org/10.1186/s12870-014-0403-3.

Samayoa, L., Cao, A., Santiago, R. et al. (2019). Genome-wide association analysis for fumonisin content in maize kernels. *BMC Plant Biology* 19 (1): 166. https://doi.org/10.1186/s12870-019-1759-1.

Sheoran, S., Jaiswal, S., Kumar, D. et al. (2019). Uncovering genomic regions associated with 36 agro-morphological traits in Indian Spring Wheat Using GWAS. *Frontiers in Plant Science* 10: 527. https://doi.org/10.3389/fpls.2019.00527.

Srivastava, R.K., Singh, R.B., Pujarula, V.L. et al. (2020). Genome-wide association studies and genomic selection in Pearl Millet: advances and prospects. *Frontiers in Genetics* 28 (10): 1389.

Su, J., Li, L., Zhang, C. et al. (2018). Genome-wide association study identified genetic variations and candidate genes for plant architecture component traits in Chinese upland cotton. *Theoretical and Applied Genetics* 131: 1299–1314.

Sukumaran, S., Dreisigacker, S., Lopes, M. et al. (2014). Genome-wide association study for grain yield and related traits in an elite spring wheat population grown in temperate irrigated environments. *Theoretical and Applied Genetics* 128: 353–363. https://doi.org/10.1007/s00122-014-2435-3.

Sun, C., Zhang, F., Yan, X. et al. (2017). Genome-wide association study for 13 agronomic traits reveals distribution of superior alleles in bread wheat from the Yellow and Huai Valley of China. *Plant Biotechnology Journal* 15: 953–969. https://doi.org/10.1111/pbi.12690.

Sun, Z., Li, H., Zhang, Y. et al. (2018). Identification of SNPs and candidate genes associated with salt tolerance at the seedling stage in cotton (*Gossypium hirsutum* L). *Frontiers in Plant Science* 9: 1011.

Swamy, B.P., Vikram, P., Dixit, S. et al. (2011). Meta-analysis of grain yield QTL identified during agricultural drought in grasses showed consensus. *BMC Genomics* 12: 319. https://doi.org/10.1186/1471-2164-12-319.

Szymczak, S., Biernacka, J.M., Cordell, H.J. et al. (2009). Machine learning in genome-wide association studies. *Genetic Epidemiology* 33: S51–S57. https://doi.org/10.1002/gepi.20473.

Thabet, S.G., Moursi, Y.S., Karam, M.A. et al. (2018). Genetic basis of drought tolerance during seed germination in barley. *PLoS One* 13 (11): e0206682.

Thirulogachandar, V., Alqudah, A.M., Koppolu, R. et al. (2017). Leaf primordium size specifies leaf width and vein number among row type classes in barley. *Plant Journal: For Cell and Molecular Biology* 91 (4): 601–612.

Thornsberry, J., Goodman, M., Doebley, J. et al. (2001). Dwarf8 polymorphisms associate with variation in flowering time. *Nature Genetics* 28: 286–289.

Tian, F., Bradbury, P., Brown, P. et al. (2011). Genome-wide association study of leaf architecture in the maize nested association mapping population. *Nature Genetics* 43: 159–162.

Tsai, H., Janss, L.L., Andersen, J.R. et al. (2020). Genomic prediction and GWAS of yield, quality and disease-related traits in spring barley and winter wheat. *Scientific Reports* **10**: 3347. https://doi.org/10.1038/s41598-020-60203-2.

Turner, A., Beales, J., Faure, S. et al. (2005). The pseudo-response regulator Ppd-H1 provides adaptation to photoperiod in barley. *Science* 310 (5750): 1031–1034.

Van Inghelandt, D., Melchinger, A.E., Martinant, J.P. et al. (2012). Genome-wide association mapping of flowering time and northern corn leaf blight (*Setosphaeria turcica*) resistance in a vast commercial maize germplasm set. *BMC Plant Biology* 12: 56. https://doi.org/10.1186/1471-2229-12-56.

Varshney, R.K., Paulo, M.J., Grando, S. et al. (2012). Genome wide association analyses for drought tolerance related traits in barley (*Hordeum vulgare* L.). *Field Crops Research* 126: 171–180.

Wang, M., Jiang, N., Jia, T. et al. (2012). Genome-wide association mapping of agronomic and morphologic traits in highly structured populations of barley cultivars. *TAG. Theoretical and Applied Genetics. Theoretische Und Angewandte Genetik* 124 (2): 233–246.

Wang, Q., Tian, F., Pan, Y. et al. (2014). A SUPER powerful method for genome wide association study. *PLoS One* https://doi.org/10.1371/journal.pone.0107684.

Wang, G., Leonard, J.M., von Zitzewitz, J. et al. (2014). Marker–trait association analysis of kernel hardness and related agronomic traits in a core collection of wheat lines. *Molecular Breeding* 34: 177–184. https://doi.org/10.1007/s11032-014-0028-0.

Wang, H., Li, K., Hu, X. et al. (2016). Genome wide association analysis of forage quality in maize mature stalk. *BMC Plant Biology* 16: 227.

Wang, R., Chen, J., Anderson, J.A. et al. (2017). Genome-wide association mapping of Fusarium head blight resistance in spring wheat lines developed in the Pacific Northwest and CIMMYT. *Phytopathology* 107: 1486–1495.

Wang, F., Longkumer, T., Catausan, S.C. et al. (2018). Genome-wide association and gene validation studies for early root vigour to improve direct seeding of rice. *Plant, Cell & Environment* 41: 2731–2743. https://doi.org/10.1111/pce.13400.

Wang, H., Wei, J., Li, P. et al. (2019). Integrating GWAS and gene expression analysis identifies candidate genes for root morphology traits in maize at the seedling stage. *Genes* 10 (10): 773. https://doi.org/10.3390/genes10100773.

Wang, L., Muhammad, A., Hu, W. et al. (2020). Uncovering genomic regions controlling plant architectural traits in hexaploid wheat using GWAS. https://doi.org/10.21203/rs.2.23079/v1. researchsquare.com

Warburton, M., Tang, J., Windham, G. et al. (2015). Genome-wide association mapping of *Aspergillus flavus* and aflatoxin accumulation resistance in maize. *Crop Science* 55: 1857–1867.

Ward, B.P., Brown-Guedira, G., Kolb, F.L. et al. (2019). Genome-wide association studies for yield-related traits in soft red winter wheat grown in Virginia. *PLoS One* 14 (2): e0208217. https://doi.org/10.1371/journal.pone.0208217.

Wessinger, C.A., Kelly, J.K., Jiang, P. et al. (2018). SNPskimming: a fast approach to map loci generating quantitative variation in natural populations. *Molecular Ecology Resources* 18: 1402–1414.

Whitlock, M.C. (2005). Combining probability from independent tests: the weighted Z-method is superior to Fisher's approach. *Journal of Evolutionary Biology* 18: 1368–1373.

Willer, C.J., Li, Y., and Abecasis, G.R. (2010). METAL: fast and efficient meta-analysis of genomewide association scans. *Bioinformatics* 26: 2190–2191.

William, H.M., Trethowan, R., and Crosby-Galvan, E.M. (2007). Wheat breeding assisted by markers: CIMMYT's experience. *Euphytica* 157 (3): 307–319.

Wojcik-Jagla, M., Fiust, A., Koscielniak, J. et al. (2018). Association mapping of drought tolerance-related traits in barley to complement a traditional biparental QTL mapping study. *TAG. Theoretical and Applied Genetics. Theoretische Und Angewandte Genetik* 131 (1): 167–181.

Wu, Y., Huang, M., Tao, X. et al. (2016). Quantitative trait loci identification and meta-analysis for rice panicle-related traits. *Molecular Genetics and Genomics: MGG* 291 (5): 1927–1940. https://doi.org/10.1007/s00438-016-1227-7.

Xing, A., Gao, Y., Ye, L. et al. (2015). A rare SNP mutation in Brachytic2 moderately reduces plant height and increases yield potential in maize. *Journal of Experimental Botany* 66: 3791–3802.

Xu, X., Sharma, R., Tondelli, A. et al. (2018). Genome-wide association analysis of grain yield-associated traits in a Pan-European Barley Cultivar Collection. *The Plant Genome* 11 (1): 170073.

Yan, J., Kandianis, C., Harjes, C. et al. (2010). Rare genetic variation at Zea mays crtRB1 increases b-carotene in maize grain. *Nature Genetics* 42: 322–327.

Yang, Q., Li, Z., Li, W. et al. (2013). CACTA-like transposable element in ZmCCT attenuated photoperiod sensitivity and accelerated the postdomestication spread of maize. *Proceedings of the National Academy of Sciences* 110: 16969–16974.

Yang, N., Lu, Y., Yang, X. et al. (2014). Genome wide association studies using a new nonparametric model reveal the genetic architecture of 17 agronomic traits in an enlarged maize association panel. *PLoS Genetics* 10: e1004573.

Yano, K., Yamamoto, E., Aya, K. et al. (2016). Genome wide association study using whole-genome sequencing rapidly identifies new genes influencing agronomic traits in rice. *Nature Genetics* 48: 927–934. https://doi.org/10.1038/ng.3596.

Youssef, H.M., Eggert, K., Koppolu, R. et al. (2017). VRS2 regulates hormone-mediated inflorescence patterning in barley. *Nature Genetics* 49 (1): 157–161.

Yu, J., Zhao, W., Tong, W. et al. (2018). A genome-wide association study reveals candidate genes related to salt tolerance in rice (*Oryza sativa*) at the germination stage. *International Journal of Molecular Sciences* 19: https://doi.org/10.3390/ijms19103145.

Zegeye, H., Rasheed, A., Makdis, F. et al. (2014). Genome-wide association mapping for seedling and adult plant resistance to stripe rust in synthetic hexaploid wheat. *PLoS One* 9 (8): e105593. https://doi.org/10.1371/journal.pone.0105593.

Zhang, X., Warburton, M., Setter, T. et al. (2016). Genome-wide association studies of drought-related metabolic changes in maize using an enlarged SNP panel. *Theoretical and Applied Genetics* 129: 1449–1463.

Zhao, K., Tung, C.-W., Eizenga, G.C. et al. (2011). Genome-wide association mapping reveals a rich genetic architecture of complex traits in *Oryza sativa*. *Nature Communications* 2: 467.

Zhao, Y., Zhang, H., Xu, J. et al. (2018). Loci and natural alleles underlying robust roots and adaptive domestication of upland ecotype rice in aerobic conditions. *PLoS Genetics* 14: e1007521.

Zhao, J., Sauvage, C., Zhao, J. et al. (2019). Meta-analysis of genome-wide association studies provides insights into genetic control of tomato flavor. *Nature Communications* 10 (1): 1–12.

Zhao-Ming, Q., Ya-Nan, S., Qiong, W. et al. (2011). A meta-analysis of seed protein concentration QTL in soybean. *Canadian Journal of Plant Science* 91: 221–230.

Zhu, C., Gore, M., Buckler, E.S. et al. (2008). Status and prospects of association mapping in plants. *The Plant Genome* 1: 5–20.

Zhu, X.-M., Shao, X.-Y., Pei, Y.-H. et al. (2018). Genetic diversity and genome-wide association study of major ear quantitative traits using high-density SNPs in maize. *Frontiers in Plant Science* 9: 966.

# 6

# Genotyping of Seeds While Preserving Their Viability

*Vinaykumar Rachappanavar[1,2], Arushi Padiyal[2], and Jitender Kumar Sharma[3]*

[1]*Department of Agriculture, MS Swaminathan School of Agriculture, Shoolini University, Solan, Himachal Pradesh, India*
[2]*Department of Seed Science and Technology, Dr. Yashwant Singh Parmar University of Horticulture and Forestry, Solan, Himachal Pradesh, India*
[3]*Department of Agriculture, School of Agriculture, Baddi University of Emerging Sciences & Technology, Baddi, Himachal Pradesh, India*

## 6.1 Introduction

Most of the policy makers estimated that world population count may cross nine billion by middle of the twenty-first century. This may become the biggest challenge for the world food production and supply (Godfray et al. 2010; Cleland 2013). A seed has an important role in food supply system, and it fulfills more than 70% of the world daily calories demand. To remove the world from hungry bowl and provide the sufficient balanced diet for growing population, researchers have to concentrate on new breeding approaches to double the agriculture productivity by 2050 (Tilman et al. 2011; Ray et al. 2013; Becker et al. 2014). In the past two decades, marker-assisted selection (MAS) has become more common in various breeding programs (William et al. 2007). The ultimate aim of MAS is to develop breeding products by combining desirable genes/genotypes and alleles or linking disequilibrium blocks and optimize the new gene network (Xu et al. 2012). For this, leaf tissue-based genomic DNA extraction protocols are developed and recommended to avoid the secondary metabolites contamination, although the collection of young leaves from field followed by labeling and tracking is more tedious for genotyping (Xu and Crouch 2008). If live plants are absent, leaf DNA-based genotyping requires seed germination followed by plant growth for leaf sample collection (Kang et al. 1998; Gao et al. 2008). Even some seeds have dormancy and they need additional treatment for germination. To avoid all these constrains, seed DNA-based genotyping is selected as an alternate to the leaf-based genotyping. Seed DNA-based genotyping saves money and time and dramatically improves the genotyping efficiency by allowing breeders to select individuals before planting (Von Post et al. 2003; Gao et al. 2008). Seed DNA-based genotyping become very simple and more economical when required to analyze large population (Arbelaez et al. 2019).

*Genotyping by Sequencing for Crop Improvement*, First Edition.
Edited by Humira Sonah, Vinod Goyal, S.M. Shivaraj, and Rupesh K. Deshmukh.

To maximize the efficacy of the results, in-depth study is required to know the exact source of origin of different tissues of seed before DNA isolation. It is universally known that fertilization of female gametophyte by male leads to the seed development. Seed can be distinctly divided into embryo, endosperm, and seed coat based on their functionalities (Ohad et al. 1999). In the fertilization event, the zygotic embryo is formed by the fusion of sperm cells and egg cells, whereas the endosperm results from the fertilization of diploid central cell with the second sperm cell, and it supports the embryo throughout its growth until complete seedling establishment. Finally, the seed coat originates from maternal tissue, and it serves to protect the embryo and endosperm throughout the seed development (Becker et al. 2014). Brown et al. (2003) further divided the endosperm into micropylar endosperm (MCE, proximal to the embryo), the peripheral endosperm (PEN), and the chalazal endosperm (CZE) subregions based on their position of deposition in seed. Most of the researchers recommend to use the CZE for the seed DNA extraction and genotyping (Becker et al. 2014).

Seed DNA isolation method is considered as one of the nondestructive methods in which a small portion of the endosperm is utilized for the DNA isolation. Further, the same seed can be utilized for field trial, but it needs aseptic growing condition to avoid the pest/disease infection at the seed damage site caused during endosperm isolation. In the past, several researchers utilized the same seed for DNA extraction method for genetic testing as well as transgenic detection in rice (Kang et al. 1998), maize (Sangtong et al. 2001), wheat (Abd-Elsalam et al. 2011), soybean (Chunwongse et al. 1993), barley (Von Post et al. 2003), and peanuts (Chenault et al. 2007) and found comparable results as that of leaf-based DNA extraction method (Papazova et al. 2005). Seed DNA extraction method has several advantages over other tissue-specific DNA isolation methods. Few of them are:

1) Same sampled seed can be utilized for complete plant development;
2) No need of extra seeds for genetic evaluation;
3) It helps to know the genetic purity of the seed before planting;
4) More feasible for those plants that produce very less number of seeds in both the natural and artificial environmental conditions;
5) No need for an additional progeny test to know the seed zygosity condition/health;
6) Large size population can be handled in both laboratory and field conditions;
7) Based on the off-season marker evaluation, selected seeds are only utilized for further studies.

In plant-breeding programs, molecular markers have ample opportunities. Among them, polymerase chain reaction (PCR)-based markers are frequently utilized in genome mapping, pedigree, and phylogenetic analysis. Because they are more efficient as compared to non-PCR-based markers, they need very small amount of pure DNA for genotyping. PCR, sequencing, cloning, and restriction analysis require pure and quality DNA (Sonmezoglu and Ates 2015). On the other hand, DNA extraction from seeds has its own disadvantages, which makes this method less important for genotyping. In general, well-developed mature endosperm consists of carbohydrates, lipids, proteins, oils, mucilage, polyphenols, and other secondary metabolites, which affects the quality and storability of the extracted DNA (Porebski et al. 1997; Aboul-Maaty and Oraby 2019). During seed DNA isolation process, reserve materials present in the cell come in contact with the nuclei.

Polysaccharides present in the cells precipitate with genomic DNA, whereas polyphenols react with the nucleic acids and proteins; thus, brown-colored DNA is produced by the action of restriction enzymes, polymerases, and ligases (Bhattacharjee et al. 2004). Other than this, seed morphological issues should be considered before the seed DNA-based genotyping (Xu and Crouch 2008).

To extract quality DNA, the seed DNA isolation protocol must contain all the following necessary characteristics to eliminate contamination. Firstly, the seed size must be large enough to provide sufficient amount of endosperm for both DNA isolation and seedling establishment. Secondly, the seed structure should be able to tolerate dehydration treatment without affecting seed germination. Finally, the pericarp contamination should be minimum to low during endosperm excision. So, in order to make seed DNA-based genotyping, the protocol must meet three important challenges. First, the complete protocol should be easily automated for high-throughput applications. Next, a required amount of DNA must be isolated from single seed with very low amount of contamination. This will help to analyze large number of markers during genotyping. Finally, the DNA quality must be adequate for genotyping without any further protocol modification (Gao et al. 2008).

### 6.1.1 Genotyping

Molecular markers are prerequisite for gene mapping and tagging, genetic diagnosis forensic examination, segregation analysis, and phylogenetic analysis (Sonah et al. 2011). But most of the markers have their own limitations. Several types of molecular markers have been developed and used for decades for plant molecular breeding. The restriction fragment length polymorphism (RFLP) is the first molecular marker which is effectively used in plant genotyping (Botstein et al. 1980). It is a non-PCR technique, which is mostly used in linkage map construction. But it has major drawbacks such as complicated hybridization process, it involves handling radioactive reagents, and low availability of probes for hybridization (Bernatsky and Tanksley 1986). Later, with further advancement in biotechnology, less-complex, cost-effective PCR-based markers such as random amplification of polymorphic DNA (RAPD; Williams et al. 1990); amplified fragment length polymorphisms (AFLPs; Vos et al. 1995), cleaved amplified polymorphic sequences (CAPSs; Konieczny and Ausubel 1993), sequence-characterized amplified region (SCAR; Paran and Richard 1993), simple sequence repeats (SSRs; Salimath et al. 1995), single-nucleotide polymorphism (SNP; Wang et al. 1998), and direct amplification of length polymorphisms (DALPs; Desmarais et al. 1998) were developed and effectively utilized in plant-breeding activities. These methods are relatively less expensive and less time consuming as compared to non-PCR-based markers. Among PCR-based molecular markers, SNPs are most suitable for analysis on a wide range of genomic scales (Rafalski et al. 2008). Advancement in SNP-based marker technology has increased the genotyping efficiency as compared to previous marker systems. Fluorescent-based SNP-specific hybridization techniques using TaqMan, molecular beacons and invaders (Ragoussis et al. 2006), sequenom homogeneous mass extend (hME), and increased plexing efficiency and flexibility for massarray system (iPLEX) genotyping systems generate the hundreds to few thousand SNPs information per sample per day (Procunier et al. 2009). Newly developed Illumina bead-based assay systems such as Golden Gate allow to screen a large number of samples using a single multiplexed assay (Deschamps et al. 2012), whereas Illumina's Infinium generates four million SNPs on a single sample. The Affimetrix GeneChip system is extensively used for SNPs detection by sequencing through hybridization (SbH) (Adler et al. 2013).

Bulked segregation analysis (BSA) and genome-wide associate studies (GWAS) are the major intensive applications that are used in breeding activities (Schneeberger and Weigel 2011). As compared to linkage mapping, GWAS produces improved resolution in mapping by exploiting the natural diversity which is generated by multigeneration recombination (Hansen and Ostermeier 2001). Lack of knowledge regarding the structure and degree of linkage disequilibrium (LD) distribution obstructs the utilization of GWAS in plants. Epistasis, degree of selfing, and admixture also affect the LD distribution (Flint-Garcia et al. 2005). To achieve optimal resolution in a genome-wide scan, millions of markers are needed to correct the population structure which increases the total cost of genotyping. For this, candidate-gene association analysis is used to minimize the cost of GWAS. Near isogenic lines (NILs) are created to resolve the complex traits. Even inbred lines, double haploids, and recombinant inbred lines (RILs) are also used to conduct GWAS (Zhu et al. 2016). Cost- and time-effective BSA can be used to identify the phenotype-specific markers without using any linkage map and large population for sampling. In BSA, only those plants present in the extreme ends of phenotype range are selected. This method helps to detect QTLs (quantitative trait loci) easily in large populations when widely distributed large number of markers are available (Wolyn et al. 2004). Finally, cost-effective genomic or genome-wide selection (GS) has been proposed as a breeding method for trait involving low heritable multiple QTLs (Bhat et al. 2016). In this method, arrays of markers are used to directly predict the traits without establishing association with any traits. For that, only genotypic data are needed to select individuals with the best breeding values (Kim et al. 2020).

### 6.1.2 Genotyping-by-Sequencing

In modern genetics, polymorphism is effectively utilized for the genotype identification and connecting them to phenotypic expression. Restriction enzyme digestion and multiplex enrichment PCR technique are extensively utilized for the identification of species-specific markers (He et al. 2014). Elshire et al. (2011) developed cost-effective simple and new sequencing protocol, called genotyping-by-sequencing (GBS), which can afford millions of markers across the genome, and they can be applied for large genomes (Shendure and Ji 2008). GBS is more efficient as compared to phenotypic selection or MAS for genotyping a variety of species (He et al. 2014).

DNA sequencing has a major role in gene cloning, breeding programs, finding unknown or foreign gene, and comparative and evolutionary studies. In the past three decades, genome sequencing technology has undergone tremendous evolution. Rosche (2008) proposed the first next-generation sequencing (NGS) technology, 454 sequencing. It is a highly efficient *in vitro* DNA amplification method which utilizes the PCR emulsion for sequencing and bypasses the cloning steps completely. Later, pyrosequencing methodology was developed which uses sequencing-by-synthesis (SBS) technique that measures pyrophosphate release during sequencing process (De Barba et al. 2017). Subsequently, sequencing by oligonucleotide ligation and detection (SOLiD) method was developed by Agencourt. It is a massively parallel sequencing system that can determine the underlying nucleotides sequence using florescent nucleotides. Initially, the read length of SOLiD was 35 bp, and the output data per run was 3Gb. It can be used for whole-genome resequencing and epigenome and transcriptome research works (Deurenberg et al. 2017). Solexa (2006) developed and released SBS-based genomic analyzer (GA) (Bennett 2004). Initially, the GA output was 1Gb per run; later, the output of GA increased to 20Gb per run by improvement through upgrading polymerase, buffer, flow cell, and software. Later, Illumina launched HiSeq 2000 in 2010 with the same GA futures with the output 200Gb per run initially. It needs HiSeq control software for program control and consensus assessment of sequence and variation (CASAVA) for secondary analysis. They can calculate the position and number of clusters based on their output and quality. After advancement in sequencing, all NGS followed similar strategies for DNA template preparation, where universal adopters were used to ligate randomly sheared DNA fragments. For Illumina NGS sequencers, DNA molecules are polymerized with the help of primers on solid surfaces. These cloned DNA are immobilized on solid surface to generate several billions of sequences in a parallel. Later, sequencing is performed in a systematic manner, where sequencers are used to detect the incorporation of every single illuminous reversible terminator base (Metzker 2010). A camera present in the sequencer captures images of the fluorescently labeled nucleotides. Finally, to allow the next cycle of polymerization, terminal 3′ blocker is chemically removed from the DNA (Mardis 2008). Both SNP assay and diversity array technology (DArT) provide same markers density, but SNP assay provides more precise and well-characterized gene prediction as compared to DArT array assay (Oliver et al. 2013).

To avoid the cost and improve the genotyping efficiency, GBS was introduced. In this method, restriction enzymes are used to digest the genomic DNA followed by the ligation of barcode adapter, PCR amplification, and sequencing of the amplified DNA. Later, various bioinformatics pipelines are used to analyze and interpret GBS datasets. In simple, sequencing is performed in a series of cycles consisting of single reversible fluorescent terminator deoxyribonucleotides extension and florescent detection to know which base pair incorporated followed by cleavage of both the fluorescent label and the chemical moieties to allow for the next cycle to occur (Deschamps et al. 2018). GBS can be utilized for genome characterization, genomic selection (Poland et al. 2012), saturating existing genetic map (Spindel et al. 2013), genetic ordering of draft genome sequence (Mascher et al. 2013), and germplasm characterization (Lu et al. 2013).

## 6.2 Genotyping-by-Sequencing with Minimum DNA

As the size of tissue decreases, it reduces the yield of DNA, which hampers the genotyping analysis. The size of the targeted tissue and its subsequent DNA quality and content are therefore crucial for genomic analysis. Laser-assisted microdissection (LAM) to quantitative real-time reverse transcription–polymerase chain (RT-PCR) and then to DNA chips has increased the efficiency of DNA utilization, and thus, these approaches are utilized in genotyping programs where DNA is isolated from seed (Becker et al. 2014). As the NGS requires small amount of nucleic acid for genotyping, it has the ability to detect low abundance of transcripts, increased sensitivity, more reproducible result, and ability to compare distinctly related organisms (Mastrokolias et al. 2012).

## 6.3 DNA Extraction from Half Grain

In the past, several research works were conducted to standardize the seed DNA extraction protocols (Kang et al. 1998; Von Post et al. 2003; Chenault et al. 2007). In certain studies, relatively small number of markers need to be run on relatively large number of individuals/lines to directly determine the linkage of localized regions, selection for rare recombination events, fingerprinting of plant or seed material, and large-scale selection against transgenes. For large number of DNA extraction, the laboratories require time and trained workforces. On the other hand, DNA extraction from single seed affects the seed germination as well as the DNA extraction efficiency. Even the presence of coprecipitated polysaccharides inhibits the polymerase activity (Vibhuti et al. 2015). The poor quality seed DNA is incompatible in PCR and fails to amplify products larger than 500 bp (Liang et al. 2015). This is why most of the quality DNA extraction protocols recommend to use meristem tissue even if they require hazardous and costly chemicals and enzymes (Liang et al. 2015; Vibhuti et al. 2015). Therefore, the rapid DNA sampling method from ungerminated seed or seedling is required for PCR analysis. DNA extracted from these seeds must be directly used for the genotyping.

For DNA extraction from seed, a portion of the endosperm or cotyledon should be mined by drilling as reported in maize (Abdel-Latif and Osman 2017); however, this leads to seed overheating. To minimize the overheating problem, dehydration followed by scalping method is frequently recommended by various researchers for orthodox seeds. However, water soaking leads to seed pregermination and biochemical changes, which affects the quality DNA isolation. To avoid these problems, diluted ABA solution is recommended for seed soaking (Bewley 1997). The sectioned or drilled seeds are more susceptible to fungal and bacterial infection (Zheng et al. 2015a). To avoid this, low doses of seed protectants such as fungicides and bactericides are recommended. In certain conditions, direct contact of seed with highly concentrated protectants may cause severe damage. Therefore, seed coating, a promising technique, is recommended instead of direct exposing of dissected seed to hazardous chemicals (Rao et al. 2017). Maternal seed tissues such as pericarp and seed coat contain only female parental alleles or genes. They may result in detection of false heterozygosity in segregating population. Even DNA extracted from endosperm can produce false result by heterofertilization (Sprague 1932), although this false result cannot be evaded in seed DNA extraction method (Robertson 1985). The large sized seeds can provide enough nutrition for seedling establishment even after a portion of endosperm is utilized for DNA extraction (Figure 6.1).

DNA extracted from the seeds is effectively used for genotyping by PCR. The amount of DNA produced through the nondestructive seed method is sufficient for 100–1000 reactions. Initially DNA extracted in strongly basic buffer is later neutralized into normal condition before genotyping and storage. This method helps to extract DNA from hundreds of thousand seeds or seedling samples at a very low cost (Zheng and Rosenfeld 2015). In recent years, seed DNA-based genotyping is fully automated from endosperm sampling to MAS. In general, seed-based genotyping is used for recombinant individual identification from large population and chromosome-specific fine mapping using specific PCR-based markers. Dayteg et al. (2007) developed automated PCR-based genotyping, and it could be used for crops that produce large-sized seeds, and this method can replace the leaf DNA-based genotyping. Benito et al. (1993) initially developed a standard DNA isolation protocol

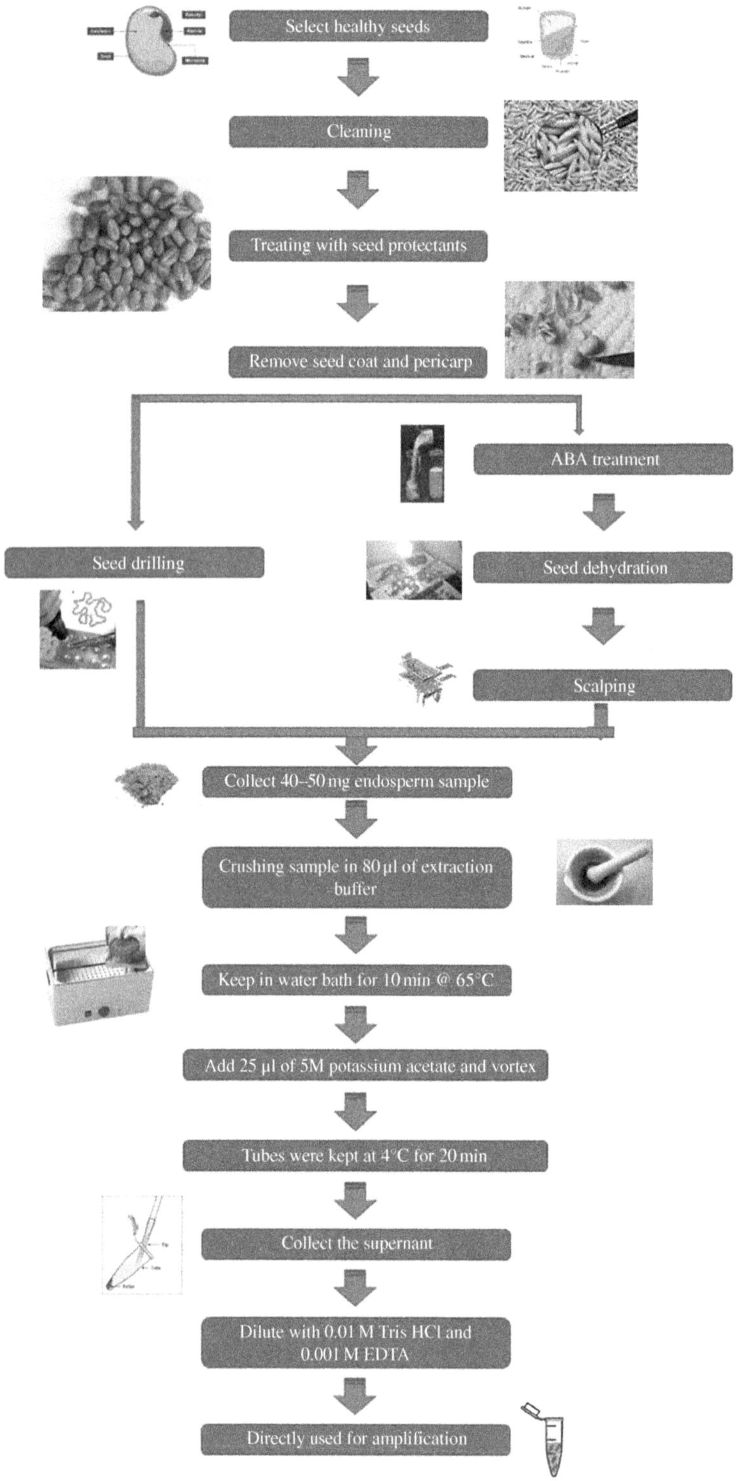


**Figure 6.1** DNA extraction from seed endosperm. **Source:* Amenic181/Adobe Stock Photos.

for seed sample in which they selected healthy seeds and imbibed them for 16 h in sterilized water and kept them at room temperature to remove excess water. Later, DNA isolation was done by the following steps: initial crushing of 40–50 mg hydrated seed in 80 μl extract buffer (2× the amount of seed used) with the help of a spatula or a metal rod. The extraction buffer contained 0.1 M tris–HCl (pH 8), 0.05 M ethylenediamine tetraacetic acid (EDTA) (pH 8), and 0.5 M NaCl; 8 μl of 20% sodium dodecyl sulfate (SDS) was added to the PCR tubes after complete crushing. The tubes were kept at water bath at 65 °C for 10 min. The tubes were later cooled, and 27 μl of 5 M potassium acetate was added to all the tubes and vortexed thoroughly. The tubes were maintained at 4 °C for 20 min. The supernatant was extracted after centrifugation at 4 °C for 20 min at 14 000g. Later, the supernatant was diluted with 0.01 M tris–HCl and 0.001 M EDTA and directly used for amplification. This protocol was widely followed to extract DNA from all crops. Later, Sharma et al. (2013) used the Kang et al. (1998) and Wulff et al. (2002) DNA isolation protocol with the following modifications for seed DNA extraction to reduce the DNA isolation time by reducing the number of steps from standard protocol:

i) Instead of the sample being freeze dried before crushing, the seeds were preheated at 65 °C for 30 min;
ii) SDS was used instead of cetyltrimethylammonium bromide (CTAB) as the detergent for the preparation of the extraction buffer;
iii) PEG (40 000 mol. wt.) was added in the extraction buffer.

It can extract the required quantity of quality DNA from 100 mg of seed, and this protocol does not depend on liquid nitrogen. This protocol was successfully used to isolate DNA from banana, yam, okra, and mango from which the isolation of nuclear DNA was found to be difficult due to the presence of excess amount of polyphenols and proteins (Cassago et al. 2002). DNA isolation from legume seeds is more complex as compared to cereals. Legume seeds contain excess polyphenols, phenolic terpenoids, tannins, and polysaccharides which cannot be easily separated from DNA. These biomolecules bind to RNA and DNA after cell lysis. DNA can be isolated from legume seeds, but they need special treatment before amplification. Some authors recommend isopropanol–ammonium acetate precipitation treatment to improve the quality of isolated DNA. Here, we compiled some crop-specific rapid seed DNA extraction methods suitable for genotyping. These modifications consistently produce pure DNA which can be directly utilized for amplification without any additional treatments (Agbagwa et al. 2012).

### 6.3.1 DNA Extraction from Rice Seeds

Paddy seeds have several advantages and disadvantages over leaves as a source of DNA. However, the presence of 80–90% starch in seed obstructs the quality of DNA extraction by inhibiting DNA polymerase activity (Duan et al. 2015). There are limited number of seed DNA isolation protocols present for rice seeds (Thomas and Dominic 2015). Liang et al. (2015) extracted the DNA from rice seed using CTAB, phenol, chloroform, and α-amylase. However, the quantity of DNA using single seed is very low. Later, Mutou et al. (2014) described the revised seed DNA extraction protocol, but they utilized the hazardous chemicals. That protocol is unfit for rice seed DNA extraction in common labs.

To resolve this problem, Sajib et al. (2017) developed a very simple and highly affordable nondestructive seed DNA extraction protocol based on Chelex-100. In this, 10–12 mg of endosperm is removed from the seed and mechanically crushed to a coarse powder. Then, 200 μl of 0.9% NaCl solution and 100 μl of 20% Chelex®-100 (C7901, Sigma) are added to the tube containing grain powder and completely vortexed for 10–15 s. Incubate this mixture in a water bath at 95 °C for 10–20 min and then centrifuge at 13 000 rpm for 10 min. Transfer the supernatant to a new microcentrifuge tube and add an equal volume of isopropanol. Centrifuge the mixture again at 13 000 rpm for 10 min, wash the pellets with ethanol twice, and store at 4 °C in TE buffer after RNAse treatment.

### 6.3.2 DNA Extraction from Wheat and Barley Seeds

Freeze the seed material at −80 °C and grind the extracted seed-storage tissue portion at 1500 strokes per min in Spex Geno Grinder for 2 min or grind the seed storage tissue in silica gel using Spex Geno Grinder for 5–10 min at 500 strokes per min. Add 500 μl of the extraction buffer (90 mM tris HCL (pH 7.5) + 630 mM NaCl + 45 mM EDTA (pH 8.0) + 1% Sarcosyl + 0.9% SDS) to the microcentrifuge tube containing grounded seed material and incubate at 65 °C in water bath for 1 h. After incubation, cool at room temperature and add 250 μl of 6 M ammonium acetate (ice chilled) to precipitate the protein and plant tissues. Transfer the supernatant into a new 1.5-ml microcentrifuge tube containing 360 μl of isopropanol. Keep the tubes at −20 °C overnight. Then after, follow the Doyle and Doyle isolation protocol to complete the DNA extraction and purification method (Gao et al. 2008).

### 6.3.3 DNA Extraction from Maize Seeds

Choose large-sized maize seeds and soak them in water for approximately 24 h. Dry the seed to remove the excess water and place the seed with embryo side up on a plastic board. Separate a small portion of endosperm from the seed with the help of scalpel without damaging the embryo and limit the amount of endosperm extraction to 20–60 mg. The remaining sectioned seed can be stored for more than one month without losing viability under freeze condition (Gao et al. 2008). The extracted endosperm portion later is utilized for DNA extraction as recommended for leaf tissue DNA extraction method with slight modification. Care should be taken before grinding to reduce overheating risk. For this, whole samples containing plate are kept at deep freeze (−20 °C) for at least half an hour just before grinding. The sampled seeds are allowed to dry for at least two days at room temperature before phenotypic screening under controlled condition (Abdel-Latif and Osman 2017).

In addition, NaOH tris method was developed by slight modification of Von Post et al.'s (2003) method for nondestructive seed DNA extraction. Mix approximately 30 mg of endosperm powder well in 200 μl of 0.15 M NaOH solution and heat twice in a 700-W microwave oven for at least 1 min or incubate at 55 °C for 10 min. Add a total of 750 μl containing 0.03 M tris–HCl (pH 8.0) and 1 mM EDTA (pH 8.0) and keep the sample overnight (14 h) to settle. Centrifuge the sample at 3500 rpm for 10 min and collect 800 μl of supernatant and transfer into a storage plate.

### 6.3.4 DNA Extraction from Soybean Seeds

According to Motokazu and Tadahiko (2003), individual soybean seeds were drilled at the center of the cotyledon to collect 10–30 mg of cotyledon powder without damaging the hypocotyl. This seed powder is directly used for DNA isolation without any further treatments. The seed powder is digested with DNA extraction buffer containing 10 mM tris–HCl (pH 7.8), 0.5% SDS, 5 mM EDTA and NP-40, 0.5% Tween 20, and 16 μg of proteinase K in a water bath at 55 °C for 20 min. After incubation, add 0.2 ml of phenol/chloroform/isoamyl alcohol (25:24:1). After this, follow the standard DNA isolation and purification protocol to obtain quality DNA for genotyping.

### 6.3.5 DNA Extraction from Cotton Seeds

DNA isolation from cotton (*Gossypium* spp.) seed is more complicated by the presence of excess of secondary metabolites such as phenolic compounds (Paterson et al. 1993). They can oxidize and irreversibly bind with the proteins and nucleic acids during tissue homogenization process (Loomis 1974), which is hard to separate during DNA purification and render the DNA unsuitable for genotyping (Porebski et al. 1997). For this, several leaf-based DNA extraction protocols were developed but did not address the practical method to genotype the ungerminated seed or seedlings. To solve these problems, Zheng et al. (2015a) developed seed DNA extraction protocol for cotton using sand drum. Seeds are treated with 60-grit sanding bands just for exposing the cotyledonary tissue and the seed coat covering removed. These treated seeds are placed in physically modified PCR plates (#MPS499; Phenix Research, Candler, NC) with the sanded end facing the bottom. The seeded plate is covered with any adhesive element to prevent the seeds from falling out of the wells. The PCR plate is then inverted, and approximately 3.5 mg of particulate endospermic or cotyledonary fine tissue particles is obtained by drilling at slow speed with the help of an engraving cutter (Dremel #107). After each seed abrasion, the cutting head is cleaned by drilling a hole into a common pencil eraser. The still-inverted modified PCR plate is physically placed on another PCR plate in an upright down position to transfer the drillings (seed particles). The DNA is extracted from the grounded tissue using different basic and acidic buffers as described by Xin et al. (2003). Buffer A (100 mM NaOH and 2% Tween 20) and buffer B (100 M tris–HCl, 2 mM EDTA, adjusted to pH 2.0) are freshly prepared just before use. After genotyping, the seeds are selected and germinated in "ragdolls" of germination paper that had been presoaked with sterile water plus Banrot. These seeds can be stored for at least 12 months in modified PCR plates under ambient room temperature without losing viability (Zheng et al. 2015b).

### 6.3.6 DNA Extraction from Papaya Seeds

Some papaya cultivars produce intermediate storable small-sized seeds. Tissue removal or sectioning can damage the embryo, so it can affect the germination process after sampling. For nondestructive seed DNA-based genotyping, the sectioned seeds need to be stored for at least six days to complete the sex-specific DNA molecular marker analysis. So, these sectioned seeds are more susceptible for fungal and bacterial infestation. Generally, fungicides are recommended to treat at sectioned site to obtain a good germination rate (Chunwongse et al. 1993; Zheng et al. 2015a).

### 6.3.7 DNA Extraction from Watermelon Seeds

Based on the size and dimension of the watermelon seeds, Meru and McGregor (2013) developed two different seed DNA extraction protocols. Sampling is done by removing ⅓ or ½ of the distal portion of the seeds using sharp blade, and the remaining seeds are kept in cells containing sterile 3B soil amended with Osmocote Classic (1.38 g/kg N, 1.38 g/kg P, and 1.38 g/kg K) for germination under controlled environmental condition. The seed samples are placed in individual tubes on a 96-well plate for seed DNA extraction.

Protocol-1: (modified from Gao et al. 2008) Incubate the seed sample in buffer solution containing 450 μl of 1% CTAB, 1% SDS, or 1% sarcosyl in a water bath at 65 °C for 30 min. After incubation, add an equal volume of chloroform–isoamyl alcohol (24:1) and centrifuge at 2465g for 10 min at room temperature. Transfer the supernatant to a new tube and add 0.6× of cold isopropanol (where X is the volume of the supernatant). DNA is allowed to precipitate for 30 min at −20 °C and separated by centrifugation at 2465g for 10 min. Wash the DNA pellets twice in 70% ethanol and completely dry before storing in 50 μl tris–EDTA (Meru et al. 2013).

Protocol-2: (modified from Abd-Elsalam et al. 2011): Incubate the grounded sample in an extraction buffer containing 450 μl 0.5% SDS, 1% PVP, 200 mM tris–HCl, pH 7.5, 250 mM NaCl, 25 mM EDTA, pH 8, and 0.2% (v/v) β-mercaptoethanol in 65 °C water bath for 30 minutes. Later, add NaOAc (0.3×) to each sample tube and then centrifuge at 2465g for 30 min. Transfer the supernatant to a new tube and add equal amount of ice-cold isopropanol and continue the purification process as previously described in protocol-I (Meru et al. 2013).

## 6.4 GBS with Half Seed

Seed DNA-based genotyping allowed to determine the genetic constitution of seeds before planting in field condition. Theoretically, DNA extracted from seeds is not as pure as DNA extracted from any meristematic plant tissues. The tissues are made up of a group of differentiated specialized cell types. Each has its own anatomical and physiological expression. The lack of precision makes it nearly impossible to isolate only the endosperm without contamination from the neighboring tissues. However, PCR-based markers used for MAS can produce the same result as DNA extracted from leaves (Zhang et al. 2006), which do not demand quality DNA in large amount as reported in maize (Gao et al. 2008), papaya (Chen et al. 2019), watermelon (Meru et al. 2013), cotton (Zheng et al. 2015a), rice and wheat (Chunwongse et al. 1993). DNA extracted from 30 mg of seed material is sufficient for 200 PCRs. With the advancement of chip-based SNP genotyping, DNA extracted from single seed can be used for the evaluation of millions of SNPs. Seed DNA-based genotyping provides an opportunity to select individuals from population for phenotypic evaluation with minimum errors (Blair and Corrigan 2019). To conduct study related to seed fertilization including apomixis, self-incompatibility, male sterility, and abnormal fertilization (Belicuas et al. 2007).

The lack of precision in manual tissue extraction protocols leads to select other approaches such as isolation of nuclei tagged in specific cell types (INTACT), laser-capture microdissection (LCM), and fluorescence-activated cell sorting (FACS) of fluorescently labeled cells or nuclei (Nelson et al. 2006; Frumkin et al. 2008). Of these techniques, LCM and FACS can be used for seed DNA isolation, but both methods have their own limitations such as they require highly sophisticated and costly instruments and often produce less-than-optimal yield and purity of the targeted cell type.

Laser micro dissection (LMD) technology is successfully used to dissect a selected region from seed without contamination from the neighboring tissues (Khan et al. 2014) (Figure 6.2). It permits to isolate the specific tissue composed of heterogeneous cells (Emmert-Buck et al. 1996). The extraction of a portion of the endosperms for genotyping is not easy. LMD has proven to be an effective technique to isolate endosperms for transcriptome profiling (Day et al. 2008). LMD can allow to isolate tissue-specific nucleic acids and other metabolites to elucidate the associated molecular events (Isenberg et al. 1976; Nelson et al. 2006; Ramsay et al. 2006; Balestrini and Bonfante 2008). The laser microdissection apparatus extracts the homogeneous cell population with the help of light microscope and computer-based dissection instruments. LCM and laser cutting are the two microscope-based laser dissection methods (Nelson et al. 2006). In LCM, pulsed infrared laser is used to melt and fuse the thermo cap onto the selected cells. When the cap is removed, the ligated cells are separated from the surrounding tissue (Emmert-Buck et al. 1996). These cells are separated from the thermocap and directly used for the GBS, whereas laser cutting uses a ultraviolet (UV) light to excise the targeted region without damaging the nucleic acids and other biomolecules. The excised fragment is catapulted upward using laser pressure catapulting (LPC) (Tang et al. 2006). For LCM, the initial portion of the tissue is dissected, and then the targeted cell groups are isolated without contamination. For this, sample preparation is required before dissection to preserve enough visual detail to target and extract the tissue and improve the recovery of nucleic acids from the harvested cells. The sample preparation has been clearly illustrated in several research articles and reviews (Asano et al. 2002; Nelson et al. 2006; Balestrini and Bonfante 2008; Gomez and Harrison 2009). Two methods are generally used for sectioning before LM, which are paraffin sectioning and cryosectioning. Cryosectioning is mostly preferred for its work speed, maintaining sample cold throughout the procedure, and better preserve intact biomolecules. During cryosectioning, the freezing process may form ice crystals in cell vacuole. To avoid this, sucrose-like cryoprotectants are used to maintain the cell histology (Asano et al. 2002; Casson et al. 2005). As an alternative, samples are embedded in paraffin after fixation. Although this protocol provides excellent morphology, it reduces the nucleic acid yield as compared with that from frozen samples (Tang et al. 2006). Although some researchers performed DNA extraction using LM (Scutt et al. 1997; Hobza and Vyskot 2007), most of the plant researchers recommend LM method for transcriptome studies (Balestrini and Bonfante 2008), since it produces sufficient amount of RNA for molecular analysis such as RT-PCR and global profiling analysis (Ithal et al. 2007).

To avoid the mechanical cell purification, transgenic nuclei tagging is done in targeted cell types to avoid the mechanical cell purification. For this strategy, INTACT system is

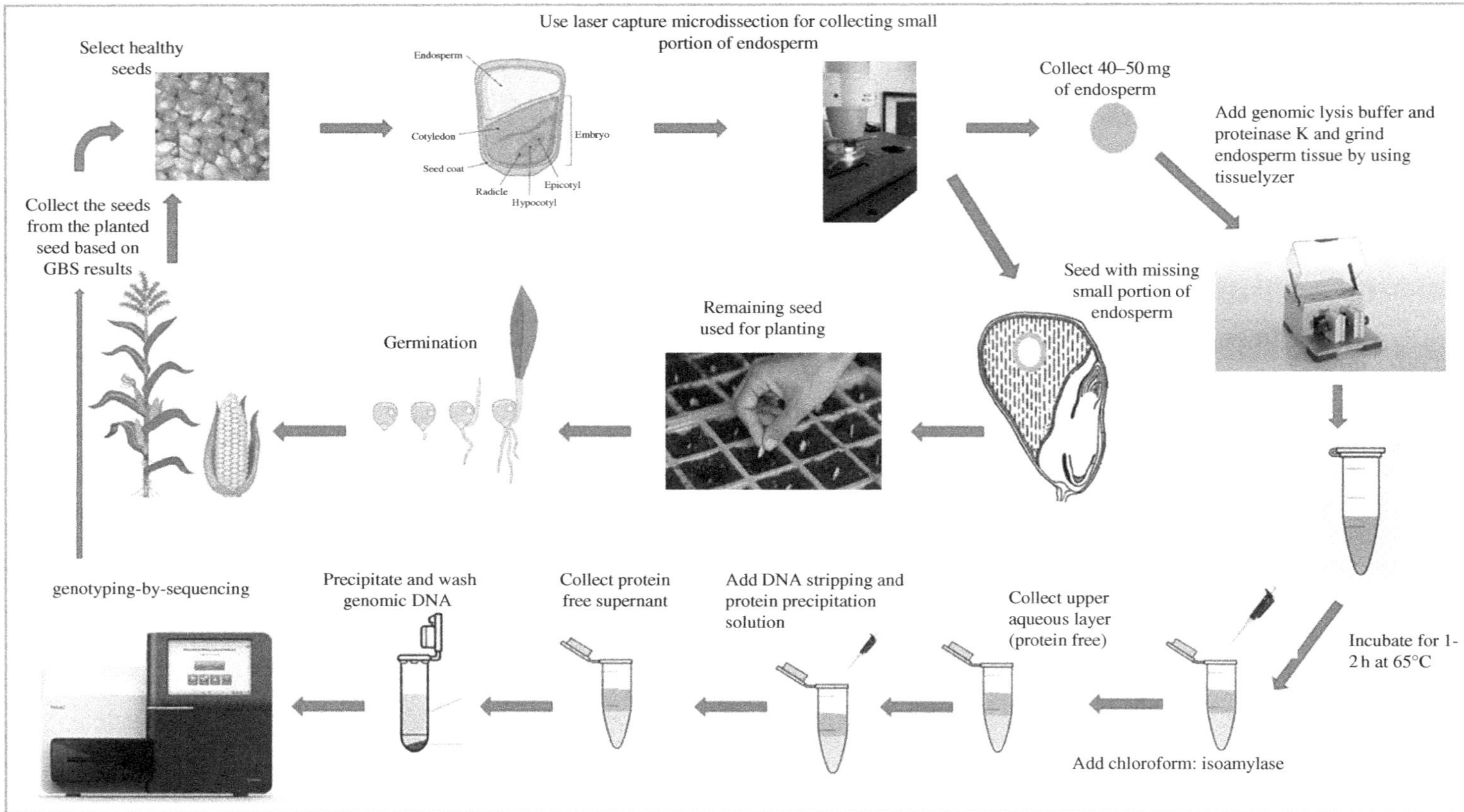


**Figure 6.2** Seed DNA-based genotyping-by-sequencing using laser microdissection. *Source:* Based on Khan et al. (2014); Illumina, Inc.

preferred to isolate pure nuclei for genomic and chromatin profiling (Deal and Henikoff 2010). It provides 100% pure DNA from desired cells. This method is more rapid as compared to the previous methods and does not require hazardous chemical, and anyone can handle the whole processes with little knowledge. This method can be used for any organism without any restriction for targeted nuclei isolation.

## 6.5 Applications of GBS as Diagnostic Tool

### 6.5.1 Germplasm Conservation and Quality Control

The exploration and exploiting of biodiversity of cultivated crops and their ancestors and wild relatives is required to improve the food security. Control and understanding of germplasm identity and purity are considered as fundamental activities for germplasm management (Kaga et al. 2012; Gaafar and Saker 2006). Seed protection becomes one of the major key points to safeguard the seed industry and also provide quality assurance to the purchaser. Quality control is considered as one of the critical components of breeding and germplasm conservation activities.

Most of the national and international organizations present in the world deal with seed quality standards. Generally, DUS (distinctness, uniformity, and stability) characterization is used to determine the varietal identity and genetic purity during new variety registration according to national and international rules and regulations (EU Commission Directive 2016/1914/EU). Characterization of candidate varieties through DUS protocol is time consuming, expansive, and requires highly trained persons to minimize human errors. To minimize the error occurred during morphological characterization, molecular testing is recommended along with morphological characterization for variety registration and certification processes (Jones et al. 2013). The combination of improved sampling strategy and molecular method has high potential to determine the purity of the variety. Even UPOV (International Union for the Protection of New Varieties of Plants) supports the use of trait-specific molecular markers, but it does not replace the DUS characterization (UPOV 2011). In Europe, various molecular techniques are used to manage the reference varieties (Reid et al. 2011). In OECD (Organization for Economic Co-operation and Development) and ISTA (International Seed Testing Association), a separate working group was created to standardize the biochemical and molecular techniques.

Selection of high-quality markers that can easily distinguish varieties is an important decision for breeding activities. The number of markers used for research work is decided based on the level of resolution required for genotyping to minimize the cost. Limited number of markers are used to avoid mislabeling of seed containers or plots used for accession regeneration, whereas large number of markers are required to estimate the heterozygosity and identification of each line in RIL and NIL populations (Alonso-Blanco et al. 2003; Fletcher et al. 2013). In recent years, adaptation of sequencing technologies has significantly increased the availability of SNPs for genotyping (Chia et al. 2012). SNP-based KASP (Kompetitive allele-specific PCR) genotyping is used for molecular analysis (He et al. 2014). Allele-specific primers are used to achieve Bi-allelic discrimination. This genotyping method is most cost-effective, simple, and easy to run. Further, correlation of specific SNPs to DUS characteristics can be done through this assay (Cockram et al. 2012). Other than KASP assay, the recently evolved DArTSeq NGS is used for genotyping. This methodology is used when high marker density is required for genome-wide association studies. DArTseq approach has much higher coverage with lower marker density and missing data in comparison to Cornell GBS system. DArTseq has the ability to directly score heterozygosity, so they can be directly used for diversity-related and genomic selection activities (Troggio et al. 2008; Semagn et al. 2014). Various factors affect the cost of genotyping l. Among them, the number of markers used for genotyping directly influences the cost per sample. As compared to the array type and sequenced-based platforms, KASP, a single plex system, has more influence on the cost factor. The accuracy of quality control may differ based on the stages of breeding and germplasm bank activities. Optimization of sampling strategy and marker density are the key factors that determine the genotyping costs (Patterson et al. 2017; Majeed et al. 2019; Grewal et al. 2020).

### 6.5.2 Tracking Crop Varieties

Commonly collecting the name and morphological plant descriptions from farmers is the main approach to accurately identify the crop cultivars. Also, environmental condition affects the morphological descriptions (Elhoumaizi et al. 2002). To overcome these challenges, molecular markers are used for varietal identification, and their expression will not be affected by any environmental factor. Genome-wide markers such as SSR and SNPs not only differentiate the varieties but also provide the information regarding the relative portion of ancestries derived from various founder genotypes of the currently grown varieties (Yoon et al. 2007). This ancestry interference can help to select the parental material in breeding programs. In the past, SSR, RFLP, and RAPD markers are utilized in DNA-based fingerprinting applications (Ercisli et al. 2011). However, limited multiplexing ability, their low frequency in genome, and high cost made them to be replaced by SNPs. GBS is simple and inexpensive to genotype large population of individuals (Glaubitz et al. 2014). Rabbi et al. (2015) used the SNP markers for varietal identification and tracking the notified and released varieties in farmer's field. It is also possible to know the farmers interest toward new varieties and varietal adaptation to the varying climatic conditions.

### 6.5.3 Sex Determination

In regard to sexual systems, plant kingdom has remarkable diversity (Barrett 2002), and most of the seed-bearing plants are "cosexual" in which both male and female reproductive organs are present in the same flower (hermaphrodite) or in separate flowers (Lloyd and Bawa 1984). Sex determination was usually identified with the physical separation of female and male plants from each other (Heikrujam et al. 2015). The sex of the flower is decided by the interaction of the plant genetic content and the environmental factors such as the soil nutrient status, climatic condition during flower-establishing stage, and the stress condition.

In deciduous plants, sex determination is a tedious job, especially before flowering and during early stages of plant development. Plant breeders give most priority for these plants for crop improvement programs because they need less attention during the hybridization step. There are several cultivated species where hermaphrodite flowers have more importance than female flowers and least importance is given for male flowers. These include *Actinidia deliciosa* (Shirkot et al. 2002), *Borassus flabellifer* (George et al. 2007), *Carica papaya* (Parasnis et al. 2000), *Eucommia ulmoides* (Xu et al. 2004), *Hippophae rhamnoides* (Persson et al. 1998; Sharma et al. 2010), *Myristica fragrans* (Shibu et al. 2000), *Phoenix dactylifera* (Younis et al. 2008), and *Simmondsia chinensis* (Agrawal et al. 2007). Sex chromosomes have been determined across all the eukaryotic lineages in deciduous species. Dioecy has evolved hundreds of times independently from hermaphroditic lineages (Renner 2014). Even evolutionary transition appeared several times in between monoecy and dioecy. It happened by the evaluation of an X/Y or Z/W sex chromosome pairs by mutation in two closely linked antagonistic sexual genes (Charlesworth and Charlesworth 1978; Charlesworth 2018). In some families, inconstant males also produce fruits if dominant suppressor of femaleness ($Su^F$) is absent (Lloyd and Bawa 1984). Several sex determination genes were identified in garden asparagus (Irish and Nelson 1989), persimmon (Akagi et al. 2018), and kiwifruit (Akagi et al. 2019). In most of the crops, sex determination region is present (SDR) on the Y chromosomes to determine the sex of plant by suppressing the female and promoting the male characters (Renner and Müller 2021).

Sex determination in papaya and date palm like dioecious plants at seed stage is a practice of great importance. For decades, several attempts were made to know the exact sex of the plant at seedling or early stage of the plant growth to maintain the higher female plants ratio as compared to male (El Hadrami and El Hadrami 2009). Till classification of plant sex phenotypically remains mysterious. Normally, 3–5 seedlings or plants per hill are maintained till the flowering stage to decide the sex of the plants. Later, excess male and all female plants are discarded from the field to get high fruit yield. Hermaphrodite plants produce elongated fruits with small internal cavity, which fetches higher market value in both domestic and international markets (Dantas and Pausas 2013). A wide range of biochemical and molecular techniques were developed to determine the sex of the dioecious crop plants (Elmeer and Mattat 2012; Al-Ameri et al. 2016). In recent years, the inter simple sequence repeat (ISSR) markers are gaining more importance for plant sex determination in plants and have several advantages over the other dominant markers but rarely used in dioecious plants (Heikrujam et al. 2014).

In addition to ISSRs, several molecular markers are discovered to distinguish the papaya plants at seedling stage (Costa et al. 2011; Liao et al. 2017), but they require all seeds to be planted for leaf tissue collection for DNA isolation (Vashisht et al. 2016). In papaya, the male and hermaphrodite gender is decided by the male-specific region (MSY) present on the Y and hermaphrodite-specific region (HSY) on $Y^h$ chromosomes (Ming et al. 2007). Among all the tested MSY region-specific markers, PS1 and PS2 are the only markers that can find the exact sex of the seed. Amplification of both the primers occurs when DNA is isolated from XX (female) seed, whereas hermaphrodite seed DNA forms heterozygote bands after amplification (Oliveira et al. 2007; Pirovani et al. 2018). But endospermic (triploid tissue) DNA from hermaphrodite seed can produce amplification of $XXY^h$, which favors the exact determination of the sex of the seed as compared to the leaf DNA-based sex determination (Borghetti and Ferreira 2004).

After papaya, date palm is an important ornamental, perennial, dioecious, and oldest cultivated plantation crop (Mathew et al. 2015). It plays a major socioeconomic role in dry tropical regions around the world by providing food, fiber, and shelter (Jain et al. 2011; Srivastava et al. 2013). For commercial use, 90:10 female and male ratio should be maintained to get higher economic return. In date palm, Younis et al. (2008) used the ISSR and RAPD markers, and Cherif et al. (2013) used SSR and SCAR markers for plant sex determination. Subsequently, GBS was successfully utilized for the discovery of sex-specific markers (Elshire et al. 2011; Davey et al. 2011). Sequencing is the only tool that helps to detect the sex-determining region (SRY) genes in plants. In date palm, male-specific primers are developed by Mohei et al. (2019) by sequencing. It yields a 350–400 bp PCR product as same as in human (Drobnic et al. 2006) and papaya (Yu et al. 2008) after amplification. A number of researchers (Foote et al. 1992; Devlin et al. 1998; Shibata et al. 1999; Okada et al. 2001; Ishizaki et al. 2002) have reported several Y chromosome-specific unique repeat sequences in various crops, which are similar to human and mice *SRY* genes (Skaletsky et al. 2003; Hawley 2003; Rozen et al. 2003; Tanurdzic and Banks 2004).

### 6.5.4 Transgenic Detection

Recently, transgenic-based breeding strategies are adopted for trait-specific crop improvement programs. It requires sophisticated equipment for distinguishing homozygous and hemizygous individuals in a segregating population (Geng et al. 2017). Quantitative real-time PCR (qPCR) is a single-step protocol used for detection as well as amplification of the targeted DNA segment (Heid et al. 1996). But qPCR can detect only targeted known sequence, which can limit the use of qPCR techniques for transgenic identification. Under such circumstance, NGS is the only solution for the identification and characterization of both known and unknown genetically modified organisms (GMOs) in seed without the need of Southern blot analysis. However, uneven coverage of the genome restricts the use of NGS technology for transgenic identification. Initially, whole-genome sequencing is used to detect the inserted sequences. But this methodology is inefficient to detect transgenic when it is present at very low amount. To solve this problem, amplicons by sequence were proposed, but their performance was lower than qPCR (Cain-Hom et al. 2017; Zhang et al. 2018; Debode et al. 2019; Kim et al. 2021). Later, combining NGS with DNA walking strategies and site-finding PCR was recommended to differentiate between transgenic and nontransgenic plants even at low percentages of genetic modifications (Fraiture et al. 2017).

### 6.5.5 Detection of Seed-borne Diseases

Seed-borne plant diseases are caused by various organisms, which ultimately limit the crop production and substantially cause economic losses. Advance identification and detection of foreign pathogenic agents in plants is considered as an imperative strategy for disease management (Chalupowicz et al. 2018). All living organisms have developed a strong defensive system against any pest infections that are triggered by initial identification of foreign invaders. Plants use the pathogen-associated molecular pattern (PAMP)-triggered immune (PTI) responses such as signaling cascades by changing the host gene expression, which eventually modify the gene expression of the hosts. The disease incident in plants depends on the defense level of the host (Jones and Dangl 2006). For this, accurate and rapid diagnosis tools including serological and molecular methods are required to minimize the impact of the infection (De Boer and López 2012; Martinelli et al. 2015). Recently, loop-mediated isothermal amplification has been most commonly employed for the detection of pathogens (Parida et al. 2008). It can easily detect any fungal seed-borne disease with high accuracy (Endo and Komori 2004; Ohori et al. 2006; Uemura et al. 2008).

The development of NGS in recent years has created new ways for detection and identification of pathogens (Van Dijk et al. 2014). Habitually, NGS analysis of 16S amplicons is used for microbiome analysis (Niu et al. 2017). Transcriptome (total RNA sequencing) is done for the detection of both DNA and RNA viruses present in the seed (Adams 2009; Luria et al. 2019; Santala and Valkonen 2018). NGS-based technologies are most commonly used for pathogenic viruses as compared to bacteria and fungi for the identification and characterization (Al Rwahnih et al. 2009; Cantu et al. 2011). The major sequencing platforms used for the detection of seed-borne pathogens are 454 sequencing, AB sequencing by oligonucleotide ligation and detection (AB SOLiD) technology, Solexa/Illumina sequencing, and Oxford Nanopore DNA sequencing (Mancini et al. 2016). However, Oxford Nanopore DNA sequencing like platforms are frequently utilized for seed-borne disease identification. It is a single molecule sequencing platform which can generate high-throughput data by sequencing long reads in a short time (Chalupowicz et al. 2018).

## 6.6 Summary

Mass sequencing of genome and transcriptomes has revolutionized the field of plant genetics. GBS is carried out in most of the cultivated crops to know the critical information which is necessary for breeding programs. However, the utilization of GBS in plant-breeding activities is still in its nascent stage. Developed countries possess advanced laboratories to utilize the seed DNA for genotyping, whereas, in the developing countries, several bottlenecks limit the use of seed DNA-based genotyping technology. As a result, seed DNA-based genotyping can be effectively utilized in plant-breeding programs when cost-effective, high-throughput techniques and analytical pipelines for are developed.

## References

Abdel-Latif, A. and Osman, G. (2017). Comparison of three genomic DNA extraction methods to obtain high DNA quality from maize. *Plant Methods* 13 (1): 1–9.

Abd-Elsalam, K., Bahkali, A., Moslem, M. et al. (2011). An optimized protocol for DNA extraction from wheat seeds and loop-mediated isothermal amplification (LAMP) to detect *Fusarium graminearum* contamination of wheat grain. *International Journal of Molecular Sciences* 12 (6): 3459–3472.

Aboul-Maaty, N.A.F. and Oraby, H.A.S. (2019). Extraction of high-quality genomic DNA from different plant orders applying a modified CTAB-based method. *Bulletin of the National Research Centre* 43 (1): 25.

Adams, J. (2009). The role of time perspective in smoking cessation amongst older English adults. *Health Psychology* 28 (5): 529.

Adler, A.J., Wiley, G.B., and Gaffney, P.M. (2013). Infinium assay for large-scale SNP genotyping applications. *Journal of Visualized Experiments* 81: e50683.

Agbagwa, I.O., Datta, S., Patil, P.G. et al. (2012). A protocol for high-quality genomic DNA extraction from legumes. *Genetics and Molecular Research* 11: 4632–4639.

Agrawal, V., Sharma, K., Gupta, S. et al. (2007). Identification of sex in *Simmondsia chinensis* (Jojoba) using RAPD markers. *Plant Biotechnology Reports* 1 (4): 207–210.

Akagi, T., Henry, I.M., Ohtani, H. et al. (2018). A Y-encoded suppressor of feminization arose via lineage-specific duplication of a cytokinin response regulator in kiwifruit. *The Plant Cell* 30 (4): 780–795.

Akagi, T., Pilkington, S.M., Varkonyi-Gasic, E. et al. (2019). Two Y-chromosome-encoded genes determine sex in kiwifruit. *Nature Plants* 5 (8): 801–809.

Al-Ameri, A.A., Al-Qurainy, F., Gaafar, A.Z. et al. (2016). Male specific gene expression in dioecious *Phoenix dactylifera* (Date palm) tree at flowering stage. *Pakistan Journal of Botany* 48 (1): 131–135.

Al Rwahnih, M., Daubert, S., Golino, D. et al. (2009). Deep sequencing analysis of RNAs from a grapevine showing Syrah decline symptoms reveals a multiple virus infection that includes a novel virus. *Virology* 387 (2): 395–401.

Alonso-Blanco, C., Bentsink, L., Hanhart, C.J. et al. (2003). Analysis of natural allelic variation at seed dormancy loci of *Arabidopsis thaliana*. *Genetics* 164 (2): 711–729.

Arbelaez, J.D., Tandayu, E., Reveche, M.Y. et al. (2019). Methodology: ssb-MASS: a single seed-based sampling strategy for marker-assisted selection in rice. *Plant Methods* 15 (1): 1–11.

Asano, T., Masumura, T., Kusano, H. et al. (2002). Construction of a specialized cDNA library from plant cells isolated by laser capture microdissection: toward comprehensive analysis of the genes expressed in the rice phloem. *The Plant Journal* 32 (3): 401–408.

Balestrini, R. and Bonfante, P. (2008). Laser microdissection (LM): applications to plant materials. *Plant Biosystems* 142 (2): 331–336.

Barrett, S.C. (2002). The evolution of plant sexual diversity. *Nature Reviews Genetics* 3 (4): 274–284.

Becker, M.G., Hsu, S.W., Harada, J.J., and Belmonte, M.F. (2014). Genomic dissection of the seed. *Frontiers in Plant Science* 5: 464.

Belicuas, P.R., Guimarães, C.T., Paiva, L.V. et al. (2007). Androgenetic haploids and SSR markers as tools for the development of tropical maize hybrids. *Euphytica* 156 (1): 95–102.

Benito, C., Figueiras, A.M., Zaragoza, C. et al. (1993). Rapid identification of *Triticeae* genotypes from single seeds using the polymerase chain reaction. *Plant Molecular Biology* 21 (1): 181–183.

Bennett, S. (2004). Solexa ltd. *Pharmacogenomics* 5 (4): 433–438.

Bernatsky, R. and Tanksley, S.D. (1986). Toward a saturated linkage map in tomato based on isozymes and random cDNA sequences. *Genetics* 112: 887–898.

Bewley, J.D. (1997). Seed germination and dormancy. *The Plant Cell* 9 (7): 1055.

Bhat, J.A., Ali, S., Salgotra, R.K. et al. (2016). Genomic selection in the era of next generation sequencing for complex traits in plant breeding. *Frontiers in Genetics* 7: 221.

Bhattacharjee, R., Kolesnikova-Allen, M., Aikpokpodion, P. et al. (2004). An improved semiautomated rapid method of extracting genomic DNA for molecular marker analysis in cocoa, *Theobroma cacao* L. *Plant Molecular Biology Reporter* 22 (4): 435–436.

Blair, E.O. and Corrigan, D.K. (2019). A review of microfabricated electrochemical biosensors for DNA detection. *Biosensors and Bioelectronics* 134: 57–67.

Borghetti, F. and Ferreira, A.G. (2004). *Germinação: Do Básicoaoaplicado*, 209–222. Porto Alegre: Artmed.

Botstein, D., White, R.L., Skolnick, M., and Davis, R.W. (1980). Construction of a genetic linkage map in man using restriction fragment length polymorphisms. *American Journal of Human Genetics* 32 (3): 314.

Brown, J., Enright, N.J., and Miller, B.P. (2003). Seed production and germination in two rare and three common co-occurring *Acacia* species from south-east Australia. *Austral Ecology* 28 (3): 271–280.

Cain-Hom, C., Splinter, E., van Min, M. et al. (2017). Efficient mapping of transgene integration sites and local structural changes in Cre transgenic mice using targeted locus amplification. *Nucleic Acids Research* 45 (8): 62.

Cantu, D., Govindarajulu, M., Kozik, A. et al. (2011). Next generation sequencing provides rapid access to the genome of *Puccinia striiformis* f. sp. *tritici*, the causal agent of wheat stripe rust. *PLoS One* 6 (8): e24230.

Cassago, A., Panepucci, R.A., Baião, A.M.T. et al. (2002). Cellophane based mini-prep method for DNA extraction from the filamentous fungus *Trichoderma reesei*. *BMC Microbiology* 2 (1): 14.

Casson, S., Spencer, M., Walker, K., and Lindsey, K. (2005). Laser capture microdissection for the analysis of gene expression during embryogenesis of Arabidopsis. *The Plant Journal* 42 (1): 111–123.

Chalupowicz, L., Dombrovsky, A., Gaba, V. et al. (2018). Diagnosis of plant diseases using the nanopore sequencing platform. *Plant Pathology* 68 (2): 229–238.

Charlesworth, B. and Charlesworth, D. (1978). A model for the evolution of dioecy and gynodioecy. *The American Naturalist* 112 (988): 975–997.

Charlesworth, D. (2018). Does sexual dimorphism in plants promote sex chromosome evolution. *Environmental and Experimental Botany* 146: 5–12.

Chen, W., Lv, H., Nie, F. et al. (2019). i6mA-Pred: identifying DNA N6-methyladenine sites in the rice genome. *Bioinformatics* 35 (16): 2796–2800.

Chenault, K.D., Gallo, M., Seib, J.C. et al. (2007). A non-destructive seed sampling method for PCR-based analyses in marker assisted selection and transgene screening. *Peanut Science* 34 (1): 38–43.

Cherif, E., Zehdi, S., Castillo, K. et al. (2013). Male-specific DNA markers provide genetic evidence of an XY chromosome system, a recombination arrest and allow the tracing of paternal lineages in date palm. *New Phytologist* 197 (2): 409–415.

Chia, J.M., Song, C., Bradbury, P.J. et al. (2012). Maize HapMap2 identifies extant variation from a genome in flux. *Nature Genetics* 44 (7): 803–807.

Chunwongse, J., Martin, G.B., Tanksley, S.D. et al. (1993). Pre-germination genotypic screening using PCR amplification of half-seeds. *Theoretical and Applied Genetics* 86 (6): 694–698.

Cleland, A.B. (2013). *Food Knowledge–What is It and Where Does It Come from: A Study of Culinary Management Students at William Angliss Institute*. University of Melbourne.

Cockram, J., Jones, H., Norris, C., and O'Sullivan, D.M. (2012). Evaluation of diagnostic molecular markers for DUS phenotypic assessment in the cereal crop, barley (*Hordeum vulgare* ssp. *vulgare* L.). *Theoretical and Applied Genetics* 125 (8): 1735–1749.

Costa, F.R.D., Pereira, T.N.S., Gabriel, A.P.C. et al. (2011). ISSR markers for genetic relationships in *Caricaceae* and sex differentiation in papaya. *Crop Breeding and Applied Biotechnology* 11 (4): 352–357.

Dantas, V.D.L. and Pausas, J.G. (2013). The lanky and the corky: fire-escape strategies in savanna woody species. *Journal of Ecology* 101 (5): 1265–1272.

Davey, J.W., Hohenlohe, P.A., Etter, P.D. et al. (2011). Genome-wide genetic marker discovery and genotyping using next-generation sequencing. *Nature Reviews Genetics* 12 (7): 499–510.

Day, R.C., Herridge, R.P., Ambrose, B.A., and Macknight, R.C. (2008). Transcriptome analysis of proliferating Arabidopsis endosperm reveals biological implications for the control of syncytial division, cytokinin signaling, and gene expression regulation. *Plant Physiology* 148 (4): 1964–1984.

Dayteg, C., Tuvesson, S., Merker, A. et al. (2007). Automation of DNA marker analysis for molecular breeding in crops: practical experience of a plant breeding company. *Plant Breeding* 126 (4): 410–415.

De Barba, M., Miquel, C., Lobréaux, S. et al. (2017). High-throughput microsatellite genotyping in ecology: improved accuracy, efficiency, standardization and success with low-quantity and degraded DNA. *Molecular Ecology Resources* 17 (3): 492–507.

De Boer, S.H. and López, M.M. (2012). New grower-friendly methods for plant pathogen monitoring. *Annual Review of Phytopathology* 50: 197–218.

Deal, R.B. and Henikoff, S.A. (2010). Simple method for gene expression and chromatin profiling of individual cell types within a tissue. *Developmental Cell* 18: 1030–1040.

Debode, F., Hulin, J., Charloteaux, B. et al. (2019). Detection and identification of transgenic events by next generation sequencing combined with enrichment technologies. *Scientific Reports* 9: 1–9.

Deschamps, S., Llaca, V., and May, G.D. (2012). Genotyping-by-sequencing in plants. *Biology* 1 (3): 460–483.

Deschamps, S., Zhang, Y., Llaca, V. et al. (2018). A chromosome-scale assembly of the sorghum genome using nanopore sequencing and optical mapping. *Nature Communications* 9 (1): 1–10.

Desmarais, E., Lanneluc, I., and Lagnel, J. (1998). Direct amplification of length polymorphisms (DALP), or how to get and characterize new genetic markers in many species. *Nucleic Acids Research* 26 (6): 1458–1465.

Deurenberg, R.H., Bathoorn, E., Chlebowicz, M.A. et al. (2017). Application of next generation sequencing in clinical microbiology and infection prevention. *Journal of Biotechnology* 243: 16–24.

Devlin, R.H., Stone, G.W., and Smailus, D.E. (1998). Extensive direct-tandem organization of a long repeat DNA sequence on the Y chromosome of chinook salmon (*Oncorhynchus tshawytscha*). *Journal of Molecular Evolution* 46 (3): 277–287.

Drobnic, M., Radosavljevic, D., Ravnik, D. et al. (2006). Comparison of four techniques for the fixation of a collagen scaffold in the human cadaveric knee. *Osteoarthritis and Cartilage* 14 (4): 337–344.

Duan, C.G., Wang, X., Tang, K. et al. (2015). MET18 connects the cytosolic iron-sulfur cluster assembly pathway to active DNA demethylation in Arabidopsis. *PLoS Genetics* 11 (10): e1005559.

El Hadrami, I. and El Hadrami, A. (2009). Breeding date palm. In: *Breeding Plantation Tree Crops: Tropical Species* (ed. S.M. Jain and P.M. Priyadarshan), 191–216. New York: Springer.

Elhoumaizi, M.A., Saaidi, M., Oihabi, A., and Cilas, C. (2002). Phenotypic diversity of date-palm cultivars (*Phoenix dactylifera* L.) from Morocco. *Genetic Resources and Crop Evolution* 49 (5): 483–490.

Elmeer, K. and Mattat, I. (2012). Marker-assisted sex differentiation in date palm using simple sequence repeats. *Biotech* 2 (3): 241–247.

Elshire, R.J., Glaubitz, J.C., Sun, Q. et al. (2011). A robust, simple genotyping-by-sequencing (GBS) approach for high diversity species. *PLoS One* 6: e19379.

Emmert-Buck, M.R., Bonner, R.F., Smith, P.D. et al. (1996). Laser capture microdissection. *Science* 274 (5289): 998–1001.

Endo, S. and Komori, T. (2004). Detection of gp43 of *Paracoccidioides brasiliensis* by the loop-mediated isothermal amplification (LAMP) method. *FEMS Microbiology Letters* 234 (1): 93–97.

Ercisli, S., Ipek, A., and Barut, E. (2011). SSR marker-based DNA fingerprinting and cultivar identification of olives (*Olea europaea*). *Biochemical Genetics* 49 (9): 555–561.

Fletcher, R.S., Mullen, J.L., Yoder, S. et al. (2013). Development of a next-generation NIL library in *Arabidopsis thaliana* for dissecting complex traits. *BMC Genomics* 14 (655): 1–13.

Flint-Garcia, S.A., Thuillet, A.C., Yu, J. et al. (2005). Maize association population: a high-resolution platform for quantitative trait locus dissection. *The Plant Journal* 44 (6): 1054–1064.

Foote, S., Vollrath, D., Hilton, A. et al. (1992). The human Y chromosome: overlapping DNA clones spanning the euchromatic region. *Science* 258 (5079): 60–66.

Fraiture, M.A., Herman, P., Papazova, N. et al. (2017). An integrated strategy combining DNA walking and NGS to detect GMOs. *Food Chemistry* 232: 351–358.

Frumkin, D., Wasserstrom, A., Itzkovitz, S. et al. (2008). Amplification of multiple genomic loci from single cells isolated by laser micro-dissection of tissues. *BMC Biotechnology* 8 (1): 1–16.

Gaafar, R.M. and Saker, M.M. (2006). Monitoring of cultivars identity and genetic stability in strawberry varieties grown in Egypt. *World Journal of Agricultural Sciences* 2 (1): 29–36.

Gao, S., Martinez, C., Skinner, D.J. et al. (2008). Development of a seed DNA-based genotyping system for marker-assisted selection in maize. *Molecular Breeding* 22 (3): 477–494.

Geng, L., Deng, D.D., Wubben, M.J. et al. (2017). A high-throughput standard PCR-based genotyping method for determining transgene zygosity in segregating plant populations. *Frontiers in Plant Science* 8: 1252.

George, J., Karun, A., Manimekalai, R. et al. (2007). Identification of RAPD markers linked to sex determination in palmyrah (*Borassus flabellifer* L.). *Current Science* 93 (8): 1075–1077.

Glaubitz, J.C., Casstevens, T.M., Lu, F. et al. (2014). TASSEL-GBS: a high capacity genotyping by sequencing analysis pipeline. *PLoS One* 9 (2): e90346.

Godfray, H.C.J., Beddington, J.R., Crute, I.R. et al. (2010). Food security: the challenge of feeding 9 billion people. *Science* 327 (5967): 812–818.

Gomez, S.K. and Harrison, M.J. (2009). Laser microdissection and its application to analyze gene expression in *Arbuscular mycorrhizal* symbiosis. *Pest Management Science: Formerly Pesticide Science* 65 (5): 504–511.

Grewal, S., Hubbart-Edwards, S., Yang, C. et al. (2020). Rapid identification of homozygosity and site of wild relative introgressions in wheat through chromosome-specific KASP genotyping assays. *Plant Biotechnology Journal* 18 (3): 743–755.

Hansen, N. and Ostermeier, A. (2001). Completely derandomized self-adaptation in evolution strategies. *Evolutionary Computation* 9 (2): 159–195.

Hawley, R.S. (2003). The human Y chromosome: rumors of its death have been greatly exaggerated. *Cell* 113 (7): 825–828.

He, J., Zhao, X., Laroche, A. et al. (2014). Genotyping-by-sequencing (GBS), an ultimate marker-assisted selection (MAS) tool to accelerate plant breeding. *Frontiers in Plant Science* 5: 484.

Heid, C.A., Stevens, J., Livak, K.J. et al. (1996). Real time quantitative PCR. *Genome Research* 6 (10): 986–994.

Heikrujam, M., Sharma, K., Kumar, J., and Agrawal, V. (2014). Generation and validation of unique male sex-specific sequence tagged sites (STS) marker from diverse genotypes of dioecious Jojoba-*Simmondsia chinensis* (Link) Schneider. *Euphytica* 199 (3): 363–372.

Heikrujam, M., Sharma, K., Prasad, M. et al. (2015). Review on different mechanisms of sex determination and sex-linked molecular markers in dioecious crops: a current update. *Euphytica* 201 (2): 161–194.

Hobza, R. and Vyskot, B. (2007). Laser microdissection-based analysis of plant sex chromosomes. *Methods in Cell Biology* 82: 433–453.

Irish, E.E. and Nelson, T. (1989). Sex determination in monoecious and dioecious plants. *The Plant Cell* 1 (8): 737.

Isenberg, G., Bielser, W., Meier-Ruge, W., and Remy, E. (1976). Cell surgery by laser micro-dissection: a preparative method. *Journal of Microscopy* 107 (1): 19–24.

Ishizaki, K., Shimizu-Ueda, Y., Okada, S. et al. (2002). Multicopy genes uniquely amplified in the Y chromosome-specific repeats of the liverwort *Marchantia polymorpha*. *Nucleic Acids Research* 30 (21): 4675–4681.

Ithal, N., Recknor, J., Nettleton, D. et al. (2007). Parallel genome-wide expression profiling of host and pathogen during soybean cyst nematode infection of soybean. *Molecular Plant-Microbe Interactions* 20 (3): 293–305.

Jain, S.M., Al-Khayri, J.M., and Johnson, D.V. (ed.) (2011). *Date Palm Biotechnology*. Springer Science & Business Media.

Jones, J.D. and Dangl, J.L. (2006). The plant immune system. *Nature* 444 (7117): 323–329.

Jones, H., Norris, C., Smith, D. et al. (2013). Evaluation of the use of high-density SNP genotyping to implement UPOV Model 2 for DUS testing in barley. *Theoretical and Applied Genetics* 126 (4): 901–911.

Kaga, A., Shimizu, T., Watanabe, S. et al. (2012). Evaluation of soybean germplasm conserved in NIAS genebank and development of mini core collections. *Breeding Science* 61 (5): 566–592.

Kang, H.W., Cho, Y.G., Yoon, U.H. et al. (1998). A rapid DNA extraction method for RFLP and PCR analysis from a single dry seed. *Plant Molecular Biology Reporter* 16 (1): 90.

Khan, D., Millar, J.L., Girard, I.J. et al. (2014). Transcriptional circuitry underlying seed coat development in Arabidopsis. *Plant Science* 219: 51–60.

Kim, K.D., Kang, Y., and Kim, C. (2020). Application of genomic big data in plant breeding: past, present, and future. *Plants* 9 (11): 1454.

Kim, S.J., Park, J.S., Shin, Y.H., and Park, Y.D. (2021). Identification and validation of genetic variations in transgenic Chinese cabbage plants (*Brassica rapa* ssp. pekinensis) by next-generation sequencing. *Genes* 12 (5): 621.

Konieczny, A. and Ausubel, F.M. (1993). A procedure for mapping Arabidopsis mutations using co-dominant ecotype-specific PCR-based markers. *The Plant Journal* 4 (2): 403–410.
Liang, P., Xu, Y., Zhang, X. et al. (2015). CRISPR/Cas9-mediated gene editing in human tripronuclear zygotes. *Protein and Cell* 6 (5): 363–372.
Liao, Z., Yu, Q., and Ming, R. (2017). Development of male-specific markers and identification of sex reversal mutants in papaya. *Euphytica* 213 (2): 53.
Lloyd, D.G. and Bawa, K.S. (1984). Modification of the gender of seed plants in varying conditions. In: *Evolutionary Biology* (ed. M.K. Hecht, B. Wallace and G.T. Prance), 255–338. Berlin: Springer-Verlag.
Loomis, W.D. (1974). Overcoming problems of phenolics and quinones in the isolation of plant enzymes and organelles. In: *Methods in Enzymology*, vol. 31, 528–544. Academic Press.
Lu, B., Lee, J., Nie, X. et al. (2013). Phosphorylation of human TFAM in mitochondria impairs DNA binding and promotes degradation by the AAA+ Lon protease. *Molecular Cell* 49 (1): 121–132.
Luria, N., Smith, E., Sela, N. et al. (2019). Insights into a watermelon virome contribute to monitoring distribution of whitefly-borne viruses. *Phytobiomes Journal* 3 (1): 61–70.
Majeed, U., Darwish, E., Rehman, S.U., and Zhang, X. (2019). Kompetitive allele specific PCR (KASP): a singleplex genotyping platform and its application. *Journal of Agricultural Science* 11 (1): 11–20.
Mancini, V., Murolo, S., and Romanazzi, G. (2016). Diagnostic methods for detecting fungal pathogens on vegetable seeds. *Plant Pathology* 65 (5): 691–703.
Mardis, E.R. (2008). Next-generation DNA sequencing methods. *Annual Review of Genomics and Human Genetics* 9: 387–402.
Martinelli, F., Scalenghe, R., Davino, S. et al. (2015). Advanced methods of plant disease detection. *Annual Review of Agronomy for Sustainable Development* 35 (1): 1–25.
Mascher, M., Richmond, T.A., Gerhardt, D.J. et al. (2013). Barley whole exome capture: a tool for genomic research in the genus Hordeum and beyond. *The Plant Journal* 76 (3): 494–505.
Mastrokolias, A., den Dunnen, J.T., van Ommen, G.B. et al. (2012). Increased sensitivity of next generation sequencing-based expression profiling after globin reduction in human blood RNA. *BMC Genomics* 13: 1–9.
Mathew, L.S., Seidel, M.A., George, B. et al. (2015). A genome-wide survey of date palm cultivars supports two major subpopulations in *Phoenix dactylifera*. *Genes, Genomes, Genetics* 5 (7): 1429–1438.
Meru, G., McDowell, D., Waters, V. et al. (2013). A non-destructive genotyping system from a single seed for marker-assisted selection in watermelon. *Genetics and Molecular Research* 11 (1): 702–709.
Meru, G. and McGregor, C. (2013). Genetic mapping of seed traits correlated with seed oil percentage in watermelon. *Horticulture Science* 48 (8): 955–959.
Metzker, M.L. (2010). Sequencing technologies-the next generation. *Nature Reviews Genetics* 11 (1): 31–46.
Ming, R., Wang, J., Moore, P.H. et al. (2007). Sex chromosomes in flowering plants. *American Journal of Botany* 94 (2): 141–150.
Mohei, E.L., Mohasseb, H.A.A., Al-Khateeb, A.A. et al. (2019). Identification and sequencing of Date-SRY gene: a novel tool for sex determination of date palm (*Phoenix dactylifera* L.). *Saudi Journal of Biological Sciences* 26 (3): 514–523.

Motokazu, K. and Tadahiko, K. (2003). Rapid DNA extraction method from soybean seeds. *Breeding Science* 53: 277–279.

Mutou, C., Tanaka, K., and Ishikawa, R. (2014). DNA extraction from rice endosperm (including a protocol for extraction of DNA from ancient seed samples). In: *Cereal Genomics*, 7–15. Totowa, NJ: Humana Press.

Nelson, T., Tausta, S.L., Gandotra, N., and Liu, T. (2006). Laser microdissection of plant tissue: what you see is what you get. *Annual Review of Plant Biology* 57: 181–201.

Niu, B., Paulson, J.N., Zheng, X. et al. (2017). Simplified and representative bacterial community of maize roots. *Proceedings of the National Academy of Sciences of the United States of America* 11 (12): 2450–2459.

Ohad, N., Yadegari, R., Margossian, L. et al. (1999). Mutations in FIE, a WD polycomb group gene, allow endosperm development without fertilization. *The Plant Cell* 11 (3): 407–415.

Ohori, A., Endo, S., Sano, A. et al. (2006). Rapid identification of *Ochroconis gallopava* by a loop-mediated isothermal amplification (LAMP) method. *Veterinary Microbiology* 114 (3–4): 359–365.

Okada, S., Sone, T., Fujisawa, M. et al. (2001). The Y chromosome in the liverwort *Marchantia polymorpha* has accumulated unique repeat sequences harboring a male-specific gene. *Proceedings of the National Academy of Sciences of the United States of America* 98 (16): 9454–9459.

Oliveira, R.R.D., Feldberg, E., Anjos, M.B.D. et al. (2007). Karyotype characterization and ZZ/ZW sex chromosome heteromorphism in two species of the catfish genus Ancistrus Kner, 1854 (Siluriformes: Loricariidae) from the Amazon basin. *Neotropical Ichthyology* 5 (3): 301–306.

Oliver, R.E., Tinker, N.A., Lazo, G.R. et al. (2013). SNP discovery and chromosome anchoring provide the first physically-anchored hexaploid oat map and reveal synteny with model species. *PLoS One* 8 (3): 58068.

Papazova, N., Malef, A., Degrieck, I. et al. (2005). DNA extractability from the maize embryo and endosperm–relevance to GMO assessment in seed samples. *Seed Science and Technology* 33 (3): 533–542.

Paran, I. and Richard, W.M. (1993). Development of reliable PCR-based markers linked to downy mildew resistance genes in lettuce. *Theoretical and Applied Genetics* 85 (8): 985–993.

Parasnis, A.S., Gupta, V.S., Tamhankar, S.A., and Ranjekar, P.K. (2000). A highly reliable sex diagnostic PCR assay for mass screening of papaya seedlings. *Molecular Breeding* 6 (3): 337–344.

Parida, M., Sannarangaiah, S., Dash, P.K. et al. (2008). Loop mediated isothermal amplification (LAMP): a new generation of innovative gene amplification technique; perspectives in clinical diagnosis of infectious diseases. *Reviews in Medical Virology* 18 (6): 407–421.

Paterson, A.H., Brubaker, C.L., and Wendel, J.F. (1993). A rapid method for extraction of cotton (*Gossypium* spp.) genomic DNA suitable for RFLP or PCR analysis. *Plant Molecular Biology Reporter* 11 (2): 122–127.

Patterson, E.L., Fleming, M.B., Kessler, K.C. et al. (2017). A KASP genotyping method to identify northern watermilfoil, Eurasian watermilfoil, and their interspecific hybrids. *Frontiers in Plant Science* 8: 752.

Persson, H.A., Lundquist, K., and Nybom, H. (1998). RAPD analysis of genetic variation within and among populations of Turk's-cap lily (*Lilium martagon* L.). *Hereditas* 128 (3): 213–220.

Pirovani, A.A.V., Ramos, H.C.C., Santa-Catarina, R. et al. (2018). A hermaphrodite genotype in dioecious papaya progeny: sex reversal or contamination? *Euphytica* 214 (12): 227.

Poland, J.A., Brown, P.J., Sorrells, M.E. et al. (2012). Development of high-density genetic maps for barley and wheat using a novel two-enzyme genotyping-by-sequencing approach. *PLoS One* 7 (2): 32253.

Porebski, S., Bailey, L.G., and Baum, B.R. (1997). Modification of a CTAB DNA extraction protocol for plants containing high polysaccharide and polyphenol components. *Plant Molecular Biology Reporter* 15 (1): 8–15.

Procunier, J.D., Prashar, S., Chen, G. et al. (2009). Rapid ID technology (RIDT) in plants: high-speed DNA fingerprinting in grain seeds for the identification, segregation, purity, and traceability of varieties using labautomation robotics. *Journal of the Association for Laboratory Automation* 14 (4): 221–231.

Rabbi, I.Y., Kulakow, P.A., Manu-Aduening, J.A. et al. (2015). Tracking crop varieties using genotyping-by-sequencing markers: a case study using cassava (*Manihot esculenta* Crantz). *BMC Genetics* 16 (1): 1–11.

Rafalski, A., Tingey, S., and Henry, R. (2008). *SNPs and Their Use in Maize. Plant Genotyping II: SNP Technology*, 30–43. Wallingford, UK: CAB International.

Ragoussis, J., Elvidge, G.P., Kaur, K. et al. (2006). Matrix-assisted laser desorption/ionisation, time-of-flight mass spectrometry in genomics research. *PLoS Genetics* 2 (7): 100.

Ramsay, K., Jones, M.G., and Wang, Z. (2006). Laser capture microdissection: a novel approach to microanalysis of plant–microbe interactions. *Molecular Plant Pathology* 7 (5): 429–435.

Rao, N.K., Dulloo, M.E., and Engels, J.M. (2017). A review of factors that influence the production of quality seed for long-term conservation in gene banks. *Genetic Resources and Crop Evolution* 64 (5): 1061–1074.

Ray, A., Gupta, S.D., and Ghosh, S. (2013). Evaluation of anti-oxidative activity and UV absorption potential of the extracts of *Aloe vera* L. gel from different growth periods of plants. *Industrial Crops and Products* 1 (49): 712–719.

Reid, A., Hof, L., Felix, G. et al. (2011). Construction of an integrated microsatellite and key morphological characteristic database of potato varieties on the EU common catalogue. *Euphytica* 182 (2): 239.

Renner, S.S. (2014). The relative and absolute frequencies of angiosperm sexual systems: dioecy, monoecy, gynodioecy, and an updated online database. *American Journal of Botany* 101 (10): 1588–1596.

Renner, S.S. and Müller, N.A. (2021). Plant sex chromosomes defy evolutionary models of expanding recombination suppression and genetic degeneration. *Nature Plants* 7: 1–11.

Robertson, D.S. (1985). A possible technique for isolating genic DNA for quantitative traits in plants. *Journal of Theoretical Biology* 117 (1): 1–10.

Rosche, L., Semroc, J.M., and Gilbert, L.K. (2008). *Dragonflies and Damselflies of Northeast Ohio*. Cleveland Museum of Natural History.

Rozen, S., Skaletsky, H., Marszalek, J.D. et al. (2003). Abundant gene conversion between arms of palindromes in human and ape Y chromosomes. *Nature* 423 (6942): 873–876.

Sajib, A.A., Bhuiya, M.A.I., and Huque, R. (2017). A simple, efficient and rapid method for good quality DNA extraction from rice grains. *Rice Science* 24 (2): 119–122.

Salimath, S.S., Oliveira, A.C.D., Bennetzen, J.L. et al. (1995). Assessment of genome origins and genetic diversity in the genus Eleusine with DNA markers. *Genome* 38 (4): 757–763.

Sangtong, V., Mottl, E.C., Long, M.J. et al. (2001). Serial extraction of endosperm drillings (seed)—a method for detecting transgenes and proteins in single viable maize kernels. *Plant Molecular Biology Reporter* 19 (2): 151–158.

Santala, J. and Valkonen, J. (2018). Sensitivity of small RNA-based detection of plant viruses. *Frontiers in Microbiology* 9: 939.

Schneeberger, K. and Weigel, D. (2011). Fast-forward genetics enabled by new sequencing technologies. *Trends in Plant Science* 16 (5): 282–288.

Scutt, C.P., Li, Y., Robertson, S.E. et al. (1997). Sex determination in dioecious *Silene latifolia* (effects of the Y chromosome and the parasitic smut fungus (*Ustilago violacea*) on gene expression during flower development). *Plant Physiology* 114 (3): 969–979.

Semagn, K., Babu, R., Hearne, S. et al. (2014). Single nucleotide polymorphism genotyping using Kompetitive allele specific PCR (KASP): overview of the technology and its application in crop improvement. *Molecular Breeding* 33 (1): 1–14.

Sharma, A., Zinta, G., Rana, S. et al. (2010). Molecular identification of sex in *Hippophae rhamnoides* L. using isozyme and RAPD markers. *Forestry Studies in China* 12 (2): 62–66.

Sharma, R., Masaki, J., and Agarwal, A. (2013). Sperm DNA fragmentation analysis using the TUNEL assay. In: *Spermatogenesis* (ed. D.T. Carrell and K.I. Aston), 121–136. Totowa, NJ: Humana Press.

Shendure, J. and Ji, H. (2008). Next-generation DNA sequencing. *Nature Biotechnology* 26 (10): 1135–1145.

Shibata, F., Hizume, M., and Kuroki, Y. (1999). Chromosome painting of Y chromosomes and isolation of a Y chromosome-specific repetitive sequence in the dioecious plant *Rumex acetosa*. *Chromosoma* 108 (4): 266–270.

Shibu, M.P., Ravishankar, K.V., Lalitha, A. et al. (2000). Identification of sex-specific DNA markers in the dioecious tree, nutmeg (*Myristica fragrans* Houtt.). *Plant Genetic Resources Newsletter* 121: 59–61.

Shirkot, P., Sharma, D.R., and Mohapatra, T. (2002). Molecular identification of sex in *Actinidia deliciosa* var. deliciosa by RAPD markers. *Scientia Horticulturae* 94 (1–2): 33–39.

Skaletsky, H., Kuroda-Kawaguchi, T., Minx, P.J. et al. (2003). The male-specific region of the human Y chromosome is a mosaic of discrete sequence classes. *Nature* 423 (6942): 825–837.

Solexa, I. (2006). Illumina: protocol for whole genome sequencing using solexa technology. In: *BioTechniques Protocol Guide*, 29. New York: BioTechniques.

Sonah, H., Deshmukh, R.K., Sharma, A. et al. (2011). Genome-wide distribution and organization of microsatellites in plants: an insight into marker development in Brachypodium. *PLoS One* 6 (6): 21298.

Sonmezoglu, O. and Ates, K. (2015). Determination of genetically modified corn and soy in processed food products. *Journal of Applied Biology & Biotechnology* 3: 32–37.

Spindel, J., Wright, M., Chen, C. et al. (2013). Bridging the genotyping gap: using genotyping by sequencing (GBS) to add high-density SNP markers and new value to traditional bi-parental mapping and breeding populations. *Theoretical and Applied Genetics* 126 (11): 2699–2716.

Sprague, G.F. (1932). The nature and extent of hetero-fertilization in maize. *Genetics* 17 (3): 358.

Srivastava, A., Raj, S.K., Kumar, S., and Snehi, S.K. (2013). New record of Papaya leaf curl virus and Ageratum leaf curl beta satellite associated with yellow vein disease of aster in India. *New Diseases Reports* 28 (6): 2044-0588.

Tang, C., Li, X., and Du, J. (2006). Hydrogen sulfide as a new endogenous gaseous transmitter in the cardiovascular system. *Current Vascular Pharmacology* 4 (1): 17–22.

Tanurdzic, M. and Banks, J.A. (2004). Sex-determining mechanisms in land plants. *The Plant Cell* 16 (1): 61–71.

Thomas, J. and Dominic, V.J. (2015). Isolation of high quality DNA from pulverized rice grain samples by tri-phase method. *International Journal of Science Research* 4 (11): 524–527.

Tilman, D., Balzer, C., Hill, J. et al. (2011). Global food demand and the sustainable intensification of agriculture. *Proceedings of the National Academy of Sciences of the United States of America* 108 (50): 20260–20264.

Troggio, M.I.C.H.E.L.A., Malacarne, G., and Vezzulli, S. (2008). Comparison of different methods for SNP detection in grapevine. *Vitis* 47 (1): 21–30.

Uemura, N., Makimura, K., and Onozaki, M. (2008). Development of a loop-mediated isothermal amplification method for diagnosing *Pneumocystis pneumonia*. *Journal of Medical Microbiology* 57 (1): 50–57.

UPOV (2011). Examining stability. TGP/11 (ed. International Union for the Protection of New Varieties of Plants), 1–3. http://www.upov.int/edocs/tgpdocs/en/tgp_11.pdf (accessed 2 March 2017).

Van Dijk, E.L., Auger, H., Jaszczyszyn, Y. et al. (2014). Ten years of next-generation sequencing technology. *Trends in Genetics* 30 (9): 418–426.

Vashisht, I., Pal, T., Sood, H., and Chauhan, R.S. (2016). Comparative transcriptome analysis in different tissues of a medicinal herb, *Picrorhiza kurroa* pinpoints transcription factors regulating picrosides biosynthesis. *Molecular Biology Reports* 43 (12): 1395–1409.

Vibhuti, C.S., Bargali, K., and Bargali, S.S. (2015). Seed germination and seedling growth parameters of rice (*Oryza sativa* L.) varieties as affected by salt and water stress. *Indian Journal of Agricultural Sciences* 85 (1): 102–108.

Von Post, R., Von Post, L., Dayteg, C. et al. (2003). A high-throughput DNA extraction method for barley seed. *Euphytica* 130 (2): 255–260.

Vos, P., Hogers, R., Bleeker, M. et al. (1995). AFLP: a new technique for DNA fingerprinting. *Nucleic Acids Research* 23: 4407–4414.

Wang, D.G., Fan, J.B., Siao, C.J. et al. (1998). Large-scale identification, mapping, and genotyping of single-nucleotide polymorphisms in the human genome. *Science* 280 (5366): 1077–1082.

William, H.M., Trethowan, R., and Crosby-Galvan, E.M. (2007). Wheat breeding assisted by markers: CIMMYT's experience. *Euphytica* 157 (3): 307–319.

Williams, J.G., Kubelik, A.R., Livak, K.J. et al. (1990). DNA polymorphisms amplified by arbitrary primers are useful as genetic markers. *Nucleic Acids Research* 18 (22): 6531–6535.

Wolyn, D.J., Borevitz, J.O., Loudet, O. et al. (2004). Light-response quantitative trait loci identified with composite interval and extreme array mapping in *Arabidopsis thaliana*. *Genetics* 167 (2): 907–917.

Wulff, E.G., Torres, S., and Vigil, E.G. (2002). Protocol for DNA extraction from potato tubers. *Plant Molecular Biology Reporter* 20 (2): 187.

Xin, Z., Velten, J.P., Oliver, M.J. et al. (2003). High-throughput DNA extraction method suitable for PCR. *Biotechniques* 34 (4): 820–826.

Xu, W.J., Wang, B.W., and Cui, K.M. (2004). RAPD and SCAR markers linked to sex determination in *Eucommia ulmoides* Oliv. *Euphytica* 136 (3): 233–238.

Xu, Y. and Crouch, J.H. (2008). Marker-assisted selection in plant breeding: from publications to practice. *Crop Science* 48 (2): 391–407.

Xu, Y., Lu, Y., Xie, C. et al. (2012). Whole-genome strategies for marker-assisted plant breeding. *Molecular Breeding* 29 (4): 833–854.

Yoon, M.S., Song, Q.J., Choi, I.Y. et al. (2007). BARCSoySNP23: a panel of 23 selected SNPs for soybean cultivar identification. *Theoretical and Applied Genetics* 114 (5): 885–899.

Younis, R.A., Ismail, O.M., and Soliman, S.S. (2008). Identification of sex-specific DNA markers for date palm (*Phoenix dactylifera* L.) using RAPD and ISSR techniques. *Research Journal of Agriculture and Biological Sciences* 4 (4): 278–284.

Yu, Q., Hou, S., and Feltus, F.A. (2008). Low X/Y divergence in four pairs of papaya sex-linked genes. *The Plant Journal* 53 (1): 124–132.

Zhang, X., Yazaki, J., and Sundaresan, A. (2006). Genome-wide high-resolution mapping and functional analysis of DNA methylation in Arabidopsis. *Cell* 26 (6): 1189–1201.

Zhang, S., Sun, S., Sun, J. et al. (2018). Identification of transgene insertions in two genetically modified soybeans using high throughput next generation sequencing. *Oil Crop Science* 3 (2): 92.

Zheng, Y. and Rosenfeld, D. (2015). Linear relation between convective cloud base height and updrafts and application to satellite retrievals. *Geophysical Research Letters* 42 (15): 6485–6491.

Zheng, X., Hoegenauer, K.A., Maeda, A.B. et al. (2015a). Non-destructive high-throughput DNA extraction and genotyping methods for cotton seeds and seedlings. *Biotechniques* 58 (5): 234–243.

Zheng, Y., Wang, Y., and Wu, J. (2015b). A dynamic evolutionary and functional landscape of plant phased small interfering RNAs. *BMC Biology* 13 (1): 32.

Zhu, Z., Zhang, F., Hu, H. et al. (2016). Integration of summary data from GWAS and eQTL studies predicts complex trait gene targets. *Nature Genetics* 48 (5): 481–487.

# 7

# Genomic Selection: Advances, Applicability, and Challenges

*Naina Garewal, Riya Joon, Ravneet Kaur, and Kashmir Singh*

*Department of Biotechnology, Panjab University, Chandigarh, India*

## 7.1 Introduction

The science propelling the major plant breeding experiments primarily concerns increasing the overall productivity along with the quality of major food, feed as well as commercial crop plants. Two main approaches established in plant breeding were that of conventional and marker-assisted breeding. The conventional breeding was carried out by hybridizing two superior trait parent varieties followed by cycles of subsequent selections over the later generations for generating elite cultivars. This approach presents several limitations as it requires long time (5–12 years) for crop development, the selection is based on phenotype, presents with higher environmental noise, and is also not very effective when dealing with complex and low heritability associated traits. Molecular-assisted breeding, on the other hand, used molecular markers to indirectly select traits of interest in crops. Marker-assisted breeding was primarily introduced to combat the limitations of conventional breeding, as it initially requires atleast the information on the phenotype during its training phase. The marker-assisted breeding in turn comprises marker-assisted selection (MAS) and genomic selection (GS). MAS uses molecular markers to identify phenotypes containing traits of interest thereby selecting plants containing the alleles with desirable traits. Traits associated with fewer quantitative trait loci (QTLs) are identified using MAS, whereas complex traits governed by minor QTLs in larger numbers still face difficulty with selection by using MAS. GS offers a better solution in handling complex traits. Meuwissen et al. (2001) first proposed this technique where it incorporates the total additive genetic variance effects along with the use of molecular markers from the entire genome. It was later applied for molecular breeding where handling of difficult quantitative traits was involved. This technique combined both the pedigree predictions used in traditional breeding relationship estimates along with genetic effect prediction with the use of various molecular markers. GS prediction model utilizes both the genotypic and phenotypic training population data to estimate genomic estimated breeding values (GEBV) for constitutive members of the breeding population (BP). The GEBV values, in turn, help us determine the performance of

*Genotyping by Sequencing for Crop Improvement*, First Edition.
Edited by Humira Sonah, Vinod Goyal, S.M. Shivaraj, and Rupesh K. Deshmukh.

a cultivar in difficult and challenging environments (Poland and Rutkoski 2016). Cross-validation involves bisecting a set of lines that are to be genotyped and phenotyped into a training set and a validation set. The training set estimates GEBV values for the validation set. This allows correlation of GEBV values with observed phenotypic values of validation set and helps determine predictive ability. Increased correlation coefficient increases the predictive ability and leads to an increase in success of prediction (Stewart-Brown et al. 2019).

With the recent reduction in the cost of next-generation sequencing (NGS) platforms, single nucleotide polymorphisms (SNP) discovery has become easy and less time consuming which has in turn given birth to genotyping by sequencing (GBS). GBS is a simple technique with a multitude of applications. It was developed at Buckler lab under Illumina NGS platform. Since then it has been applied for a large number of crop species with the objective of crop improvement.

## 7.2 Natural Selection

The threat of climate change reduces productivity in the cultivated crop varieties, it presents a major challenge for food security. There is an immediate need for climate-resilient crop varieties capable of surviving extreme climate changes. Natural wild populations of plants are easily capable of coping with these drastic climatic fluctuations owing to their adaptability through the process of natural selection. This adaptation gets passed down generations of plants over time and can be traced between different plant populations growing in differing conditions under various environmental gradients. Thus the roadmap to understanding how to adapt crop varieties to these varying climates lies in exploring how different plant populations adapt to environmental changes as well as variations as a result of natural selection. Also, the analysis of plants undergoing adaptation from different environmental conditions will eventually help identify the correlation between genetic variations associated with diverse environmental effects. The correlation between the growth of plants and local adaptations might help in uncovering different genetic loci as well as mechanisms involved in the adaptation of these plants in all the concerned environments. In some cases adapting the crops to different biotic environments becomes crucial than adapting them to varied temperature fluctuations (Henry and Nevo 2014).

The plant breeding programs began with the underlying motive of utilizing and identification of genome-wide adaptation genes pointing toward the important genetic loci to be selected while working with breeding for the generation of new varieties adaptable for changing climates. This identification can come from analyzing the wild gene pools of plant populations that undergo natural selection, thereby generating adaptive paths important in the evolution of plant genomes. Evolutionary patterns of convergent or parallel evolution can also be determined based on adaption to natural selection with different climatic conditions. Local adaptation to climatic variations might be a result of either a parallel mutation where the same mutations get selected independently or because of convergent evolution where only mutations having the same impact on the phenotype get chosen and thrive with time. There is, however, also a need to prioritize the conservation of wild populations along with avoiding the gene flow from domesticated to wild crop varieties in order to avoid any anomalies while studying natural selection (Henry and Nevo 2014).

Identifying the key adaptation genes can aid in targeted selection facilitating rapid gene introduction from wild populations into domesticated crop varieties. This can be useful in ensuring resilience from wild crops to be retained while undergoing selection for domestication traits. The breeding of crops with better resilience to differing climates can also be achieved by focussing on genome plasticity genes of the plant, as they can help in radical phenotype alteration in plants in response to changing environments. Whole-genome epigenome analysis also plays an important role in understanding adaptability as a result of natural selection and should therefore be explored. Mechanisms governing the hybridization and performance of polyploidy plant of wild populations can help demonstrate the contribution of heterosis or polyploidy to adaptive variations in climates (Henry and Nevo 2014).

## 7.3 Breeding Selection

With the advent of classical breeding during the green revolution of the 1970s, there were drastic crop improvements seen with respect to the development of high-yielding and nutrient-responsive breeding varieties. It has played a major role in the generation of most of the modern-day cultivars of almost all the major crop varieties with significant gains in crop productivity. The yearly genetic gain was pushed by 1% which was still not sufficient to keep up with the growing demand of 2% genetic gain required per year (Bhat et al. 2016).

Conventional breeding involves selection of cultivars based on their phenotypes where every individual plant was required to be phenotyped using pedigree-based methods such as best linear unbiased prediction (BLUP) in order to predict individual breeding values (BV) based on their pedigree. Conventional selection has been demerited due to slow progression in the generation of breeding lines as well as the need for extensive phenotypic evaluations (Hawkins and Yu 2018). It is also less effective for low heritability complex multigenic traits such as yield, quality, biotic, and abiotic stresses with a huge influence from the environment based on Gene × Environment (G × E) effects. In addition, the phenotypic selection is time-consuming, laborious, cost-ineffective, requires large land area, population size-dependent along with low precision as well as reliability. Thus the development of more efficient, reliable, and less time-consuming systems was the need of the hour for developing climate-resistant and high-yielding varieties. Therefore to overcome these challenges marker-based selection strategies were introduced in plant breeding. MAS and eventually GS was developed, and they relied on reduced phenotyping and selection has been carriedout based on marker or phenotypic profiling. GEBV calculated in GS is derived by combining various useful loci and serves as a direct estimation for the likelihood of generating a superior phenotype (Bhat et al. 2016).

In comparison to conventional phenotypic breeding GS leads to shorter generation cycles along with increasing the genetic gain per year for the concerned crop varieties. Moreover, for complex trait analysis usually requiring longer generation times, GS offers a cheaper and easier alternative and can help in analyzing and characterizing more candidates thereby providing improved selection intensity. Besides these advantages GS based high-density genetic markers are available at a comparable cost to that of conventional phenotyping-based breeding selection (Bhat et al. 2016).

## 7.4 Marker-assisted Selection

MAS typically identifies a small marker set with maximum gravity towards the trait of interest and further selects candidates based on fixation of single trait associated with the desirable alleles (Hawkins and Yu 2018). Commonly used MAS strategies include marker-assisted backcrossing with both foreground and background selection, enrichment of favorable alleles in the earlier generations, quantitative trait selection on multiple loci using multiple markers as well as selection across multiple cycles (Bassi et al. 2016).

Thus MAS is a technique where phenotypic selection is carried out based on the genotype of a marker. MAS helps to overcome difficulties faced in conventional plant breeding and has upgraded the standards governing cultivar selection. MAS is widely used for the identification of dominant or recessive alleles along with plant generations to identify the most favorable progeny among various segregating populations. Important steps in MAS constitute population development, phenotyping, genotyping, QTL mapping, and linkage map construction, QTL validation, and marker validation (Nadeem et al. 2018).

MAS-based schemes are primarily categorized into marker-assisted backcrossing, gene pyramiding, marker-assisted recurrent selection, and GS. Marker-assisted backcrossing is backcrossing utilizing molecular markers which are selected based on their genome distribution, haplotype diversity, and polymorphism indices, their association with candidate genes, and agronomical traits. It occurs in three levels with the first level involving foreground selection where gene or QTL are screened in combination or as a substitute by using markers. The second level includes recombinant selection where backcross progeny possessing target or QTL genes is selected for performing recombination by linking flanking markers and target locus thereby reducing the size of the donor chromosome segment. The third level involves background selection where backcross progeny possessing a large amount of recurrent parental genome gets selected using markers unlinked to the target locus (Nadeem et al. 2018).

Marker-assisted recurrent selection endorses the application of molecular markers in each generation thereby targeting all the traits of interest. In this process, selected individuals are crossed for performing crossing at every crossing and selection cycle. This technique primarily focuses on the improvement of F2 generation by undergoing one cycle of MAS (utilizing phenotypic data with marker scores) followed by 2–3 cycles performing marker-based selections. This technique does not require any prior knowledge of QTL's and its selection depends on associations established between a marker and a trait. Maker-assisted gene pyramiding, however, involves the transfer of multiple QTL genes for one or multiple traits into a cultivar lacking those traits. This technique mainly increases the level of resistance to specific diseases and insects, as its selection focuses on two or more genes simultaneously being selected (Nadeem et al. 2018).

Therefore, we can categorize traits that can be addressed by MAS in plant breeding into four broad catogories i.e. (a) traits of difficult management by phenotypic selection since they are expensive, time-consuming, or have low penetrance and complex inheritance; (b) traits with their selection depending on specific environments or developmental stages influencing the expression of their target phenotypes; (c) maintaining recessive alleles during backcrossing or trying to speed up the backcross breeding; (d) pyramiding of multiple monogenic traits (Xu and Crouch 2008). Many breeding programs require a combination of traits to be introgressed or pyramided into the crops to generate crops possessing diverse resistances along with improved crop performance and yields derived in a shorter time frame than in conventional breeding. MAS in contrast to conventional breeding offers the potential to accumulate target characteristics in the same genotype with fewer unintentional losses and fewer selection cycles (Xu and Crouch 2008). In MAS, the opportunities for upgrading and working with complex traits such as abiotic stress tolerance are limited by low heritability, multiple gene involvement with unpredictable epistasis along additional impacts from diverse environmental factors. These traits are still challenging and periodically requires continuous customizations (Xu and Crouch 2008).

## 7.5 Genomic Selection

GS encompasses the utilization of a wide range of dense markers to predict the BV of an individual. The deployment of GS includes setting up a statistical model consisting of a test population (TP) by exercising the knowledge of its genotypic and phenotypic data. Calibration of the statistical model will further help in the prediction of genotype or phenotype values pertaining to the selection of nonphenotyped candidates. These candidates with known genotypes but unknown phenotypes comprise the BP and therefore TP is used for analyzing BP trait performance based on the phenotypic identity among the allelic loci. GEBV is determined based on BP as it confers a direct estimation of the phenotypic superiority of a breeding variety which in turn helps in the selection of parents for a new breeding cycle (Figure 7.1).

The GEBV estimation, in turn, reduces the duration of a breeding cycle as it relies only on a plant's genotype which can be estimated when the plant is young and thus it removes the waiting time on phenotyping of quantitative traits in the late filial generations. Validation population (VP) is a population that gets both genotyped and phenotyped to procure a GEBV value which in turn dictates the accuracy of a GS model on the pretext of its correlation with the actual phenotypic value (Bassi et al. 2016). Whilst selection for breeding can be done based on GEBVs, but commercially selling these varieties calls in need for a face value as an estimation of their market value. This is in turn evaluated via the

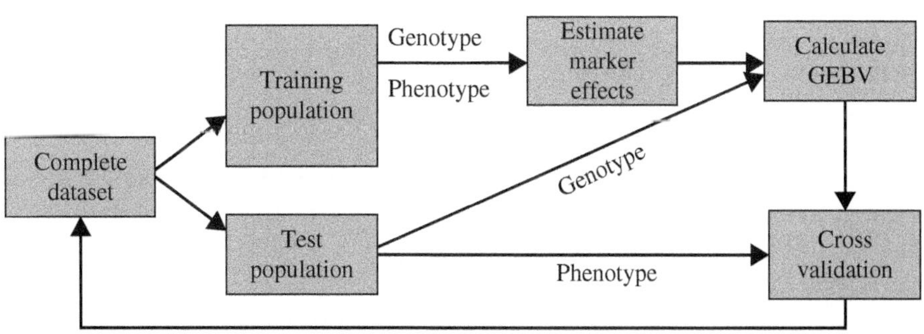


**Figure 7.1** Overview of estimate marker effects in order to get a genomic estimated breeding value (GEBV) of lines in the test population under genomic selection. *Source:* The figure is reproduced from Robertsen et al. (2019) available with CC-BY-4.0.

genetic gain which evaluates the performance surges generated via selection programs in the breeding varieties. Thus, the expected genetic gain from GS per unit time will be:

$$\Delta G = i\sigma_A r_{MG} / t$$

where $\Delta G$ is the expected genetic gain, $i$ is the intensity of the selection, $\sigma_A$ is genetic SD (the square root of additive genetic variance), $r_{MG}$ is the selection accuracy (measured by correlating GEBVs and the BVs).

Genetic gain can, in turn, be inflated by several means, i.e. increasing selection intensity, raising the selection accuracy or heritability, amping up the genetic SD, and narrowing of the breeding cycle timing (Xu et al. 2020).

GS uses all the molecular markers for genomic prediction in contrast to QTL and association mapping. The main advantages of using GS over phenotypic selection are that it reduces costs by up to 50% along with saving the time required for selecting the lines as it bypasses the selection process directly to stage II during a variety development (Crossa et al. 2017).

BVs generated for candidate selection with GS are predicted firstly by calculating the parental average derived from assessing the average of the mean BV of both parents along with performance deviation of the progeny from this value as a result of Mendelian sampling. The Mendelian sampling consequently accounts for the quantification of progeny tested in multienvironmental field trials to address the within-family variations. With GS the Mendelian sampling gets quantified as it utilizes dense markers thereby avoiding extensive phenotyping of the progeny, which in turn saves time, enhances selection response per unit time as well as the expected genetic gain thereby saving resources. GS can be used both to improve complex traits with low heritability as well as increasing accuracy for simple traits with heritability higher than complex traits. Two limiting factors of GS, i.e., cost of genotyping and unclear guidelines concerning the use of GS in a breeding program. GS primarily focuses on the prediction of additive effects in early generations to establish a rapid selection cycle in a short time frame; since only BV prediction is the primary goal in GS, linear additive models summarizing the effects of markers are sufficient in this case. But GP involves the prediction of complete genetic values with both the additive and nonadditive effects, for attesting toward the commercial value of the cultivars (Crossa et al. 2017).

GS presumes the additive effects as mediators behind the genetic variance of a trait. Fisher's infinite model was foundational behind this assumption that multiple genes contribute toward a trait variance in a population. Genomic best linear unbiased prediction (G-BLUP) models are hinged on fisher's infinite model and are extensively used in GS. G-BLUP modeling also contributes to significantly improve the prediction accuracy by integrating G × E interactions. With fast-paced improvements, new computationally efficient prediction models based on Bayesian methods and Markov chain Monte Carlo were developed to incorporate nonadditive effects, combined multitrait effect on multiple environments, and the conditional distribution of trait parameters (He and Li 2020).

For better prediction accuracy of GS, a large reference population encompassing whole-genome high-density markers is crucial. Both the reference population and the subsequent prediction models should be adaptive and updated with respect to the density of the markers and their corresponding phenotypes (He and Li 2020).

GS uses a diverse variety of statistical models to estimate its BVs. Based on the type of trait, assessment required models are either used for analyzing quantitative or qualitative inheritance. G-BLUP and Ridge-regression best linear unbiased prediction (RR-BLUP) are the two most popularly used quantitative inheritance models. G-BLUP model is a mixed linear model, with individuals as random effects, and covariance is assumed to be proportional to the genomic relationship matrix generated using genome-wide markers. In this model, close relatives are weighed more heavily. G-BLUP is modified BLUP where instead of genomic relationships, pedigree was used. Both RR-BLUP and G-BLUP assume the conditioning of a large number of additive loci. Wimmer et al. (2013) compared the four of the commonly used GS models: RR-BLUP and BayesB (Meuwissen et al. 2001), LASSO (Tibshirani 1996), and the elastic net (Zou and Hastie 2005). RR-BLUP was highly recommended for crop species since it was easy to use and quick comparatively since Bayesian selection models assume reduction in several markers to help explain genetic variance about traits. Bayesian models such as BayesA, BayesB (Hayes and Goddard. 2001), BayesCπ (Habier et al. 2011), and Bayesian LASSO (Park and Casella 2008). Bayes A assumes each marker to have a unique variance, Bayes B assumes unique variance but allows some markers to have no effect. BayesCπ assumes equal variance of all markers but with a certain proportion of markers with no effect. In Bayesian LASSO markers have equal variance but shrinkage affects smaller loci more than those of larger loci (Poland and Rutkoski 2016).

Nonlinear models are those which consider the nonadditive effects of genes. These are Reproducing Kernel Hilbert Space (RKHS), Random Forest (RF), and neural networks-based models. RKHS is similar to G-BLUP, but its relationship matrix is defined as squared Euclidean distance divided by parametric scale value influencing how quickly genomic covariance decays with distance. RF on the other hand is a machine learning program and uses a series of regression trees as model-training set (Poland and Rutkoski 2016).

## 7.6 Genotyping for Genomic Selection

GBS is employed for genomic breeding by identification of various SNP markers (from a TP) which are in turn used to identify SNP and marker–trait associations as this helps predict GEBV values in GS. Currently, GBS is the most sought-after reduced representation approach for GS. This approach takes advantage of restriction enzymes such as methylation-sensitive enzymes to trim the genome at fixed regions thereby deploying to genotyping of the fragmented restriction sequenced ends. GBS incorporates genotyping of multiple polymorphic sites, it not only incorporates all marker–trait loci but also helps identify a large number of little-effect loci (He and Li 2020).

As depicted by Elshire et al. 2011 a typical GBS includes Genomic DNA extraction of each individual under study, digestion of DNA using one or two of the commonly used restriction enzymes i.e. ApeKI and EcoT22I depending on the required coverage. A unique barcoded adaptor is ligated to the ends of individual digested samples prior to de-multiplexing these samples for in silico analysis. Following sample combination and PCR amplification, the samples are sequenced using a high-throughput NGS platform such as Illumina's HiSeq platform which uses sequencing primer binding site common with adaptors. Single-ended 100 bp sequencing is the most common. Further various polymorphic loci consisting of SNP's and multiple nucleotide polymorphisms (MNP's) are identified called genotypes based on the coverage. The coverage of 48–60× defines gene dosage at the side but a call of only 15× is needed to determine its presence or absence. This data is further utilized for GS using phenotypic data of sampled individuals to identify variant loci linked to traits of interest. Due to lesser genomic coverage, most polymorphisms are more likely to be in linkage disequilibrium (LD) with the causal loci. Since linkage is used to estimate the BVs, throughout the breeding period, it will also lead to recombinations among the causal and marker loci which will, in turn, reduce the efficacy of the prediction models. Thus there is a perpetual requirement for upgradation throughout the selection procedure. GBS-based marker sets along with associated linkage maps offer a better option than transcriptome-based sets as they are not limited to transcribed regions. Although less economical whole-genome sequencing will still provide more information in comparison (Hawkins and Yu 2018).

Genotyping by sequencing is a powerful tool that allows rapid, high-resolution mapping of genetic variation for important agronomical traits. High-throughput genotyping helped in target-oriented variant enrichment, which in turn led to rapid phenotypic prediction using genotyping of a large number of hybrid/inbred lines. Important agronomical traits undergo targeted variant enrichment using a reduced representation sequencing approach. Reduced-representation sequencing not only foreshortens the sequencing cost but also facilitates the screening of a large number of phenotypes for the desired trait. Reduced-representation sequencing comprises exome capture or transcriptome sequencing enabling targeted recognition of desired SNP's from the protein-coding regions of the genome. Region-specific targeting of chromosome regions can be done using custom-designed capture probes that will target specific regions bearing candidate genes for the desired traits. These reduced representation sequencing strategies may, however, miss a few genetic variations since they do not incorporate intergenic regions which have an essential role in regulating trait variations (He and Li 2020).

Soybean was the first crop where GS was utilized to upgrade its agronomical aspects using genotyping by sequencing in the breeding program (Sonah et al. 2015), subsequently, other crops reported utilizing GBS were pea (Annicchiarico et al. 2017), soybean (Jarquín et al. 2014), chickpea (Verma et al. 2015), groundnut (Ho et al. 2017), and pigeon pea (Saxena et al. 2017).

GBS studies in some crop plants along with the prediction models used and their prediction accuracy are shown in Table 7.1.

**Table 7.1** Table showing different crop plants where GBS was used (Adapted from Bhat et al. 2016).

| S. No. | Name of crop | Traits analyzed | Prediction accuracy | Statistical model (software used) | References |
|---|---|---|---|---|---|
| 1. | Rice | Grain yield, flowering time | 0.31–0.63 | RR-BLUP (R package rrBLUP) | Spindel et al. (2015) |
| 2. | Wheat | Grain yield | 0.28–0.45 | BLUP (ASReml 3.0) | Poland et al. (2012) |
| 3. | Wheat | Grain yield, plant height, heading date, preharvest sprouting | 0.54 | BLUP (R package rrBLUP) | Heslot et al. (2013) |
| 4. | Wheat | Stem rust resistance | 0.61 | G-BLUP B (R package GAPIT) | Rutkoski et al. (2014) |
| 5. | Wheat | Yield and yield-related traits | 0.20–0.59 | BLUP (rrBLUP version 4.2) | Isidro et al. (2015) |

**Table 7.1** (Continued)

| S. No. | Name of crop | Traits analyzed | Prediction accuracy | Statistical model (software used) | References |
|---|---|---|---|---|---|
| 6. | Wheat | Fusarium head blight resistance | 0.4–0.90 | RR-BLUP (R-package GAPIT) | Arruda et al. (2016) |
| 7. | Wheat | Grain yield, protein content, protein yield | 0.19–0.51 | RR-BLUP (R-package rrBLUP) | Michel et al. (2016) |
| 8. | Wheat | Grain yield | 0.50 | G-BLUP (R-package rrBLUP) | Lado et al. (2016) |
| 9. | Wheat | Grain yield, Fusarium head blight resistance, softness equivalence, and flour yield | 0.35–0.62 | BLUP (BGLR R-package) | Hoffstetter et al. (2016) |
| 10. | Wheat | Heat and drought stress | 0.18–0.65 | G-BLUP (BGLR R-package) | Crossa et al. (2016) |
| 11. | Maize | Grain yield, anthesis date, anthesis-silking interval | 0.62 | PGBLUP, PRKHS (R software) | Crossa et al. (2013) |
| 12. | Maize | Drought stress | 0.40–0.50 | G-BLUP (BGLR R-package) | Zhang et al. (2015) |
| 13. | Soybean | Yield and other agronomic traits | 0.43–0.64 | G-BLUP (MissForest R package, TASSEL 5.0) | Jarquín et al. (2014) |
| 14. | Alfa alfa | Biomass yield | 0.66 | BLUP (R package, TASSEL software) | Li et al. (2015) |
| 15. | Alfa alfa | Biomass yield | 0.50 | SVR (R package rrBLUP, R package BGLR, R package RandomForest) | Annicchiarico et al. (2015) |
| 16. | Switch grass | Biomass yield | 0.52 | BLUP (glmnet R package, R package rrBLUP) | Lipka et al. (2014) |
| 17. | Grapevine | Yield and related traits | 0.50 | RR-BLUP (R package BLR, R package rrBLUP) | Fodor et al. (2014) |
| 18. | Intermediate Wheatgrass | Yield and other agronomic traits | 0.67 | RR-BLUP (R package rrBLUP, BGLR R-package) | Zhang et al. (2016) |
| 19. | Perennial ryegrass | Plant herbage dry weight and days-to-heading | 0.16–0.56 | RR-BLUP (R software) | Faville et al. (2016) |
| 20. | Pea | Grain yield under severe terminal drought | 0.5 | BL/RR-BLUP (R-package) | Annicchiarico et al. (2017) |

*Source:* Adapted from Bhat et al. (2016).

## 7.7 Integration of Genomic Selection in MAS Program

MAS led to the initial incorporation of molecular markers into the traditional breeding programs. MAS deploys a selection of simple quantitative traits based on quantitative trait loci associated markers only showcasing major effects. It is difficult to find the same QTL across diverse environments due to interactions between QTL and environment or in candidates with dissimilar genetic backgrounds. This has led to unsuccessful outcomes while dealing with the improvement of complex traits. The identification and quantification of rare economically important trait QTLs with small effects but higher environmental impacts pose a major challenge for association mapping and QTL detection studies. However, with the reduction in cost prices of SNP assays, identification of more QTLs became feasible using high-density SNP arrays. These, in turn, drove breeders to generate improved statistical models with better prediction accuracy concerning marker–trait associations (Crossa et al. 2017).

GS on the other hand is an advanced form of MAS as it predicts the genetic values of selected candidates based on their GEBV values. GEBV values are predicted using the high-density markers which are distributed across the genome. GEBV prediction model combines both phenotypic and pedigree data to increase its prediction accuracy. In comparison to MAS, GEBV prediction utilizes all markers including minor and major marker effects. This technique selects markers covering the entire genome and incorporates that all QTLs in LD with at least a single marker. High-throughput phenotyping and GS of complex traits have revolutionized this technique by increasing the selection accuracy (Nadeem et al. 2018). The comparison of the MAS and the GS has been illustrated in Figure 7.2.

## 7.8 The Efficiency of Genomic Selection for Complex Traits

The multitrait GS comprises four main strategies: (i) tandem selection (independently selects different traits in different generations); (ii) independent culling (involves simultaneous truncated selections on multiple traits with independent thresholds); (iii) index selection (involves multiple trait selection based on an index constructed by a linear combination of multiple traits); (iv) mate selection (involves the selection of multiple traits simultaneously, based on Pareto optimal solutions of mate selection index). Tandem selection lacks simultaneous multiple trait selection, it helps in early generational trait selections. Whereas independent culling can achieve simultaneous selections, it exhibits sensitivity toward truncation points of different traits. Also while index selection is a widely used and important method for the generation of superior phenotypes, mate selection considers different constraints and breeding goals for multiple traits and evaluates those criteria based on a proposed set of matings. Thus the changes in the optimum selection index with each generation will vary based on the number of traits under study (Moeinizade et al. 2020). Genomic complexity for gene and environment interactions concerning multiple traits should use statistical-genetic models exploiting multitrait, variance-covariance of multiple environments, and genetic environmental correlations between individual traits as well as between environments and traits, simultaneously (Crossa et al. 2017).

GS has also been used in the selection of simultaneously agronomical and essential but negatively correlated traits such as grain yield and protein content and it helped to generate

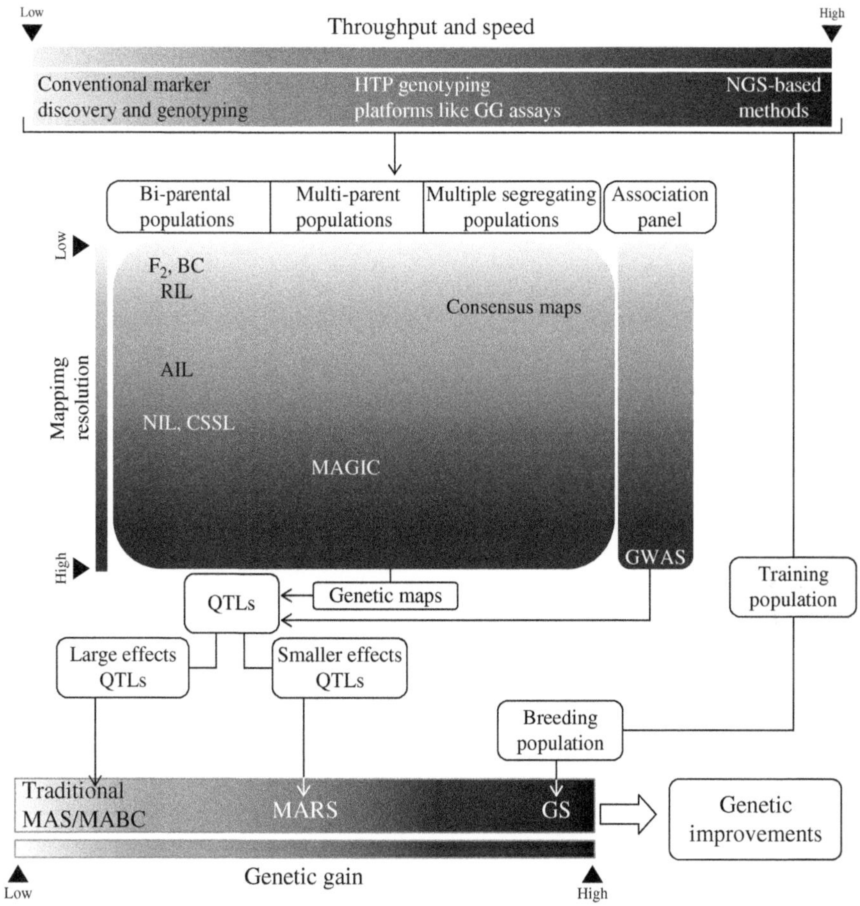


**Figure 7.2** The methodology involved in marker-assisted section (MAS) and genomic selection (GS). MARS, marker-assisted recurrent selection; MABC, marker-assisted backcrossing; GWAS, genome-wide association study. *Source:* The figure is reproduced from Bohra et al. (2014) available with CC-BY-4.0.

varieties displaying superior characteristics in wheat. GS also exhibits selective advantage for improving the accuracy of phenotypic traits with low heritability such as grain yield, in contrast to traits with high heritability, such as grain weight, disease resistance as well as plant height. This is primarily due to marker profiling cost to genetic-gain ratio being too low in comparison to direct phenotypic selection (He and Li. 2020).

An efficient and accurate GS model requires an establishment of a large, adaptable, and updatable reference population that is genetically diverse in desired traits so that it lies within the variation patterns of genotypes and phenotypes to be analyzed. A large germplasm population can not only serve as a training population for prediction models but also aids in the characterization of large-scale germplasm accessions to uncover elite genetic variants for the generation of superior cultivars. Higher accuracy in GS can be achieved by a large germplasm collection along with maximizing the phenotypic variance in the training population (He and Li 2020).

Phenotypic selection using multiple agronomic traits has been a limiting bottleneck as it limits the power and scale of breeding. High-throughput phenotyping platforms (HTTPs) offer a significant breakthrough in regards to it as they allow precise phenotypic variation measurements of various developmental stages in a nondestructive way (He and Li 2020).

## 7.9 Integration of Genomic Selection in the Varietal Trial Program

Improvement of desirable traits such as higher yield and biotic and abiotic stress tolerance using conventional as well as MAS is challenging. GS thus provides new opportunities for increasing the efficiency of crop breeding programs, since it fixes all genetic variation and can accurately select individuals with higher BVs without the requirement of collecting all phenotypes of individuals. Thus reducing the breeding cycle and early-stage introgression of desired traits and is believed to help in combating the global hunger crisis. GS has been integrated into several crop breeding programs such as rice (Xu et al. 2018; Zeng et al. 2017), wheat (Haile et al. 2018; Larkin et al. 2019), maize (Mastrodomenico et al. 2019; Beyene et al. 2015; Allier et al. 2020), barley (Bhatta et al. 2020), millets (Liang et al. 2018), banana (Nyine et al. 2018), cacao (McElroy et al. 2018), carrot (Corak et al. 2019), cassava (de Oliveira et al. 2012), soybean (Jarquín et al. 2014; Xavier et al. 2016), strawberry (Gezan et al. 2017), etc. Some of these have been compared, see Table 7.2.

**Table 7.2** Table showing various GS studies carried out in different crops (Adapted from Lin et al. 2014).

| S. No. | Name of crop | Traits studied | Prediction accuracy | Statistical model | Markers used | References |
|---|---|---|---|---|---|---|
| 1. | Alfalfa | Biomass yield | 0.28–0.66 | RR-BLUP | SNPs | Li et al. (2015) |
| 2. | Apple | Flowering time, astringency | 0.43 | RR-BLUP, Bayes-LASSO | SNPs | Kumar et al. (2012) |
| 3. | *Arabidopsis thaliana*, Barley, Maize | Root lodging, flowering time, dry matter yield, grain yield, stalk lodging | 0.74, 0.88, 0.94, 0.6, 0.26 | GBLUP, Bayes-A,B | SSR's, SNP's, RFLP's | Lorenzana and Bernardo (2009) |
| 4. | *Arabidopsis thaliana* | Plant width | ~0.45 | LASSO, the elastic net, Bayes-B, RR-BLUP | SNP's | Wimmer et al. (2013) |
| 5. | Barley | Soluble solids, deoxynivalenol | $0.89^2$, 0.5–0.7 | RR-BLUP, Bayes-$C_{II}$, LASSO | SNP's | Lorenz et al. (2012) |
| 6. | Barley | Grain yield | ~0.6 | Bayes-LASSO | SNP's | Heslot et al. (2013) |
| 7. | Canola | Flowering time | 0.64 | RR-BLUP | DArTs | Raman et al. (2016) |
| 8. | Cassava | Dry matter content, shoot weight | 0.67, 0.83 | RR-BLUP | SNP's | de Oliveira et al. (2012) |

**Table 7.2** (Continued)

| S. No. | Name of crop | Traits studied | Prediction accuracy | Statistical model | Markers used | References |
|---|---|---|---|---|---|---|
| 9. | Eucalyptus | Kernel weight, pulp yield | 0.3, 0.55 | RR-BLUP | DArTs | Resende et al. (2012a, b, c) |
| 10. | Loblolly pine | Height growth, diameter breast height, total height | 0.79, 0.65–0.75 | RR-BLUP | SNP's | Resende et al. (2012a, b, c) |
| 11. | Loblolly pine | Lignin | 0.17 | RR-BLUP, Bayes A, $C_{II}$, LASSO | SNP's | Resende et al. (2012a, b, c) |
| 12. | Maize | Grain yield, male-flowering time | 0.42, 0.79 | RR-BLUP, RKHS, Bayes-LASSO | SNP's | Crossa et al. (2010) |
| 13. | Maize | Dry yield matter, dry matter content | 0.5-0.7, 0.6-0.8 | GBLUP | SNP's | Albrecht et al. (2011) |
| 14. | Maize | Northern corn leaf blight, grey leaf spot | ~0.5 | Bayes-LASSO, RKHS | SNP's | Crossa et al. (2011) |
| 15. | Maize | Grey leaf spot, male flowering time | 0.2–0.5, 0.82 | Bayes-LASSO, RBFNN, RKHS | SNP's | González-Camacho et al. (2012) |
| 16. | Maize | Anthesis-silking interval | 0.26 | RR-BLUP, Bayes-A, B | SNP's | Guo et al. (2012) |
| 17. | Maize | Sugar content | 0.69 | RR-BLUP | SNP's | Riedelsheimer et al. (2012) |
| 18. | Maize | Kernels per row, Kernel rows | 0.3–0.6, 0.7–0.9 | GBLUP | SNP's | Riedelsheimer and Melchinger (2013) |
| 19. | Maize | Northern corn leaf blight | 0.7 | GBLUP | SNP's | Technow et al. (2013) |
| 20. | Maize | Grain yield | 0.5-0.6 | RR-BLUP | SNP's | Zhao et al. (2013) |
| 21. | Maize | Protein content, oil content | 0.29, 0.42 | GBLUP | SNP's | Guo et al. (2014) |
| 22. | Maize | Ear rot disease resistance | 0.25–0.59 | RR-BLUP | DArTs | dos Santos et al. (2016) |
| 23. | Miscanthus | Phenology, biomass, cell wall composition traits | 0.57 | BLUP | RADs | Slavov et al. (2014) |
| 24. | Oats | Yield, β-glucan | 0.15–0.3, 0.35–0.5 | RR-BLUP, Bayes-$C_{II}$ | DArTs | Asoro et al. (2011) |
| 25. | Rice | Flowering time, seed length | 0.46–0.59, 0.75–0.78 | LASSO, the elastic net, Bayes-B, RR-BLUP | SNP's | Wimmer et al. (2013) |
| 26. | Rice | Yield, apiculus color | 0.16, 0.98 | LASSO | SNP's | Xu. (2013) |
| 27. | Rice | Oil content, protein content, seed width | 0.42, 0.44, 0.84 | GBLUP | SNP's | Guo et al. (2014) |

*(Continued)*

**Table 7.2** (Continued)

| S. No. | Name of crop | Traits studied | Prediction accuracy | Statistical model | Markers used | References |
|---|---|---|---|---|---|---|
| 28. | Rice | Grain yield, plant height | 0.54 | GBLUP, RR-BLUP | DArTs | Grenier et al. (2015) |
| 29. | Rice | Grain yield per plant, thousand-grain weight, panicle number per plant, plant height, secondary branch number, grain number per panicle, panicle length, and primary branch number | ~0.5 | GBLUP, LASSO, BayesB, PLS, SVM, and RKHS | SNP's | Xu et al. (2018) |
| 30. | Sugar beet | Nitrogen content, white sugar yield | 0.5, 0.9 | RR-BLUP | SNP's | Würschum et al. (2013) |
| 31. | Wheat | Grain yield | 0.48–0.61 | RR-BLUP, RKHS, Bayes-LASSO | DArTs | Crossa et al. (2010) |
| 32. | Wheat | Grain yield | 0.43–0.79 | Bayes-LASSO, RKHS | DArTs | Crossa et al. (2011) |
| 33. | Wheat | Grain yield, flour yield | 0.48, 0.76 | RR-BLUP, Bayes-A,B, $C_{II}$ | DArTs | Heffner et al. (2011) |
| 34. | Wheat | Grain yield | 0.5–0.6 | Bayes-LASSO | DArTs | Burgueño et al. (2012) |
| 35. | Wheat | Date to heading, grain yield | 0.5–0.6, 0.6–0.7 | RR-BLUP, Bayes-A, LASSO, RKHS, RBFNN, BRNN | DArTs | Pérez-Rodríguez et al. (2012) |
| 36. | Wheat | Grain yield, 1000-kernel weight | 0.2–0.4, ~0.3 | GBLUP | DArTs, SNP's | Poland et al. (2012) |
| 37. | Wheat | Heading date, Fusarium head blight | ~0.3, ~0.6 | RR-BLUP, Bayes-LASSO, RKHS, RF | DArTs, SSR's | Rutkoski et al. (2012) |
| 38. | Wheat | Stem rust, stripe rust | 0.2–0.4 | GBLUP, Bayes-R | SNP's | Daetwyler et al. (2013) |
| 39. | Wheat | Grain yield | ~0.2 | GBLUP | SNP's | Dawson et al. (2013) |
| 40. | Wheat | Yield, date to heading | 0.43–0.51, 0.60–0.66 | LASSO, the elastic net, Bayes-B, RR-BLUP | SNP's | Wimmer et al. (2013) |
| 41. | Wheat | Grain yield | 0.3–0.6 | RR-BLUP, Bayes-A, B, C, $C_{II}$ | SNP's | Zhao et al. (2013) |
| 42. | Wheat | Grain yield | 0.27–0.36 | G-BLUP | DArTs | Saint Pierre et al. (2016) |

*Source:* Adapted from Lin et al. (2014).

## 7.10 Cost Comparison of GS vs MAS

Molecular marker technology has the potential to reshape breeding programs and facilitate rapid gains from the selection. MAS is effective to carry out manipulation of large-effect alleles with known association to a marker; however, many alleles with small effects and no substantial reliable effects are not identified using MAS. One of the major weaknesses of using traditional MAS is the way MAS splits the task into two components, i.e. first identifying QTL and then estimating their effects. On the other hand, the scope of GS in breeding has been very high in recent years. GS has the potential for higher rates of genetic gain over traditional selection. The cost involvement in GS is reduced considerably due to advancement in SNP genotyping. As the accuracy of selection and the increase in genetic gain is high, the cost encountered by GS research work is nominal. "It is this incredible development of the genotyping technology that makes GS feasible" (Meuwissen, 2007). In other words, the most important criteria for GS to work is the accessibility of low-cost, reasonable, and high-density marker systems. The development of NGS technologies has resulted in sequencing of an increased number of genomes hence providing low-cost and high-density SNP markers for genotyping. These advanced marker technologies have aided in the determination of population structure of training as well as validation set, and have also enhanced the selection accuracy of GS. Furthermore, the genomic sequencing cost is expected to further decrease in the future, which will make the whole genome sequencing a cost-effective process leading to increased prediction accuracy of GS. Till that time, targeted sequencing seems to be a more cost-effective option for large-scale marker discoveries and GS, particularly for those cases which have larger and un-decoded genomes. Several targeted marker discovery techniques which utilize NGS are reduced-representation library generation, complexity reduction of polymorphic sequences. restriction-site associated DNA sequencing (RAD-seq), double digest RADSeq (ddRADSeq), 2bRADSeq, ezRADSeq, DArTSeq, sequence-based polymorphic marker technology, multiplexed shotgun genotyping (MSG), genome reduction on restriction site conservation (GR-RSC), genotyping by sequencing (GBS), molecular inversion probe, solution hybrid selection, and microarray-based GS can be utilized even in the absence of prior WGS knowledge (Bhat et al. 2016). Since these NGS-based technologies provide for genome-wide coverage with higher precision at a very low cost per data point they are an efficient choice over MAS and other conventional breeding programs.

## References

Albrecht, T., Wimmer, V., Auinger, H.J. et al. (2011). Genome-based prediction of testcross values in maize. *Theoretical and Applied Genetics* 123 (2): 339.

Allier, A., Teyssèdre, S., Lehermeier, C. et al. (2020). Genomic prediction with a maize collaborative panel: identification of genetic resources to enrich elite breeding programs. *Theoretical and Applied Genetics 133* (1): 201–215.

Annicchiarico, P., Nazzicari, N., Li, X. et al. (2015). Accuracy of genomic selection for alfalfa biomass yield in different reference populations. *BMC Genomics* 16 (1): 1020.

Annicchiarico, P., Nazzicari, N., Pecetti, L. et al. (2017). GBS-based genomic selection for pea grain yield under severe terminal drought. *The Plant Genome 10* (2): 1–13.

Arruda, M.P., Brown, P., Brown-Guedira, G. et al. (2016). Genome-wide association mapping of Fusarium head blight resistance in wheat using genotyping-by-sequencing. *The Plant Genome 9* (1): 1–14.

Asoro, F.G., Newell, M.A., Beavis, W.D. et al. (2011). Accuracy and training population design for genomic selection on quantitative traits in elite North American oats. *The Plant Genome 4* (2): 132–144.

Bassi, F.M., Bentley, A.R., Charmet, G. et al. (2016). Breeding schemes for the implementation of genomic selection in wheat (Triticum spp.). *Plant Science 242*: 23–36.

Beyene, Y., Semagn, K., Mugo, S. et al. (2015). Genetic gains in grain yield through genomic selection in eight bi-parental maize populations under drought stress. *Crop Science 55* (1): 154–163.

Bhat, J.A., Ali, S., Salgotra, R.K. et al. (2016). Genomic selection in the era of next generation sequencing for complex traits in plant breeding. *Frontiers in Genetics 7*: 221.

Bhatta, M., Gutierrez, L., Cammarota, L. et al. (2020). Multi-trait genomic prediction model increased the predictive ability for agronomic and malting quality traits in Barley (*Hordeum vulgare* L.). *G3: Genes, Genomes, Genetics 10* (3): 1113–1124.

Bohra, A., Pandey, M.K., Jha, U.C. et al. (2014). Genomics-assisted breeding in four major pulse crops of developing countries: present status and prospects. *Theoretical and Applied Genetics* 127 (6): 1263–1291.

Burgueño, J., de los Campos, G., Weigel, K. et al. (2012). Genomic prediction of breeding values when modeling genotype× environment interaction using pedigree and dense molecular markers. *Crop Science 52* (2): 707–719.

Corak, K.E., Ellison, S.L., Simon, P.W. et al. (2019). Comparison of representative and custom methods of generating core subsets of a carrot germplasm collection. *Crop Science 59* (3): 1107–1121.

Crossa, J., de Los Campos, G., Pérez, P. et al. (2010). Prediction of genetic values of quantitative traits in plant breeding using pedigree and molecular markers. *Genetics 186* (2): 713–724.

Crossa, J., Pérez, P., de los Campos, G. et al. (2011). Genomic selection and prediction in plant breeding. *Journal of Crop Improvement 25* (3): 239–261.

Crossa, J., Beyene, Y., Kassa, S. et al. (2013). Genomic prediction in maize breeding populations with genotyping-by-sequencing. *G3: Genes, Genomes, Genetics 3* (11): 1903–1926.

Crossa, J., Jarquín, D., Franco, J. et al. (2016). Genomic prediction of gene bank wheat landraces. *G3: Genes, Genomes, Genetics 6* (7): 1819–1834.

Crossa, J., Pérez-Rodríguez, P., Cuevas, J. et al. (2017). Genomic selection in plant breeding: methods, models, and perspectives. *Trends in Plant Science 22* (11): 961–975.

Daetwyler, H., Hayden, M., Bansal, U. et al. (2013). Genomic selection for disease and morphological traits in diverse wheat landraces. *Plant and Animal Genomes 21*: 11–16.

Dawson, J.C., Endelman, J.B., Heslot, N. et al. (2013). The use of unbalanced historical data for genomic selection in an international wheat breeding program. *Field Crops Research 154*: 12–22.

dos Santos, J.P.R., Pires, L.P.M., de Castro Vasconcellos, R.C. et al. (2016). Genomic selection to resistance to Stenocarpella maydis in maize lines using DArTseq markers. *BMC Genetics* 17 (1): 86.

Elshire, R.J., Glaubitz, J.C., Sun, Q. et al. (2011). A robust, simple genotyping-by-sequencing (GBS) approach for high diversity species. *PloS one* 6 (5): e19379.

Faville, M.J., Ganesh, S., Moraga, R. et al. (2016). Development of genomic selection for perennial ryegrass. In: *Breeding in a World of Scarcity*, 139–143, Cham. Springer.

Fodor, A., Segura, V., Denis, M. et al. (2014). Genome-wide prediction methods in highly diverse and heterozygous species: proof-of-concept through simulation in grapevine. *PLoS One* 9 (11): e110436.

Gezan, S.A., Osorio, L.F., Verma, S. et al. (2017). An experimental validation of genomic selection in octoploid strawberry. *Horticulture Research 4* (1): 1–9.

González-Camacho, J.M., de Los Campos, G., Pérez, P. et al. (2012). Genome-enabled prediction of genetic values using radial basis function neural networks. *Theoretical and Applied Genetics 125* (4): 759–771.

Grenier, C., Cao, T.V., Ospina, Y. et al. (2015). Accuracy of genomic selection in a rice synthetic population developed for recurrent selection breeding. *PloS One* 10 (8): e0136594.

Guo, Z., Tucker, D.M., Lu, J. et al. (2012). Evaluation of genome-wide selection efficiency in maize nested association mapping populations. *Theoretical and Applied Genetics 124* (2): 261–275.

Guo, Z., Tucker, D.M., Basten, C.J. et al. (2014). The impact of population structure on genomic prediction in stratified populations. *Theoretical and Applied Genetics 127* (3): 749–762.

Habier, D., Fernando, R.L., Kizilkaya, K. et al. (2011). Extension of the Bayesian alphabet for genomic selection. *BMC Bioinformatics* 12 (1): 186.

Haile, J.K., N'Diaye, A., Clarke, F. et al. (2018). Genomic selection for grain yield and quality traits in durum wheat. *Molecular Breeding* 38 (6): 75.

Hawkins, C. and Yu, L.X. (2018). Recent progress in alfalfa (Medicago sativa L.) genomics and genomic selection. *The Crop Journal 6* (6): 565–575.

Hayes, B.J. and Goddard, M.E. (2001). Prediction of total genetic value using genome-wide dense marker maps. *Genetics 157* (4): 1819–1829.

He, T. and Li, C. (2020). Harness the power of genomic selection and the potential of germplasm in crop breeding for global food security in the era with rapid climate change. *The Crop Journal.*

Heffner, E.L., Jannink, J.L., and Sorrells, M.E. (2011). Genomic selection accuracy using multifamily prediction models in a wheat breeding program. *The Plant Genome 4* (1): 65–75.

Henry, R.J. and Nevo, E. (2014). Exploring natural selection to guide breeding for agriculture. *Plant Biotechnology Journal 12* (6): 655–662.

Heslot, N., Rutkoski, J., Poland, J. et al. (2013). Impact of marker ascertainment bias on genomic selection accuracy and estimates of genetic diversity. *PloS One* 8 (9): e74612.

Ho, W.K., Chai, H.H., Kendabie, P. et al. (2017). Integrating genetic maps in bambara groundnut [Vigna subterranea (L) Verdc.] and their syntenic relationships among closely related legumes. *BMC Genomics* 18 (1): 192.

Hoffstetter, A., Cabrera, A., and Sneller, C. (2016). Identifying quantitative trait loci for economic traits in an elite soft red winter wheat population. *Crop Science 56* (2): 547–558.

Isidro, J., Jannink, J.L., Akdemir, D. et al. (2015). Training set optimization under population structure in genomic selection. *Theoretical and Applied Genetics 128* (1): 145–158.

Jarquín, D., Kocak, K., Posadas, L. et al. (2014). Genotyping by sequencing for genomic prediction in a soybean breeding population. *BMC genomics* 15 (1): 740.

Kumar, S., Chagné, D., Bink, M.C. et al. (2012). Genomic selection for fruit quality traits in apple (Malus× domestica Borkh.). *PloS One* 7 (5): e36674.

Lado, B., Barrios, P.G., Quincke, M. et al. (2016). Modeling genotype × environment interaction for genomic selection with unbalanced data from a wheat breeding program. *Crop Science 56* (5): 2165–2179.

Larkin, D.L., Lozada, D.N., and Mason, R.E. (2019). Genomic selection—considerations for successful implementation in wheat breeding programs. *Agronomy* 9 (9): 479.

Li, X., Wei, Y., Acharya, A. et al. (2015). Genomic prediction of biomass yield in two selection cycles of a tetraploid alfalfa breeding population. *The Plant Genome 8* (2): 1–10.

Liang, Z., Gupta, S.K., Yeh, C.T. et al. (2018). Phenotypic data from inbred parents can improve genomic prediction in pearl millet hybrids. *G3: Genes, Genomes, Genetics 8* (7): 2513–2522.

Lin, Z., Hayes, B.J., and Daetwyler, H.D. (2014). Genomic selection in crops, trees and forages: a review. *Crop and Pasture Science 65* (11): 1177–1191.

Lipka, A.E., Lu, F., Cherney, J.H. et al. (2014). Accelerating the switchgrass (*Panicum virgatum* L.) breeding cycle using genomic selection approaches. *PloS one* 9 (11): e112227.

Lorenz, A.J., Smith, K.P., and Jannink, J.L. (2012). Potential and optimization of genomic selection for Fusarium head blight resistance in six-row barley. *Crop Science 52* (4): 1609–1621.

Lorenzana, R.E. and Bernardo, R. (2009). Accuracy of genotypic value predictions for marker-based selection in biparental plant populations. *Theoretical and Applied Genetics 120* (1): 151–161.

Mastrodomenico, A.T., Bohn, M.O., Lipka, A.E. et al. (2019). Genomic selection using maize ex-plant variety protection germplasm for the prediction of nitrogen-use traits. *Crop Science 59* (1): 212–220.

McElroy, M.S., Navarro, A.J., Mustiga, G. et al. (2018). Prediction of cacao (Theobroma cacao) resistance to Moniliophthora spp. diseases via genome-wide association analysis and genomic selection. *Frontiers in Plant Science* 9: 343.

Meuwissen, T. (2007). Genomic selection: marker assisted selection on a genome wide scale. *Journal of Animal Breeding and Genetics 124* (6): 321–322.

Meuwissen, T.H., Hayes, B.J., and Goddard, M.E. (2001). Prediction of total genetic value using genome-wide dense marker maps. *Genetics* 157 (4): 1819–1829.

Michel, S., Ametz, C., Gungor, H. et al. (2016). Genomic selection across multiple breeding cycles in applied bread wheat breeding. *Theoretical and Applied Genetics 129* (6): 1179–1189.

Moeinizade, S., Kusmec, A., Hu, G. et al. (2020). Multi-trait genomic selection methods for crop improvement. *Genetics 215* (4): 931–945.

Nadeem, M.A., Nawaz, M.A., Shahid, M.Q. et al. (2018). DNA molecular markers in plant breeding: current status and recent advancements in genomic selection and genome editing. *Biotechnology and Biotechnological Equipment 32* (2): 261–285.

Nyine, M., Uwimana, B., Blavet, N. et al. (2018). Genomic prediction in a multiploid crop: genotype by environment interaction and allele dosage effects on predictive ability in banana. *The Plant Genome 11* (2): 1–16.

de Oliveira, E.J., de Resende, M.D.V., da Silva Santos, V. et al. (2012). Genome-wide selection in cassava. *Euphytica 187* (2): 263–276.

Park, T. and Casella, G. (2008). The bayesian lasso. *Journal of the American Statistical Association 103* (482): 681–686.

Pérez-Rodríguez, P., Gianola, D., González-Camacho, J.M. et al. (2012). Comparison between linear and non-parametric regression models for genome-enabled prediction in wheat. *G3: Genes, Genomes, Genetics 2* (12): 1595–1605.

Poland, J. and Rutkoski, J. (2016). Advances and challenges in genomic selection for disease resistance. *Annual Review of Phytopathology 54*: 79–98.

Poland, J., Endelman, J., Dawson, J. et al. (2012). Genomic selection in wheat breeding using genotyping-by-sequencing. *The Plant Genome 5* (3): 103–113.

Raman, H., Raman, R., Coombes, N. et al. (2016). Genome-wide association analyses reveal complex genetic architecture underlying natural variation for flowering time in canola. *Plant, Cell and Environment 39* (6): 1228–1239.

Resende, M.F.R., Munoz, P., Acosta, J.J. et al. (2012a). Accelerating the domestication of trees using genomic selection: accuracy of prediction models across ages and environments. *New Phytologist 193* (3): 617–624.

Resende, M.D., Resende, M.F., Sansaloni, C.P. et al. (2012b). Genomic selection for growth and wood quality in Eucalyptus: capturing the missing heritability and accelerating breeding for complex traits in forest trees. *New Phytologist 194* (1): 116–128.

Resende, M.F., Muñoz, P., Resende, M.D. et al. (2012c). Accuracy of genomic selection methods in a standard data set of loblolly pine (*Pinus taeda* L.). *Genetics 190* (4): 1503–1510.

Riedelsheimer, C. and Melchinger, A.E. (2013). Optimizing the allocation of resources for genomic selection in one breeding cycle. *Theoretical and Applied Genetics 126* (11): 2835–2848.

Riedelsheimer, C., Czedik-Eysenberg, A., Grieder, C. et al. (2012). Genomic and metabolic prediction of complex heterotic traits in hybrid maize. *Nature Genetics* 44 (2): 217.

Robertsen, C.D., Hjortshøj, R.L., and Janss, L.L. (2019). Genomic selection in cereal breeding. *Agronomy* 9 (2): 95.

Rutkoski, J., Benson, J., Jia, Y. et al. (2012). Evaluation of genomic prediction methods for Fusarium head blight resistance in wheat. *The Plant Genome 5* (2): 51–61.

Rutkoski, J.E., Poland, J.A., Singh, R.P. et al. (2014). Genomic selection for quantitative adult plant stem rust resistance in wheat. *The Plant Genome 7* (3): 1–10.

Saint Pierre, C., Burgueño, J., Crossa, J. et al. (2016). Genomic prediction models for grain yield of spring bread wheat in diverse agro-ecological zones. *Scientific Reports 6*: 27312.

Saxena, R.K., Singh, V.K., Kale, S.M. et al. (2017). Construction of genotyping-by-sequencing based high-density genetic maps and QTL mapping for fusarium wilt resistance in pigeonpea. *Scientific Reports 7* (1): 1–11.

Slavov, G.T., Nipper, R., Robson, P.R., et al. (2014). Genome-wide association studies and prediction of traits related to phenology, biomass and cell wall composition in Miscanthus sinensis. *Plant & Animal Genome XXII Conference*, San Diego, USA.

Sonah, H., O'Donoughue, L., Cober, E. et al. (2015). Identification of loci governing eight agronomic traits using a GBS-GWAS approach and validation by QTL mapping in soya bean. *Plant Biotechnology Journal 13* (2): 211–221.

Spindel, J., Begum, H., Akdemir, D. et al. (2015). Genomic selection and association mapping in rice (Oryza sativa): effect of trait genetic architecture, training population composition, marker number and statistical model on accuracy of rice genomic selection in elite, tropical rice breeding lines. *PLoS Genetics* 11 (2): e1004982.

Stewart-Brown, B.B., Song, Q., Vaughn, J.N. et al. (2019). Genomic selection for yield and seed composition traits within an applied soybean breeding program. *G3: Genes, Genomes, Genetics 9* (7): 2253–2265.

Technow, F., Bürger, A., and Melchinger, A.E. (2013). Genomic prediction of northern corn leaf blight resistance in maize with combined or separated training sets for heterotic groups. *G3: Genes, Genomes, Genetics 3* (2): 197–203.

Tibshirani, R. (1996). Regression shrinkage and selection via the lasso. *Journal of the Royal Statistical Society: Series B (Methodological) 58* (1): 267–288.

Verma, S., Gupta, S., Bandhiwal, N. et al. (2015). High-density linkage map construction and mapping of seed trait QTLs in chickpea (*Cicer arietinum* L.) using Genotyping-by-Sequencing (GBS). *Scientific Reports* 5: 17512.

Wimmer, V., Lehermeier, C., Albrecht, T. et al. (2013). Genome-wide prediction of traits with different genetic architecture through efficient variable selection. *Genetics 195* (2): 573–587.

Würschum, T., Reif, J.C., Kraft, T. et al. (2013). Genomic selection in sugar beet breeding populations. *BMC Genetics 14* (1): 1–8.

Xavier, A., Muir, W.M., and Rainey, K.M. (2016). Assessing predictive properties of genome-wide selection in soybeans. *G3: Genes, Genomes. Genetics 6* (8): 2611–2616.

Xu, S. (2013). Genetic mapping and genomic selection using recombination breakpoint data. *Genetics 195* (3): 1103–1115.

Xu, Y. and Crouch, J.H. (2008). Marker-assisted selection in plant breeding: From publications to practice. *Crop Science 48* (2): 391–407.

Xu, Y., Wang, X., Ding, X. et al. (2018). Genomic selection of agronomic traits in hybrid rice using an NCII population. *Rice 11* (1): 1–10.

Xu, Y., Liu, X., Fu, J. et al. (2020). Enhancing genetic gain through genomic selection: from livestock to plants. *Plant Communications* 1 (1): 100005.

Zeng, D., Tian, Z., Rao, Y. et al. (2017). Rational design of high-yield and superior-quality rice. *Nature Plants 3* (4): 1–5.

Zhang, X., Pérez-Rodríguez, P., Semagn, K. et al. (2015). Genomic prediction in biparental tropical maize populations in water-stressed and well-watered environments using low-density and GBS SNPs. *Heredity 114* (3): 291–299.

Zhang, X., Sallam, A., Gao, L. et al. (2016). Establishment and optimization of genomic selection to accelerate the domestication and improvement of intermediate wheatgrass. *The Plant Genome 9* (1): 1–18.

Zhao, Y., Zeng, J., Fernando, R. et al. (2013). Genomic prediction of hybrid wheat performance. *Crop Science 53* (3): 802–810.

Zou, H. and Hastie, T. (2005). Regularization and variable selection via the elastic net. *Journal of the Royal Statistical Society: Series B (Statistical Methodology) 67* (2): 301–320.

# 8

# Analytical Pipelines for the GBS Analysis

*Vinaykumar Rachappanavar*[1,2], *Arushi Padiyal*[2], *Jitender Kumar Sharma*[3], *and Narender Negi*[4]

[1] *Department of Agriculture, MS Swaminathan School of Agriculture, Shoolini University, Solan, Himachal Pradesh, India*
[2] *Department of Seed Science and Technology, Dr. Yashwant Singh Parmar University of Horticulture and Forestry, Solan, Himachal Pradesh, India*
[3] *Department of Agriculture, School of Agriculture, Baddi University of Emerging Sciences & Technology, Baddi, Himachal Pradesh, India*
[4] *Department of Fruit Science, ICAR-NBPGR Regional Station, Shimla, Himachal Pradesh, India*

## 8.1 Introduction

The Sanger has developed more efficient DNA sequencing techniques, and it has undergone various modifications and refinement in intervening years. Such as the use of base-specific fluorescent dyes instead of radioactive dyes (Smith et al. 1986) and complex gel electrophoresis has been replaced by automated capillary electrophoresis (Luckey et al. 1990). After all these improvements, Sanger sequencing is widely considered the gold standard for clinical diagnostic use (Moorthie et al. 2011). However, this technique remained expensive for routine sequencing. To address this shortcoming, new sequencing techniques are developed, they have improved run capacity, processing speed, low cost, and affordability for whole-genome sequencing (Pettersson et al. 2009; Voelkerding et al. 2009). These technologies are referred to as next-generation sequencing (NGS) technologies. All NGS strategies follow a similar protocol for template preparation and ligation but use different platforms based on their working principles. In addition, they depend upon the cyclic interrogation of millions of clonally amplified DNA molecules immobilized on artificial surface to generate up to billions of sequences in a massively parallel style (Metzker 2010). Most of the NGS systems are capable of generating reliable sequences and show close to ideal coverage behavior on guanine–cytosine (GC)-rich, neutral, and moderately on adenine–thymine (AT)-wealthy genomes (Quail et al. 2012). NGS enabled sequencing of ancient DNA samples, resequencing to generate a large number of single-nucleotide polymorphism (SNPs) for exploring diversity, and metagenomics analysis of environmentally influenced samples (Mardis 2008; Elshire et al. 2011). Multiplex sequencing has also been done by way of tagging randomly shared DNA fragments with particular, short DNA collection and pooling

*Genotyping by Sequencing for Crop Improvement*, First Edition.
Edited by Humira Sonah, Vinod Goyal, S.M. Shivaraj, and Rupesh K. Deshmukh.

samples into sequencing channels (Craig et al. 2008). This approach has been rapidly used in the discovery and mapping of SNPs in crops (Huang et al. 2012). The evolution of NGS is explained in Figure 8.1.

Genotype-by-sequencing (GBS) allows millions of bases to be sequenced in a single round which increased capabilities far beyond that of traditional Sanger sequencing (Sanger et al. 1977). It is cost-effective method to quickly identify the genetic variation present in the population by using specific restriction enzymes to divide the whole genome into fragments whose ends are sequenced using short-red sequencing platforms (Wickland et al. 2017). The development of next-generation of improved sequencing techniques has revolutionized biological research including agriculture and medicinal sectors. Whole-genome sequencing identifies millions of SNPs that serve as markers to analyze the association between genotypes and phenotypes. These SNPs identified using reduced-representation approaches involve a subset of locations from the genome. The restriction site-associated DNA (RAD) sequencing is the earliest reduced sequencing method that use allied high efficient restriction enzymes to break the whole genome into DNA fragments and these sheared DNAs were sequenced on next-generation platforms (Miller et al. 2007; Baird et al. 2008; Scheben et al. 2017).

GBS gained popularity in genotyping because of its high-throughput and low-cost, it involves reduced sample handling, fewer polymerase chain reaction (PCR) and purification steps, no size fractionation, and no reference sequence limits. It has been widely used in genomic selection (GS), gene mapping, and genome-wide associated studies (GWAS) in various crop species (Liu et al. 2014; Sonah et al. 2015). It offers simplified library construction which is amenable to work on a large number of lines (Elshire et al. 2011). It also depends on restriction enzymes to generate a reduced representation, but it has fewer steps for library preparation protocol than RAD sequencing and it mainly lacks a size selection step (Elshire et al. 2011) but two-enzyme digestion step was incorporated (Poland et al. 2012). In pseudo-testcross progeny, GBS combined with genome-independent imputation is a simple and most-efficient method for genetic map construction (Ward et al. 2013). To reduce the cost per sample sequencing, Sonah et al. (2013) described the modified library preparation protocol. In which selective amplification is recommended to improve both the number of SNPs and their depth of coverage.

Two unique GBS techniques were advanced with the Ion personal genome machine (PGM) device. Those are (i) restriction enzyme digestion, in which sample DNA is digested with one or more selected restriction enzymes before ligation, and no specific SNPs have been identified in this system. (ii) Multiplex enriched PCR, in which area-specific PCR primers are used for amplification and a set of SNPs has been defined for a selection of the genome. Baird et al. (2008) demonstrated the value of RAD (restriction site-associated genomic DNA) sequencing for high-density SNP discovery and genotyping. Multiplex sequencing strategy that uses an inexpensive barcoding system to reduce the cost and improve efficiency. To eliminate the need for a second Illumina sequencing read, barcodes are included in one of the adaptor sequences just upstream of the restriction site in genomic DNA. This strategy is similar to RAD however modulation of barcode nucleotide composition and length affects in fewer series phasing errors (Baird et al. 2008). GBS is substantially less complicated compared to the RAD method, straight forward, fragments are not size selected, single-well digestion of genomic DNA followed by adopter ligation, fewer DNA

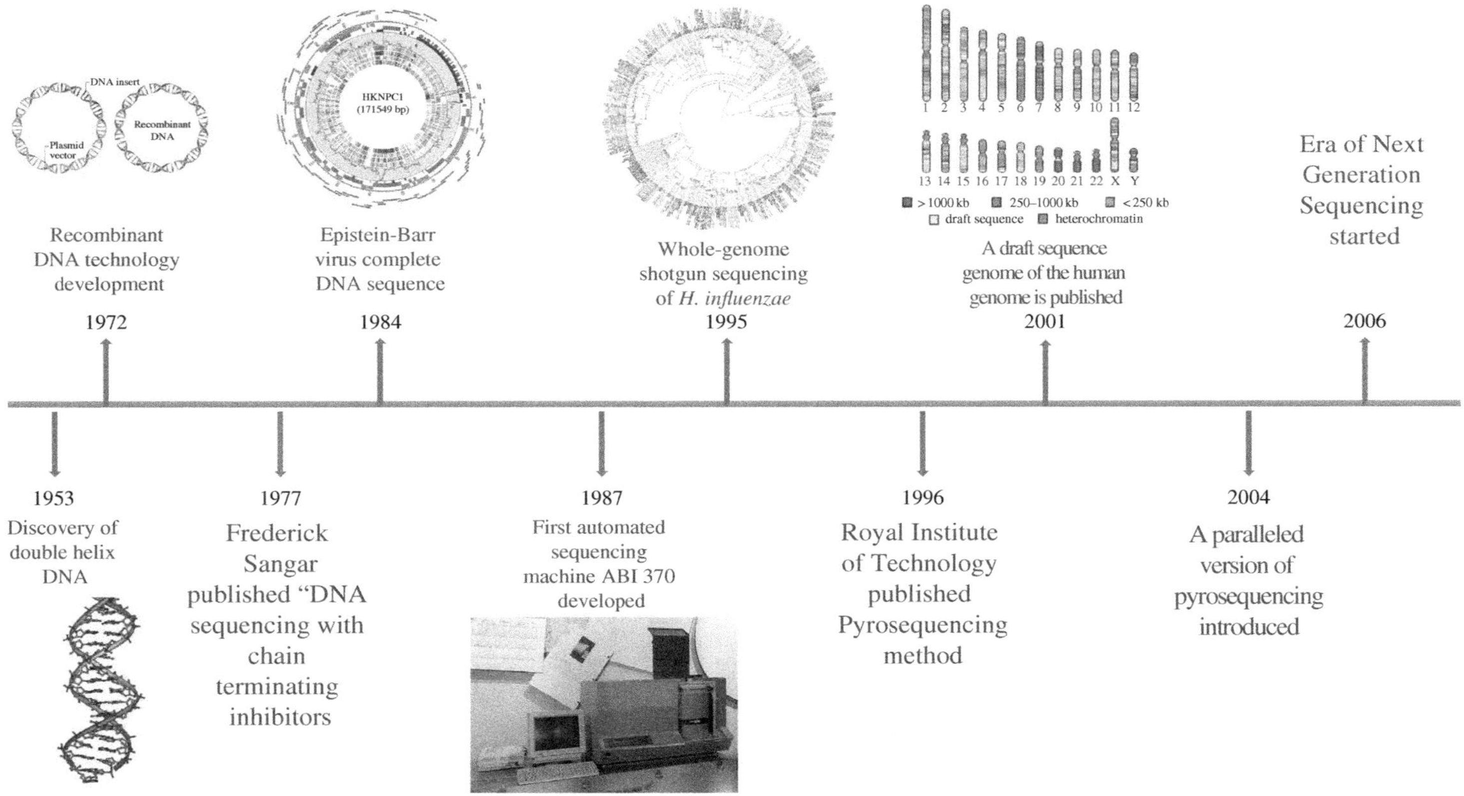


**Figure 8.1** Evolution of next-generation sequencing. *Source:* Giac83/Wikimedia Commons/CC BY SA 2.5

purification steps, and shallow genome sequencing followed by an accusation of missing SNPs in haplotype blocks to further reduce the cost (He et al. 2014).

## 8.2 Applications of NGS

NGS has several applications such as DNA-sequencing and assembly of the produced data to determine the unknown genome. These technologies are most commonly utilized in mutation identification rather than other studies. Initially, short reads with lengths about 50–250 bp are mapped to the reference genome, and after that variation is detected from the mapped data. While most of the NGS platforms concentrated on the identification of SNPs or small insertion–deletion mutations (INDELs), bigger INDELs, structural variation (SV) including translocations and copy number variation (CNV) (Wadapurkar and Vyas 2018).

## 8.3 NGS Sequencing Platforms

There are several NGS platforms are available which use different sequencing technologies. However, all these platforms can sequence millions of small DNA fragments parallelly. NGS has created a vast amount of opportunities as well as challenges to bioinformaticians, computational biologists, and end-users endeavoring to assemble a large amount of data generated by its different tools. NGS became commercial around 2005 with the introduction of Solexa sequencing technology. Since then, several advanced sequencing techniques have been developed. These techniques can be broadly grouped into three main types: (i) sequencing by synthesis, (ii) sequencing by ligation, and (iii) single-molecule sequencing.

### 8.3.1 Sequencing by Synthesis

Same as Sanger sequencing, NGS determines the base composition of DNA segment through the detection of chemiluminescence created by fluorescent nucleotide incorporation during DNA polymerase activity. In sequencing by synthesis, DNA strands are split into appropriately sized fragments, ligated to adaptor sequences, and clonally amplified to promote fluorescent signals. Later templets are separated and immobilized for the next step that is flow-cell cycles. Three major platforms follow sequencing by synthesis which differs by reading length and how templets are amplified and immobilized on a solid surface.

#### 8.3.1.1 Roche 454 Pyrosequencing

In Roche 454 pyrosequencing, a microbead is used to adhere to a single, primed DNA template and amplified using emulsion PCR. Each bead is then placed in a well of PicoTiterPlate where it is incubated with different reagents such as DNA polymerase, luciferase, adenosine triphosphate (ATP) sulfurylase, and apyrase. For every incorporation of appropriate deoxynucleotide triphosphate (dNTP) into a new strand, pyrophosphate is released and it is transformed into ATP in the presence of ATP sulfurylase. Later fluorescence was produced by the formation of oxyluciferin from luciferin in the presence of ATP (Ronaghi et al. 1998; Nyrén 2007). Later all unused, excess ATP and nucleotides are then removed by the apyrase wash. This procedure is repeated until the complete DNA has elongated. A camera is used to detect the fluorescence produced by nucleotide incorporation and analyzed to produce information regarding DNA sequence (Egan et al. 2012).

#### 8.3.1.2 Illumina

Solexa developed Illumina Genomic Analyzer which uses solid-phase bridge amplification. In this 3′ and 5′ adaptors are attached to each end of the DNA template and one end of the fragment is attached to the substrate to provide the stability for amplification (http://www.Illumina.com). These adaptors create a bridge by hybridizing to immobilize primers and facilitate amplification. Clusters of identical templets formed by amplicons, which can enhance chemiluminescent production. The first being the addition of all four nucleotide combinations and each modified as 3′-*O*-azidomethyl reversible terminators followed by series of cyclic washes to complete the extension (Bateman and Quackenbush 2009).

#### 8.3.1.3 Ion Torrent

Ion Torrent is differing from others in the detection of the sequencing process. It measures the $H^+$ ion concentration upon every nucleotide incorporation rather than fluorescence using specialized semiconductor technology (Rothberg et al. 2011). The machine can detect sequentially adding specific nucleotides based on pH change. Two systems that use this technique are (i) Ion PGM, for small scale application and it can produce over 100 Mb of sequence with 11.1 million sensors. (ii) Ion Proton which provides higher throughput.

### 8.3.2 Sequencing by Ligation

Sequencing by ligation improves the sequencing efficiency by eliminating the mismatch sensitivity of DNA ligase (Landegren et al. 1988). In these methods, varying length florescent tagged oligonucleotides are used based on nucleotides to be determined. Short, known sequence anchors are used to prime the fragmented DNA templates. Different sets of probes are used repeatedly to query the DNA template and assess the sequence of newly incorporated nucleotides.

#### 8.3.2.1 SOLiD

Support oligonucleotide ligation detection (SOLiD) platform created by Applied Biosystems (ABI) that utilizes sequencing by ligation to determine DNA sequence. Emulsion PCR on microbeads is used to enrich fragmented primed libraries. A set labeled with a different fluorophore probe is added to the flow cell. A corresponding 1.2 fluorescent probes are ligated by DNA ligase to the primer. Following phosphorothiolate link cleaved by silver ions leads to 5′ phosphate group regeneration for subsequent ligation. After sufficient elongation by repetition, the extended product is removed. This method is often used in transcriptome resequencing studies or genome sequencing (Shulaev et al. 2011).

#### 8.3.2.2 Polonator

Dr. George Church's laboratory developed this system at Harvard University in collaboration with Dover systems. It uses emulsion PCR for amplification of templates used for library preparation. It is an automated polymerase colony machine (Shulaev et al. 2011) which is an expensive but highly accurate sequencing technique. It has the potential to read millions of immobilized DNA sequences in parallel. The read length of this system is 26 bp per amplicon and 13 bp per tag with 4–5 bp gap for each tag. It worked mainly on paired-end tags. Tag library consists of 135 base-paired DNA templates with two 17–20 base-paired genomic tags. The fluorescence microscope and a computer-controlled flow-cell were used in these sequencing instruments. Very flexible and inexpensive instrumentation makes this technique very useful. It can be used for bacterial artificial chromosome and bacterial genome resequencing, SAGE (serial analysis of gene expression) tag, and barcode sequencing. Although 786 GB raw data will be collected from this technique only 0.00001–0.0001% bits are useful. Another problem with this technique is the nonuniform implication which lowers the efficiency of this technique.

### 8.3.3 Single-Molecule Sequencing

Single-molecule sequencing allows examining billions of DNA or RNA molecules in parallel. This method can be used to sequence the whole genome. It is often termed as "third-generation sequencing" technology, these methods use chemiluminescence to produce detectable signals for every nucleotide incorporation. It can sequence low concentrated or moderately degraded initial material without any complications (Orlando et al. 2011) and avoid the biases introduced during PCR amplification which are commonly observed in the above sequencing technologies. It offers both sequencing and quantification by eliminating sample preparation which evade complex ligation and PCR for amplification steps (Hayden 2009).

#### 8.3.3.1 Helicos

Helicos single-molecule sequencing is an integrated system that minimizes human interference in genome biology and provides both high accurate quantification and sequencing information. In this, the DNA molecules were shared into small pieces, tailed by attaching poly-A followed by hybridization to oligo-dT present on the flow cell surface. This process utilizes very few components as compared to previous generation sequencing platforms. There is no restriction on the specific size range of target molecules. This technique allows detecting the variation in more precise order. Along with this, it has a wide array of applications such as small RNA studies, CNV, ChIP-Seq using picogram quantities of pure DNA, and ancient and low to completely degraded DNA (Thompson and Steinmann 2010).

#### 8.3.3.2 Pacific Bioscience

Pacific Biosciences is a single molecule, real-time sequencing technique developed by Pacific Biosciences. The *de novo* assemblies help to advance the characterization of SV by narrowing the distance between reference assemblies. Longer reads generated by this technique in shorter run times facilitate the identification of novel genes and annotate gene isoforms effortlessly. Several hybridization sequencing strategies developed by using this Pacific Bioscience technique are more reliable and easier to use in small-sized laboratories. It is used in the discovery of short tandem repeats (*STRs*), fragile X mental retardation

(*FMR1*) mutations, and cyclin-dependent kinase (*CDK2A*) deletion which are associated with brain diseases, cancer, and respiratory diseases. It also helps to identify methylation like base modification (Rhoads and Au 2015).

#### 8.3.3.3 ChIP-Sequencing

ChIP-sequencing combines chromatin immunoprecipitation and parallel sequencing which lowers the cost and time of sequencing. Protein-specific DNA sequences were identified by immunoprecipitation. In immunoprecipitation, protein-specific chromatins are fixed by forming a covalent linkage between DNA and DNA-binding proteins. Then these fragments are separated, amplified by using PCR, and sequenced using ChIP-sequencing technologies. The reads generated are then mapped to the relative genome (Shin et al. 2013). It allows identifying DNA binding proteins which influence phenotypic expression, TFs (transcription factors), or microarray expression which helps to an in-depth study of biological processes. This technique is mainly used in epigenetic studies which give insights into the epigenetic mechanism (De Sá et al. 2018).

## 8.4 Tools for NGS Data Analysis

The primary bottleneck in sequencing is the data analysis which is generated by different sequencing instruments. Not all researchers have access to bioinformatics and most of the programs are written in LINUX or UNIX background. Most of the systems are costly and need incorporated algorithms to perform data analysis, e.g. GenomeQuest, CLC Genomics Workbench, DNAnexus, and Galaxy. Galaxy provides easy to use platform that incorporates community Linux command and popular open-source. It comprises a variety of workflows for NGS data analysis. It uses open-source tools such as RNA-Seq (Tophat, Cufflinks), ChIP-Seq (model-based analysis of ChIP-Seq (MACS), GeneTrack indexer, Peak predictor), and finding small deletion, insertion, and SNPs using sequence alignment/map (SAM) tools. The Galaxy cloud-based program does not need expensive high performing computers. The new generation of parallel DNA sequence data needs software tools that can align billions of short DNA sequence reads more accurately and rapidly to a large reference genome. To align reads generated by sequencing to the entire genome is grossly impractical and costly. For this, many algorithms have been developed that reduced both costs and to search target for alignment. We have discussed some tools which are routinely used in NGS data analysis.

a) **Mapping quality (MAQ):** Many sequencing platforms produce very short reads (10–100 bp). For this, new algorithms and software are required to efficiently align them to the reference genome. For this, MAQ is used. It is accurate, versatile, efficient, and free to use from http://maq.sourceforge.net. It uses mate-pair information and the error probability of individual read alignment. It is commonly used to build assemblies by mapping shotgun short reads to a reference genome (Li et al. 2008).
b) **FASTX Tool Kit:** It is a group of command-line tools for Short-Reads FASTA (fast-all)/FASTQ file processing. Most of the NGS machines produce the data in FASTA or FASTQ files which contain information of multiple short-read sequences. It aligns the sequence

to reference genome or other databases using particular programs such as LastZ, basic local alignment search tool (BLAST), MAQ, and many others. It can most commonly use to process the raw data FASTA or FASTQ files before mapping the sequence to the genome to produce better results.

c) **Genome analysis toolkit (GATK):** It is an efficient and robust analysis toolkit for NGS which uses the functional programming philosophy of MapReduce. It provides a rich but small set of data access patterns that encompass the needs of most of the analytical tools. It separates specific analysis calculations from data management infrastructure which enables shared and distributed memory parallelization which improves the central processing unit (CPU) and memory efficiency. It has been used in large-scale sequencing projects like complete genomes projects of different species (McKenna et al. 2010).
d) **Genome-wide complex trait analysis (GCTA):** GWAS identifies the hundreds of SNPs conferring the genetic variation. Despite this, most trait-associated SNPs explain a small fraction of heritability and there is no consensus on the clarification of "missing heritability" (Manolio et al. 2009; Maher 2008). For this, GCTA was developed which estimates the total amount of phenotypic variance and has a function to estimate the inbreeding coefficient from all SNPs data. It uses a mixed linear model to fit the effect of all the SNPs as random effects. It is a universal tool to estimate and partition complex trait variation with large GWAS data sets (Yang and Li 2011).
e) **SAM Tools:** It consists of parallel software components that can work efficiently on single-node multicore large memory systems. It is 156–186 times faster in marking duplicate reads on a single node system as compared to do facto standard tools. It has pleasing features like conversion of file format SAM to binary alignment/map (BAM), analyzing, and duplicate marking. It could reduce the runtime to 9-minute from 20 hours by using 711 GB of memory (Ogasawara et al. 2016).
f) **BLAT-like fast accurate search tool (BFAST):** Homer and his colleagues have developed the BFAST. It can align giga-scale short reads with higher accuracy and sensitivity for deletion or insertion to a reference genome. It can be customized for sensitivity and for a given time. It is a two-step method in which first step utilizes multiple indexes of the reference genome to identify the candidate alignment locations (CALs) followed by further local alignment to identify the best match. Most of the other tools do not perform a gapped local alignment. For this, BFAST scarifies some speed in the CAL identification to perform gapped local alignment (Homer et al. 2009).
g) **Burrows–Wheeler Aligner (BWA):** It is the first algorithm designed to analyze up to 100 bp Illumina sequence reads. Under this many algorithms developed. Those are BWA-backtrack (Li and Durbin 2009), BWA-SW (Smith–Waterman) (Li and Durbin 2010) and BWA-MEM (maximal exact match) (Li 2013), bowie2 (Langmead and Salzberg 2012), cushew. BWA-MEM is a faster and more accurate single-node tool in genome alignment so it can be preferred over BWA-SW. Later BWASpark (GATK) was developed which integrates BWA-MEM. Instead of calling a whole BWA program, it creates BWA threads with java native interface (JNI). It processes only part of data loaded in sam file format to BWA-MEM because only original sequence information is required to perform an alignment. To generate the complete result, the alignment result is integrated with the original data file, and the result is then stored in hadoop distributed file system (HDFS) (Zhang et al. 2019).

h) **Bowtie:** Second-generation sequencers increased throughput creating a challenge in computational alignment. Bowtie is the first step for many workflows, including digital gene expression and variant detection. It aligns reads with the aid of an index of the reference genome to achieve both speed and memory efficiency. The Bowtie index is an improved version of the Ferragina and Manzini (FM) Index (Ferragina and Manzini 2000), which in turn uses the Burrows–Wheeler transform (Burrows and Wheeler 1994). The users must build or obtain an appropriate index before reads can align. Once an index is built can be queried several times without any restriction.

i) **VariationHunter:** A soft clustering strategy developed to assign each read to multiple clusters without any clinch (Hormozdiari et al. 2009). It processes both Variation Hunter-SC (set cover) and Variation Hunter-Pr (Probabilistic) algorithms to implement soft clustering of the discarded reads it maps each paired end reads to a particular SV to minimize the total number of genomic clusters (Zhao et al. 2013).

j) **Micro-red fast alignment search tool (mrFAST):** It can easily allow for the prediction of absolute copy-number variation of duplicated genes and segments. mrFAST helps to calculate accurate read depth and return all possible single-nucleotide differences within duplicated and unique genome portions. It is mainly designed for short sequence reads (>25 bp). Same as BLAST, it uses the seed-and-extend method and creates indices of the reference genome by implementing a hash table (Alkan et al. 2009).

k) **Short oligonucleotide alignment program (SOAP):** It has a special benefit over others. It allows either a continuous gap or a certain number of mismatches. The best hit of each read has a smaller gap or a minimal number of mismatches. By default, this program will allow a maximum of two mismatches. The SNPs in-between two haplotypes is much higher than that of small deletion or addition, so un-gapped hits have precedence over gapped hits. For gapped alignment, 1–3 bp continues only one gap is acceptable and to avoid ambiguous gaps no mismatches are permitted. SOAP can iteratively trim a small portion of 3′-end to minimize the sequencing errors which sometimes make them unalignable to the reference sequences. Additionally, SOAP provides special modules for small RNA discovery with size between 18 and 26 bp and messenger RNA (mRNA) tag profiling analysis (Ruiqiang et al. 2008).

l) **SOAP Denovo:** This tool is mainly used for transcriptome studies where a reference genome is not available. Assemblers like Cufflinks (Trapnell et al. 2010), enhanced read analysis of gene expression (ERANGE) (Mortazavi et al. 2008), and Scripture (Guttman et al. 2010) can reconstruct the transcripts from very short reads by aligning the reads to a reference genome. When the genome is unavailable or polyploid or unusually large, *de novo* assembly is required. For this, Velvet (Zerbino and Birney 2008), assembly by short sequences (ABySS) (Simpson et al. 2009), and SOAPdenovo (Li et al. 2010), were developed. These programs were prepared to recover the known size of genomes with a defined number of chromosomes. In contrast, transcriptome assembler must recover tens of thousands of sequences. Further, the length of the transcriptome sequence is in kilobases whereas, chromosomes can be hundreds of Megabases in length.

m) **RNA-Seq:** It has a wide range of applications. Different pipelines can be used based on the project work. It plays a key role between genomics and proteomics by transcript identification and quantification of gene expression in a single high-throughput sequencing assay. Scientists chose the specific analysis strategy based on research goals and the organism being studied. RNA-seq can be joined with other assays to analyze RNA–protein binding, RNA–RNA interaction and RNA structure, and many other aspects of RNA biology (Conesa et al. 2016).

n) **Celera Assembler/(Celera Assembler with best overlap graph (CABOG):** It is a Sanger era assembler and it discovers overlaps using compressed seed. Now it is revised and improved to assemble 454 data. Celera leverage the strengths of two or more sequencing platforms (Hybrid sequencing) for specific read type combinations (Hall 2007). Celera Assembler was more sensitive to the different read lengths between platforms and sensitive to the higher coverage typical of lower-cost pyrosequencing. It reduces the repeats of single letters that are reduced homopolymer runs to overcome homopolymer run-length uncertainty in data. CABOG evades initial more susceptible repeat-induced false overlaps to improve efficacy. Instead of fixing the read, it adjusts the tabulated error rates in overlaps spanning the incidental error and it applies a user-decided threshold for error rates (Miller et al. 2010).

o) **BreakDancer:** It consists of two complementary algorithms. The first one is BreackDancerMax, which provides genome-wide identification of addition, deletion, inversion, and inter and extrachromosomal translocation. The second one is BreackDanceMini, specially designed for detecting 10–100bp INDELs that are not routinely detected by BreackDancerMax. In population-based investigation, pooling improves the detection of common variation whereas, both BreackDancerMax and BreackDanceMini algorithms are suitable for pooled data analysis by eliminating inherited variation (Chen et al. 2009). BreakDancer run in two steps. First, The first few thousand reads in BAM input file by "bam2cfg.pl" and product configuration. Later these files were used by "BreackDancerMax" algorithm to identify and cluster DRPs (discordant read-file), ultimately yielding a list of putative SNVs (single nucleotide variants).

p) **PEMer:** High resolution and massive paired-end mapping (PEM) has the upper hand over other SNP genotyping-based algorithms and microarray-based approaches by allowing high-sensitivity and resolution SV reconstruction by eliminating cross-hybridization in repeat-rich regions. PEM is more economical than split read or assembly comparison analysis (Korbel et al. 2009).

q) **VarScan:** It is an independent mutation caller platform for exome, targeted, and whole-genome resequencing data generated by different NGS instruments. The new version of VarScan 2 is written in Java, so it can work in most of the computer operating systems (Linux, Mac, operating system X (OSX), UNIX, even windows) through the Java Virtual Machine. It can be used to detect germline variants, multi-sample variants, somatic mutations, and somatic copy number alterations (http://dkoboldt.github.io/varscan). It can effectively call variants with data from multiple sequencing platforms, in both pooled and individual samples. VarScan offers a platform-independent, central platform tool for variant detection (Koboldt et al. 2009).

r) **Allpaths-LG:** It implements graph topology-based error correction to improve the speed, counting, and avoid misassembly. They assemble short read-pair data to resolve short repeats. Long contingents produced by joining several read pairs extended them along the available path by using different algorithms (Henson et al. 2012). It cannot run without a longer insert library and one overlapping read pair library and it will now accept PacBio reads to complete the task (Illumina® 2020, www.Illumina.com.).

s) **Velvet:** This assembler was designed to quickly build *contigs*, out of short-read database produced by different NGS technologies. This function is very advantageous when a new organism for which there is no reference genome data is available. It builds *Bruijn* graph with no errors (Zerbino and Birney 2008). Velvet uses *velveth* and *velvetg* programs together. *VelvetOptimiser* was used to tune some parameters to automate the process. *Velveth* builds a dictionary of all words of length k using a sequence file. Whereas, *Velvetg* builds a de Bruijn graph from dictionary produced by *Velveth* and finally produces an error-free, simplified graph and resolves repeats based on the parameters provided by the user (Zerbino 2010).

t) **Sequence search and alignment by hashing algorithm (SSAHA):** It is used for performing fast searches on the database containing multiple Gb of DNA. Initially, the sequence databases are divided into consecutive $k$-tuples of $k$ contiguous bases and then the hash table is used to store the position of occurrence of each $k$-tuple. Later, searching for query sequence is done by obtaining from the hash table the "hits" for each $k$-tuple. This algorithm is used for very large-scale assembly and high-throughput SNP detection (Ning et al. 2001).

u) **String graph assembler (SGA):** It reconstructs a target sequence from a set of sequence reads in three phases. Those are overlap, layout, and consensus. Overlap graph is prepared by matching suffix and prefix of all pairs of sequence reads. In the layout phase, the locations of the reads are determined and in the consensus phase, the base for each position is selected to reconstruct the target sequence. It is a time-consuming process. Therefore, Myers (1995) proposed an assembly string graph as an alternative to an overlap graph. It has two most important advantages over the Brujin graph, at first, it does not require splitting the reads database into $k$-mers and the second one is, it always maintains read coherence. SGA uses an algorithm based on the BWT to compute suffix and prefix matches so that a string graph directly be constructed (Gonnella and Kurtz 2012).

v) **IMPUTATION:** It aims to fill in missing variant calls by comparing reference genomes using haplotypic information to detect similar haplotypes between two data sets (Marchini and Howie 2010). It can compare variants present in phylogenetic near neighbor to impute mutation for a set of haploid nonrecombining samples. It also corrects erroneous variants which can appear on the phylogenetic tree as reversion. But the appearance of reversions higher than expected rate may indicate sequence errors. Imputation can operate on any given haploid data without using separate reference data (Jobin et al. 2018).

## 8.5 Generalized Procedure for NGS Data Analysis

NGS enables sequencing of genome or transcriptome at a low-cost, high-throughput manner. Illumina, Roche, ABI/SOLiD like NGS platforms used for wet lab analysis whereas BWA, Galaxy, Bowtie, SanGenix like computational tools are used for dry-lab analysis (Wadapurkar and Vyas 2018). There are presently three generations of sequencing technologies are available. In first-generation, sequencing technology is developed based on the principle of DNA chain termination. Where dideoxynucleotide triphosphate (ddNTPs) with four fluorochromes were used and the sequence of base pairs is identified by the fluorescence detection device. This method produces long contiguous sequence reads (>500 bp). It is a time-consuming method and at present, it is frequently used for validating results of NGS studies. Second-generation sequencing is also known as NGS or high-throughput sequencing which can create millions of short reads within a few hours with more exactness (Thomas et al. 2007). These have been created by Roche 454 (Margulies et al. 2005), Illumina (Bennett 2004; Bentley et al. 2008), and SOLiD (Metzker 2010). Finally, third-generation sequencing can sequence the whole genome within a matter of hours at little or no human efforts. But presently, it is under the developmental phase.

In DNA sequencing, targeted regions such as exomes and intromes or whole chromosome or whole genome are targeted based on demand. This can be performed in two distinct ways (i) reference genome dependent and (ii) *de novo* (when reference sequence not available). There are three most important fundamental steps behind the reference genome approach. (i) DNA molecules are randomly fragmented by using restriction enzymes or mechanical forces, (ii) library of known insert sizes are prepared from such fragments, and (iii) these fragments are sequenced and mapped back to a known related reference sequence present in any database (Walter et al. 2012). In *de novo*, the small sequence fragments are assembled into contigs and a consensus sequence is obtained from these contigs (Figure 8.2).

NGS platforms produce a huge amount of unprocessed data to be further processed. The analysis of this raw data is done in five different steps. (i) determination of the quality of the raw data, (ii) alignment of reads to reference genome, (iii) variant identification, (iv) annotation of the variants, and (v) data visualization. The steps involved in NGS data analysis are presented in Figure 8.3.

### 8.5.1 Assessment of Quality

In this step the quality of NGS reads is evaluated for quality, trim, or remove the reads which are not meeting the required standards. Tools like FASTQC are used to assess base calling errors, poor quality reads, etc., with the calculation of quality scores (Dai et al. 2010).

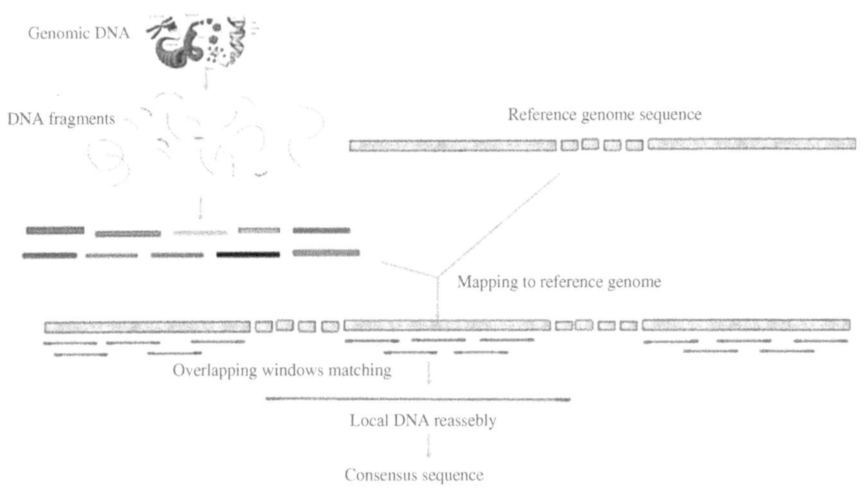


**Figure 8.2** Sequencing and assembly of DNA.

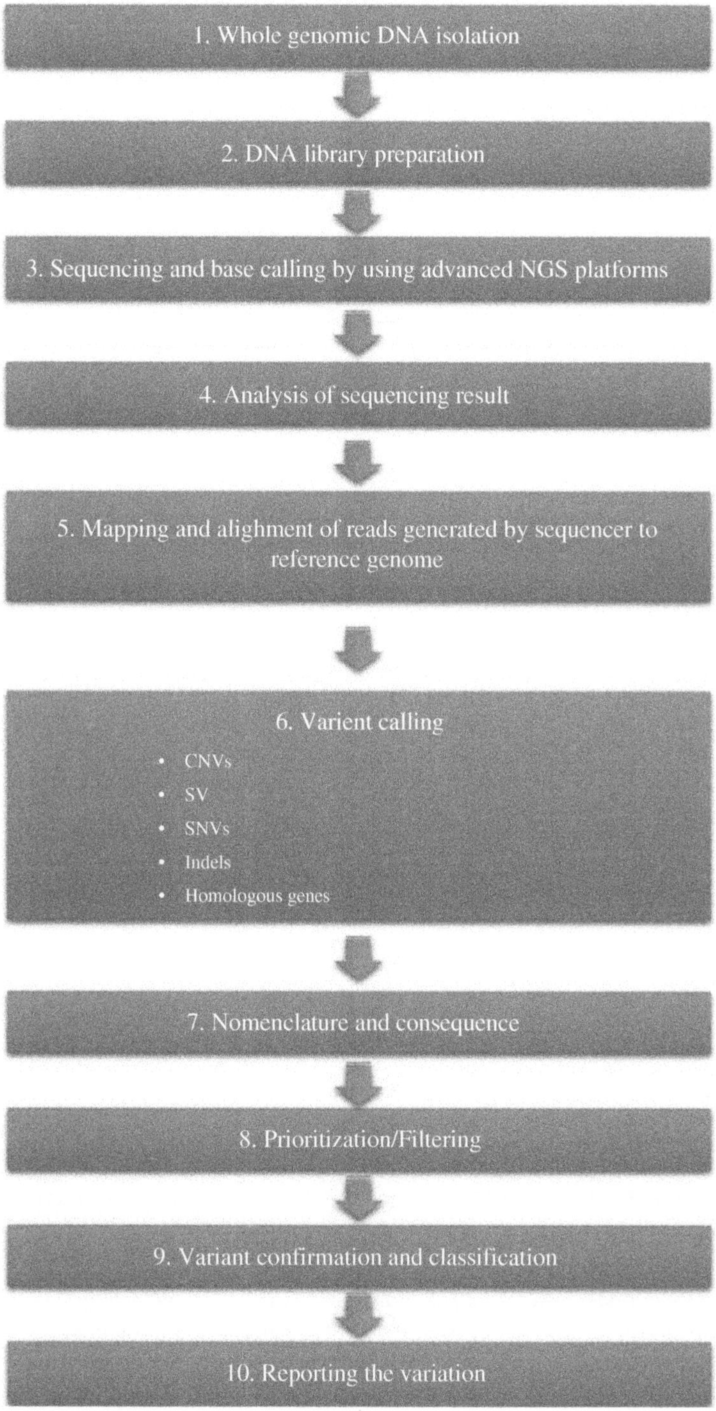


**Figure 8.3** The workflow showing the steps involved in NGS data analysis. *Source:* The figure is adapted after modification from Wadapurkar and Vyas (2018).

### 8.5.2 Aligning Sequences

After the assessment and improvement of read quality, the reads are aligned to the reference genome. For example, GRC (Genome Reference Consortium) and UCSC (University of Santa Cruz) are used as a source of the human reference genome (Nielsen et al. 2011; Rosenbloom et al. 2010). Ambiguity in mapping short reads to the reference genome is resolved by considering paired-end reads (Lee and Schatz 2012). The reads with many mismatches must be removed to improve the accuracy of prediction.

### 8.5.3 Identification of Variants

It is an important part of the NGS data analysis system which involves the identification of mutations that should be supported by several reads (Neuman et al. 2013). Variant identification tools are classified into four groups: (i) germline callers, (ii) somatic callers, (iii) CNVs, and (iv) Structural Variants Identification. The list of variant identification tools is provided in Table 8.1.

**Table 8.1** Different variant identification tools.

| S. No | Tools | Input file format | Output fileformat | Identifies | Use |
|---|---|---|---|---|---|
| 01. | Galaxy platform | BAM/SAM | VarScan, VSC, VCF | INDEL, SNP | Germline caller and Somatic caller |
| 02. | SanGenix platform | BAM/SAM | VarScan, VSC, VCF | INDEL, SNP | Germline caller and Somatic caller |
| 03. | VarScan2 | Pileup/mpileup | VarScan, VSC, VCF | INDEL, SNP | Germline caller and Somatic caller |
| 04. | SNVer | BAM/SAM | VCF | INDEL, SNP | Germline caller |
| 05. | GATK | BAM/SAM | VCF | INDEL, SNP | Germline caller and Somatic caller |
| 06. | SAMtools | FASTA, BAM/SAM | VCF | INDEL, SNP | Germline caller and Somatic caller |
| 07. | ExomeCNV | Pile+bed+FASTA, BAM/SAM | CSV | LOH, CNV | CNV identification |
| 08. | CNVnator | FASTA, BAM/SAM | CSV | CNV | CNV identification |
| 09. | CONTRA | FASTA, BAM/SAM | CSV, VCF | CNV | CNV identification |
| 10. | RDXplorer | FASTA, BAM/SAM | CSV | CNV | CNV identification |
| 11. | GASVPro | BAM/SAM | Cluster file | INV, INDEL, TRANS | SV identification |
| 12. | CLEVER | FASTA, BAM/SAM | CLEVER format | INDEL | SV identification |
| 13. | BreakDancer | Config file, BAM/SAM | CSV, BED | CNV, INDEL, INV | SV identification |
| 14. | Breakpointer | BAM/SAM | GFF | INDEL | SV identification |
| 15. | SVMerge | FASTA, BAM/SAM | BED | INV, INDEL, CNV | SV identification |

## 8.6 Variant Annotation

It involves the addition of secondary knowledge and metadata to quality-filtered variant calls to enhance the assessment of variants. While information regarding known allele frequencies, phenotypic information, and predicted lethal effects are commonly added (Sefid Dashti and Gamieldien 2017). Computer annotation tools provide annotation of INDELs and SNPs. The various annotation tools are listed in Table 8.2.

### 8.6.1 Visualization of NGS Data

Various genome browsers and visualization tools are used to visualize the NGS data. By this, we get information regarding varients, such as MAQ, aligned readers and annotation information, etc. (Nielsen et al. 2010).

## 8.7 Role of NGS Informatics in Identifying Variants

Presently, genomics playing important role in innovations, discoveries, and applications (Metzker 2010; Edwards et al. 2013; Kilpinen and Barrett 2013). The combination of NGS and reduced representation methods of genotyping has made it possible to simultaneously perform

**Table 8.2** Tools for variant annotation.

| S. No | Tools | Input file format | Output file format | Variants |
|---|---|---|---|---|
| 01. | Galaxy platform | Pileup/TXT, VCF | TXT, VCF, HTML overview | INDEL, SNP |
| 02. | SanGeniX platform | Pileup/TXT, VCF | TXT, VCF, HTML overview | INDEL, SNP |
| 03. | VARIANT | BED, GFF2, VCF | Web report, TXT | INDEL, SNP, CLI, Web |
| 04. | snpEff | VCF | HTML, VCF, overview TXT | INDEL, SNP, CLI |
| 05. | NGS-SNP | VCF, MAQ, pileup, diBAYes, TXT | TXT | CLI, SNP |
| 06. | VEP | Pileup, HGVS, VCF, TXT, variant identifiers | TXT | INDEL, SNP, CNV, CLI |
| 07. | ANNOVA-R | GFF3-SOLiD, VCF, SOAPsnp, pileup, MAQ, CASAVA, complete genomics | TXT | INDEL, CNV, CLI, SNP |
| 08. | AnnTools | Pileup, VCF, TXT | VCF | INDEL, CNV, CLI, SNP |
| 09. | SettleSeq | GATK, BED, VCF, custom, CASAVA, MAQ | SeattleSeq, VCF | INDEL, SNP, Web |
| 10. | SVA | SV.events file, VCF, BCO | CSV | SNP, CNV, INDEL, GUI,CLI |

SNPs discovery and genotyping in a single step even in species with large genomes (Davey et al. 2011) by using several related methods such as CRoPS < RAD-seq, double-digest RAD-seq, GBS, and type IIb restriction site–associated DNA (2bRAD) (van Orsouw et al. 2007; Wang et al. 2012; Etter et al. 2007). These various methods make it possible to solve the problems related to population genetics, ecological genetics, molecular breeding, and evolution by using millions of genetic markers in a wide array of species (Davey et al. 2011). GBS is a simple, high-throughput, low-cost, attractive complexity reduction method for genotyping in both model and nonmodel species (Elshire et al. 2011). Because of the ease to use, the GBS method of Elshire et al. (2011) and other closely related methods have been already applied by numerous scientists in many organisms (Lu et al. 2013; Morris et al. 2013).

## 8.8 Genotyping by Sequencing

GBS is a cost-effective powerful method to obtain genome-wide variability information for a population (Crossa et al. 2013; Morris et al. 2013; Spindel et al. 2015). In brief, the GBS protocol starts with restriction enzyme-mediated DNA digestion aiming to reduce the genome size. Then, all these fragments are ligated to adopters followed by amplification, and sequencing is done through the specialized platform (Bentley et al. 2008). The low cost and its flexibility make this an excellent tool for many applications in genetics and plant breeding activities which allows for the genotyping of thousands of millions of SNPs and improves the probability of identifying traits linked to SNPs (Kumar et al. 2012). Barcodes are used when multiple samples to be sequenced in one single lane (Elshire et al. 2011). The sequenced output data reads can then be demultiplexed and either aligned to a reference genome or analyzed *de novo*. Variation present in the population can be identified by using bioinformatics analysis pipelines (ICGMC 2015). GBS has several applications in plant breeding such as estimation of breeding values in GS (Spindel et al. 2015), analysis of population dynamics (Morris et al. 2013), and genetic mapping of complex traits through GWAS (Romay et al. 2013). Bioinformatics pipeline is the key component of any GBS protocol to obtain the polymorphic information by analyzing reads. Even though advancement in sequencing platforms produces thousands of sequence reads per run, data analysis becomes more complex due to the use of restriction enzymes, sample multiplexing, and variable read length (Davey et al. 2011). Advanced analysis pipeline for GBS data analysis must include filtration step to filter out poor-quality reads, shorting of reads by an individual or pooled samples based on sequence barcode and align the sequence data to discover polymorphism and often score genotypes for each individual included in the study (Bradbury et al. 2007). Currently, different GBS platforms are available and offer different advantages to determine the variation among individual genotypes.

## 8.9 Analytical Pipelines for GBS

The accurate and efficient computational processing, genotype, or variant calling of large-scale NGS data are the bottleneck in genomics. To overcome these problems several bioinformatics pipelines have been developed (Bradbury et al. 2007). All of them follow the same protocol from (i) obtaining raw sequence data, (ii) demultiplexed pooled sequence read data, (iii) filtering out low-quality reads, (iv) aligning or assembling reads to (v) discovering polymorphic loci. Each step has its specific challenges and ambiguities (Nielsen et al. 2011). On the other hand, technical factors such as the degree of DNA quality, sample multiplexing, read length, the total number of reads produced per sample, and the sequence error rate interacts with these biological factors (Lynch 2009). Therefore, the selection of appropriate parameters (quality of read mapping, depth of coverage, etc.) is necessary for successful mapping of the reads. At present, several GBS bioinformatics pipelines are available to process raw GBS sequence data to identify SNPs (Torkamaneh et al. 2016). In general, bioinformatics pipelines are categorized into *de novo* and reference-based GBS data handling. Up to now, several genome references based GBS analysis pipelines have been developed. The most commonly used reference-based GBS analysis pipelines are Stacks, trait analysis by association evolution and linkage-genotyping-by-sequencing (TASSEL-GBS) (v1 and v2), and IBIS genotyping-by-sequencing tool (IGST) (Sonah et al. 2013).

a) **TASSEL-GBS:** To minimize the several bottlenecks in bioinformatics, researchers developed TASSEL-GBS analysis pipeline in Java script. This pipeline can accept nearly all restriction enzymes and barcoding approaches. Currently, it accepts 15 restriction enzymes and easily new enzymes can be added without any major modification (Mascher et al. 2013). As compared to pipelines developed in the past. TASSEL-GBS can handle large quantities of data from a large number of samples. TASSEL-GBS is designed for both species with reference genome and pseudo-reference genome where incomplete genome assemblies consisting of numerous contigs are available (Lu et al. 2013). This TASSEL-GBS specially designed (i) to accelerate high-throughput plant breeding, (ii) accommodate the high-throughput diversity, and (iii) provide a platform that can run with modest computational resources in many contexts.
b) **Stacks:** It is a restriction enzyme-based GBS data analysis pipeline which can work with any restriction enzyme without any difficulty. It is computationally flexible and more robust and it can align data to a reference genome or it can assemble short reads *de novo*. Stack software can incorporate a Michael Widenius's structured query language (MySQL) database and web front end for efficient data visualization, management, and modification. Initially, this software was developed for genetic mapping (Catchen et al. 2011) and later updated for ecological and evolutionary genomic analysis. Stacks can handle GBS raw data in FASTQ or FASTA format to align reads against reference genome in SAM and to identify loci *de novo* (Li et al. 2009). Stacks uses *k-mer* search algorithm to merge alleles into loci. The Stacks pipeline follow a specified procedure in five major stages: (i) raw input sequence data/reads are demultiplexed and cleaned, (ii) data from each sample are grouped into loci and identify the polymorphic sites, (iii) catalog is written by grouping loci across the individual, (iv) allelic sites at each locus in each individual is determined by matching loci from each loci to catalog, and (v) allelic states are either subjected to population genetic statistics via populations or converted into a set of mappable genotypes (Catchen et al. 2013a). For *de novo* stack formation, *k-mer* search algorithm is used to merge alleles into loci. First, the hashing algorithm is used to form stacks by using exactly matching reads, and subsequently, these stacks are decomposed to *k-mers* (Catchen et al. 2013b). While pstacks program is used for stacks formation when a reference genome is available. Pstacks can use data

from any alignment program that can produce an output file in SAM or BAM format and has been extensively tested with BWA (Li and Durbin 2009), Bowtie (Langmead et al. 2009), and genomic short-read nucleotide alignment program (GSNAP) (Wu and Nacu 2010).

Bounded-error model is used to identify the SNPs. Stacks deploy a multinomial-based likelihood model in the diploid organism for SNPs identification (Catchen et al. 2011). This model works by estimating the maximum-likelihood value of the sequence error at each base pair for each possible genotype for both stacks and pstacks and then calculating the likelihood of the two most frequently observed genotypes (Hohenlohe et al. 2011). Finally, chi-square distribution and one-degree freedom are used to perform a standard likelihood ratio to test the two hypotheses. Stacks population also by default reports which haplotypes are present in each individual in the analysis, and it uses letters or some other meaningful scheme to encode these haplotypes. Furthermore, to produce longer haplotypes from randomly sequenced paired ends by increasing read lengths of common sequencing platforms (Etter et al. 2011).

c) **IGST-GBS:** This pipeline was developed in pearl programming language, and it uses BWA, FASTAX, etc. tools for sequencing data analysis. Whereas, SNP detection is done based on missing data, read depth, and minor allele frequencies (MAF) (Sonah et al. 2013). Very little information is available w.r.t IGST-GBS in the public domain.
d) **Fast-GBS:** This analysis platform is available without any restriction and can be run with different types of sequencing data and modern computational resources (Torkamaneh et al. 2017a). This pipeline was developed by assimilating public packages with internally developed tools. The public packages include (i) Sabre for demultiplexing (Torkamaneh et al. 2017b), (ii) Cutadapt for read trimming and cleaning (Martin 2011), (iii) BWA for read mapping (Li 2011), (iv) SAMtools for file conversion and indexing (Li 2011), and (v) Platypus for postprocessing of reads, haplotype construction, and variant calling (Rimmer et al. 2014).

Fast-GBS require raw input data in FASTQ format. The parameter files were prepared for Fast-GBS. This file contains information regarding the type of sequence, the adapter sequence, sequencing technology, path to the FASTQ files, barcodes, and reference genome. In this file, we can define critical filtering options such as a minimal number of reads required to call a genotype, minimal quality scores for reads, and maximum amount of missing data should be allowed. This file also comes with the Fast-GBS pipeline. Fast-GBS uses Sabre (Torkamaneh et al. 2017b) to demultiplex barcode reads into separate files since resulting pooled reads will need to be demultiplexed before SNP calling. It separates the reads into appropriate barcodes based on comparing 5′ end of each read with barcodes. Those mismatches with barcodes are put into an "unknown" file. After demultiplexing, Fast-GBS pipeline uses Cutadapt to find and trim or clean primers, adapter sequences, and other types of unwanted sequences from high-throughput sequencing read (Martin 2011).

Fast-GBS pipeline uses the MEM algorithm that works by seeding alignment followed by extending seed similar to SW algorithm (Li and Durbin 2010). This algorithm can perform parallel alignment which improves the speed of the analysis. After initial alignment, Fast-GBS pipeline further process the initial aligned data by using Platypus (28) to improve the specificity and sensitivity of variant calling. Platypus improves the quality by re-examination and classifies the poorly mapped reads into three distinct classes: (i) reads with numerous mismatches, (ii) reads mapping to multiple locations in the genome, and (iii) remaining adopter sequence or adaptor sequences and are highlighted using BadReads flag. For haplotype construction and variant calling, Fast-GBS uses Platypus. Platypus helps to exploit information from multiple samples by using multi-sample variant calling. This tactic reduces the errors around large variants (multiple nucleotide polymorphism (MNPs)) and INDELs. Local assembler first read a few kb (small window) followed by generation of Bruijn graph using all the reads and then it generates an exhaustive list of haplotypes by using all candidate variant. Expectation-maximization (EM) algorithm utilized in haplotype frequencies estimation (Torkamaneh et al. 2017a).

The main output file generated contains detailed information regarding each of the variants in vcf file format (Danecek et al. 2011) and Fast-GBS also generates a simple text file containing only the genotype data. In case the running of the Fast-GBS pipeline prematurely terminates due to some error, it generates a log file which contains the information regarding at which step analysis stopped and the reason. After error correction, analysis can be started at any point on the existing intermediate file (Torkamaneh et al. 2017a).

e) **Universal Network Enabled Analysis Kit (UNEAK):** It is a part of TASSEL 3.0 which is independent of the reference sequence for SNP discovery (*de novo*). In UNEAK, identical reads are trimmed to 64 bp and are collapsed into tags. The pairwise alignment is utilized to recognize base pair mismatches. There are several sequencing errors, paralogs and repeats are present in raw sequencing data which are rejected by using network filters. Those 64 bp TAGs present multiple times across all samples are used for TAG pair identification. UNEAK rejects half of TM SNPs automatically and maintains 1 SNP per 64 bp. Among all *de novo* pipelines, UNEAK requires three times more memory space but completes the task in 1/3rd time of other pipelines (Torkamaneh et al. 2016). However, this pipeline works when the input file is in BAM format (Berthouly-Salazar et al. 2016).
f) **NGS experience platform (NGSEP):** Deconvolution process is a novel functionality implemented in NGSEP to distribute the reads in one Illumina lane and produces a separate file for each sample. This method can be applied to any type of sequencing where samples are marked by suitable barcodes. NGSEP receives lane information containing fastq file and a text file describing the link between samples and barcodes. This method is very important in various GBS experiments where 96–384 samples are sequenced per lane to improve cost efficiency (Perea et al. 2016). NGSEP can be deployed as a standalone command-line program, using the extensible markup language (XML) scripts or a graphical user interface (GUI) for integration in the galaxy environment (Goecks et al. 2010). This pipeline depends on the libraries jsci-core for statistics, sam-1.68 to handle sam/bam files, the SortSam package of Picard to sort alignments, and Xchart to produce graphical interference (Torkamaneh et al. 2016). NGSEP reads and writes BAM, virtual contact file (VCF), or general feature format (GFF) standard formats for genomic analysis.

In this NGSEP pipeline, one step wizard for parallel multisample analysis is introduced to greatly reduce the amount of work and time needed by researchers to obtain variation datasets. For error control, the user has an option to skip analysis in any step of the problematic sample analysis process and finish the process including only the successful samples or abort the whole process at any step. In this module, most of the filtering strategies are introduced to maintain the balance between genotype quality and variant genome coverage which are required for different applications. Genotype calls can be filtered based either on total read coverage or genome quality. This helps to focus on a particular region of the genome and removes markers in undesired regions. SNPs can be selected based on several criteria's such as MAF, number of individual genotypes, and distance between markers and functional annotations. This GBS pipeline has a conversion module to translate genotype data into input format needed by several tools such as DARwin to build distance-based dendrogram (DaRwin 2016), the R package (ridge regression (RR), which is equivalent to best linear unbiased prediction (BLUP) (rrBLUP)) to perform estimation and cross-validation of breeding values for GS (Spindel et al. 2015) and JoinMap for construction of genetic maps (Van Ooijen 2011).

GBS has one major drawback compared to other high-throughput genotyping platforms. It produces missing data by the random distribution of reads coverage both across genomic sites as well as across samples. Although several imputation platforms are developed to deal with this problem. For this, NGSEP was also implemented with a haploid version of the hidden Markov model (HMM) in fastPHAST to perform imputation in the population of related or unrelated members and inbred lines.

## 8.10 Comparison of GBS Pipelines

Here, we compared the two *de novo* pipelines (UNEAK and Stacks) that do not require a reference genome and five pipelines requiring a reference genome (Stacks, TASSEL-GBS, IGST, and Fast-GBS). Among them, Fast-GBS can call the greatest number of polymorphisms and yield the highest accuracy for Illumina sequence data (Torkamaneh et al. 2016). TASSEL-GBS v1 is becoming the fastest running pipeline among reference-based pipelines (almost 1 h 45 min), whereas IGST took 13 h to complete the same task and it became the slowest one among them. Among *de novo* pipelines, UNEAK completes task almost three times faster than Stacks and become one of the fastest among all pipelines. Regarding memory requirement, IGST requires 17.1-fold more memory (240 GB) than stacks in among the reference-based pipelines whereas, STACKS required almost three times less memory space compared to UNEAK in among *de novo* pipelines (Torkamaneh et al. 2016). For the accuracy and efficiency of GBS bioinformatics pipelines, GBS-called SNPs were performed at the same level of tolerance for missing data (≤80%) and MAF (≥0.05) on all SNPs. Among *de novo* pipelines, the proportion of missing data was less variable compared to reference-based pipelines, ranging from 39.4% in Stacks to 41.3% in UNEAK. Whereas, in reference-based pipelines, the missing data varied from 28% in TASSEL GBS v1 to as much as 57.3% in Stacks. When compared *de novo* pipelines (UNEAK and Stacks), UNEAK greatly outperformed Stacks by generating 82% more SNPs but the similarity was observed in terms of genotypic accuracy (~94%) and missing data (~40%) (Lu et al. 2013). Among reference-based pipelines, Fast-GBS can produce the highest genotyping accuracy for a very large population in a short period. TASSEL-GBS v1 is one of the tested pipelines which stood out from the rest of the group in terms of the number of SNP loci called (50–100% more than others) but whose genotype calls were accurate in less than 90% of cases and it is not easy to distinguish true from false genotypes. So, TASSEL-GBSv1 is inefficient to be used on its own. According to many previous studies, estimates of genotypic accuracy were often achieved by performed on a very small subset of SNP loci (Sonah et al. 2013) or by indirect measurement (Lu et al. 2013).

## References


Alkan, C., Kidd, J.M., Marques-Bonet, T. et al. (2009). Personalized copy number and segmental duplication maps using next-generation sequencing. *Nature Genetics* 41: 1061–1067.

Baird, N.A., Etter, P.D., Atwood, T.S. et al. (2008). Rapid SNP discovery and genetic mapping using sequenced RAD markers. *PLoS One* 3: e3376.

Bateman, A. and Quackenbush, J. (2009). Bioinformatics for next generation sequencing. *Bioinformatics* 25: 429.

Bennett, S. (2004). Solexa ltd. *Pharmacogenomics* 5: 433–438.

Bentley, D.R., Balasubramanian, S., Swerdlow, H.P. et al. (2008). Accurate whole human genome sequencing using reversible terminator chemistry. *Nature* 456: 53–59.

Berthouly-Salazar, C., Mariac, C., Couderc, M. et al. (2016). Genotyping-by-sequencing SNP identification for crops without a reference genome: using transcriptome based mapping as an alternative strategy. *Frontiers in Plant Science* 7: 777.

Bradbury, P.J., Zhang, Z., Kroon, D.E. et al. (2007). TASSEL: software for association mapping of complex traits in diverse samples. *Bioinformatics* 23: 2633–2635.

Burrows, M. and Wheeler, D.J. (1994). *A Block Sorting Lossless Data Compression Algorithm*. Palo Alto, CA: Digital Equipment Corporation Technical Report 124.

Catchen, J.M., Amores, A., Hohenlohe, P. et al. (2011). Stacks: building and genotyping loci *de novo* from short-read sequences. *G3: Genes, Genomes, Genetics* 1: 171–182.

Catchen, J., Bassham, S., Wilson, T. et al. (2013a). The population structure and recent colonization history of Oregon three-spine stickleback determined using RAD-seq. *Molecular Ecology* 22: 2864–2883.

Catchen, J., Hohenlohe, P.A., Bassham, S. et al. (2013b). Stacks: an analysis tool set for population genomics. *Molecular Ecology* 22: 3124–3140.

Chen, K., Wallis, J.W., McLellan, M.D. et al. (2009). BreakDancer: an algorithm for high-resolution mapping of genomic structural variation. *Nature Methods* 6: 677–681.

Conesa, A., Madrigal, P., Tarazona, S. et al. (2016). A survey of best practices for RNA-seq data analysis. *Genome Biology* 17: 13.

Craig, D.W., Pearson, J.V., Szelinger, S. et al. (2008). Identification of genetic variants using bar-coded multiplexed sequencing. *Nature Methods* 5: 887–893.

Crossa, J., Beyene, Y., Kassa, S. et al. (2013). Genomic prediction in maize breeding populations with genotyping-by-sequencing. *G3: Genes, Genomes, Genetics* 11: 1903–1926.

Dai, M., Thompson, R.C., Maher, C. et al. (2010). NGSQC: cross-platform quality analysis pipeline for deep sequencing data. *BMC Genomics. BioMed Central* 11: 1–9.

Danecek, P., Auton, A., Abecasis, G. et al. (2011). The variant call format and VCFtools. *Bioinformatics* 27: 2156–2158.

DARwin (2016). Dissimilarity analysis and representation for windows. http://darwin.cirad.fr/ (accessed 26 February 2021).

Davey, J.W., Hohenlohe, P.A., Etter, P.D. et al. (2011). Genome-wide genetic marker discovery and genotyping using next-generation sequencing. *Nature Reviews Genetics* 12: 499–510.

de Sá, P.H., Guimarães, L.C., das Graças, D.A. et al. (2018). Next-generation sequencing and data analysis: strategies, tools, pipelines and protocols. In: *Omics Technologies and Bio-Engineering*, 191–207. Academic Press.

Edwards, D., Batley, J., and Snowdon, R.J. (2013). Accessing complex crop genomes with next-generation sequencing. *Theory of Applied Genetics* 126: 1–11.

Egan, A.N., Schlueter, J., and Spooner, D.M. (2012). Applications of next-generation sequencing in plant biology. *American Journal of Botany* 99: 175–185.

Elshire, R.J., Glaubitz, J.C., Sun, Q. et al. (2011). A robust, simple genotyping-by-sequencing (GBS) approach for high diversity species. *PLoS One* 6: e19379.

Etter, P.D., Atwood, T.S., Currey, M.C. et al. (2007). Rapid SNP discovery and genetic mapping using sequenced RAD markers. *PLoS One* 3: e3376.

Etter, P.D., Bassham, S., Hohenlohe, P.A. et al. (2011). SNP discovery and genotyping for evolutionary genetics using RAD sequencing. *Methods in Molecular Biology* 772: 157–178.

Ferragina, P. and Manzini, G. (2000). Opportunistic data structures with applications. *Proceedings 41st Annual Symposium on Foundations of Computer Science*, 390–398. IEEE.

Goecks, J., Nekrutenko, A., and Taylor, J. (2010). Galaxy: a comprehensive approach for supporting accessible, reproducible, and transparent computational research in the life sciences. *Genome Biology* 11: 86.

Gonnella, G. and Kurtz, S. (2012). Readjoiner: a fast and memory efficient string graph-based sequence assembler. *BMC Bioinformatics* 13: e82.

Guttman, M., Garber, M., Levin, J.Z. et al. (2010). *Ab initio* reconstruction of cell type–specific transcriptomes in mouse reveals the conserved multi-exonic structure of lincRNAs. *Nature Biotechnology* 28: 503–510.

Hall, N. (2007). Advanced sequencing technologies and their wider impact in microbiology. *Journal of Experimental Biology* 210: 1518–1525.

Hayden, E. (2009). Genome sequencing: the third generation. *Nature* 457: 768–769.

He, J., Zhao, X., Laroche, A. et al. (2014). Genotyping-by-sequencing (GBS), an ultimate marker-assisted selection (MAS) tool to accelerate plant breeding. *Frontiers in Plant Science* 5: 484.

Henson, J., Tischler, G., and Ning, Z. (2012). Next-generation sequencing and large genome assemblies. *Pharmacogenomics* 13: 901–915.

Hohenlohe, P.A., Amish, S.J., Catchen, J.M. et al. (2011). Next-generation RAD sequencing identifies thousands of SNPs for assessing hybridization between rainbow and westslope cutthroat trout. *Molecular Ecology Resources* 11: 117–122.

Homer, N., Merriman, B., and Nelson, S.F. (2009). BFAST: an alignment tool for large scale genome resequencing. *PLoS One* 4: e7767.

Hormozdiari, F., Alkan, C., Eichler, E.E., and Sahinalp, S.C. (2009). Combinatorial algorithms for structural variation detection in high-throughput sequenced genomes. *Genome Research* 19: 1270–1278.

Huang, W., Li, L., Myers, J.R., and Marth, G.T. (2012). ART: a next-generation sequencing read simulator. *Bioinformatics* 28: 593–594.

Illumina® (2020). www.Illumina.com (accessed 1 August 2020).

International Cassava Genetic Map Consortium (ICGMC) (2015). High-resolution linkage map and chromosome-scale genome assembly for Cassava (*Manihot esculenta* Crantz) from ten populations. *G3: Genes, Genomes, Genetics* 5: 133–144.

Jobin, M., Schurz, H., and Henn, B.M. (2018). IMPUTOR: phylogenetically aware software for imputation of errors in next-generation sequencing. *Genome Biology and Evolution* 10: 1248–1254.

Kilpinen, H. and Barrett, J.C. (2013). How next-generation sequencing is transforming complex disease genetics. *Trends in Genetics* 29: 23–30.

Koboldt, D.C., Chen, K., Todd, W. et al. (2009). VarScan: variant detection in massively parallel sequencing of individual and pooled samples. *Bioinformatics* 25: 2283–2285.

Korbel, J.O., Abyzov, A., Mu, X.J. et al. (2009). PEMer: a computational framework with simulation-based error models for inferring genomic structural variants from massive paired-end sequencing data. *Genome Biology* 10: p23.

Kumar, S., Banks, T.W., and Cloutier, S. (2012). SNP discovery through next-generation sequencing and its applications. *International Journal of Plant Genomics* 2012: 1–15.

Landegren, U., Kaiser, R., Caskey, C.T., and Hood, L. (1988). DNA diagnostics--molecular techniques and automation. *Science* 242: 229–237.

Langmead, B., Trapnell, C., Pop, M., and Salzberg, S.L. (2009). Ultrafast and memory-efficient alignment of short DNA sequences to the human genome. *Genome Biology* 10: 25.

Langmead, B. and Salzberg, S.L. (2012). Fast gapped-read alignment with Bowtie 2. *Nature Methods* 9: 357–359.

Lee, H. and Schatz, M.C. (2012). Genomic dark matter: the reliability of short read mapping illustrated by the genome mappability score. *Bioinformatics* 28: 2097–2105.

Li, H., Ruan, J., and Durbin, R. (2008). Mapping short DNA sequencing reads and calling variants using mapping quality scores. *Genome Research* 18: 1851–1858.

Li, H. and Durbin, R. (2009). Fast and accurate short read alignment with Burrows–Wheeler transform. *Bioinformatics* 25: 1754–1760.

Li, H., Handsaker, B., Wysoker, A. et al. (2009). The sequence alignment/map format and SAMtools. *Bioinformatics* 25: 2078–2079.

Li, H. and Durbin, R. (2010). Fast and accurate long-read alignment with Burrows-Wheeler transform. *Bioinformatics* 26: 589–595.

Li, R., Zhu, H., Ruan, J. et al. (2010). *De novo* assembly of human genomes with massively parallel short read sequencing. *Genome Research* 20: 265–272.

Li, H. (2011). A statistical framework for SNP calling, mutation discovery, association mapping and population genetical parameter estimation from sequencing data. *Bioinformatics* 27: 2987–2993.

Li, H. (2013). Aligning sequence reads, clone sequences and assembly contigs with BWA-MEM. arXiv preprint arXiv:1303.3997.

Liu, H., Bayer, M., Druka, A. et al. (2014). An evaluation of genotyping by sequencing (GBS) to map the Breviaristatum-e (ari-e) locus in cultivated barley. *BMC Genomics* 15: 104.

Lu, F., Lipka, A.E., Elshire, R.J. et al. (2013). Switchgrass genomic diversity, ploidy and evolution: novel insights from a network-based SNP discovery protocol. *PLoS Genetics* 9: e1003215.

Luckey, J.A., Drossman, H., Kostichka, A.J. et al. (1990). High speed DNA sequencing by capillary electrophoresis. *Nucleic Acids Research* 18: 4417–4421.

Lynch, M. (2009). Estimation of allele frequencies from high coverage genome-sequencing projects. *Genetics* 182: 295–301.

Maher, B. (2008). Personal genomes: the case of the missing heritability. *Nature* 456: 18–21.

Manolio, T.A., Collins, F.S., Cox, N.J. et al. (2009). Finding the missing heritability of complex diseases. *Nature* 461: 747–753.

Marchini, J. and Howie, B. (2010). Genotype imputation for genome-wide association studies. *Nature Review of Genetics* 11: 499–511.

Mardis, E.R. (2008). Next-generation DNA sequencing methods. *Annual Review of Genomics and Human Genetics* 9: 387–402.

Margulies, M., Egholm, M., Altman, W.E. et al. (2005). Genome sequencing in microfabricated high-density picolitre reactors. *Nature* 437: 376–380.

Martin, M. (2011). Cutadapt removes adapter sequences from high-throughput sequencing reads. *European Molecular Biology Network* 17: 10–12.

Mascher, M., Wu, S., Amand, P.S. et al. (2013). Application of genotyping-by-sequencing on semiconductor sequencing platforms: a comparison of genetic and reference-based marker ordering in barley. *PLoS One* 8: e76925.

McKenna, A., Hanna, M., Banks, E. et al. (2010). The genome analysis toolkit: a MapReduce framework for analyzing next-generation DNA sequencing data. *Genome Research* 20: 1297–1303.

Metzker, M.L. (2010). Sequencing technologies - the next generation. *Nature Review of Genetics* 11: 31–46.

Miller, M., Dunham, J., Amores, A. et al. (2007). Genotyping using restriction site associated DNA (RAD) markers. *Genome Research* 17: 240–248.

Miller, J.R., Koren, S., and Sutton, G. (2010). Assembly algorithms for next-generation sequencing data. *Genomics* 95: 315–327.

Moorthie, S., Mattocks, C.J., and Wright, C.F. (2011). Review of massively parallel DNA sequencing technologies. *The HUGO Journal* 5: 1–12.

Morris, G.P., Ramu, P., Deshpande, S.P. et al. (2013). Population genomic and genome-wide association studies of agroclimatic traits in sorghum. *Proceedings of the National Academy of Sciences of the United States of America* 110: 453–458.

Mortazavi, A., Williams, B.A., McCue, K. et al. (2008). Mapping and quantifying mammalian transcriptomes by RNA-Seq. *Nature Methods* 5: 621–628.

Myers, E.W. (1995). Toward simplifying and accurately formulating fragment assembly. *Journal of Computational Biology* 2: 275–290.

Neuman, J.A., Isakov, O., and Shomron, N. (2013). Analysis of insertion–deletion from deep-sequencing data: software evaluation for optimal detection. *Briefings in Bioinformatics* 14: 46–55.

Nielsen, C.B., Cantor, M., Dubchak, I. et al. (2010). Visualizing genomes: techniques and challenges. *Nature Methods* 7: S5–S15.

Nielsen, R., Paul, J.S., Albrechtsen, A., and Song, Y.S. (2011). Genotype and SNP calling from next-generation sequencing data. *Nature Review of Genetics* 12: 443–451.

Ning, Z., Cox, A.J., and Mullikin, J.C. (2001). SSAHA: a fast search method for large DNA databases. *Genome Research* 11: 1725–1729.

Nyrén, P. (2007). The history of pyrosequencing®. In: *Pyrosequencing® Protocols*, 1–13. Humana Press.

Ogasawara, T., Cheng, Y., and Tzeng, T.H.K. (2016). Sam2bam: high-performance framework for NGS data preprocessing tools. *PLoS One* 11: e0167100.

Orlando, L., Ginolhac, A., Raghavan, M. et al. (2011). True single-molecule DNA sequencing of a Pleistocene horse bone. *Genome Research* 21: 1705–1719.

Perea, C., De La Hoz, J.F., Cruz, D.F. et al. (2016). Bioinformatic analysis of genotype by sequencing (GBS) data with NGSEP. *BMC Genomics* 17: 539–551.

Pettersson, E., Lundeberg, J., and Ahmadian, A. (2009). Generations of sequencing technologies. *Genomics* 93: 105–111.

Poland, J.A., Brown, P.J., Sorrells, M.E., and Jannink, J.L. (2012). Development of high density genetic maps for barley and wheat using a novel two-enzyme genotyping-by-sequencing approach. *PLoS One* 7: e32253.

Quail, M.A., Smith, M., Coupland, P. et al. (2012). A tale of three next generation sequencing platforms: comparison of Ion Torrent, Pacific Biosciences and Illumina MiSeq sequencers. *BMC Genomics* 13: 341.

Rhoads, A. and Au, K.F. (2015). PacBio sequencing and its applications. *Genomics, Proteomics & Bioinformatics* 13: 278–289.

Rimmer, A., Phan, H., Mathieson, I. et al. (2014). Integrating mapping, assembly and haplotype-based approaches for calling variants in clinical sequencing applications. *Nature Genetics* 46: 912–918.

Romay, M.C., Millard, M.J., Glaubitz, J.C. et al. (2013). Comprehensive genotyping of the USA national maize inbred seed bank. *Genome Biology* 14: 55.

Ronaghi, M., Uhlén, M., and Nyrén, P. (1998). A sequencing method based on real-time pyrophosphate. *Science* 281: 363–365.

Rosenbloom, K.R., Dreszer, T.R., Pheasant, M. et al. (2010). ENCODE whole-genome data in the UCSC Genome Browser. *Nucleic Acids Research* 38: 620–625.

Rothberg, J.M., Hinz, W., Rearick, T.M. et al. (2011). An integrated semiconductor device enabling non-optical genome sequencing. *Nature* 475: 348–352.

Ruiqiang, L., Yingrui, L., Karsten, K., and Jun, W. (2008). SOAP: short oligonucleotide alignment program. *Bioinformatics* 24: 713–714.

Sanger, F., Nicklen, S., and Coulson, A.R. (1977). DNA sequencing with chain terminating inhibitors. *Proceedings of the National Academy of Sciences of the United States of America* 74: 5463–5467.

Scheben, A., Batley, J., and Edwards, D. (2017). Genotyping-by-sequencing approaches to characterize crop genomes: choosing the right tool for the right application. *Plant Biotechnology Journal* 15: 149–161.

Sefid Dashti, M.J. and Gamieldien, J. (2017). A practical guide to filtering and prioritizing genetic variants. *Biotechniques* 62: 18–30.

Shin, H., Liu, T., Duan, X. et al. (2013). Computational methodology for ChIP-seq analysis. *Quantitative Biology* 1: 54–70.

Shulaev, V., Sargent, D.J., Crowhurst, R.N. et al. (2011). The genome of woodland strawberry (*Fragaria vesca*). *Nature Genetics* 43: 109–116.

Simpson, J.T., Wong, K., Jackman, S.D. et al. (2009). ABySS: a parallel assembler for short read sequence data. *Genome Research* 19: 1117–1123.

Smith, L.M., Sanders, J.Z., Kaiser, R.J. et al. (1986). Fluorescence detection in automated DNA sequence analysis. *Nature* 321: 674–679.

Sonah, H., Bastien, M., Iquira, E. et al. (2013). An improved genotyping by sequencing (GBS) approach offering increased versatility and efficiency of SNP discovery and genotyping. *PLoS One* 8: e54603.

Sonah, H., O'Donoughue, L., Cober, E. et al. (2015). Identification of loci governing eight agronomic traits using a GBS-GWAS approach and validation by QTL mapping in soya bean. *Plant Biotechnology Journal* 13: 211–221.

Spindel, J., Begum, H., Akdemir, D. et al. (2015). Genomic selection and association mapping in rice (*Oryza sativa*): effect of trait genetic architecture, training population composition, marker number and statistical model on accuracy of rice genomic selection in elite, tropical rice breeding lines. *PLoS Genetics* 11: e1004982.

Thomas, R.K., Baker, A.C., DeBiasi, R.M. et al. (2007). High-throughput oncogene mutation profiling in human cancer. *Nature Genetics* 39: 347–351.

Thompson, J.F. and Steinmann, K.E. (2010). Single molecule sequencing with a HeliScope genetic analysis system. *Current Protocols in Molecular Biology* 92: 7–10.

Torkamaneh, D., Laroche, J., and Belzile, F. (2016). Genome-wide SNP calling from genotyping by sequencing (GBS) data: a comparison of seven pipelines and two sequencing technologies. *PLoS One* 11: e0161333.

Torkamaneh, D., Laroche, J., Bastien, M. et al. (2017a). Fast-GBS: a new pipeline for the efficient and highly accurate calling of SNPs from genotyping-by-sequencing data. *BMC Bioinformatics* 18: 5.

Torkamaneh, D., Laroche, J., Bastien, M. et al. (2017b). Fast-GBS: a new pipeline for the efficient and highly accurate calling of SNPs from genotyping-by-sequencing data. *BMC Bioinformatics* 18: 1–7.

Trapnell, C., Williams, B.A., Pertea, G. et al. (2010). Transcript assembly and quantification by RNA-Seq reveals unannotated transcripts and isoform switching during cell differentiation. *Nature Biotechnology* 28: 511–515.

Van Ooijen, J.W. (2011). Multipoint maximum likelihood mapping in a full-sib family of an outbreeding species. *Genetics Research* 93: 343–349.

van Orsouw, N.J., Hogers, R.C.J., Janssen, A. et al. (2007). Complexity reduction of polymorphic sequences (CRoPS): a novel approach for large-scale polymorphism discovery in complex genomes. *PLoS One* 2: e1172.

Voelkerding, K.V., Dames, S.A., and Durtschi, J.D. (2009). Next-generation sequencing: from basic research to diagnostics. *Clinical Chemistry* 55: 641–658.

Wadapurkar, R.M. and Vyas, R. (2018). Computational analysis of next generation sequencing data and its applications in clinical oncology. *Informatics in Medicine Unlocked* 11: 75–82.

Walter, M.J., Shen, D., Ding, L. et al. (2012). Clonal architecture of secondary acute myeloid leukemia. *New England Journal of Medicine* 366: 1090–1098.

Wang, S., Meyer, E., McKay, J.K., and Matz, M.V. (2012). 2b-RAD: a simple and flexible method for genome-wide genotyping. *Nature Methods* 9: 808–810.

Ward, J.A., Bhangoo, J., Fernández-Fernández, F. et al. (2013). Saturated linkage map construction in *Rubus idaeus* using genotyping by sequencing and genome-independent imputation. *BMC Genomics* 14: 2.

Wickland, D.P., Battu, G., Hudson, K.A. et al. (2017). A comparison of genotyping-by-sequencing analysis methods on low-coverage crop datasets shows advantages of a new workflow, GB-eaSy. *BMC Bioinformatics* 18: 1–12.

Wu, T.D. and Nacu, S. (2010). Fast and SNP-tolerant detection of complex variants and splicing in short reads. *Bioinformatics* 26: 873–881.

Yang, X. and Li, L. (2011). miRDeep-P: a computational tool for analyzing the microRNA transcriptome in plants. *Bioinformatics* 27: 2614–2615.

Zerbino, D.R. and Birney, E. (2008). Velvet: algorithms for *de novo* short read assembly using de Bruijn graphs. *Genome Research* 18: 821–829.

Zerbino, D.R. (2010). Using the velvet *de novo* assembler for short-read sequencing technologies. *Current Protocols in Bioinformatics* 31: 11–15.

Zhang, L., Liu, C., and Dong, S. (2019). PipeMEM: a framework to speed up BWA-MEM in spark with low overhead. *Genes* 10: 886.

Zhao, M., Wang, Q., Wang, Q. et al. (2013). Computational tools for copy number variation (CNV) detection using next-generation sequencing data: features and perspectives. *BMC Bioinformatics* 14: 1–16.

# 9

# Recent Advances and Applicability of GBS, GWAS, and GS in Maize

*Anshuman Tiwari, Shalu Choudhary, Jayendra Padiya, Abhijit Ubale, Venugopal Mikkilineni, and Bharat Char*

*Mahyco Research Centre, Mahyco Private Limited, Jalna, Maharashtra, India*

## 9.1 Introduction

Maize (*Zea Mays* L.) originated from Central America from where it spread to the different parts of the world except for Antarctica. Maize is one of the top ten produced commodities in the world which shows its importance in food security as food and feed. Among the three major cereals (Rice, Wheat, and Maize), maize stands first in terms of world production with more than 1140 million tons which was cultivated on 193 million hectares of agricultural land globally. America and Asia are two major continents that share 52.5 and 29.1% of world maize production respectively (FAOSTAT 2018).

Maize can be grown in a wide range of environments which makes it the preferred crop to meet the demand from rapidly increasing population along with changing climate. Maize has greater economic value too, as it not only serves as food for human consumption but also as feed for livestock and raw material for industrial supply. The presence of micronutrients, phytochemicals, carotenoids, anthocyanins, dietary fiber, and lipids in the maize kernel makes it a choice of grain as food. Maize and its products are a staple food and a major source of nutrients, dietary fiber, and protein supply in the developing countries of Africa, America, and Asia (Prasanna et al. 2019). Daily intake of maize as food ranges from 50 to >300 g per person per day in Sub Saharan Africa (SSA) and Latin America (Prasanna et al. 2019). As a feed, Maize plays a very important role in the livestock-to-meat cycle across the world. Since people from developed countries can afford milk, egg, and meat hence maize demand as feed is more in developed countries as compared to developing nations. Quality protein maize (QPM) is enriched with two essential amino acids lysine and tryptophan thus increasing its nutritional value as food and feed. Maize is also providing raw materials to a variety of industries like glue, paint, alcoholic beverages, sweetener, insecticides, toothpaste, rubber tires, rayon, and molded plastics. Maize is also a source of ethanol for biofuel production (FAO (Food and Agriculture Organization) 1992). All these

*Genotyping by Sequencing for Crop Improvement*, First Edition.
Edited by Humira Sonah, Vinod Goyal, S.M. Shivaraj, and Rupesh K. Deshmukh.

uses of maize make it one of the top traded crops across the world and one of the most profitable grain crops.

## 9.2 Maize Genetics

Grass family Poaceae is a very large family which includes more than 10 000 species (Kellogg 1998). Maize is a member of the Panicoideae subfamily under the grass family *Gramineae* (*Poaceae*). It evolved from its closest wild relative Teosinte. Although the morphology and inflorescence architecture of both maize and teosinte are very distinct, teosinte is the progenitor of maize. Teosinte has many tillers while modern maize does not have any tillers. But both maize and teosinte are very much similar at the genomic level. Both species have the same number of chromosomes and identical chromosomal morphology. Maize and teosinte are compatible for crossing and develop fully fertile $F_1$ (Doebley 2004). According to the teosinte hypothesis, maize is the domesticated/cultivated form of teosinte evolved due to continuous human selection over thousands of years (Beadle 1939).

Maize is a diploid with chromosome number of $2n = 2x = 20$. Like the other large genome plant species, maize consists of up to 85% repetitive sequences which cause the variation in the genome size of maize inbred lines ranging from 2.3 to 2.7 Gb. On the other hand, the unique region of the maize genome does not show significant diversity/variation among the maize genotypes (Rayburn et al. 1985, 1993; Bennetzen 2009; Schnable et al. 2009; Zhou et al. 2009). The repetitive region of the maize genome includes transposable elements (TEs), high copy short tandem repeats, and ribosomal deoxyribonucleic acid (rDNA) located mostly at telomeres, centromeres, and heterochromatin knobs (McClintock 1931; Peacock et al. 1981; Ananiev et al. 1998; Morgante 2006). TEs are also termed selfish sequences of DNA and can move from their original site to other parts of the genome. Based on their transposition intermediate, RNA or DNA is classified into two classes; Class I (retrotransposons) and Class II (DNA transposons), respectively (Feschotte et al. 2002).

Maize has been a model crop for genetic research for many years. Maize has many features at morphological and genome level which makes it an ideal crop for a research study. Approximately 2.4 Gb of genome content of haploid maize is distributed and packed into 10 chromosomes which makes their size large enough that they can be studied under a light microscope easily. Because of this, many cytogenetic studies have been conducted on maize that helped to understand the maize genome in a better way. Maize is a cross-pollinated crop and has a separate male and female flower, tassel, and silk respectively which facilitate controlled pollination. This makes maize an ideal crop to study the heterosis which is the primary reason behind the success of maize as a commercial crop (Nannas and Dawe 2015). Availability of a large number of seeds per successful pollination help plant breeder to study rare events of kernel color and segregation ratio. Cross-pollination nature and availability of hundreds of seeds per cross help breeder in developing mapping populations and genetically dissect the complicated traits of maize and their associated regions in the genome. This helped to hasten the marker development in maize and advanced genome analysis such as genome-wide association study (GWAS) and genomic selection (GS).

## 9.3 Importance of Genomics and Genotyping-based Applications in Maize Breeding Programs

As maize is the main source of food and feed in developing nations in Africa, Asia, and Latin America, maize breeders are under pressure to develop high-yielding hybrids and varieties to meet the food demand of the increasing population. The utilization of advanced molecular breeding approaches like GWAS and GS is an effective way to speed up the genetic improvement of maize in less time. The ultra-high-throughput and cost-effective genotyping platforms and the advanced statistical methods are prerequisite for successful application and implementation of GS and GWAS.

The molecular breeding methods GWAS and GS provide more insight into complex traits as these methods account for the effects of all the markers distributed across the genome hence capture the genes with minor effects also (Heffner et al. 2009; Xu et al. 2012). GWAS enables analysis of the genetic architecture of complex traits. GWAS offers higher resolution and greater ability to identify favorable genetic loci responsible for the trait of interest while saving cost and time. Linkage disequilibrium (LD) decay is rapid in maize due to its highly diverse nature. Therefore, a large number of polymorphic SNPs are required to ensure complete coverage of the genome. Genotyping-by-sequencing (GBS) generates millions of SNPs at affordable cost. GS provides more genetic gain per unit time by shortening the breeding cycle and by reducing the phenotyping of the test population and allowing the selection of lines based on only genotyping (Heffner et al. 2010; Xu et al. 2017).

Under marker-assisted selection (MAS), we use trait/gene linked marker for parental line screening hence its accuracy is high and correlates with phenotype data but in GS entire marker set is used for screening the lines without phenotyping. Accuracy of GS ($r_{MG}$) is measured in terms of Pearson's correlation ($r$) between true breeding value (actual phenotyping value) and genomic estimated breeding value (GEBV) of the candidate individuals of the test population. Many studies have been conducted in maize to understand the factors affecting the prediction accuracy so that it can be improved and applied in maize breeding programs (Zhang et al. 2017a).

Identification of region [QTL (quantitative trait locus) or gene] and its linked marker is a traditional way of marker–trait association study. Two maize inbred lines H427 and H761 were crossed to develop $F_2$ population in 1986 to create the first linkage map using RFLP markers (Helentjaris et al. 1986). Since then, many researchers have used other molecular markers such as AFLP, SSR, and EST to construct linkage maps and identify the QTL for several traits in maize. But all these markers have limitations in terms of a low number of polymorphic markers between parental lines used to develop the mapping populations. Low genomic coverage by a limited number of markers has affected the output of QTL mapping experiments. Advances in genotyping platforms including GBS have resolved the issue of low coverage and provided a large number of SNP markers at an affordable cost. Several studies have been conducted in maize using GBS for QTL mapping of complex traits like plant architecture (Zhou et al. 2016), plant height (PH) (Zhou et al. 2016; Liu et al. 2015), yield and their related traits (Su et al. 2017; Wang et al. 2018a; Zhao and Su 2019). There are several examples in maize where, GBS, is the preferred genotyping method in molecular breeding programs to study different types of traits starting from simple traits like PH, flowering time to a complicated trait-like grain yield (GY). In this chapter, we are presenting the recent research work done for the improvement of maize using advance molecular breeding approaches with a focus on GBS-based GWAS and GS.

## 9.4 GBS-based QTL Mapping in Maize

Indirect selection for a trait of interest with the help of a DNA marker is known as MAS. QTL mapping is one of the common methods to study marker–trait relationships and establish marker–trait associations. The outcome of the QTL mapping study depends upon many factors and one of them is marker density on a linkage map. Many of the plant traits such as PH, flowering time, biotic and abiotic resistance, and yield are polygenic and they are regulated by several gene(s) or QTL with minor effects (Sham et al. 2002). A high-density linkage map is required to dissect these complex traits. Advancements in genotyping platforms, specifically GBS and other high-throughput genotyping techniques explained earlier in this chapter help to populate large genomes such as maize, with thousands of markers in a cost-effective manner. In maize where GBS has been applied extensively, high-density linkage map has been developed ranging from 1471 cM (Semagn et al. 2013) to 11 198 cM (Liu et al. 2020a, 2015) with an average distance between adjacent markers from 6.1 to 1.7 cM, respectively (Table 9.1). Bin map in QTL mapping is another approach being used which increases the resolution and improves the accuracy of QTL mapping. A bin is defined as perfect LD block within which all SNPs segregate together. Statistical software like PlabMQTL (Utz 2012), QTLNetwork v2.1 (Yang et al. 2008), R/QTL package (Broman et al. 2003), QTL IciMapping v3.3 (www.isbreeding.net), compliment the use of advance genotyping platforms in QTL mapping projects. This section discusses the application of next-generation sequencing (NGS) techniques, especially GBS for QTL mapping in maize.

**Table 9.1** Linkage map developed in maize using genotyping by sequencing (GBS).

| References | No of bin markers | cM | Average distance between adjacent markers cM | Total markers |
|---|---|---|---|---|
| Liu et al. (2015, 2020a) | 6618 | 11198.5 | 1.7, 0.48 | 1151856 SNPs, 2916 SSRs |
| Wu et al. (2020) | 3876 | 2413.25 | 0.62 | 31486 SNPs |
| Su et al. (2017), Zhao and Su (2019) | 3305 | 2236.66 | 0.68 | 29927 SNPs |
| Wang et al. (2018a) | 16769 | 2508.111 | 0.152 | 49640 |
| Wang et al. (2018b) | 2157 | 1568.28 | 1.4 | 45212/2157 SNPs |
| Jiménez-Galindo et al. (2017) | 285 | 2372.1 | 8.6 | 955690/1145 SNPs |
| Zhou et al. (2016) | 4183 | 1545.65 | 0.37 | 88268 SNPs |
| Nair et al. (2015) | 156 | | 0.4 | 156 SNPs |
| Semagn et al. (2013) | 430 | 1471 | 6.1 | |
| Li et al. (2015) | 4932 (CN-NAM) | 1700.44 | 0.34 | 39620 SNPs |
| | 5296 (US-NAM) | 1456.68 | 0.28 | 109104 SNPs |

In Maize, commercially important traits are yield and yield-related traits (Su et al. 2017), insect resistance (Jiménez-Galindo et al. 2017), drought resistance, and traits related to plant architectures (Wang et al. 2018a). QTL mapping experiments were conducted to identify the genomic regions associated with these traits. Zhou et al. (2016) genotyped the Ye478 × Qi319 derived RIL population of 314 progenies using GBS. RIL progenies were sequenced with an average of 357 376 reads per individual, which provided an equivalent coverage of 0.07 fold of maize B73 reference genome. A total of 88 268 SNPs were categorized into the 4183 bin markers which were used for developing high-density linkage map of 1545 cM length across 10 maize chromosomes. This linkage map was used to identify the QTL of three maize plant architecture traits namely PH, ear height (EH), and internode number (IN). A total of 35 QTLs identified across the three environments, 14 for PH, 14 for EH, and 7 for IN traits. Nine out of thirty-five QTL were found to be stable QTL as they were identified in multiple environments and six QTL were pleiotropic (pQTLs) as the same QTL region was found for more than one trait. pQTL10 identified in maize chromosome 10 between marker mk4012 and mk4037 was found influencing all three traits and accounted for 9.35% of phenotypic variation for PH (Table 9.2).

In another experiment, the two well-studied maize lines B73 and Mo17 were used to generate RIL population to study flowering time and PH (Liu et al. 2015). To obtain a higher genetic resolution, a ten-generation inter-mated B73 × Mo17 (IBM) doubled haploid mapping population (IBM Syn10 DH Population) was used for QTL mapping of the above traits. A total of 6618 bin markers along with 1 151 865 high-quality SNP markers were used to create a very high-density genetic linkage map using 280 progenies of Syn10 population. Using this linkage map, 35 QTL for flowering time and 34 QTL for PH were detected (Table 9.2). The genotyping data generated in this study is available in iPlant Discovery Environment (DE) and provides an excellent resource for all future research in maize. The linkage map developed in this study was also used by Liu et al. (2020a) to study the kernel size.

Maize kernel size, which includes kernel length (KL), kernel thickness (KT), and kernel width (KW), plays an important role in determining kernel weight (KWt) and thus influences yield. Several studies have been conducted on these traits and genomic regions were identified. Kernel size-associated QTL was identified on different maize chromosomes. Liu et al. (2017) have identified 213 QTL using 10 RIL populations (Table 9.2), Zhang et al. (2017b) found 24 QTL, and Liu et al. (2014) detected 40 QTL based on $F_2$ population. Unlike the previous studies, Liu et al. (2020a) have used two different populations for identifying the QTL region and associated marker for kernel size in maize, where the IBM Syn10 DH population of 265 progenies and second natural population of 310 inbred lines were used. The combined use of linkage mapping and association mapping (AM) increased the power to identify QTL regions and their associated marker for the trait. To identify a stable QTL (a QTL which is found in either more than one environment or in the different genetic background) both populations were phenotyped at multiple locations, Syn10 population was phenotyped at seven locations while association panel of 310 inbred lines was phenotype in three different environments. A linkage map of 6618 bin markers was used for identifying QTL in IBM Syn10 population using windows QTL Cartographer software version 2.5 (Wang et al. 2012). A total of 50 QTL were identified in this experiment for three kernel size traits of which, 29 QTL were common and the remaining 21 QTLs were novel and unique

**Table 9.2** List of GBS-based QTL mapping studies in maize.

| References | Trait | Population | Type of population | No. of QTLs | Chromosome | Minimum LOD | Maximum LOD | Minimum PV | Maximum PV |
|---|---|---|---|---|---|---|---|---|---|
| Wu et al. (2020) | Pericarp thickness (PT) | M03×M08 | BC4F3 | 14 | 1, 5, 6, 7, 10 | 2.1 | 7.13 | 3.22 | 35.38 |
| Liu et al. (2020a) | Kernel length (KL) | B73×Mo17 | RIL (Syn10 IBM) | 15 | 1, 2, 3, 4, 5, 6, 7, 9 | 2.21 | 5.4 | 3.48 | 10.11 |
| | Kernel width (KW) | B73×Mo18 | RIL (Syn10 IBM) | 21 | 1, 2, 3, 4, 5, 7, 8, 9, 10 | 2.47 | 4.99 | 3.44 | 8.43 |
| | Kernel thickness (KT) | B73×Mo19 | RIL (Syn10 IBM) | 9 | 1, 3, 4, 6, 7, 8, 10 | 2.51 | 9.16 | 3.38 | 15.04 |
| Zhao and Su (2019) | Grain weight per ear (GWP) | SG-5×SG-7 | F2:3 | 11 | 2, 3, 5, 7, 9 | 3.3 | 6.2 | 2.98 | 13.3 |
| | Kernel row number (KRN) | SG-5×SG-7 | F2:3 | 39 | 1, 2, 3, 4, 5, 7, 8, 9 | 3.3 | 19.4 | 0.33 | 15 |
| | 100-Kernel weight (KWt) | SG-5×SG-7 | F2:3 | 7 | 3, 7, 8, 9 | 3.1 | 6.7 | 2.67 | 10.5 |
| | Cob diameter (CD) | SG-5×SG-7 | F2:3 | 13 | 1, 2, 3, 4, 5, 6, 7 | 3.5 | 8.9 | 1.71 | 16.8 |
| | Ear diameter (EAD) | SG-5×SG-7 | F2:3 | 6 | 1, 2, 4, 10 | 3 | 7.4 | 1.82 | 12 |
| | Ear length (EAL) | SG-5×SG-7 | F2:3 | 5 | 1, 2, 3, 5 | 3 | 5.1 | 6.4 | 25.6 |
| | Ear width (EW) | SG-5×SG-7 | F2:3 | 6 | 2, 5, 7, 9 | 3.4 | 7.3 | 2.28 | 15 |
| Wang et al. (2018a) | Plant height (PH) | Zheng58×Chang7-2 | RIL (Zhengdan958) | 8 | 1, 2, 3, 5, 6, 9 | 3.9 | 21.83 | 1.28 | 9.15 |
| | Ear height (EH) | Zheng58×Chang7-2 | RIL (Zhengdan958) | 9 | 1, 2, 3, 5, 6, 7, 8 | 3.45 | 15.34 | 0.81 | 7.28 |
| | Tassel branch number (TBN) | Zheng58×Chang7-2 | RIL (Zhengdan958) | 6 | 1, 2, 4, 7, 10 | 3.82 | 5.84 | 1.02 | 7.78 |
| | Tassel length (TL) | Zheng58×Chang7-2 | RIL (Zhengdan958) | 14 | 1, 2, 3, 4, 5, 6, 8 | 4.14 | 10.58 | 0.05 | 3.9 |
| | Upper leaf number (ULN) | Zheng58×Chang7-2 | RIL (Zhengdan958) | 6 | 1, 2, 3, 5, 7, 10 | 4.66 | 18.18 | 1.78 | 7.71 |
| | Relative ear height (EH/PH) | Zheng58×Chang7-2 | RIL (Zhengdan958) | 8 | 1, 2, 3, 6, 8, 9 | 3.88 | 11.8 | 0.22 | 5.03 |

*(Continued)*

**Table 9.2** (Continued)

| References | Trait | Population | Type of population | No. of QTLs | Chromosome | Minimum LOD | Maximum LOD | Minimum PV | Maximum PV |
|---|---|---|---|---|---|---|---|---|---|
| Wang et al. (2018b) | Plant height (PH) | CML220×1091-AF | F2 | 2 | 8 | 4.04 | 4.3 | 14.34 | 16.57 |
| | Ear height (EH) | CML220×1091-AF | F2 | 5 | 4, 5, 8, 10 | 4.07 | 11.81 | 8.62 | 28.53 |
| | Kernel row number (KRN) | CML220×1091-AF | F2 | 2 | 1 | 3.87 | 5.84 | 16.63 | 22.78 |
| | Relative ear height (REH) | CML220×1091-AF | F2 | 3 | 4, 8, 10 | 4.78 | 9.84 | 11.45 | 25.93 |
| | Angle of tassel branch (AOT) | CML220×1091-AF | F2 | 2 | 2 | 4.56 | 4.76 | 14.77 | 15.47 |
| | Leaf width (LW) | CML220×1091-AF | F2 | 2 | 1, 9 | 4.54 | 6.22 | 14.77 | 21.03 |
| | Stem cross-sectional area (AA) | CML220×1091-AF | F2 | 2 | 2 | 3.84 | 3.96 | 13.02 | 13.46 |
| | Nodal root number (RN) | CML220×1091-AF | F2 | 2 | 1, 2 | 4.65 | 6.98 | 12 | 18.74 |
| | Color of anther (COA) | CML220×1091-AF | F2 | 4 | 1, 2, 6, 8 | 4.86 | 16.82 | 6.04 | 26.17 |
| | Color of glum (COG) | CML220×1091-AF | F2 | 1 | 10 | | 5.82 | | 18.63 |
| | Color of kernel (COK) | CML220×1091-AF | F2 | 2 | 6, 10 | 4.37 | 10.47 | 11.57 | 31.9 |
| | Ear diameter in Beijing (EDBJ) | CML220×1091-AF | F2 | 1 | 1 | | 4.97 | | 19.75 |
| Su et al. (2017) | Ear length (EAL) | SG5×SG7 | F2 | 4 | 1, 6, 10 | 4.1 | 4.1 | 8.1 | 8.2 |
| | Ear diameter (EAD) | SG5×SG7 | F2 | 5 | 1, 4, 7 | 4 | 7 | 6.4 | 11.6 |
| | Cob diameter (CD) | SG5×SG7 | F2 | 2 | 1, 2 | 4.6 | 9.1 | 7.7 | 18 |
| | Kernel row number (KRN) | SG5×SG7 | F2 | 1 | 8 | | 4.1 | | 7.6 |
| | Corn grains per row (CGR) | SG5×SG7 | F2 | 3 | 4, 6 | 4 | 5.4 | 8.2 | 10 |
| | Ear weight (EW) | SG5×SG7 | F2 | 5 | 4, 7 | 4.2 | 5.2 | 8.2 | 19.7 |
| | Grain yield per plant (GWP) | SG5×SG7 | F2 | 5 | 4, 6, 7 | 4.4 | 5.3 | 8.1 | 11.8 |
| | 100-Kernel weight (KWt) | SG5×SG7 | F2 | 4 | 3, 4, 6, 8 | 4.5 | 7.5 | 7.4 | 13 |

| | | | | | | | | | |
|---|---|---|---|---|---|---|---|---|---|
| Jiménez-Galindo et al. (2017) | Tunnel length | A637×A509 | RIL | 3 | 5, 8, 10 | 3.92 | 6.92 | 5.57 | 10.4 |
| | Kernel resistance | A637×A509 | RIL | 1 | 9 | | 5.44 | | 7.2 |
| | Stalk damage (%) | A637×A509 | RIL | 1 | 10 | | 5.5 | | 13.1 |
| | Yield (Kg ha$^{-1}$) | A637×A509 | RIL | 1 | 8 | | 8.23 | | 15.9 |
| | Silking (days) | A637×A509 | RIL | 3 | 1, 8, 9 | 8.22 | 9.38 | 8.27 | 23.9 |
| | Plant height | A637×A509 | RIL | 3 | 2, 4, 8 | 4.11 | 8.27 | 5.08 | 9.12 |
| Liu et al. (2017) | 100-Kernel weight (HKWt) | B73, BY804, BK, YU87-1, K22, CI7, DAN340, KUI3, B77, ZONG3, DE3, BY815, ZHENG58, SK | 10 RIL | 90 | 1, 2, 3, 4, 5, 7, 8, 9, 10 | 2.53 | 12.78 | 3.2 | 18.35 |
| | Kernel length (KL) | B73, BY804, BK, YU87-1, K22, CI7, DAN340, KUI3, B77, ZONG3, DE3, BY815, ZHENG58, SK | 10 RIL | 61 | 1, 2, 3, 4, 5, 7, 8, 9, 10 | 2.51 | 11.41 | 3.27 | 18.57 |
| | Kernel thickness (KT) | B73, BY804, BK, YU87-1, K22, CI7, DAN340, KUI3, B77, ZONG3, DE3, BY815, ZHENG58, SK | 10 RIL | 63 | 1, 2, 3, 4, 5, 7, 8, 9, 10 | 2.52 | 9.21 | 3.52 | 15.09 |
| | Kernel test weight (KTW) | B73, BY804, BK, YU87-1, K22, CI7, DAN340, KUI3, B77, ZONG3, DE3, BY815, ZHENG58, SK | 10 RIL | 70 | 1, 2, 3, 4, 5, 7, 8, 9, 10 | 2.52 | 11.96 | 3.36 | 19.43 |
| | Kernel width (KW) | B73, BY804, BK, YU87-1, K22, CI7, DAN340, KUI3, B77, ZONG3, DE3, BY815, ZHENG58, SK | 10 RIL | 89 | 1, 2, 3, 4, 5, 7, 8, 9, 10 | 2.51 | 9.71 | 2.91 | 15.97 |
| Zhou et al. (2016) | Plant height (PH) | Ye478×Qi319 | RIL | 14 | 1, 2, 4, 5, 8, 10 | 3.53 | 10.63 | 2.6 | 12.01 |
| | Ear height (EH) | Ye478×Qi319 | RIL | 14 | 1, 3, 5, 6, 7, 8, 10 | 3.68 | 14.63 | 3.43 | 15.68 |
| | Internode number (IN) | Ye478×Qi319 | RIL | 7 | 1, 3, 8, 10 | 6.26 | 13.08 | 6.48 | 18.73 |
| Liu et al. (2015) | Plant height (PH) | B73×Mo17 | RIL (Syn10 IBM) | 34 | 1, 3, 4, 8, 9 | 2.59 | 10.31 | 3.3 | 14.5 |
| | Flowering time (FT) | B73×Mo17 | RIL (Syn10 IBM) | 35 | 1, 8, 9 | 2.57 | 17.89 | 2.82 | 24.06 |
| Nair et al. (2015) | *Maize Streak virus1 (Msv1)* | CML206×CML312 | F2 | 3 | 1, 3, 10 | 3.46 | 11.76 | 8.6 | 67.36 |

to this experiment only. Of the 50 QTL, 15 QTL identified for KL, 9 QTL for KT, 21 QTL for KW, and 5 QTL for more than one trait (Table 9.2). These QTLs were distributed across the 10 chromosomes and phenotypic variation explained (PVE) by these QTL ranged from 3.48 to 10.11% for KL, from 3.4 to 8.43% for KW, and from 3.38 to 15.04% for KT (Table 9.2). For KT, stable and major QTL qKT-4.02 with more than 10%, phenotypic variation was found on chromosome 4 (Liu et al. 2020a). In earlier studies too, QTL cluster qGW4.05 was found on chromosome 4 which showed a pleiotropic effect on kernel size-related traits (KL, width, and KWt). Efforts were made to fine map this QTL to a 279.6 kb region and six polymorphic sites found within candidate gene *GRMZM2G039934* that were significantly associated with KWt and kernel size (Chen et al. 2016a). In genome-wide association mapping study by Liu et al. (2020a), 39 354 SNPs were used for genotyping the AM panel of 310 inbred lines. Analysis with three GWAS methods, (i) the compressed mixed linear model (CMLM) in GAPIT (Lipka et al. 2012; Zhang et al. 2010) with the kinship matrix and population structure (Q matrix); (ii) the mixed linear model (MLM) in TASSEL 5.0 (Bradbury et al. 2007) with the kinship matrix and Q matrix; and (iii) fixed and random model circulating probability unification (FarmCPU; Liu et al. 2016a) with population structure (PCA, principal component analysis) as covariates. A total of 56 SNPs were found significantly associated with three kernel size traits and located within intervals of the 18 QTL identified in this study. The top 21 significant SNPs by GWAS analysis led to 17 candidate genes involved in regulating the kernel development in maize. Further study on these findings will facilitate the identification of markers for kernel size in the future.

In maize, the yield depends on associated traits such as KWt, ear length (EAL), ear diameter (EAD), cob diameter (CD), kernel row number (KRN), corn grains per row (CGR), ear weight (EW), and grain yield per plant (GWP) and is affected by several loci with minor effects (Su et al. 2017). To detect the minor QTL, 199 $F_2$ progenies along with their parental lines SG5 and SG7 were sequenced on Illumina HiSeq$^{TM}$ sequencer. A total of 29 927 SNPs were identified after filtering for quality and were used to construct 2236.66 cM long linkage map consisting 3305 bins classified on 10 maize chromosomes (Table 9.1). Phenotype data were collected for KWt, EAL, EAD, CD, KRN, CGR, EW on the $F_2$ mapping population. In this study, QTL mapping was conducted using two approaches; composite interval mapping (CIM) in which a total of 28 QTL were detected for above mentioned seven yield-related traits, and 29 QTLs were identified using LASSO method. A total 16 QTLs were commonly identified between both analysis methods (Table 9.2). Based on this study, many candidate genes were identified which are involved in regulating the yield-related traits, that include an F-box family protein gene (*GRMZM2G139872*), three genes found in the intervals of qKW-2 QTL, *GRMZM2G180811*, *GRMZM5G828139*, and *GRMZM5G873194* are the WD40-repeat protein genes. Another important candidate gene *GRMZM2G019183* located in the intervals of qGWP-4 QTL is the UDP-Glycosyl transferase gene. To validate these QTL, an advanced generation of the same population was used by Zhao and Su (2019), where one generation advanced $F_3$ population was used for phenotyping while $F_2$-based linkage map was used for QTL mapping. The approach followed by Zhang and Xu (2004) was used to avoid the phenotypic segregation issue of $F_{2:3}$ population. A total of 33 QTL were identified using CIM while 54 QTL were found in genome-wide composite interval mapping (GCIM). Both $F_2$-based QTL and $F_{2:3}$-based QTL regions were compared and those QTLs which shared the same overlapping QTL interval are regarded as confirmed or stable QTLs. A total of 10 QTLs (*qKW-7, qEW-9, qEW-10, qGWP-6, qGWP-8, qGWP-10, qGWP-11, qGWP-12, qGWP-13*, and *qGWP-14*) were found as stable QTL associated with grain weight. To predict the candidate genes for this trait, transcriptome profiling of grains sampled at three different developmental stages was used and differentially expressed genes (DEGs) were obtained between parental lines. Of the several candidate genes identified in both the experiments, eight candidate genes fall between the physical interval of ten confirmed QTLs for grain size and weight in maize. These eight candidate genes are involved in different pathways such as the phytohormone signaling, transcription factor pathways, and ubiquitin-proteasome degradation pathways (Verma et al. 2004). Further study on the identified ten stable QTL across the different generations and their associated candidate genes will be useful to develop a reliable marker for grain size and grain weight for MAS in maize.

Developing the insect resistance genotype is another way to improve the maize yield as in some cases corn borer caused the yield loss of up to 30% (Meissle et al. 2010). QTL mapping experiment conducted by Jiménez-Galindo et al. (2017) is different in a way that instead of working on individual characters, this group decided to use the index which includes more than one trait. In this study, they worked both on yield and Mediterranean corn borer (MCB), *Sesamia nonagrioides* jointly. Based on standardized data, four indexes were calculated, (i) Y + kr − tl (yield plus kernel resistance minus tunnel length); (ii) Y − tl (yield minus tunnel length); (iii) Y + kr (yield plus kernel resistance); and (iv) Kr − tl (kernel resistance minus tunnel length) and QTL analysis for these indexes were carried out using PlabMQTL (Utz 2012). QTL found for individual traits under this study (Table 9.2) were found by many other scientists earlier also but in this experiment index QTL was identified which can improve both yields as well as resistance to MCB simultaneously.

To accurately map the QTL region with small effects and identify markers associated with it requires many markers along with a large mapping population to accurately identify recombinational breakpoints (Chen et al. 2014; Song et al. 2016). To identify the minor genes affecting the plant architecture-related traits (PATs) in maize, RIL population ($F_7$ generation) with 1147 progenies were developed by crossing two maize genotypes Zheng58 and Chang7-2, and phenotyped for six PATs those are silk color (SC), EH, PH, tassel branch number (TBN), tassel length (TL), and upper leaf number (ULN). In his study, Wang et al. (2018a) found a total of 51 QTL for these six PATs spread across the ten chromosomes of maize (Table 9.2). As expected, all the identified QTL were with minor effect and explained phenotypic variation between 0.05 and 9.15%. A high-density linkage map of 2508.11 cM length with 16 769 recombinant bins was used in this study (Table 9.1). This provided a finer interval for all the QTL than the previous studies and was able to capture many genes with known function. About 40 out of 51 QTLs were found in less than 5 Mb interval range and hence capturing genes with known functions like *zfl2* for TL, *td1*, *na1,* and *d3* for PH and *ra1* for TBN.

GBS approaches have limitation like low sequencing depth leading to a high genotype error rate at the heterozygous site and missing data. To overcome these limitations, Wang et al. (2018b) applied the sliding window approach in their $F_2$ population, where 153 progenies with a window size of 25 SNPs and a step size of 2 SNPs (Huang et al. 2009) were used. In each sliding window, a ratio of 18/25 of SNP alleles from two parental lines was used as a threshold to determine the individual genotype and classified as homozygous or heterozygous. Any $F_2$ progeny containing the 18/25 SNP alleles from either 1091-AF or CML220 parental lines were scored as homozygous otherwise categorized as heterozygous. Then blocks were formed by combining the adjacent widows with the same genotype and their boundaries were defined as recombination breakpoints with different genotypes. In this study, only blocks with more than 5 Mb were considered for further analysis. Genotypic maps of individual $F_2$ progenies were used to generate bin maps and consecutive 100 kb intervals without any recombination on the same chromosome across the $F_2$ population defined as a single bin (Huang et al. 2009). These bins were used to create high-density linkage map using R/qtl package (Broman et al. 2003) in this experiment. A total of 29 QTLs were found for 12 traits includes PH, EH, KRN, relative ear height (REH), angle of tassel branch (AOT), leaf width (LW), stem cross-sectional area (AA), nodal root number (RN), color of anther (COA), color of glume (COG), color of kernel (COK), and ear diameter in Beijing (EDBJ) (Table 9.2). A significant outcome of this study is that the QTL region of EH4, a major QTL on chromosome 8 for trait EH, is narrowed down from 11 to 3.23 Mb using a modified fine-mapping approach.

The maize streak virus (MSV) is considered as one of the most important biotic stress of maize, causes severe economic loss in SSA region (Pratt and Gordon 2005). Several studies have identified bin 1.05 on chromosome 1 as a major QTL region *Msv1* for MSV which explains the phenotypic variation up to 92% (Kyetere et al. 1999; Welz et al. 1998; Pernet et al. 1999a,b; Lagat et al. 2008). But due to low coverage of markers and small population size, the physical interval of this major QTL region ranged between 16 and 133 Mb on the short arm of chromosome one. To increase the accuracy of MAS program for MSV resistance, Nair et al. (2015) fine mapped this region to 0.87 cM which is equal to 7 Mb physical distance according to RefGen_V2 of B73. Two QTLs, QTL3 and QTL10 on chromosomes 3 and 10, respectively, were identified along with *Msv1* on chromosome one. To fine map *Msv1* in the presence of two minor QTL/gene, QTL isogenic recombinant (QIR) strategy was used. A large $F_{2:3}$ population was used where MSV resistant parent CML206 was crossed with MSV susceptible parent CML312 and $F_2$ population with 4725 progenies was developed. $F_2$ seeds were used for genotyping with nondestructive seed DNA isolation method suggested by Gao et al. (2008), while the $F_3$ population was used for replicated phenotyping.

The pericarp of maize kernel is composed of nondigestible cellulose. Pericarp thickness affects the crispness, tenderness, and residue percentage of sweet corn. Wu et al. (2020) started a QTL mapping experiment with M03 thinner pericarp thickness line as recurrent parent and M08 as thicker pericarp thickness parental lines as a donor. A $BC_4F_3$ population with 148 progenies was developed by crossing these two sweet corn parental lines. Genotyping was done using a specific length amplified fragment (SLAF) approach. A total of 3876 SLAF tags were used to construct the high-density genetic linkage map of 2413.25 cM lengths with an average distance of 0.62 cM (Table 9.1). The phenotyping of the $BC_4F_3$ population along with its parental lines was done in three different seasons from 2014 to 2016. QTL mapping analysis revealed a total of 14 QTLs across the three different seasons which are distributed on maize chromosomes 1, 5, 6, 7, and 10. A stable QTL qPT10-5 on chromosome 10 was identified in all three seasons which explained 7.8 to 35% phenotypic variations across different years (Table 9.2). To further refine the analysis, transcriptome profiling of the contrasting parental lines was conducted, and based on DEGs, five candidate genes were found in the 901.2 kb region of qPT10-5 QTL. Out of five genes, *GRMZM2G143352*, *GRMZM2G143402*, and *GRMZM2G143389* are considered as candidate genes for pericarp thickness in maize kernel. GBS-based QTL mapping studies on maize is helping breeder to identify the regions responsible for the inheritance of traits and their role. Markers identified in different mapping projects will speed up their breeding program and develop genotypes with improved traits.

## 9.5 GBS Protocols and Analytical Pipelines for Maize

In large and complex plant genomes, GBS is a challenge. To increase the depth of sequencing and to represent a larger portion of the genome, target enrichment and complexity reduction are required. Reducing genome complexity using restriction enzymes (REs) leading to reduction representation sequencing (RRS) is a common approach used to reduce repetitive regions for sequencing and to capture low copy regions of the genome. Reduce representation of the genome with specific RE focuses on sequencing fragments at ends of restriction sites. Selecting a candidate REs or RE combinations is a critical first step of RSS methods that determines genome sampling efficiency and in turn has led to development of different GBS protocols/RRS methods (Scheben et al. 2017). The earliest method is restriction site-associated DNA (RAD) sequencing (Baird et al. 2008), which has been applied in maize for inbred characterization (Tamaki et al. 2016). Elshire et al. (2011), simplified RAD seq method, optimized a GBS protocol in maize using a single methylation-sensitive type II RE, *Ape*KI, and demonstrated in maize IBM RIL population where 20 000 sequence tags were mapped. Generally, RE that leaves 2 to 3 bp overhang and does not cut frequently in the major repetitive fraction of the genome is selected (Elshire et al. 2011). For maize and close relatives (teosinte), *Ape*KI has been used routinely for GBS which creates a 5′ overhang (3 bp), has relatively few recognition sites in the major classes of maize retrotransposons, and is partially methylation sensitive (Elshire et al. 2011). This protocol is simple, robust that involves restriction digestion of genomic DNA followed by ligation with two barcode adapters with variable length. The barcode adaptors consist of a 4-8bp barcode on the 3′ end immediately upstream to the compatible sticky ends to tag unique sample and "common adapter." Ligation was followed by DNA pooling, amplification of pooled fragments, purification, and running on Illumina sequencing flow cell. The protocol with a few modifications has been used by maize research community for germplasm characterization (Romay et al. 2013; Gouesnard et al. 2017), generation of high-density linkage maps, and mapping major and minor trait QTLs (Chen et al. 2014; Song et al. 2016; Wang et al. 2018a,b), GWASs (Rashid et al. 2020) and genomic prediction (GP) (Crossa et al. 2013). The protocols have also been found to be suitable for other species such as barley (Elshire et al. 2011) and soybean (Sonah et al. 2013).

Using a combination of two enzymes that includes one rare cutter *PstI* and one frequent cutter *MspI*, a modified protocol of Elshire et al. (2011), GBS known as two-enzyme GBS (Poland et al. 2012) has been developed. Paulsmeyer et al. (2018) constructed GBS library in maize mapping population and discovered a candidate gene "Anthocyanin Acyltransferase1 (AAT1)" using this method. The two enzyme protocol provides a greater degree of complexity reduction and uniform library for sequencing than the original protocol (Sonah et al. 2013; Fu et al. 2016a). Using specific-locus amplified fragment sequencing GBS method where predesigned RRL scheme is used to optimize marker efficiency, Liu et al. (2016b) genotyped 300 $F_2$ plants and two parental lines to understand genetic architecture of maize KRN.

In other genotyping studies in maize, *insilico* assessment on the training data/reference genome was performed for the selection of restriction enzymes. Su et al. (2017) evaluated *insilico* enzymes and size and distribution of restriction fragments and selected Res Mse I and Hae III for GBS libraries construction of the maize $F_2$ lines. Continuous efforts are in progress to identify more effective enzyme combinations that could increase genome sampling and address missing data, one of the limitations of GBS method. In this direction, for *insilico* evaluation of genome coverage, Fu et al. (2016b) developed a software called IgCoverage and proposed a new combination (HinfI + HpyCH4IV) for 12 species including maize.

As GBS has been extensively applied in maize research program for different applications, several modified protocols originated from the initial publication (Elshire et al. 2011) and were validated in maize populations. To improve the flexibility and scalability of original GBS protocol, several modifications were incorporated by Heffelfinger et al. (2014) and a modified protocol was tested in maize $F_2$ population. The key change was of replacing custom enzyme-specific adapters with standard Illumina adapters, multiplexing using dual barcoding system, and bead-based library preparation that minimized the front-end costs. To address the missing genotype data problem with GBS especially when multiple libraries are made from large samples. Manching et al. (2017) developed a GBS protocol based on double digestion technique in maize populations ($F_2$ population, inbred lines, and their $F_1$ hybrids).

Besides enzyme-based reduced representation sequencing strategies, other commonly used target enrichment methods based on PCR or hybridization-based approaches have also been used in maize to discover genomic variants. Jamann et al. (2017) performed amplicon-based enrichment coupled with semiconductor sequencing (referred as Ampliseq), designed multiplexed panel of 319 PCR assays to study 20 candidate loci associated with photoperiod sensitivity in diverse maize materials. In another study, 29 Mb genomic regions targeting 4648 genes possibly affecting biomass production in 21 diverse inbred maize lines (7 flints, 14 dents) were sequenced through designing sequence capture array (Muraya et al. 2015).

With the development of different GBS protocols and high-throughput sequencing platforms, an effective bioinformatics pipeline is required to analyze millions of reads and to identify genomic variants within the sequenced population. Custom-designed all-in-one software package, Tassel-GBS (Glaubitz et al. 2014), optimized for maize, is the most common pipeline widely used by maize researchers for analyzing both GBS and genetic data using a stand-alone PC. The pipeline specifically developed for analyzing reads generated by GBS technologies is implemented in Java Program TASSEL version 4 (Bradbury et al. 2007) and has capacity to analyze large amount of data from large number of samples [Glaubitz et al. (2014) analyzed 45 000 maize samples]. The pipeline is specifically designed for species with reference genome and is tailored to the GBS protocols of Elshire et al. (2011) or Poland et al. (2012). However, the pipeline could be extended to any of the restriction enzymes and barcoding approaches and is not strictly limited to these protocols. Beside

Tassel-GBS, tools and algorithms commonly applied to whole-genome sequence data analysis have been used to analyze GBS data in maize (Elshire et al. 2011; Song et al. 2016; Wang et al. 2018b) like BWA (Li and Durbin 2009) or BOWTIE2 (Langmead and Salzberg 2012), SAMTOOLS (Li et al. 2009), or genome analysis toolkit (GATK) (Mckenna et al. 2010). For scoring phased genotypes, Manching et al. (2017) extended LocHap program for use with GBS and performed read-based haplotyping in maize NAM population, a feature still not included in many GBS pipelines.

## 9.6 Maize Genome Sequencing and Resequencing

In 2002, a project was undertaken to decode the genome of an important maize variety B73 to understand the genome content and organization. The project led by a US-based consortium of researchers in 2009 (Schnable et al. 2009) released the first version of the maize reference genome (B73 RefGen_v1) through Sanger sequencing of bacterial artificial chromosomes and FOSMIDS. Subsequently, in 2017, an improved version 4 was released by Jiao et al. (2017) using the single-molecule real-time sequencing technologies. Further in maize, additional reference genomes for many commercial lines have been generated to understand high level of intraspecific genome diversity, identify structural variations contributing to diverse phenotypes among maize inbred lines. Whole-genome assembly of many temperate maize lines, Mo17 (Yang et al. 2017; Sun et al. 2018), W22 (Springer et al. 2018), and HZS (Li et al. 2019) were generated. In addition to temperate lines, reference genome was built for maize tropical lines SK (Yang et al. 2019) that has low hundred-kernel weight (HKWt) value and *mexicana* (Yang et al. 2017), a highland teosinte *Zea mays* subspecies *Mexicana* to capture vast genetic diversity in the species. B73, Mo17, W22, and SK genomes have high-quality assemblies and details of their genome assemblies are tabulated in Liu et al. (2020b). Generation of new diverse maize reference genomes is a valuable resource for constructing the pan genome of *Zea mays* and accelerating the maize improvement process.

### 9.6.1 Maize Resequencing

In maize, depending on the purpose, several inbred lines, landraces, and populations have been resequenced and mapped generally against the reference genome B73. Whole genome-resequencing (WGR), RNA-based (RNA-seq), and exome sequencing are the common approaches available for resequencing. Through the WGR approach, four-row Chinese Wax landrace (Liu et al. 2016c) were resequenced to capture genetic variation associated to reduced kernel row phenotype. In another study, 15 maize inbred lines (Xu et al. 2014) were resequenced through WGR approach and identified nonsynonymous SNPs (nsSNPs) and corresponding candidate genes for drought tolerance. Since the release of maize first genome sequence, several maize inbred lines have been resequenced through WGR using GBS methods (Chia et al. 2012; Hufford et al. 2012; Romay et al. 2013) and the generated data have been used in the building of first-generation HapMap (Gore et al. 2009), second-generation HapMap (Chia et al. 2012), and third-generation HapMap (Bukowski et al. 2018).

RNA-based sequencing is an effective approach to reduce genome size. Several diverse maize lines have been resequenced through RNA-seq approach for studying the extent of sequence variation in maize genes, transcriptional analysis, and identifying novel transcript sequences. RNA-seq approach has been widely used in several maize studies for transcriptome analysis. Here, we are focused on covering studies where large-scale genotypic variation captured through RNA-seq approach and/or their association with phenotype is determined. From whole seedlings, RNA-seq data were generated for (i) 21 maize inbred lines (Hansey et al. 2012) representing diverse North American and exotic germplasm for genome diversity analysis, (ii) 503 Wisconsin Diversity panel [WiDiv-503] (Hirsch et al. 2014) for building maize pan-genome and pan-transcriptome, and (iii) 453 maize inbred lines (Mazaheri et al. 2019) part of expanded WiDiv panel-942 and identified genes linked to stalk biomass and stalk anatomy through GWAS analysis. From developing kernels (15 days after pollination), Fu et al. (2013) sequenced 368 inbred lines (Maize association mapping panel) to identify SNPs and candidate genes related to kernel development and yield. An et al. (2019) sequenced N531 and SLN [Chinese maize landrace] and two derived introgression line from RNA isolated from 5-mm stage young ear. RNA-seq datasets of B73 were used to study population structure to understand the level of relatedness among B73 samples from around the globe (Liang and Schnable 2016).

Bulked segregant RNA-seq (BSR-seq) approach has also been widely used in maize to profile candidate genes and map associated SNP markers for different traits such as for waterlogging tolerance, Du et al. (2017) resequenced 18 contrasting lines and for KRN, Liu et al. (2016b) resequenced 61 and 63 $F_2$ plants of low and high KRNs. Several mutant genes have been successfully cloned in maize using BSR-seq approach (Li et al. 2013; Klein et al. 2018). Ready-to-use maize exome sequencing kits (Sequence Capture EZ Developer from Roche NimbleGen) have been used for library construction to sequence exon enriched DNA fragments of maize cultivars (Liu et al. 2012) and mutants (Jia et al. 2018).

## 9.7 Genotyping-by-Sequencing-based GWAS and GS Efforts in Maize

To address the limitations of the QTL mapping approach, such as low resolution due to a limited number of recombination events in a bi-parental mapping population, the marker–trait association method called GWAS was developed which is based on natural mapping populations covers historical recombination events. GWAS for marker–trait association is suitable to study complex traits and be able to identify SNP variants responsible for phenotypic variations. GBS provides the solution for genotyping large AM panels with a larger number of markers for GWAS study at affordable genotyping cost (Elshire et al. 2011). GWAS has been used in maize to study different traits that include agronomically important traits like PH, yield, flowering time (Yuan et al. 2019; Farfan et al. 2015) and for several disease resistance traits like maize lethal necrosis (MLN) (Gowda et al. 2015; Nyaga et al. 2019) and sorghum downy mildew (SDM) (Rashid et al. 2018).

Yield is always an important trait irrespective of the crop. However, yield is a complicated trait and depends on many associated factors. Untimely rain, delayed rain, and prolonged summers are now becoming frequent during the cropping seasons thus affecting yield significantly. As maize is a sensitive crop for drought and heat stress (DHS) especially during its flowering and early grain filling stage, it is logical to study maize yield under both drought stress (DS) and heat stress (HS). Considering this, the CIMMYT has initiated a project on Drought Tolerant Maize for Africa (DTMA) and developed a collection of 300 tropical-subtropical maize inbred lines. This collection is known as Drought Tolerant Maize for Africa association mapping panel (DTMA-AM panel). These 300 inbred lines of DTMA panel were test crossed with CML539 tropical maize line and these testcross hybrids were phenotyped in four different conditions, DS, HS, well watered (WW), and a combination of DHS. These trials were taken at different field experiment stations in India, Mexico, Zimbabwe, Thailand, and Kenya between 2008 and 2011 (Cairns et al. 2013). GBS platforms was used for genotyping of DTMA-AM panel (Elshire et al. 2011) and has been used in many projects (Table 9.3) for GWAS and GS studies (Yuan et al. 2019; Gowda et al. 2015; Nyaga et al. 2019).

DTMA-AM panel was used by Yuan et al. (2019) to conduct the GWAS analysis for three traits: GY, anthesis date (AD), and anthesis and silking interval (ASI). All these traits were observed in four different conditions/environments, DS, HS, WW, and DHS. Best linear unbiased prediction (BLUP) was estimated and used for analysis. Estimated BLUP values of the observed phenotypic data can be accessed through the repository: http://hdl.handle.net/11529/10548156 (Elshire et al. 2011). GBS protocol was used to genotype all 300 inbred lines of DTMA panel and a total of 955 690 SNPs were generated. The genotyping data are available in the repository: http://hdl.handle.net/11529/10548156. After quality checking based on a missing percentage of data (greater than 25%) and minor allele frequency (MAF) (less than 1%), a total 381 165 out of 955 690 SNPs were considered for analysis. Yuan et al. (2019) used two approaches to find the marker–trait association; the first is SNP based and the second is haplotype based. SNP-based association was conducted using Genome Association and Prediction Integrated Tool (GAPIT) (Lipka et al. 2012) while haplotype-based AM was conducted in TASSEL 3.0. In both methods, $1 \times 10^{-4}$ was used as a threshold P-value to identify significant SNPs associated with traits. Significantly trait-associated SNPs located within the genes were used to build haplotypes and a total of 301 897 such haplotypes were built under this study. These haplotypes were identified on 19 674 annotated genes. In SNP-based AM approach, gene(s) falling within the 3 kb region of a significantly associated SNP is listed as candidate gene(s) for the associated trait. A total of 1661 SNPs with an average PVE of 4.33% were identified in SNP-based association analysis for all 12 trait environment combinations (Table 9.3). All the 734 genes that are located within 3 kb region of these 1661 SNPs were considered as candidate genes for further study. In haplotype-based AM approach total of 311 of 301 897 haplotypes were found associated with all trait environment combinations other than HS conditions for all three traits GY, AD, and ASI. The highest PVE value of 19.71% was found for trait–environment combination for anthesis and silking interval-drought and heat stress (ASI-DHS) while the lowest PVE value was found for ASI-WW at 6.70%. The average PVE value of haplotype-based study (10.56%) is found to be much higher than SNP-based study (4.33%). In haplotype-based GWAS study, 156 SNPs were found significantly associated with trait-environment combinations and identified 20 candidate genes in these regions. There are four candidate genes (*GRMZM2G329229*, *GRMZM2G313009*, *GRMZM2G043764*, and *GRMZM2G109651*) which are common between both studies with PVE value greater than 9%. These putative candidate genes along with other potential candidate genes observed in this study should be further validated through linkage mapping and gene function analysis.

**Table 9.3** GBS-based GWAS efforts in maize.

| References | Number of accessions | Number of markers used | Tool(s) used for analysis | Traits | Associated SNP(s) |
|---|---|---|---|---|---|
| Nyaga et al. (2019) | 1400 maize inbred lines | 215 137 SNPs | TASSEL 5.2, META-R, R-Software, Structure 2.3.4 | Maize lethal necrosis (MLN) | 32 SNPs (18 for DS and 19 for AUDPC) |
| Yuan et al. (2019) | 300 maize inbred lines | 381 165 GBS SNPs | R – language, TASSEL 3.0 | Grain yield and flowering time under drought and heat stress condition | 1661 SNPs (for all 3 – grain yield, anthesis date, anthesis silking interval traits) |
| Rashid et al. (2018) | 368 maize inbred lines | 333 039 SNPs | META-R, Genstat, TASSEL, | Downy mildew resistance | 26 SNPs across seven chromosomes; 1, 2, 3, 5, 6, 8, and 9 |
| Wallace et al. (2016) | 15 WEMA populations | ~380 ,530 GBS SNPs from GBS v2.7 | TASSEL, R – language | Plant height and flowering time in drought stress and well-watered conditions | 115 SNPs for well-watered and 108 SNPs for drought for flowering time<br>143 SNPs for well-watered and 120 SNPs for drought for plant height |
| Gowda et al. (2015) | 615 accessions | 259 000 (IMAS AM) and 264 000 (DTMA-AM) SNPs, | SAS 9.2, TASSEL 4.0 | Maize lethal necrosis (MLN) disease | 24 SNPs (18 SNPs in IMAS and 6 SNPs in DTMA-AM Panel) |
| Farfan et al. (2015) | 346 maize inbred lines | 60 000 GBS SNPs | R – software, Structure 2.93 | Drought aflatoxin resistance and agronomic traits | 6 SNPs for aflatoxin level, 20 SNPs for grain yield (ton/ha), 4 SNPs for plant height, 6 SNPs for Ear weight, 10 SNPs for days to anthesis |

GP accuracy is expressed by $r_{GM}$ which is the correlation coefficient value between predicted and observed values. Heritability depends on the genetic architecture of traits. A simple trait like AD in this study showed higher heritability as compared to the other two complex traits GY and ASI which are polygenic in nature (Cairns et al. 2013; Zhang et al. 2017a). The GP was estimated using two approaches; first in using trait-linked SNP markers in which 8 to 339 SNPs for different trait environment combinations were used for GP and in the second approach 10 108 high-quality genome-wide SNP markers were used for GP. The average prediction accuracy ($r_{GM}$) under this study for GY in WW, DS, HS, and DHS conditions were 0.75, 0.43, 0.69, and 0.58, respectively, using trait liked SNPs markers. While the $r_{GM}$ value for GY using genome-wide SNPs for the same trait was 0.59, 0.50, 0.38, and 0.35 for WW, DS, HS, and DHS conditions, respectively. Similar results were obtained for the remaining two traits AD and ASI also. This indicates the superiority of applying trait-linked markers for GP as compared to whole-genome SNPs (Yuan et al. 2019).

In another field experiment, maize inbred lines from temperate, tropical, and subtropical regions were collected and a panel of 346 inbred lines was formed for the GWAS study. Two AM panels (i) Goodman Maize association panel containing 302 lines and (ii) Williams/Warburton panel which has 300 lines were used to create this AM panel of 346 lines. All these lines were test crossed with high yielding line Tx714 which is susceptible to aflatoxin and all the test cross hybrids were evaluated at two different conditions, irrigated and non-irrigated for yield and related traits, days to anthesis, days to silking, PH, and EH. After *Aspergillus flavus* inoculation, aflatoxin content was observed in irrigated trial. *Aspergillus flavus* is a soil-borne fungus which produces the carcinogenic mycotoxin called aflatoxin. Hot and dry condition of maize field at preharvest stage is optimum for the growth of this fungus and aflatoxin content estimation in maize. Aflatoxin resistance is a complex trait and hence GWAS can be applied to identify genomic regions with minor effects (Betran et al. 2005; Williams et al. 2003; Windham and Williams 2002). Genotyping was done using GBS (Elshire et al. 2011) and 60 000 SNPs showing MAF more than 0.05 were considered for GWAS analysis (Table 9.3). A total of 5 QTL variants (QTV) were detected for GY across different analysis methods used (Farfan et al. 2015). The QTVs were located at maize chromosomes two, seven, nine, and ten. The PVE by QTVs ranged between 3 and 5%, with an allele effect of 0.14 to 0.59 ton ha$^{-1}$. Similarly, 2 QTVs for PH, one QTV for EH were identified in this study while multiple common QTVs were also found on chromosomes 3, 4, 7, and 8 for flowering time which includes days to anthesis and silking. This indicates that the same genomic regions are responsible for the inheritance of days to anthesis and silking traits. Farfan et al. (2015) did not find any significant QTV for ASI and aflatoxin resistance in this experiment. The outcome of this experiment not only validating the previous QTL mapping experiment result but also establishes the use of testcross hybrid to identify the dominant allele.

There are several examples where GBS has been used as a genotyping platform for bi-parental mapping populations. A total of 35 bi-parental populations were developed under Water Efficient Maize for Africa (WEMA) project by CIMMYT (Beyene et al. 2015; Zhang et al. 2015). These populations were developed from tropical maize lines to study the drought-related traits. A total of 15 maize populations were used by Wallace et al. (2016) to study the PH and flowering time (days to anthesis). Out of 15 populations, 11, 2, and 2 populations were advanced to $F_2$, $BC_1$, and $BC_1F_2$ generation respectively, and their test crossed hybrids were evaluated under well-watered and drought conditions in multiple environments in Kenya, Zimbabwe, and Zambia during 2010 and 2011. To get the details of these populations and their phenotyping, refer Semagn et al. (2013) and Beyene et al. (2015). GBS was used for genotyping these populations and generated a total 955 960 sites with large missing data which was addressed by three imputation methods FSFHap (Swarts et al. 2014), BEAGLE4 (Browning and Browning 2007), and FILLIN (Swarts et al. 2014). A total of 380 530 SNP markers with more than 0.05% allele frequency were used for further GWAS analysis. Heritability for studied traits (flowering time and PH) varied from 0.47 to 0.83 across populations and environments (WW and drought). Joint GWAS analysis used BLUP values calculated across environments, while Bayesian analysis used phenotypic data of each population and environment combination individually. Both methods produced similar results for both traits. Significant associations were observed on chromosomes 2, 5, 7, and 8 for flowering time. For PH, significant associations were found on chromosomes 2, 5, and 8. To further refine these results additional analysis using FarmCPU (Liu et al. 2016a) was performed. For a flowering time, 115 SNPs under well-watered condition, and 108 SNPs under water stress condition showed a significant association and explained 80 and 36% genetic variation, respectively. Strongly associated candidate genes identified in this study for flowering time are *Vgt1/ZmRap2.7*, *ZmPRR37*, *ZmPRR37.1*, *ZCN14*, and *ZCN18*. A total of 143 SNPs under well-watered conditions and 120 SNPs under drought conditions were found significantly associated with PH (Table 9.3). The genetic variation explained by these SNPs were 65 and 49% under irrigated and stress conditions, respectively. As expected, most of the PH-associated SNPs found in this study were located near auxin-related genes. Auxin is a plant growth regulator responsible for plant architecture and growth (Gallavotti 2013). Few candidate genes identified include *auxin response factor-transcription factor 20* on chromosome 5 in both conditions, *GRMZM2G108851* located on chromosome 8, *Pin11,* and *GRMZM2G479596* on chromosome 2 found under well-watered conditions. To get the information of other candidate genes showing association with flowering time and PH please refer Wallace et al. (2016). For GPs, whole-genome SNPs, model-selected SNPs, and random SNPs were used within the population and across the populations. The average GP accuracy was found to range from 0.3 to 0.4.

MLN is a devastating disease that causes yield losses from 30 to 100% in the maize growing region of SSA (Mudde et al. 2018). Synergistic interaction of two viruses, *Sugarcane mosaic virus* (SCMV) and *Maize chlorotic mottle virus* (MCMV) were responsible for MLN, an important disease of maize that affects plant growth at all stages from seedling to maturity. Gowda et al. (2015) have used 615 maize lines, which belong to two AM panels; IMAS (Improved Maize for African Soils) containing 235 lines and DTMA containing 380 maize lines. IMAS and DTMA-AM panels were genotyped with the GBS method (Elshire et al. 2011; Glaubitz et al. 2014) and after all quality checking (SNPs having heterozygosity >5%, missing value >5%, and allele frequency <2% were removed). A total of 259 000 and 264 000 SNPs were considered in respective AM panels for further analysis (Table 9.3). Optimized combination of 4 : 1 ratio of SCMV and MCMV viruses were used for artificial inoculation for phenotyping of MLN in all 615 lines at 2 locations in 3 seasons from 2012 to 2014. BLUP was estimated and used for GWAS analysis in TASSEL ver 4.0 (Bradbury et al. 2007). Twenty-four SNPs (18 SNPs in IMAS AM panel and 6 SNPs in DTMA panel) showed association with MLND trait ($p < 3 \times 10^{-5}$). Significant SNPs identified in the IMAS AM panel individually explained 8 to 10% genotypic variation, while SNPs identified in the DTMA-AM panel explained greater genotypic variation ranging from 14 to 18%. Total genotypic variation explained by all 6 SNPs together in the DTMA-AM panel is 37% while 18 SNPs identified in the IMAS panel collectively explained 30% of the total genotypic variation. This study identified 20 candidate genes, and all are involved in disease resistance. A total of 615 lines from two AM panels (IMAS and DTMA) were subjected for ridge regression best linear unbiased prediction (RR-BLUP) (Whittaker et al. 2000) for GS by Gowda et al. (2015). A set of uniformly distributed 2000 SNPs with all quality parameters described above were used for GP in both panels and accuracy was measured using fivefold cross-validation with 1000 times repetition. GP accuracy for panel IMAS and DTMA was 36 and 56% respectively while prediction accuracy was 41% across the panels. The inclusion of MLND linked SNPs along with random 2000 SNPs slightly improved the accuracy from 41 to 56% for both panels separately as well as for the combined panel. In this experiment, total of 14 lines were showing the disease score from 1.17 to 1.99 (1 to 5 scale) across the environment and can be used as donors in the MLND disease resistance breeding program.

To validate the outcome of the above experiment and further refinement, Nyaga et al. (2019) repeated the entire experiment with a much bigger mapping panel of 1400 maize lines which included the two earlier AM panels DTMA and IMAS along with WEMA AM panel (Table 9.3). Out of 1400 lines, 915 lines were genotyped using GBS (Elshire et al. 2011) and 955K SNPs were filtered with quality parameters explained earlier (Gowda et al. 2015). A total of 215 137 SNPs were used for GWAS analysis. All the 1400 lines were phenotyped for disease severity in 1 (resistance) to 9 (highly susceptible) scale and area under disease progressive curve (AUDPC). After analysis 18 and 19, significantly associated SNPs ($p < 1.0 \times 10^{-6}$) were identified for SD and AUDPC respectively with 5 SNPs common between both traits. Identification of common SNPs between this and previous studies conducted by Gowda et al. (2015) and other researchers (Sitonik et al. 2019; Gowda et al. 2018) validated the result found in the previous studies. Between the two traits, SNPs identified for AUDPC are explaining the phenotypic variation from 3 to 12% individually and 24.5% together. Similarly, SNPs identified for trait DS are explaining the phenotypic variation from 3 to 5% individually and 17.05% together. The two most significantly associated SNPs were found on chromosome 1 (S1_44539940) and chromosome 4 (S4_199711804) for DS and AUDPC, respectively.

SDM is an important leaf disease of maize that is caused by *Peronosclerospora sorghi* (Weston and Uppal) and responsible for the yield loss of up to 30%. Considering its quantitative nature several QTL mapping studies have been conducted by maize breeders to identify the genomic regions responsible for an inheritance to develop tightly linked marker with the trait (George et al. 2003; Agrama et al. 1999; Sabry et al. 2006; Jampatong et al. 2013; Lohithaswa et al. 2015). To overcome the limitations of QTL mapping based on bi-parental mapping population and to increase the mapping resolution, Rashid et al. (2018) used the CIMMYT Asia Association Mapping (CAAM) panel of 368 maize lines for GWAS analysis for SDM. CAAM is a collection of maize lines specially developed by CIMMYT and includes the tropical and subtropical maize lines adopted for the Asian climate. These lines were phenotyped during the wet season of 2012 and 2015 for SDM at Mandya, Karnataka, India being a hot spot for downy mildew. The disease was scored as per the modified method suggested by Craig et al. (1977). Best linear unbiased estimators

(BLUEs) were calculated from observed phenotyping data and used for further analysis. Isolated DNA of all the 368 lines was digested with Apek1 restriction enzyme before subjecting to sequencing on Illumina Hiseq 2000 (Elshire et al. 2011). B73 sequence was used as the reference genome for SNP calling and 955 690 SNPs were scored across all the chromosomes. To select the high-quality SNPs, different filtering criteria [call rate (CR) and MAF] were used and SNPs met with those criteria were subjected for analysis. Out of 955 690 SNPs, 333 039 SNPs were identified after applying filtering criteria of CR ≥0.7 and MAF ≥0.03. Similarly, 63 546 SNPs were identified with increased quality parameters of CR ≥0.9 and MAF ≥0.01 which were then subjected for PCA and kinship calculation. For GWAS analysis the MLM model which includes genotype, PCA, and kinship matrix was used, and 26 significant SNPs were identified across the maize genome (Table 9.3). These SNPs were located on chromosomes 1, 2, 3, 5, 6, 8, and 9 and explained phenotypic variance ranging from 5.50 to 13.07 where the highest phenotypic variance was explained by SNP S2_6154311 located at chromosome 2 and the lowest variance was explained by SNP S1_88004654 located on chromosome 1. The location of these identified SNPs is matching with bin map position identified in previous QTL mapping studies (George et al. 2003; Nair et al. 2005; Jampatong et al. 2013; Lohithaswa et al. 2015). This co-localization of identified SNPs with previously found QTL regions not only validates the earlier findings but also indicates the importance of these genomic regions for further analysis (candidate gene identification) for application in the SDM disease resistance maize breeding program.

GBS is extensively used in maize for developing trait-associated markers for a variety of traits like ear rot resistance (De jong et al. 2018), carotenoid bio-fortification (Suwarno et al. 2015), Fusarium ear rot (Chen et al. 2016b), iron homeostasis (Benke et al. 2015), forage quality in maize stalk (Wang et al. 2016), major kernel traits like KL and width, EAL and diameter, CD (Zhu et al. 2018), haploid male fertility (Ma et al. 2018), corn earworm resistance (Warburton et al. 2018), etc.

Application of GS using bi-parental population is effective in maize as bi-parental populations show high GS accuracy due to the close relationship between training and test/prediction population (Zhang et al. 2015). There are many successful examples of applying GS in bi-parental population where it is more effective than marker-assisted recurrent selection. Massman et al. (2013) developed recombinant inbred lines based on bi-parental cross between Mo17 and B73 and achieved up to 50% higher genetic gain for stover, GY, and other traits. Similarly, Beyene et al. (2015) found average genetic gain per year was three times across eight tropical maize populations as compared to pedigree-based phenotypic selection in water stress conditions. To further investigate the factors affecting the accuracy of GS, like the size of the training population, marker density, and trait heritability, Zhang et al. (2017a) used a total of 28 parental lines to develop 22 bi-parental populations. These 22 populations include 4120 inbred lines. Test cross hybrids of these bi-parental populations were phenotyped in two contrasting conditions: WW and drought condition in Zambia, Kenya, and Zimbabwe from 2010 to 2011 (Semagn et al. 2013). For genotyping TaqMan assay (http://www.appliedbiosystems.com), and KASP assay (Semagn et al. 2014) were used in this study.

To know the effect of the size of the training population and the number of markers used for prediction (marker density) on GP accuracy, four levels of training population sizes were tested at three levels of marker densities. Then six trait and environment combinations were evaluated with four levels of training populations, i.e. 16, 28, 39, and 50% of entire populations along with three levels of marker density of 50, 100, and all SNPs were used in different combinations for GP. For GY at irrigated/well water condition, GS accuracy ($r_{MG}$) mean was 0.24, 0.29, and 0.33 with a marker density of 50, 100, and all SNPs markers respectively with a training population size of 90. Similarly, for days to anthesis trait $r_{MG}$ was 0.24, 0.31, and 0.34 while for PH was 0.27, 0.34, and 0.38 for 50, 100, and all SNPs markers respectively for both traits when set of 90 progenies were used as training population. This trend is true for other trait environment combinations for all three traits. With all SNPs markers, genomic accuracy found for GY was 0.24, 0.28, 0.31, and 0.33 with a training population size of 30, 50, 70, and 90 respectively under well-watered conditions. For days to anthesis, it was 0.24, 0.29, 0.32, and 0.34 while for PH 0.27, 0.32, 0.36, and 0.38 for the four levels of training populations, respectively. This trend indicates that GP accuracy is higher for simple traits with high heritability as compared to complex traits like GY with low heritability. Genomic accuracy increases as you increase the training population size and similarly $r_{MG}$ is higher with an increasing number of markers at a constant size of the training population (Zhang et al. 2017a).

In another experiment, Zhang et al. (2017c) used the eighteen tropical maize lines to create multiparental mapping population for GS. To generate a cycle zero ($C_0$) training population, all 18 lines were intercrossed two times followed by single selfing. Then 1000 ear-to-row $C_0$ families were genotyped with 955 690 SNPs and their test crosses were phenotyped in Mexico at four different locations. The first cycle of phenotypic selection was carried out at $C_0$ to $C_1$ while the remaining three cycles of rapid cycling genomic selection (RCGS) were done from $C_1$ to $C_4$ cycles. PH, EH, days to 50% flowering, plant type, and disease tolerance traits were used for selecting the 50 families (157 ears) at the $C_0$ cycle. In the later generations ($C_1$ to $C_4$), GSs were used for selecting the best grain yielder as a parental line. In each generation, 25 plants were sown per family. A total of 25, 18, and 22 families were selected in $C_1$, $C_2$, and $C_3$ generations respectively. Genomic Best linear unbiased predictor model (GBLUP) was used for GS (VanRaden 2007, 2008). GS accuracy of 0.55 for GY was achieved after implementing the fivefold cross-validation with 100 replicates. In this study, 2 breeding cycles were taken per year and realized genetic gain for GY was estimated at 0.225 ton $ha^{-1}$ $cycle^{-1}$ for RCGS ($C_1$ to $C_4$). The genetic gain was calculated using the formula Gain $= (i \times r \times h)/t$ where $i$ is selection intensity, $r$ is selection accuracy, $h$ is the square root of narrow-sense heritability and $i$ is the time in years it takes to complete a selection cycle. The outcome of this study indicates that the RCGS method allows achieving genetic gain along with maintaining diversity in a multiparent population.

## 9.8 Summary

For successful GWAS and GS study, several markers need to be used for genotyping and the AM panel or training and test population plays a very vital role. GBS can be applied for genotyping of inbred lines, bi-parental mapping populations as well as multiparent populations. GBS provides a genotyping platform for affordable genotyping per data point for these studies. It gives the flexibility of multiplexing the genotypes from 96 to 384 based on the marker density requirements and available budget. Routine application of GBS in QTL mapping, GWAS, and GS will help to develop superior maize products in the future.

## References

Agrama, H.A., Moussa, M.E., Naser, M.E. et al. (1999). Mapping of QTL for downy mildew resistance in maize. *Theoretical and Applied Genetics* 99 (3): 519–523.

An, Y., Chen, L., Li, Y.X. et al. (2019). Candidate loci for the kernel row number in maize revealed by a combination of transcriptome analysis and regional association mapping. *BMC Plant Biology* 19: 201.

Ananiev, E.V., Phillips, R.L., and Rines, H.W. (1998). Chromosome-specific molecular organization of maize (*Zea mays* L.) centromeric regions. *Proceedings of the National Academy of Sciences of the United States of America* 95 (22): 13073–13078.

Baird, N.A., Etter, P.D., Atwood, T.S. et al. (2008). Rapid SNP discovery and genetic mapping using sequenced RAD markers. *PLoS One* 3: e3376.

Beadle, G.W. (1939). Teosinte and the origin of maize. *The Journal of Heredity* 30: 245–247.

Benke, A., Urbany, C., and Stich, B. (2015). Genome-wide association mapping of iron homeostasis in the maize association population. *BMC Genetics* 16: 1.

Bennetzen, J.L. (2009). Maize genome structure and evolution. In: *Handbook of Maize: Genetics and Genomics* (eds. J.L. Bennetzen and S. Hake), 179–199. New York, NY: Springer.

Betran, J., Odvody, G., Mayfield, K. et al. (2005). Breeding corn to reduce preharvest aflatoxin contamination. In: *Aflatoxin and Food Safety* (ed. H.K. Abbas), 353–378. CRC Press.

Beyene, Y., Semagn, K., Mugo, S. et al. (2015). Genetic gains in grain yield through genomic selection in eight bi-parental maize populations under drought stress. *Crop Science* 55 (1): 154–163.

Bradbury, P.J., Zhang, Z., Kroon, D.E. et al. (2007). TASSEL: software for association mapping of complex traits in diverse samples. *Bioinformatics* 23 (19): 2633–2635.

Broman, K.W., Wu, H., Sen, S. et al. (2003). R/qtl: QTL mapping in experimental crosses. *Bioinformatics* 19 (7): 889–890.

Browning, S.R. and Browning, B.L. (2007). Rapid and accurate haplotype phasing and missing-data inference for whole-genome association studies by use of localized haplotype clustering. *The American Journal of Human Genetics* 81 (5): 1084–1097.

Bukowski, R., Guo, X., Lu, Y. et al. (2018). Construction of the third-generation *Zea mays* haplotype map. *Gigascience* 7: 1–12.

Cairns, J.E., Crossa, J., Zaidi, P.H. et al. (2013). Identification of drought, heat, and combined drought and heat tolerant donors in maize. *Crop Science* 53: 1335. https://doi.org/10.2135/cropsci2012.09.0545.

Chen, Z., Wang, B., Dong, X. et al. (2014). An ultra-high density bin-map for rapid QTL mapping for tassel and ear architecture in a large $F_2$ maize population. *BMC Genomics* 15: 433.

Chen, L., Li, Y., Li, C. et al. (2016a). Fine-mapping of qGW4.05, a major QTL for kernel weight and size in maize. *BMC Plant Biology* 16 (1): 81.

Chen, J., Shrestha, R., Ding, J. et al. (2016b). Genome-wide association study and QTL mapping reveal genomic loci associated with Fusarium ear rot resistance in tropical maize germplasm. *G3: Genes, Genomes, Genetics* 6 (12): 3803–3815.

Chia, J.M., Song, C., Bradbury, P.J. et al. (2012). Maize HapMap2 identifies extant variation from a genome in flux. *Nature Genetics* 44: 803–807.

Craig, J., Bockholt, A.J., Frederiksen, R.A. et al. (1977). Reaction of important corn inbred lines to *Sclerospora sorghi*. *Plant Disease Report* 61: 563–564.

Crossa, J., Beyene, Y., Kassa, S. et al. (2013). Genomic prediction in maize breeding populations with genotyping-by-sequencing. *G3: Genes, Genomes, Genetics* 3: 1903–1926.

Doebley, J. (2004). The genetics of maize evolution. *Annual Review of Genetics* 38: 37–59.

Du, H., Zhu, J., Su, H. et al. (2017). Bulked segregant RNA-seq reveals differential expression and SNPs of candidate genes associated with waterlogging tolerance in maize. *Frontiers in Plant Science* 8: 1022. https://doi.org/10.3389/fpls.2017.01022.

Elshire, R.J., Glaubitz, J.C., Sun, Q. et al. (2011). A robust, simple genotyping-by-sequencing (GBS) approach for high diversity species. *PLoS One* 6 (5): e19379.

FAO (Food and Agriculture Organization) (1992). *Maize in Human Nutrition. Rome*. Italy: Food and Agriculture Organization of the United Nations.

FAOSTAT (2018). http://www.fao.org/faostat/en/#data

Farfan, I.D.B., Fuente, G.N.D.L., Murray, S.C. et al. (2015). Genome wide association study for drought, aflatoxin resistance, and important agronomic traits of maize hybrids in the sub-tropics. *PLoS One* 10 (2): e0117737.

Feschotte, C., Jiang, N., and Wessler, S. (2002). Plant transposable elements: where genetics meets genomics. *Nature Reviews Genetics* 3: 329–341.

Fu, J., Cheng, Y., Linghu, J. et al. (2013). RNA sequencing reveals the complex regulatory network in the maize kernel. *Nature Communications* 4: 2832.

Fu, L.X., Cai, C.C., Cui, Y.N. et al. (2016a). Pooled mapping: an efficient method of calling variations for population samples with low-depth resequencing data. *Molecular Breeding* 36: 48–48.

Fu, Y.B., Peterson, G.W., and Dong, Y. (2016b). Increasing genome sampling and improving SNP genotyping for genotyping-by-sequencing with new combinations of restriction enzymes. *G3: Genes, Genomes, Genetics* 6: 845–856.

Gallavotti, A. (2013). The role of auxin in shaping shoot architecture. *Journal of Experimental Botany* 64 (9): 2593–2608.

Gao, S., Martinez, C., Skinner, D.J. et al. (2008). Development of a seed DNA-based genotyping system for marker-assisted selection in maize. *Molecular Breeding* 22 (3): 477.

George, M.L.C., Prasanna, B.M., Rathore, R.S. et al. (2003). Identification of QTLs conferring resistance to downy mildews of maize in Asia. *Theoretical and Applied Genetics* 107 (3): 544–551.

Glaubitz, J.C., Casstevens, T.M., Lu, F. et al. (2014). TASSEL-GBS: a high capacity genotyping by sequencing analysis pipeline. *PLoS One* 9 (2): e90346.

Gore, M.A., Chia, J.M., Elshire, R.J. et al. (2009). A first-generation haplotype map of maize. *Science* 326: 1115–1117.

Gouesnard, B., Negro, S., Laffray, A. et al. (2017). Genotyping-by-sequencing highlights original diversity patterns within a European collection of 1191 maize flint lines, as compared to the maize USDA genebank. *Theoretical and Applied Genetics* 130 (10): 2165–2189.

Gowda, M., Das, B., Makumbi, D. et al. (2015). Genome-wide association and genomic prediction of resistance to maize lethal necrosis disease in tropical maize germplasm. *Theoretical and Applied Genetics* 128 (10): 1957–1968.

Gowda, M., Beyene, Y., Makumbi, D. et al. (2018). Discovery and validation of genomic regions associated with resistance to maize lethal necrosis in four biparental populations. *Molecular Breeding* 38 (5): 66.

Hansey, C.N., Vaillancourt, B., Sekhon, R.S. et al. (2012). Maize (*Zea mays* L.) genome diversity as revealed by RNA-sequencing. *PLoS One* 7: e33071.

Heffelfinger, C., Christopher, A., Fragoso, C.A. et al. (2014). Flexible and scalable genotyping-by-sequencing strategies for population studies. *BMC Genomics* 15: 979.

Heffner, E.L., Sorrells, M.E., and Jannink, J.-L. (2009). Genomic selection for crop improvement. *Crop Science* 49: 1–12. https://doi.org/10.2135/cropsci2008.08.0512.

Heffner, E.L., Lorenz, A.J., Jannink, J.-L. et al. (2010). Plant breeding with genomic selection: gain per unit time and cost. *Crop Science* 50: 1681–1690. https://doi.org/10.2135/cropsci2009.11.0662.

Helentjaris, T., Slocum, M., Wright, S. et al. (1986). Construction of genetic linkage maps in maize and tomato using restriction fragment length polymorphisms. *Theoretical and Applied Genetics* 72 (6): 761–769.

Hirsch, C.N., Foerster, J.M., Johnson, J.M. et al. (2014). Insights into the maize pan-genome and pan transcriptome. *Plant Cell* 26: 121–135.

Huang, X., Feng, Q., Qian, Q. et al. (2009). High-throughput genotyping by whole-genome resequencing. *Genome Research* 19: 1068–1076.

Hufford, M.B., Xu, X., Heerwaarden, V. et al. (2012). Comparative population genomics of maize domestication and improvement. *Nature Genetics* 44: 808–811.

Jamann, T.M., Sood, S., Wisser, R.J. et al. (2017). High-throughput resequencing of maize landraces at genomic regions associated with flowering time. *PLoS One* 12 (1): e0168910. https://doi.org/10.1371/journal.pone.0168910.

Jampatong, C., Jampatong, S., Jompuk, C. et al. (2013). Mapping of QTL affecting resistance against sorghum downy mildew (*Peronosclerospora sorghi*) in maize (*Zea mays* L). *Maydica* 58 (2): 119–126.

Jia, S., Morton, K., Zhang, C. et al. (2018). An exome-seq based tool for mapping and selection of candidate genes in maize deletion mutants. *Genomics, Proteomics & Bioinformatics* 16 (6): 439–450.

Jiao, Y., Peluso, P., Shi, J. et al. (2017). Improved maize reference genome with single-molecule technologies. *Nature* 546: 524–527.

Jiménez-Galindo, J.C., Ordás, B., Butrón, A. et al. (2017). QTL mapping for yield and resistance against Mediterranean corn borer in maize. *Frontiers in Plant Science* 8: 698.

de Jong, G., Pamplona, A.K.A., Von Pinho, R.G. et al. (2018). Genome-wide association analysis of ear rot resistance caused by Fusarium verticillioides in maize. *Genomics* 110 (5): 291–303.

Kellogg, E.A. (1998). Relationships of cereal crops and other grasses. *Proceedings of the National Academy of Sciences of the United States of America* 95 (5): 2005–2010.

Klein, H., Xiao, Y., Conklin, P.A. et al. (2018). Bulked-segregant analysis coupled to whole genome sequencing (BSA-seq) for rapid gene cloning in maize. *G3: Genes, Genomes, Genetics* 8 (11): 3583–3592. https://doi.org/10.1534/g3.118.200499.

Kyetere, D.T., Ming, R., McMullen, M.D. et al. (1999). Genetic analysis of tolerance to maize streak virus in maize. *Genome* 42 (1): 20–26.

Lagat, M., Danson, J., Kimani, M. et al. (2008). Quantitative trait loci for resistance to maize streak virus disease in maize genotypes used in hybrid development. *African Journal of Biotechnology* 7 (15): 2573–2577.

Langmead, B. and Salzberg, S.L. (2012). Fast gapped-read alignment with Bowtie 2. *Nature Methods* 9: 357–359. https://doi.org/10.1038/nmeth.1923.

Li, H. and Durbin, R. (2009). Fast and accurate short read alignment with Burrows-Wheeler transform. *Bioinformatics* 25: 1754–1760. https://doi.org/10.1093/bioinformatics/btp324.

Li, H., Handsaker, B., Wysoker, A. et al. (2009). The sequence alignment/map format and SAM tools. *Bioinformatics* 25: 2078–2079.

Li, L., Li, D.L., Liu, S.Z. et al. (2013). The maize glossy13 gene, cloned via BSR-seq and seq-walking encodes a putative ABC transporter required for the normal accumulation of epicuticular waxes. *PLoS One* 8: e82333.

Li, C., Song, W., Luo, Y. et al. (2019). The HuangZaoSi maize genome provides insights into genomic variation and improvement history of maize. *Molecular Plant* 12: 402–409.

Liang, Z. and Schnable, J.C. (2016). RNA-seq based analysis of population structure within the maize inbred B73. *PLoS One* 11 (6): e0157942.

Lipka, A.E., Tian, F., Wang, Q. et al. (2012). GAPIT: genome association and prediction integrated tool. *Bioinformatics* 28 (18): 2397–2399.

Liu, S., Ying, K., Yeh, C.-T. et al. (2012). Changes in genome content generated via segregation of non-allelic homologs. *The Plant Journal* 72 (3): 390–399.

Liu, Y., Wang, L., Sun, C. et al. (2014). Genetic analysis and major QTL detection for maize kernel size and weight in multienvironments. *Theoretical and Applied Genetics* 127: 1019–1037.

Liu, H., Niu, Y., Gonzalez-Portilla, P.J. et al. (2015). An ultra-high-density map as a community resource for discerning the genetic basis of quantitative traits in maize. *BMC Genomics* 16: 1078.

Liu, X., Huang, M., Fan, B. et al. (2016a). Iterative usage of fixed and random effect models for powerful and efficient genome-wide association studies. *PLoS Genetics* 12 (2): e1005767.

Liu, C., Zhou, Q., Dong, L. et al. (2016b). Genetic architecture of the maize kernel row number revealed by combining QTL mapping using a high-density genetic map and bulked segregant RNA sequencing. *BMC Genomics* 17: 915.

Liu, H., Wang, X., Wei, B. et al. (2016c). Characterization of genome-wide variation in Four-Row Wax, a waxy maize landrace with a reduced kernel row phenotype. *Frontiers in Plant Science* 7: 667. https://doi.org/10.3389/fpls.2016.00667.

Liu, J., Huang, J., Guo, H. et al. (2017). The conserved and unique genetic architecture of kernel size and weight in maize and rice. *Plant Physiology* 175 (2): 774–785.

Liu, M., Tan, X., Yang, Y. et al. (2020a). Analysis of the genetic architecture of maize kernel size traits by combined linkage and association mapping. *Plant Biotechnology Journal* 18 (1): 207–221.

Liu, J., Fernie, A.R., and Yan, J. (2020b). The past, present, and future of maize improvement: domestication, genomics, and functional genomic routes toward crop enhancement. *Plant Communications* 1: 100010.

Lohithaswa, H.C., Jyothi, K., Sunil Kumar, K.R. et al. (2015). Identification and introgression of QTLs implicated in resistance to sorghum downy mildew (Peronosclerosporasorghi (Weston and Uppal) C. G. Shaw) in maize through marker-assisted selection. *Journal of Genetics* 94 (4): 741–748.

Ma, H., Li, G., Würschum, T. et al. (2018). Genome-wide association study of haploid male fertility in maize (*Zea mays* L.). *Frontiers in Plant Science* 9: 974.

Manching, H., Sengupta, S., Hopper, K.R. et al. (2017). Phased genotyping-by-sequencing enhances analysis of genetic diversity and reveals divergent copy number variants in maize. *G3: Genes, Genomes, Genetics* 7: 2161–2170. https://doi.org/10.1534/g3.117.042036.

Massman, J.M., Jung, H.-J.G., and Bernardo, R. (2013). Genomewide selection versus marker-assisted recurrent selection to improve grain yield and stover-quality traits for cellulosic ethanol in maize. *Crop Science* 53: 58–66. https://doi.org/10.2135/cropsci2012.02.0112.

Mazaheri, M., Heckwolf, M., Vaillancourt, B. et al. (2019). Genome-wide association analysis of stalk biomass and anatomical traits in maize. *BMC Plant Biology* 19: 45.

McClintock, B. (1931). The order of the genes C, Sh and Wx in *Zea mays* with reference to a cytologically known point in the chromosome. *Proceedings of the National Academy of Sciences of the United States of America* 17 (8): 485–491.

McKenna, A., Hanna, M., Banks, E. et al. (2010). The Genome Analysis Toolkit: A Map Reduce framework for analyzing next-generation DNA sequencing data. *Genome Research* 20 (9): 1297–1303. https://doi.org/10.1101/gr.107524.110.

Meissle, M., Mouron, P., Musa, T. et al. (2010). Pests, pesticide use and alternative options in European maize production: current status and future prospects. *Journal of Applied Entomology* 134: 357–375. https://doi.org/10.1111/j.1439-0418.2009.01491.x.

Morgante, M. (2006). Plant genome organisation and diversity: the year of the junk! *Current Opinion in Biotechnology* 17 (2): 168–173.

Mudde, B., Olubayo, F.M., Miano, D.W. et al. (2018). Distribution, incidence and severity of maize lethal necrosis disease in major maize growing agro-ecological zones of Uganda. *Journal of Agricultural Science* 10 (6): p72.

Muraya, M.M., Schmutze, T., Ulpinnis, C. et al. (2015). Targeted sequencing reveals large- scale sequence polymorphism in maize candidate genes for biomass production and composition. *PLoS One* 10 (7): e0132120. https://doi.org/10.1371/journal.pone.0132120.

Nair, S.K., Prasanna, B.M., Garg, A. et al. (2005). Identification and validation of QTLs conferring resistance to sorghum downy mildew (*Peronosclerospora sorghi*) and Rajasthan downy mildew (*P. heteropogoni*) in maize. *Theoretical and Applied Genetics* 110 (8): 1384–1392.

Nair, S.K., Babu, R., Magorokosho, C. et al. (2015). Fine mapping of Msv1, a major QTL for resistance to Maize Streak Virus leads to development of production markers for breeding pipelines. *Theoretical and Applied Genetics* 128 (9): 1839–1854.

Nannas, N.J. and Dawe, R.K. (2015). Genetic and genomic toolbox of *Zea mays*. *Genetics* 199 (3): 655–669.

Nyaga, C., Gowda, M., Beyene, Y. et al. (2019). Genome-wide analyses and prediction of resistance to MLN in large tropical maize germplasm. *Genes (Basel)* 11 (1): E16.

Paulsmeyer, M.N., Brown, P.J., and Juvik, J.A. (2018). Discovery of Anthocyanin Acyltransferase1 (AAT1) in maize using genotyping-by-sequencing (GBS). *G3: Genes, Genomes, Genetics* 8: 3669–3678.

Peacock, W.J., Dennis, E.S., Rhoades, M.M. et al. (1981). Highly repeated DNA sequence limited to knob heterochromatin in maize. *Proceedings of the National Academy of Sciences of the United States of America* 78 (7): 4490–4494.

Pernet, A., Hoisington, D., Franco, J. et al. (1999a). Genetic mapping of maize streak virus resistance from the Mascarene source. I. Resistance in line D211 and stability against different virus clones. *Theoretical and Applied Genetics* 99 (3): 524–539.

Pernet, A., Hoisington, D., Dintinger, J. et al. (1999b). Genetic mapping of maize streak virus resistance from the Mascarene source. II. Resistance in line CIRAD390 and stability across germplasm. *Theoretical and Applied Genetics* 99 (3): 540–553.

Poland, J.A., Brown, P.J., Sorrells, M.E. et al. (2012). Development of high density genetic maps for barley and wheat using a novel two-enzyme genotyping-by-sequencing approach. *PLoS One* 7 (2): e32253.

Prasanna, B.M., Palacios-Rojas, N., Hossain, F. et al. (2019). Molecular breeding for nutritionally enriched maize: status and prospects. *Frontiers in Genetics* 10: 1392.

Pratt, R.C. and Gordon, S.G. (2005). Breeding for resistance to maize foliar pathogens. In: *Plant Breeding Reviews* (ed. J. Janick). https://doi.org/10.1002/9780470650349.ch3.

Rashid, Z., Singh, P.K., Vemuri, H. et al. (2018). Genome-wide association study in Asia-adapted tropical maize reveals novel and explored genomic regions for sorghum downy mildew resistance. *Scientific Reports* 8 (1): 366.

Rashid, Z., Sofi, M., Harlapur, S.I. et al. (2020). Genome-wide association studies in tropical maize germplasm reveal novel and known genomic regions for resistance to Northern corn leaf blight. *Scientific Reports* 10: 21949. https://doi.org/10.1038/s41598-020-78928-5.

Rayburn, A.L., Price, H.J., Smith, J.D. et al. (1985). C-band heterochromatin and DNA content in *Zea mays*. *American Journal of Botany* 72: 1610–1617. https://doi.org/10.1002/j.1537-2197.1985.tb08425.x.

Rayburn, A.L., Biradar, D.P., Bullock, D.G. et al. (1993). Nuclear DNA content in F1 hybrids of maize. *Heredity* 70 (3): 294–300.

Romay, M.C., Millard, M.J., Glaubitz, J.C. et al. (2013). Comprehensive genotyping of the USA national maize inbred seed bank. *Genome Biology* 14: R55. https://doi.org/10.1186/gb-2013-14-6-r55.

Sabry, A., Jeffers, D., Vasal, S.K. et al. (2006). A region of maize chromosome 2 affects response to downy mildew pathogens. *Theoretical and Applied Genetics* 113 (2): 321–330.

Scheben, A., Batley, J., and Edwards, D. (2017). Genotyping-by-sequencing approaches to characterize crop genomes: choosing the right tool for the right application. *Plant Biotechnology Journal* 15: 149–161.

Schnable, P.S., Ware, D., Fulton, R.S. et al. (2009). The B73 maize genome: complexity, diversity, and dynamics. *Science* 326 (5956): 1112–1115.

Semagn, K., Beyene, Y., Warburton, M.L. et al. (2013). Meta-analyses of QTL for grain yield and anthesis silking interval in 18 maize populations evaluated under water-stressed and well-watered environments. *BMC Genomics* 14: 313.

Semagn, K., Babu, R., Hearne, S. et al. (2014). Single nucleotide polymorphism genotyping using Kompetitive Allele Specific PCR (KASP): overview of the technology and its application in crop improvement. *Molecular Breeding* 33 (1): 1–14.

Sham, P., Bader, J.S., Craig, I. et al. (2002). DNA Pooling: a tool for large-scale association studies. *Nature Reviews Genetics* 3 (11): 862–871.

Sitonik, C., Suresh, L.M., Beyene, Y. et al. (2019). Genetic architecture of maize chlorotic mottle virus and maize lethal necrosis through GWAS, linkage analysis and genomic prediction in tropical maize germplasm. *Theoretical and Applied Genetics* 132 (8): 2381–2399.

Sonah, H., Bastien, M., Iquira, E. et al. (2013). An improved genotyping by sequencing (GBS) approach offering increased versatility and efficiency of SNP discovery and genotyping. *PLoS One* 8: e54603.

Song, W., Wang, B., Hauck, A.L. et al. (2016). Genetic dissection of maize seedling root system architecture traits using an ultra-high density bin-map and a recombinant inbred line population. *Journal of Integrative Plant Biology* 58 (3): 266–279.

Springer, N.M., Anderson, S.N., Andorf, C.M. et al. (2018). The maize W22 genome provides a foundation for functional genomics and transposon biology. *Nature Genetics* 50: 1282–1288.

Su, C., Wang, W., Gong, S. et al. (2017). High density linkage map construction and mapping of yield trait QTLs in maize (*Zea mays*) using the genotyping-by-sequencing (GBS) technology. *Frontiers in Plant Science* 8: 706.

Sun, S., Zhou, Y., Chen, J. et al. (2018). Extensive intraspecific gene order and gene structural variations between Mo17 and other maize genomes. *Nature Genetics* 50: 1289–1295.

Suwarno, W.B., Pixley, K.V., Palacios-Rojas, N. et al. (2015). Genome-wide association analysis reveals new targets for carotenoid biofortification in maize. *Theoretical and Applied Genetics* 128 (5): 851–864.

Swarts, K., Li, H., Navarro, J.A.R. et al. (2014). Novel methods to optimize genotypic imputation for low-coverage, next-generation sequence data in crop plants. *The Plant Genome* 7 (3) https://doi.org/10.3835/plantgenome2014.05.0023.

Tamaki, H., Mitsuhashi, S., Kudoh, H. et al. (2016). Genome wide molecular polymorphisms among maize (*Zea mays* L.) inbred lines found from restriction-associated DNA tag sequencing (RAD-seq) analysis as a preliminary study on 'Genome wide Selection' for breeding by Japanese public sectors. *Bulletin of National Institute of Livestock and Grassland Science* 16: 1–9.

Utz, H.F. (2012). PlabMQTL - Software for meta-QTL analysis with composite interval mapping. Version 0.9. Institute of Plant Breeding, Seed Science, and Population Genetics, University of Hohenheim. Stuttgart, Germany.

VanRaden, P.M. (2007). Genomic measures of relationship and inbreeding. *Interbull Bulletin* 37: 33–33.

VanRaden, P.M. (2008). Efficient methods to compute genomic predictions. *Journal of Dairy Science* 91 (11): 4414–4423.

Verma, R., Oania, R., Graumann, J. et al. (2004). Multiubiquitin chain receptors define a layer of substrate selectivity in the ubiquitin-proteasome system. *Cell* 118 (1): 99–110.

Wallace, J.G., Zhang, X., Beyene, Y. et al. (2016). Genome-wide association for plant height and flowering time across 15 tropical maize populations under managed drought stress and well-watered conditions in Sub-Saharan Africa. *Crop Science* 56: 2365–2378. https://doi.org/10.2135/cropsci2015.10.0632.

Wang, S., Basten, C.J., and Zeng, Z.-B. (2012). Windows QTL Cartographer 2.5. Department of Statistics, North Carolina State University, Raleigh, NC. http://statgen.ncsu.edu/qtlcart/WQTLCart.htm

Wang, H., Li, K., Hu, X. et al. (2016). Genome-wide association analysis of forage quality in maize mature stalk. *BMC Plant Biology* 16 (1): 227.

Wang, B., Liu, H., Liu, Z. et al. (2018a). Identification of minor effect QTLs for plant architecture related traits using super high density genotyping and large recombinant inbred population in maize (*Zea mays*). *BMC Plant Biology* 18 (1): 17.

Wang, J., Zhang, X., and Lin, Z. (2018b). QTL mapping in a maize F2 population using genotyping-by-sequencing and a modified fine-mapping strategy. *Plant Science* 276: 171–180.

Warburton, M.L., Womack, E.D., Tang, J.D. et al. (2018). Genome-wide association and metabolic pathway analysis of corn earworm resistance in maize. *Plant Genome* 11 (1): 1–8.

Welz, H.G., Schechert, A., Pernet, A. et al. (1998). A gene for resistance to the maize streak virus in the African CIMMYT maize inbred line CML202. *Molecular Breeding* 4 (2): 147–154.

Whittaker, J.C., Thompson, R., and Denham, M.C. (2000). Marker-assisted selection using ridge regression. *Genetics Research* 75 (2): 249–252.

Williams, W.P., Windham, G.L., and Buckley, P.M. (2003). Enhancing maize germplasm with resistance to aflatoxin contamination. *Journal of Toxicology - Toxin Reviews* 22 (2–3): 175–193.

Windham, G.L. and Williams, W.P. (2002). Evaluation of corn inbreds and advanced breeding lines for resistance to aflatoxin contamination in the field. *Plant Disease* 86 (3): 232–234.

Wu, X., Wang, B., Xie, F. et al. (2020). QTL mapping and transcriptome analysis identify candidate genes regulating pericarp thickness in sweet corn. *BMC Plant Biology* 20 (1): 117.

Xu, Y., Lu, Y., Xie, C. et al. (2012). Whole-genome strategies for marker-assisted plant breeding. *Molecular Breeding* 29 (4): 833–854.

Xu, J., Yuan, Y., Xu, Y. et al. (2014). Identification of candidate genes for drought tolerance by whole-genome resequencing in maize. *BMC Plant Biology* 14: 83. https://doi.org/10.1186/1471-2229-14-83.

Xu, Y., Li, P., Zou, C. et al. (2017). Enhancing genetic gain in the era of molecular breeding. *Journal of Experimental Botany* 68 (11): 2641–2666.

Yang, J., Hu, C., Hu, H. et al. (2008). QTLNetwork: mapping and visualizing genetic architecture of complex traits in experimental populations. *Bioinformatics* 24 (5): 721–723.

Yang, N., Xu, X.W., Wang, R.R. et al. (2017). Contributions of *Zea mays* subspecies mexicana haplotypes to modern maize. *Nature Communications* 8: 1874.

Yang, N., Liu, J., Gao, Q. et al. (2019). Genome assembly of a tropical maize inbred line provides insights into structural variation and crop improvement. *Nature Genetics* 51: 1052–1059.

Yuan, Y., Cairns, J.E., Babu, R. et al. (2019). Genome-wide association mapping and genomic prediction analyses reveal the genetic architecture of grain yield and flowering time under drought and heat stress conditions in maize. *Frontiers in Plant Science* 9: 1919.

Zhang, Y.-M. and Xu, S. (2004). Mapping quantitative trait loci in F2 incorporating phenotypes of F3 progeny. *Genetics* 166 (4): 1981–1993.

Zhang, Z., Ersoz, E., Lai, C.-Q. et al. (2010). Mixed linear model approach adapted for genome-wide association studies. *Nature Genetics* 42 (4): 355–360.

Zhang, X., Pérez-Rodríguez, P., Semagn, K. et al. (2015). Genomic prediction in biparental tropical maize populations in water-stressed and well-watered environments using low-density and GBS SNPs. *Heredity* 114 (3): 291–299.

Zhang, A., Wang, H., Beyene, Y. et al. (2017a). Effect of trait heritability, training population size and marker density on genomic prediction accuracy estimation in 22 bi-parental tropical maize populations. *Frontiers in Plant Science* 8: 1916.

Zhang, C., Zhou, Z., Yong, H. et al. (2017b). Analysis of the genetic architecture of maize ear and grain morphological traits by combined linkage and association mapping. *Theoretical and Applied Genetics* 130 (5): 1011–1029.

Zhang, X., Pérez-Rodríguez, P., Burgueño, J. et al. (2017c). Rapid cycling genomic selection in a multiparental tropical maize population. *G3 (Bethesda)* 7 (7): 2315–2326.

Zhao, Y. and Su, C. (2019). Mapping quantitative trait loci for yield-related traits and predicting candidate genes for grain weight in maize. *Scientific Reports* 9 (1): 16112.

Zhou, S., Wei, F., Nguyen, J. et al. (2009). A single molecule scaffold for the maize genome. *PLoS Genetics* 5 (11): e1000711.

Zhou, Z., Zhang, C., Zhou, Y. et al. (2016). Genetic dissection of maize plant architecture with an ultra-high density bin map based on recombinant inbred lines. *BMC Genomics* 17: 178.

Zhu, X.-M., Shao, X.-Y., Pei, Y.-H. et al. (2018). Genetic diversity and genome-wide association study of major ear quantitative traits using high-density SNPs in maize. *Frontiers in Plant Science* 9: 966.

# 10

# Recent Advances and Applicability of GBS, GWAS, and GS in Soybean

*Prashant Raghunath Shingote*[1], *Dhananjay Narayanrao Gotarkar*[2], *Ravindra Ramrao Kale*[3], *Omkar Maharudra Limbalkar*[4], *and Dhiraj Lalji Wasule*[1]

[1] *Vasantrao Naik College of Agricultural Biotechnology, Dr. Panjabrao Deshmukh Krishi Vidyapeeth, Akola, Maharashtra, India*
[2] *International Rice Research Institute, Los Baños, Philippines*
[3] *ICAR-Indian Institute of Rice Research, Hyderabad, Telangana, India*
[4] *ICAR-Devision of Genetics, Indian Agriculture Research Institute, New Delhi, India*

## 10.1 Introduction

### 10.1.1 Importance of Soybean for Global Food Security

Soybean (*Glycine max* L. Merril) is the most important legume crop, often called as the "miracle crop." It occupies close to 6% of the world's arable land. It is the most versatile crop in the world and is grown in a different soil and climatic conditions. Targeted improvement in yield-related traits and reduction in production costs have upgraded the position of soybeans among arable crops. Globally, soybean is the leading crop among oilseed crops. Currently, soybeans account for about 35% of the total harvested area devoted to annual and perennial oil seed crops and >60% of the total oilseed production. The global soybean production of 336.563 million metric tons was recorded for the year 2019–2020. Brazil is the highest soybean producer with 123 million metric tons and productivity of 3333 $kg\,ha^{-1}$ followed by America and Argentina (Figure 10.1). Soybean seed contains about 40% protein, 20% oil, and other important nutrients and food health properties as compared to most other grain legumes. It contributes 25% to global vegetable oil production. China tops the soybean oil production with 15.142 million metric tons in the year 2019–2020 followed by the United States, Argentina, and Brazil (Figure 10.2a). It contributes >70% to the total protein meal consumption about two-thirds of the world's protein concentrate. Argentina tops the global Soy meal export of 30.3 million metric tons followed by Brazil and the United States (Figure 10.2b). The widespread use of soybean oil, in particular, as edible oil is due to nutrient supplies, competitive price, neutral flavor, and stability in both unhydrogenated and partially hydrogenated form. About 95% of soybean is used for edible oil production; the remaining is utilized in industrial products such as soaps, fatty acids, and biodiesel. The soybean seed is

*Genotyping by Sequencing for Crop Improvement*, First Edition.
Edited by Humira Sonah, Vinod Goyal, S.M. Shivaraj, and Rupesh K. Deshmukh.

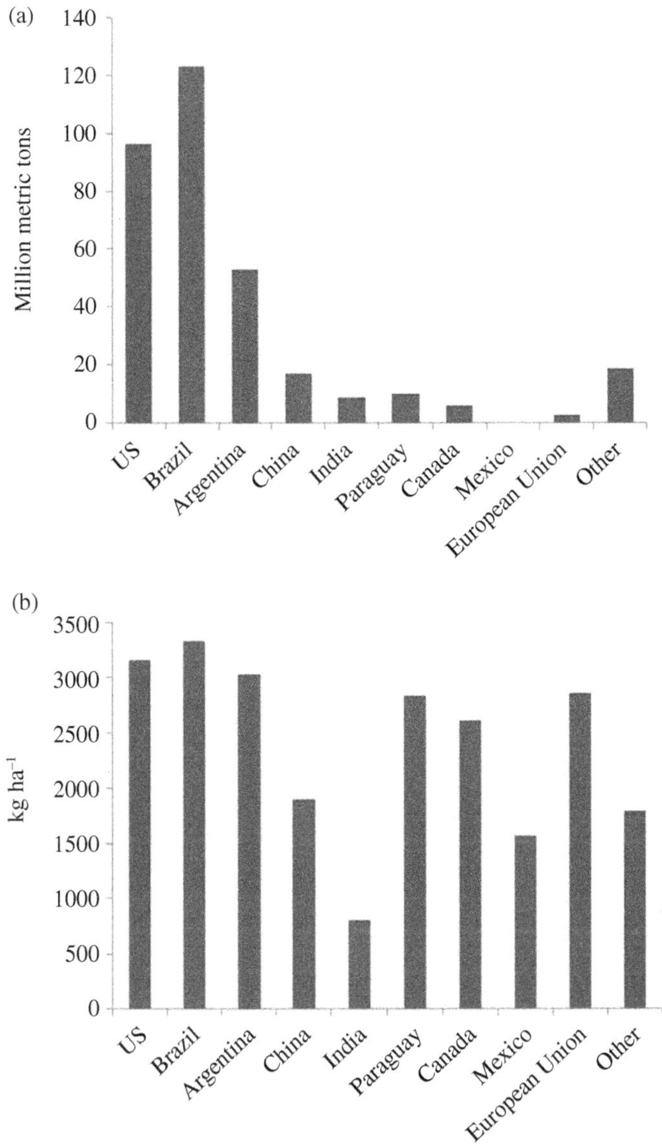


**Figure 10.1** World soybean production and productivity in 2019–2020. (a) Production in million metric tons and (b) productivity in kg $ha^{-1}$.

characterized by unique nutritional composition consisting of proteins, lipids, and carbohydrates. Other than these major nutrients soybean seeds also contain phospholipids, vitamins, and minerals. Because of its high protein content, ~98% of the soybean meal is used as a feed in livestock and aquaculture. The remaining small amount, i.e. 2%, is used to prepare soy flour and protein for human consumption. The soybean products for human consumption includes soy hamburger, tofu, and soy milk. Soybean is also used as a fresh vegetable known as "mao dou" in China, "edamame" in Japan and the United States. Several evidences showed the use of soybean as a food crop in Northeastern China around 1700–1100 BC, where the

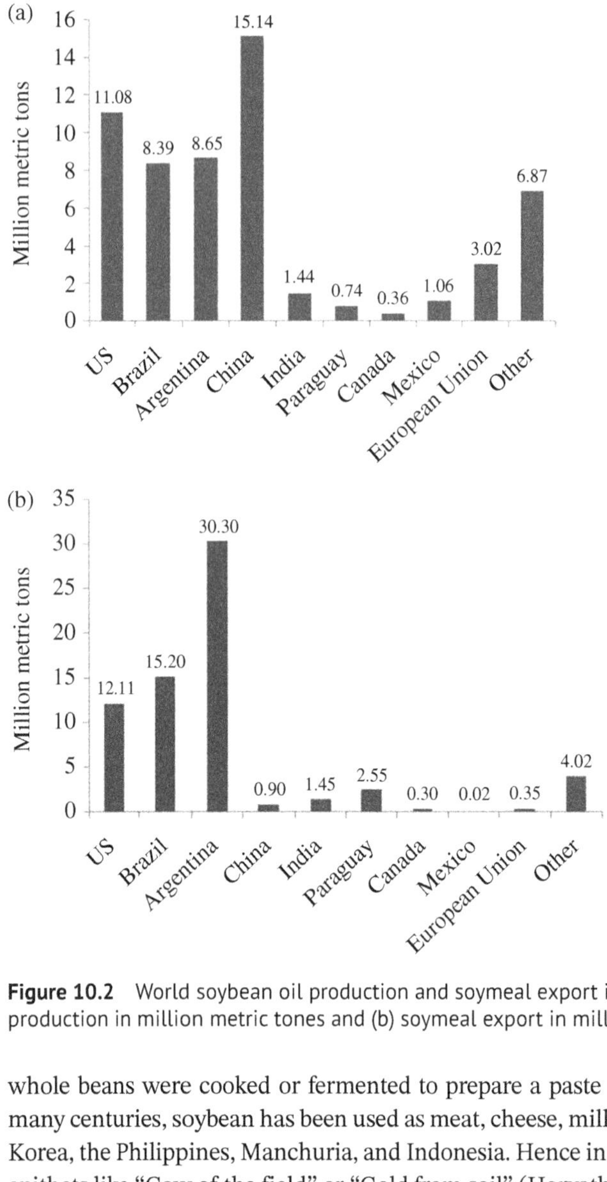


**Figure 10.2** World soybean oil production and soymeal export in the year 2019–2020. (a) Oil production in million metric tones and (b) soymeal export in million metric tones.

whole beans were cooked or fermented to prepare a paste and used in various dishes. For many centuries, soybean has been used as meat, cheese, milk, bread, and oil in China, Japan, Korea, the Philippines, Manchuria, and Indonesia. Hence in these countries, soybean earned epithets like "Cow of the field" or "Gold from soil" (Horvath 1926).

### 10.1.2 Challenges in Soybean Production

Even though soybean production has increased drastically over the decades, still several challenges are encountered to achieve the potential yield. Such challenges include the following:

1) Major soybean cultivation area is under rain-fed condition.
2) Erratic behavior of monsoon affecting planting.
3) Soil moisture stress at crucial growth stages, specifically under seed-filling stage.

4) High-temperature stress at critical growth stages.
5) Biotic factors affecting the crop growth.
6) Low level of mechanization.
7) Poor adoption of improved production technology and low risk covering ability.
8) Poor varietal diversification and mono-cropping.
9) Poor utilization in the food chain owing to the characteristic beany flavor of soybean.
10) Interference in use of soybean as pulse because of its hard-to-cook character.
11) Limited entrepreneurship for processing.
12) Inadequate availability of short duration and mechanical harvesting varieties.
13) Limited availability of resistance to pod shattering varieties.
14) Inadequate availability of resistant varieties for Charcoal Rot and Yellow mosaic.
15) Limited availability of drought and salt-tolerant varieties.

### 10.1.3 Soybean Genetic Improvement

The biotic and abiotic factors are affecting the agricultural productivity on a large scale. These include the scarcity of water, decreasing area, and degradation of land due to environmental change, pollution, the occurrence of new pathogens and pests, and climate. It is expected that the world population will grow from 7 billion to 12.3 billion (Kim et al. 2010). Plant breeding will be instrumental to increase crop productivity which helps in feeding the growing population (Bhat et al. 2016). Presently, along with the breeding for yield quantity, the quality characters are also gaining more attention. Plant breeding encompasses two major strategies, classical breeding, and molecular breeding. The classical breeding involves interbreeding among the closely related genotypes to develop novel cultivars with desirable traits. However, it requires several generations to develop improved verities which limit its ability to address the issue of increasing food demand (Tester and Langridge 2010). Molecular plant breeding utilizes molecular biology or biotechnological tools to improve crop verities. It includes two approaches, marker-assisted selection (MAS) and genetic transformation (Moose and Mumm 2008; Xu 2010). The MAS has been widely used in plant breeding to increase yield, quality, and biotic or abiotic stress tolerance. MAS involves indirect selection of the targeted traits (Collard and Mackill 2008). Recent advances in sequencing technologies like genotyping-by-sequencing (GBS) provides dense markers like SNPs which can accelerate breeding plants for novel traits.

## 10.2 GBS Efforts in Soybean

Advances in next-generation technologies have brought the costs of DNA sequencing to the point that GBS is now feasible for highly diverse and/or large genome species (Elshire et al. 2011). Next-generation sequencing technologies (NGS) employed for GBS analysis will generate high-throughput molecular marker information at a low cost. Bioinformatics pipelines have been developed to analyze the data obtained through GBS (Jarquín et al. 2014). The GBS decreases the cost through an enzyme-mediated genomic complexity reduction and use of bar-coded adapters for multiplexing (Poland and Rife 2012). Sonah et al. (2013) developed a novel GBS protocol in soybean, and reliably called 10 120 high-quality

**Table 10.1** High-throughput genomics platforms used for soybean genotyping.

| Genotyping platform | Number of SNPs/Indels | References |
|---|---|---|
| Illumina genome analyzer | 25 047 | Hyten et al. (2010) |
| Illumina GoldenGate assay | 1 536 | Vuong et al. (2010) |
| Illumina Genome Analyzer II | 205 614 | Lam et al. (2010) |
| Illumina genome analyzer/Solexa | 39 022 | Wu et al. (2010) |
| Illumina genome analyzer/RRLs | 14 550 | Varala et al. (2011) |
| Illumina Infinium SoySNP6K BeadChip | 5 376 | Akond et al. (2013) |
| Illumina GAIIx | 10 120 | Sonah et al. (2013) |
| Illumina HiSeq2000 | >47 000 | Sonah et al. (2015) |
| Illumina Genome Analyzer II | 5 102 244/707 969 | Li et al. (2013) |
| Illumina Infinium BeadChip | 52 041 | Song et al. (2013) |
| Illumina Infinium SoySNP50K BeadChip | 52 041 | Bandillo et al. (2015) |
| Illumina GoldenGate assay/Illumina BeadStation 500 | 1 536 | Phansak et al. (2016) |
| Illumina iScan platform/SoySNP8K BeadChip | 7 039 | Ping et al. (2016) |
| Illumina Infinium BeadChip | 3 343 | Patil et al. (2018) |
| llumina BeadArray platform | 1 536 | Li et al. (2019) |
| BARCSoySNP6K BeadChip | 19 | Priolli et al. (2019) |
| BARCSoySOY6K BeadChip | 3 807 | Ludke et al. (2019) |

single-nucleotide polymorphisms (SNPs) among eight diverse lines. Likewise, several researchers have used GBS to genotype soybean to develop resilient cultivars (Table 10.1).

## 10.3 High-Density Linkage Maps in Soybean

Genetic mapping involves identification of relative position and distances between markers present on a chromosome. It is also referred as linkage mapping or meiotic mapping (Semagn et al. 2006). The construction of saturated genetic maps with complete genome coverage is the initial step for certain breeding applications which are helpful in finding the location of genes or quantitative trait loci (QTL) (Tanksley et al. 1989). The principle underlying the genetic mapping is that the genes or markers segregate through chromosome recombination during meiosis, thus allowing their analysis in the progeny (Paterson et al. 2000). The success of linkage mapping depends on different factors like the type of mapping population and its size, type of molecular marker and its genetic behavior, number of markers used and their distribution, missing data, precision of phenotyping, statistical approaches, methodologies used, etc. The mapping procedure involves sequential construction of a map by adding loci one after another, starting with the most informative loci or pair. At each step, a marker is placed on the map based on linkage distance with the markers that were already mapped earlier. Map distances are equal to recombination frequencies and genetic mapping functions translate recombination distances into map distances.

The first plant polymorphism and heritability study using simple sequence repeat (SSR) markers were reported in soybean (Akkaya et al. 1992). For genomics studies of soybean precise and saturated genetic linkage map is highly necessary, which can be achieved through identification of new trait loci or understated loci including QTLs, map-based cloning, and construction of physical maps or sequencing of whole genome. The first soybean genetic map was constructed with 57 classical markers (Palmer and Kilen 1987). Subsequently, molecular maps have been incorporated with restriction-fragment length polymorphism (RFLP) markers (Lark et al. 1993; Shoemaker and Specht 1995; Mansur et al. 1996), random amplified polymorphic DNA (RAPD) markers (Ferreira et al. 2000), SSR (Akkaya et al. 1995; Cregan et al. 1999), and amplified-fragment length polymorphism (AFLP) markers (Maughan et al. 1996; Keim et al. 1997).

In recent years, the availability of molecular markers and powerful statistical tools have led to considerable progress in QTL mapping in many crops including soybean. In soybean, several genes for qualitative as well as quantitative traits have been mapped using molecular markers. Information about these QTL is available at SoyBase database (http://www.soybase.org). About 67 QTL mapping studies in soybean led to the identification of a total of 3624 QTLs for yield, yield-related traits, and domestication-related traits. Among all traits mapped, 100-seed weight (SW) accounts for the maximum number of QTLs (230), followed by plant height (PH) (223) and minimum number of QTLs for seed hardiness (02) and seeds per plant (03). Grain yield per plant, biotic stress, abiotic stresses, and others account for 172, 524, 273, and 1718 QTLs, respectively as per SoyBase database.

A breakthrough in soybean research occurred in the year 2010 with the completion of the first whole-genome sequence Glyma1.01. The sequencing of the soybean cultivar "William 82" indicated a redundant coverage equivalent to 8x obtained through whole-genome shotgun (WGS) approach (Schmutz et al. 2010). The large-scale shotgun sequencing was completed early in 2008 with ~13 million Sanger shotgun reads. Nevertheless, the assembled sequence needed an improvement since it was based on the linkage maps constructed with an inadequate number of markers and recombinant inbred lines (RILs) (Song et al. 2016). High-density linkage maps are one of the most admired tools useful for precisely localizing QTL or genes in the plant genome. Genomic information on soybean has been generated very rapidly since the initial publication of its reference genome, providing a valuable tool for the genetic enhancement of soybean. In recent years, with the advancement in sequencing techniques, investigators started reporting high-density linkage maps. An integrated high-density linkage map of soybean was constructed using an $F_2$ population derived from parents Misuzudaizu and Moshidou Gong 503 with RFLP, SSR, STS, and AFLP Markers (Xia et al. 2007). They developed and integrated a total of markers viz. 318 AFLP, 121 SSR, 108 RFLP, and 126 STS into the framework of the earlier version of the linkage map. The updated genetic map covering a map distance of 3080 cM (Kosambi function) in 20 linkage groups (LGs) was developed. The overall inter-marker distance was shortened from 5.78 to 2.41 cM and the number of gaps of more than 25 cM was reduced from 19 to 6 compared to the previously described linkage map. It also extended the coverage of the linkage map with 17 markers mapping beyond the distal ends of the previous one (Xia et al. 2007).

QTL mapping is a promising method to reveal the genetic architecture governing complex quantitative traits. But, the low density of molecular markers in genetic maps has limited the efficiency and accuracy of QTL mapping. Therefore, for the identification of consistent QTL new high-throughput genotyping and SNP discovery strategies based on NGS were used to construct a high-density linkage map in soybean. Hyten et al. (2010) developed high density integrated linkage map of soybean using SNPs where they screened three RILs populations and diverse germplasm using high-throughput GoldenGate assays for 3456 SNPs. The "Universal Soy Linkage Panel" with a comprehensive genetic map of 1536 SNP loci was developed, which can be validated in QTL mapping populations (Hyten et al. 2010). Further, Illumina Infinium BeadChip SoySNP50K was developed for high-density genotyping array in soybean (Song et al. 2013). A total of 498 921 777 reads of 35–45 bp in length were obtained from six cultivated and two wild soybean genotypes. A total of 209 903 SNPs were mapped on soybean whole-genome sequence and a total of 146 161 SNPs were identified to be ideal candidates for SoySNP50K chip design. After equalizing the distance between 60 800 SNPs (50 701 were found in euchromatic and 10 000 in heterochromatic regions), a total of 52 041 passed for the development of SoySNP50K BeadChip. Validation of the SoySNP50K associated candidate regions, 620 with domestication and 42 with recent selection. Song et al. (2013) recommended SoySNP50K BeadChip for genetic diversity analysis, identification of linkage disequilibrium (LD), and for building high-resolution linkage maps which aid in the improvement of whole-genome sequence assembly. Specific length amplified fragment sequencing (SLAF-seq) is an efficient strategy for large-scale de novo SNP discovery and genotyping in a short time. SLAF-seq was used to construct the high-density genetic map of soybean RIL population (Qi et al. 2014b). The total length of the final map was 2655.68 cM with a 0.5 cM average distance per adjacent marker with a total of 5308 markers over 20 LGs. Li et al. (2014) also constructed a high-density genetic map using SLAF-seq and studied QTL associated with isoflavone content. They obtained 116 216 high-quality SLAFs, of which 9948 were polymorphic. The final map comprised of 5785 SLAFs on 20 different LGs which spanned 2255.18 cM in size of the genome with an average distance of 0.43 cM between adjoining markers. They further, reported a high-density genetic map for soybean that exhibited high resolution and accuracy. In another study, a high-resolution genetic linkage map was constructed with BARCSoySNP6K using 357 $F_7$ RIL from a cross of "Wyandot" and PI 567301B (Lee et al. 2015). A total of 2545 SNPs were mapped across the 20 LGs with unique loci 86 to 162 per chromosome where the total length of the map was 2346 cM. The overall marker interval ranged between 0.9 and 1.3 cM and there was co-linear relationship between physical (Mb) and genetic distance (cM) of SNPs for most of the genome, showing the perfection in the updated soybean genome assembly Glyma.Wm82.a2 than that of previous assembly Glyma.Wm82.a1. This high-density linkage map showed the utility of the BARCSoySNP6K BeadChip array (Lee et al. 2015). Song et al. (2016) constructed a high-resolution genetic linkage map for improving the soybean genome sequence assembly Glyma1.01. Two high-density genetic linkage maps were constructed using 1083 RILs derived from cross Williams 82 x *G. soja* with 21 478 mapped SNPs and 922 RILs derived from Essex x Williams 82 with 11 922 mapped SNPs. Song et al. (2016) using Wm82.a2.v1 assembly updated the genomic positions of the markers exhibited in BARCSOYSSR_1.0 and the SoySNP50K BeadChip (Song et al. 2016). Li et al. (2017) constructed a high-density genetic linkage map using SLAF-seq in 200 RILs, their map spans 2534.42 cM genetic distance with an average adjacent marker distance of 0.72 cM and the map consists of 3541 markers over 20 LGs. They have identified 26 stable QTL including nine novel loci, for fatty acid composition in soybean. Cao et al. in 2017 used a combination of linkage and association mapping strategies together to elucidate the genetic basis of seed oil content of Chinese soybean in different environments (Cao et al. 2017). One RIL population (2062 specific-locus amplified fragment markers), a derived F2 population (to confirm one major QTL), and a soybean breeding germplasm population with 279 lines were used to perform a genome-wide association study using 59 845 high-quality SNP markers. Zhang et al. (2018) constructed a high-density linkage map to identify the QTLs for the yield-traits and quality-traits, using SLAF-seq. The length of the final map was 2909.46 cM with an average adjacent marker distance of 0.57 cM, they reported 14 novel QTLs and 35 QTLs distributed on 17 chromosomes for the 100-SW, PH, protein content in seeds (Protein), and oil content in seeds (Oil). Notably, found 11 QTLs with a physical distance lesser than 100 kb, indicating a direct chance to identify desirable genes. SLAF-sequencing could provide detailed information for accurate QTL localization. Watanabe et al. (2018) developed a high-density linkage map using progenies obtained from across between Japanese soybean Enrei and Chinese soybean Peking. The map based on information from 32 accessions comprising of 2177 markers with polymorphism offers a comprehensive genetic framework for these markers. The order of the marker on the linkage map showed a close similarity to a chromosome-scale assembly, Wm82.a2.v1. They reported variation in traits such as flowering, maturity, and yield mapped as 50 reproducible QTLs, and further, the QTLs with small genetic effects were evaluated as single genetic factors in a uniform genetic background. Zuo et al. (2019) analyzed the effect of segregation distortion of marker on linkage map preparation and QTL mapping. A total of 519 RILs of soybean from orthogonal and reciprocal crosses were genotyped by SLAF-seq markers. This study led to the construction of 20 LGs with 11 846 markers, including 1513 (12.77%) significantly distorted markers. Linkage maps were further validated by checking the high colinearity between marker's position on soybean reference genome and linkage maps as well as through the high consistency of 24 QTL regions. These high-density linkage maps might be beneficial to marker-assisted breeding, in addition, these could be highly useful for the localization of QTLs in soybean.

## 10.4 GBS Protocols and Analytical Pipelines for Soybean

GBS produces a huge amount of data and the key problem is calling for large-scale data. Hence bioinformatics analytical pipelines have been developed to handle such data. Appropriate bioinformatics tools are also required to extract information and reduce the complexity of raw data during library preparation before sequencing (Davey et al. 2011). Different bioinformatics pipelines have been built which follow some common data analysis steps (Figure 10.3) to troubleshoot various genomic characteristics and technical factors that make GBS data analysis challenging (Torkamaneh et al. 2017). A GBS pipeline must comprise steps from filtering poor-quality reads to scoring genotypes for each individual in the study. GBS pipelines are categorized into two groups, first one is reference-based GBS analysis pipelines which use reference sequence for analysis *viz.*, Fast-GBS, TASSEL-GBS (version1 and version2), Stacks and IGST (Sonah et al. 2013; Bradbury et al. 2007; Glaubitz et al. 2014) and the second group is de novo-based approach which does not require reference sequence for analysis viz., UNEAK,

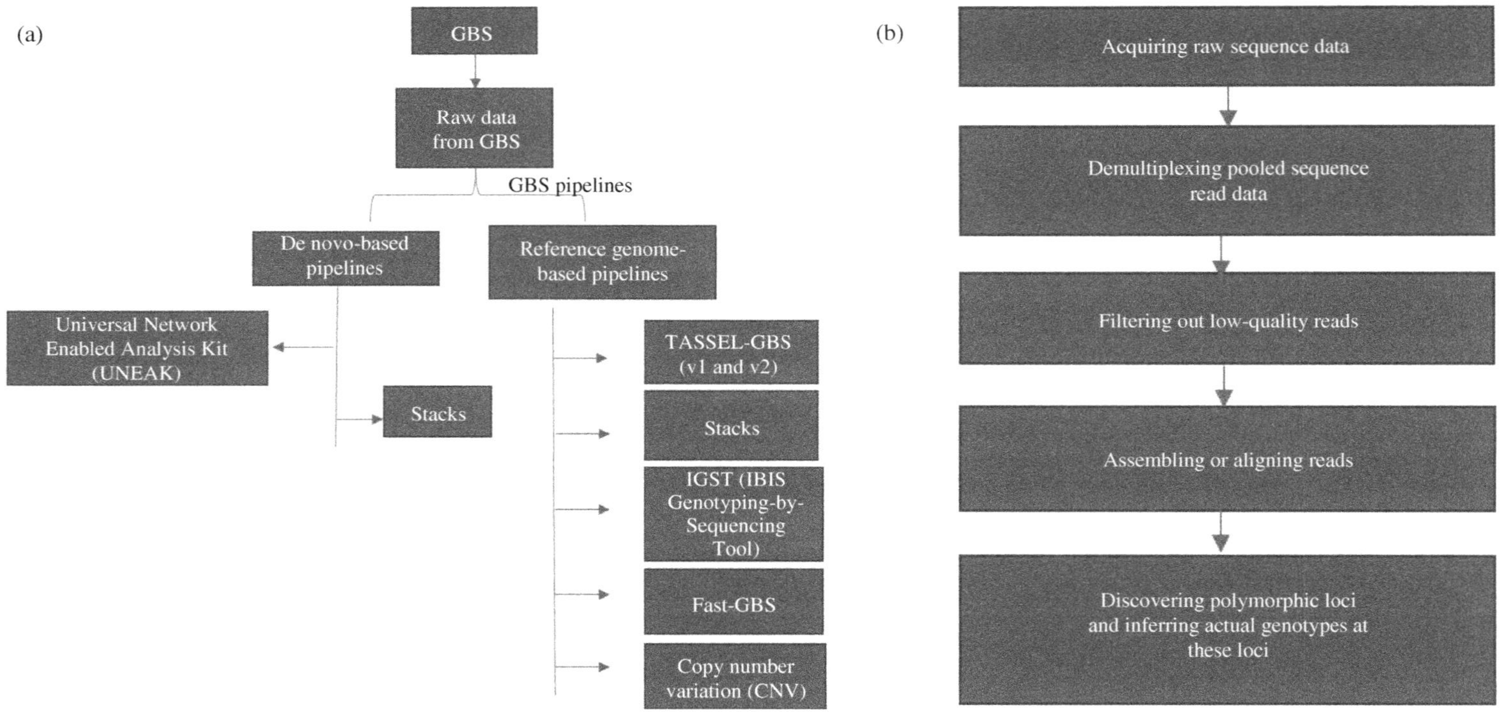


**Figure 10.3** Various pipelines and steps are involved in analyzing GBS data. (a) The different types of available GBS pipelines. (b) Common steps involved in analyzing data using GBS pipeline.

and Stacks (Catchen et al. 2013; Lu et al. 2013). It is necessary to use one or combination of pipelines as a single pipeline does not work in every situation (Torkamaneh et al. 2016).

In a study by Torkamaneh et al. (2016), GBS sequence data obtained from Illumina and Ion Torrent platform using a set of 24 resequenced soybean lines were analyzed with the help of different GBS pipelines to identify SNP. The Fast-GBS analysis pipeline enables either simultaneous processing of a large number of samples in a cluster or multiple sample serial processing on a single machine. IGST (IBIS Genotyping-by-Sequencing Tool) uses different publicly available software tools as well as some in-house tools and involved steps executed in separate shell scripts. TASSEL-GBS V1 and V2 functions similarly, requiring all reads to be trimmed to the same length (64 bp in v1, up to 92 bp in v2), and identical reads to collapse into tags that are then aligned with the reference genome, and calling SNPs from aligned tags. UNEAK (Universal Network Enabled Analysis Kit) the general design of this pipeline follows identification of single base pair mismatches which are candidate SNPs. Stacks (reference-based and de novo) pipeline uses sequenced DNA fragments from any restriction enzyme-based GBS protocol as raw input data to identify SNPs. Among all these pipelines, Fast-GBS calls for the greatest number of polymorphisms and yield the highest accuracy (98.7%). Compare to the analysis of different data from the same pipeline more overlapping SNP catalogs (79–92% overlap) have been obtained from analysis of the same data from different pipelines. Fast-GBS and TASSEL-GBS v1 proved to be the fastest running among the reference-based pipelines, whereas UNEAK was almost three times faster de novo pipelines than Stacks (Kumar et al. 2012; Torkamaneh et al. 2016)

In another study, Torkamaneh et al. (2017) found that Fast-GBS was designed to provide highly accurate genotyping, allow modest computational resources, and encourage usage. Here, they have used collections of samples from soybean, barley, and potato to study variants which include a range of different genome sizes, genome complexity, and ploidy level. Comparison of SNP genotypes derived from GBS showed estimated variants with accuracies of 98.7, 95.2, and 94% for soybean, barley, and potato, respectively. Fast-GBS is also found to be capable of detecting different kinds of variants (SNPs, MNPs, and Indels).

A genome wide occurrence of mutations in a subset of a population can be studied by Fast neutron populations. GBS data in 79 soybean mutants were used to assess the sensitivity and precision of identified copy number variations (CNVs) with the validation of results using array comparative genomic hybridization (aCGH). This study provides a proof of concept that an optimized CNVs calling pipeline on GBS data can be used as an affordable high-throughput method for assessing CNVs in mutants. The R package named delgbs has been developed for implementing the CNV calling pipeline and it is also publicly available on GitHub (github.com/malemay/delgbs) (Lemay et al. 2019)

## 10.5 GBS-based QTL Mapping Efforts in Soybean

Identification of genomic loci governing complex traits has been facilitated by the development of QTL mapping approaches (Andersen and Torp 2002). Conventionally, QTL mapping is performed using segregating bi-parental populations using low-density markers for the identification of hundreds of QTLs for different traits in different mapping populations in soybean. Typically, QTLs are distributed on all twenty chromosomes and few of these loci are common to all mapping efforts (www.soybase.org). The majority of the bi-parental QTL mapping studies with low-density marker coverage do not give the precise information regarding the number and location of the QTLs governing the traits. For accurate identification of QTLs, high-density linkage maps constructed using high-throughput markers like SNPs are advantageous. High-density linkage maps can increase the precision and provide accurate positioning of the detected QTL and may further aid in the identification of candidate genes (CGs) underlying the QTLs (Verma et al. 2015). Among the various NGS technologies, GBS is one of the NGS-based methods used for simultaneous SNP discovery and genotyping (Sonah et al. 2015).

Recently, several GBS-based QTL mapping studies have been undertaken in soybean (Table 10.2). Qi et al. (2014a) identified 15 major QTLs for 11 traits by combining *de novo* sequencing data and germplasm resequencing data to identify *GmCHX1*, a new ion transporter gene. Further, the sequence alteration in *GmCHX1* was shown to be related to the level of salt tolerance. Soybean cyst nematode (SCN) *Heterodera glycine* (HG) is the most destructive pest of soybean worldwide. Vuong et al. (2010) used a total of 927 polymorphic SSRs and SNP markers for genotyping. Further, the four QTLs linked with resistance to

**Table 10.2** List of GBS-based QTL mapping studies in soybean.

| Trait | Cross | Population | Chromosome | Genetic position/ marker interval | LOD | References |
|---|---|---|---|---|---|---|
| Seed antioxidant content | W05 X C08 | Recombinant inbreed lines | 19 | 37424658–38133607 | 4.1 | Qi et al. (2014a) |
| Seed anthocyanin content | W05 X C08 | Recombinant inbreed lines | 8 | 8019871–8609212 | 7.6 | Qi et al. (2014a) |
| Pod no per plant | W05 X C08 | Recombinant inbreed lines | 13 | 38761376–40041964 | 3.8 | Qi et al. (2014a) |
| Seed no per plant | W05 X C08 | Recombinant inbreed lines | 13 | 38761376–40041964 | 3.8 | Qi et al. (2014a) |
| Growth period | W05 X C08 | Recombinant inbreed lines | 6 | 19518049–20432589 | 4.1 | Qi et al. (2014a) |
| Salt tolerance | W05 X C08 | Recombinant inbreed lines | 3 | 40204091–41182426 | 4.3 | Qi et al. (2014a) |
| Seed coat color | W05 X C08 | Recombinant inbreed lines | 8 | 7902342–8576842 | 7.0 | Qi et al. (2014a) |
| Pod color | W05 X C08 | Recombinant inbreed lines | 19 | 37367542–37786063 | 4.6 | Qi et al. (2014a) |
| Trailing growth | W05 X C08 | Recombinant inbreed lines | 13 | 38761376–39342212 | 4.0 | Qi et al. (2014a) |
| Leaf length/ width ration | W05 X C08 | Recombinant inbreed lines | 20 | 34864294–35465896 | 3.9 | Qi et al. (2014a) |
| Nodule no per plant | W05 X C08 | Recombinant inbreed lines | 16 | 36484899–36661376 | 4.6 | Qi et al. (2014a) |

**Table 10.2** (Continued)

| Trait | Cross | Population | Chromosome | Genetic position/ marker interval | LOD | References |
|---|---|---|---|---|---|---|
| Resistance to multiple SCN HG types, 2.5.7 | Magellan X PI 567516C | Recombinant inbreed lines | 10 | Sat_038–Satt592 | 6.6 | Vuong et al. (2010) |
| Resistance to multiple SCN HG types, 1.3.5.6.7 | Magellan X PI 567516C | Recombinant inbreed lines | 10 | Sat_274–Satt592 | 9.8 | Vuong et al. (2010) |
| Resistance to multiple SCN HG types, 1.2.5.7 | Magellan X PI 567516C | Recombinant inbreed lines | 18 | Sat_210–Sat_403 | 4.5 | Vuong et al. (2010) |
| Resistance to multiple SCN HG types, 1.3.5.6.7 | Magellan X PI 567516C | Recombinant inbreed lines | 18 | Satt612–Satt191 | 3.5 | Vuong et al. (2010) |
| Resistance to multiple SCN HG types, 1.2.5.7 | Magellan X PI 567516C | Recombinant inbreed lines | 8 | Satt612–Satt191 | 4.3 | Vuong et al. (2010) |
| Average nodule dry weight | Bossier X Embrapa 20 | Recombinant inbreed lines | 13 | 26 581 863–26 934 841 | 4.66 | Grunvald et al. (2018) |
| Shoot dry weight | Bossier X Embrapa 20 | Recombinant inbreed lines | 19 | 1 637 165–3 166 049 | 3.93 | Grunvald et al. (2018) |
| Root colonization | | Diverse accession | 16 | S16_2603450 | 5.77 | Pawlowski et al. (2020) |
| Root colonization | | Diverse accession | 6 | S6_46533686 | 5.29 | Pawlowski et al. (2020) |
| Root colonization | | Diverse accession | 9 | S9_49539462 | 4.67 | Pawlowski et al. (2020) |
| Root colonization | | Diverse accession | 10 | S10_37487605 | 4.53 | Pawlowski et al. (2020) |
| Root colonization | | Diverse accession | 14 | S14_43114203 | 4.44 | Pawlowski et al. (2020) |
| Root colonization | | Diverse accession | 8 | S8_17743016 | 4.36 | Pawlowski et al. (2020) |
| Pod dehiscence | Heihe 43 X Heihe 18 | Recombinant inbreed lines | 1 | 5 149 656–5 408 961 | 3.94 | Han et al. (2019) |
| Pod dehiscence | Heihe 43 X Heihe 18 | Recombinant inbreed lines | 5 | 40 448 596–40 703 417 | 11.0 | Han et al. (2019) |
| Pod dehiscence | Heihe 43 X Heihe 18 | Recombinant inbreed lines | 8 | 35 560 353–38 482 164 | 6.18 | Han et al. (2019) |

either multiple or single SCN HG types were identified following the genetic linkage analysis, permutation tests, and composite interval mapping. Two QTLs identified in this study for resistance to multiple-SCN HG types have not been reported in any other SCN resistance sources.

In breeding programs, the economically important biological nitrogen fixation (BNF) trait for plant growth is often neglected. This was addressed by identifying two significant QTLs for nodule number and nodule dry weight (Grunvald et al. 2018). Similarly, an SNP dataset identified from a whole-genome resequencing approach led to the identification of six QTLs significantly associated with *R. intraradices* colonization which contributed for 24% of the phenotypic variance. CGs characterized from these QTL regions included the homologs to genes coding for known nodulin proteins and related to other symbiosis processes (Pawlowski et al. 2020).

The flowering trait QTL was identified rapidly and efficiently using NGS-based QTL analysis of F8:10 RILs by Kong et al. (2018), where they evaluated the traits of time to flowering, maturity, and reproductive period in the field. Han et al. (2019) has identified six new pod dehiscence QTLs mapped using inclusive composite interval mapping, which comprised of three stable QTLs (qPD01,qPD05-1andqPD08-1). These three QTLs have been validated by developing CAPS/dCAPS markers.

## 10.6 Soybean Genome Sequencing and Resequencing

In 2010, the first reference genome of the cultivated soybean has been released jointly by the soybean research community and the Department of Energy Joint Genome Institute (DOE-JGI) (Schmutz et al. 2010). The whole-genome sequence of Soybean (var. Williams 82), contained 950 Megabases (Mb) of assembled and anchored sequence, which represented about 85% of the predicted genome (1115-Mb) that was included in physical and high-density genetic maps to create a chromosome-level sequence assembly (Schmutz et al. 2010).

This achievement made the genomics-based research in soybean much easier. Further, developments in genomics and genotyping platforms for soybean had attracted many researchers to study the genetic variation among the cultivars and between wild type and domesticated soybean genotypes for agronomically important traits (Kim et al. 2010; Lam et al. 2010; Varala et al. 2011; Chung et al. 2014; Valliyodan et al. 2016), also the artificial selection during domestications leads to reduced genetic diversity in soybean cultivars was well explained by Li et al. (2013). Bandillo et al. (2015) have analyzed 14 430 genotypes to reveal the distribution of genetic variation contained within the USDA soybean collection. Apart from the genetic diversity, many scientists have taken advantage of the reference genome to look into the biotic and abiotic tolerance genes in soybean. Many have also used the genotyping-based-sequencing approach for QTL mapping which brings ease to the breeders by reducing the time and research investments.

To mention a few, a first detailed study of SCN resistance using soybean genome sequencing was performed by Patil et al. (2019) and Liu et al. (2019). Similarly, Phytophthora resistance genes were identified by whole-genome sequencing and resequencing approach by Zhong et al. (2018). A novel ion transporter was identified by resequencing of 96 genotypes to map the salt tolerance QTL by Qi et al. (2014a). Several whole-genome resequencing and reference genome efforts have been performed in soybean (Table 10.3).

**Table 10.3** Details of efforts performed for whole-genome resequencing and reference genome in soybean.

| Trait | Number of genotypes analyzed | Platform/technique used | Salient feature of the study | References |
|---|---|---|---|---|
| Genetic diversity | 1 | Illumina GA | The *Glycine soja/Glycine max* complex is at least 0.27 million years old, emerging before the relatively recent event of domestication (6000~9000 years ago) | Kim et al. (2010) |
| Genetic diversity | 31 | Illumina GAII | Provided an important resource for the analysis of wild soybeans and to aid in QTL analysis | Lam et al. (2010) |
| Genetic diversity | 4 | Illumina GAI | To estimate genetic diversity between the cultivated genotypes and with the reference Williams 82 | Varala et al. (2011) |
| Artificial domestication | 25 | Illumina | The manual selection while soybean domestication caused pronounced reduction in the genetic diversity than the use of elite cultivars over landraces | Li et al. (2013) |
| Salt tolerance | 96 | Illumina GAIII | To identify a novel ion transporter gene, *GmCHX1*, and relate its sequence alterations to salt tolerance | Qi et al. (2014a) |
| Genetic diversity | 16 | Illumina HiSeq2000 | The variation data will be valuable for breeder to identify agronomically important genes/genic regions in soybean | Chung et al. (2014) |
| Domestication | 302 | Illumina HiSeq 2000 | Soybean populations are structured geographically | Zhou et al. (2015) |
| Population structure analysis | 14430 | Illumina Infinium SoySNP50K BeadChip | To reveal the distribution of genetic variation contained within the USDA soybean collection | Bandillo et al. (2015) |
| Genetic diversity | 106 | Illumina | Analyzed the genetic variation in soybean genome, association analysis of major domestication and agronomic traits | Valliyodan et al. (2016) |
| Glyphosate tolerance | 2 | Illumina Hiseq2500 | Whole-genome sequencing is a less time consuming and cost-effective for the identification of T-DNA insertion sites and flanking sequences in soybean | Guo et al. (2016) |

*(Continued)*

**Table 10.3** (Continued)

| Trait | Number of genotypes analyzed | Platform/technique used | Salient feature of the study | References |
|---|---|---|---|---|
| Phytophthora resistance | 4 | Illumina | Integrated genomics, whole-genome resequencing, and genetic mapping to identify candidate gene, *RpsHC18* | Zhong et al. (2018) |
| Identification of astringent saponins | 1 | Illumina Hiseq 4000 platform | The mutant soybean, CWS5095 is an important resource for breeding to generating new cultivars with improved tastes | Rehman et al. (2018) |
| SCN resistant | 106 | Illumina | First report about genetic basis of broad resistance to SCN, also provided new insight into epistasis, haplotype-compatibility, CNV, promoter variation, and its effect on broad disease resistance in plants | Patil et al. (2019) |
| Resistance to soybean cyst nematode (SCN) | 6 | Illumina HiSeq X-10 | Candidate genes involved in resistance of PI 437654 to soybean cyst nematode race 4 | Liu et al. (2019) |
| Generate low-cost genotype data | 114 | Illumina NextSeq 500 | Use of the highly accurate skim sequencing along with imputation reduces the cost for obtaining dense genotypic information in soybean | Happ et al. (2019) |

## 10.7 GBS-based GWAS Efforts in Soybean

Soybean is one of the most important crops for oil and protein in the world. Many traits of agronomic importance like SW, seed protein, and oil content are shown to be inheritable. The study of the genetic nature of quantitative traits becomes easy after the availability of DNA-based molecular markers and sophisticated methods for statistical analysis. Several QTLs were mapped based on the bi-parental population and interval mapping. However, the QTLs identified in bi-parental populations are related to only that specific breeding population involving pair of parents. Therefore, the parents used in the bi-parental cross must be variable for the trait of interest. Sometimes parents used in crossing/breeding programs may or may not differ for the trait of concern then a large number of such QTLs detected through interval mapping cannot be used in actual breeding programs (Bernardo 2008) and can partly be used in MAS. Then to overcome this problem, multiline cross or multiple bi-parental populations QTL mapping has been used (Gupta et al. 2014). In the case of bi-parental mapping, the power of QTL mapping is high, but the resolution is poor because of the high level of LD, which pose difficult in fine mapping. In addition, the development of mapping population is a costly and time-consuming approach for genetic studies.

LD-based association mapping uses a diverse set of genotypes of known/unknown ancestry. This set of genotypes represents most of the genetic diversity for the trait of interest and carries multiple alleles at every locus. It is a result of hundreds of historic recombination events; hence this is providing the highest resolution for QTL mapping than bi-parental mapping (Mackay et al. 2009).

### 10.7.1 The General Procedure for Association Mapping

Singh and Singh in 2015 have briefly discussed the general procedure for genome-wide association mapping in plants but the scheme may differ according to population characteristics and methodology used for association study (Singh and Singh 2015).

1) Association Mapping (AM) Population: A diverse set of random samples from a natural population, a germplasm core collection, the breeding lines including cultivars, or a population obtained from a cross between multiparent is used for AM. The sample should be representing the maximum genetic diversity present in the population/germplasm collection. This collection of sample establishes the association mapping panel or association mapping population, or association panel.
2) Phenotyping: The selected population is assessed for several traits of interest this is known as phenotyping. Phenotyping must be based on replicated trials conducted over many locations and years for minimizing the influence of the environment. A suitable experimental design should be used to conduct a trial like a randomized block design, nested design, augmented design, etc. Accurate and reliable phenotyping is very important for the success of any mapping effort.
3) Genotyping: The samples are then genotyped using a set of molecular markers (SSR, SNP markers) which are evenly distributed over the entire genome. The markers should

be selected in such a way that they should be unlinked, i.e. located more than 40 cM distance in the genome.

4) Structure and kinship analysis: The genotypic data will be used to analyze structure and kinship in the association panel. The marker information help in the detection and estimation of the population structure of the sample using the STRUCTURE and TASSEL program.
5) Genotyping for LD Analysis: LD is the nonrandom association of alleles of two or more loci in a random mating population, this concept was first proposed by Jennings in 1917 (Jennings 1917). For its estimation, the samples should be genotyped with an appropriately large number of molecular markers. The number of markers should be very large so that it could cover the entire genome as densely as possible, and the LD between markers and the loci of interest can be detected accurately. To determine the maximum number of markers essential for covering the whole genome, a pattern of LD in the genomic region of interest and the extent of LD found among different populations of the species will be taken into consideration. The most widely used marker systems for association mapping are SSR and SNP.

Different statistical models such as general linear model (GLM), mixed linear model, and multitrait mixed model are used to find the association between the phenotype and the genotype. It will help to detect and estimate the exact LD between the markers and the genes/QTLs responsible for traits of interest. To minimize the false-positive association of marker and trait of interest the model always uses kinship and population structure as a covariate. A simplified schematic representation of the steps involved in association mapping has been given in Figure 10.4. Gupta et al. (2014) suggested three basic approaches in the analysis of association mapping, (i) one of the approaches called unexperienced where only genotypic and phenotypic data are used without considering population structure which results in the highest rate of false-positive associations. (ii) GLM: genotypic and phenotypic data are used with the population structure (GLM + Q). Q values can be obtained either by using software STRUCTURE or with the help of principal component analysis (PCA)/principal coordinate analysis (PCoA)/multidimensional scaling (MDS). But false-positive results are lesser than that in "naive" but are still fairly high. (iii) In the last approach, GLM is used along with marker-based kinship (K) with or without Q value, this model is known as the MLM (Q + K or K only), which yields the lowest rate of false-positives results.

### 10.7.2 Approaches Used for Association Studies

There are two widely used approaches for association studies, which include genome-wide association studies (GWAS) and comparative genomics (CG) approach.

GWAS

- Markers used for genotyping are distributed evenly and densely, throughout the genome.
- All loci involved in the control of all traits showing total variation.
- Considerable amount of resources and effort are necessary for reliable phenotyping of various traits.

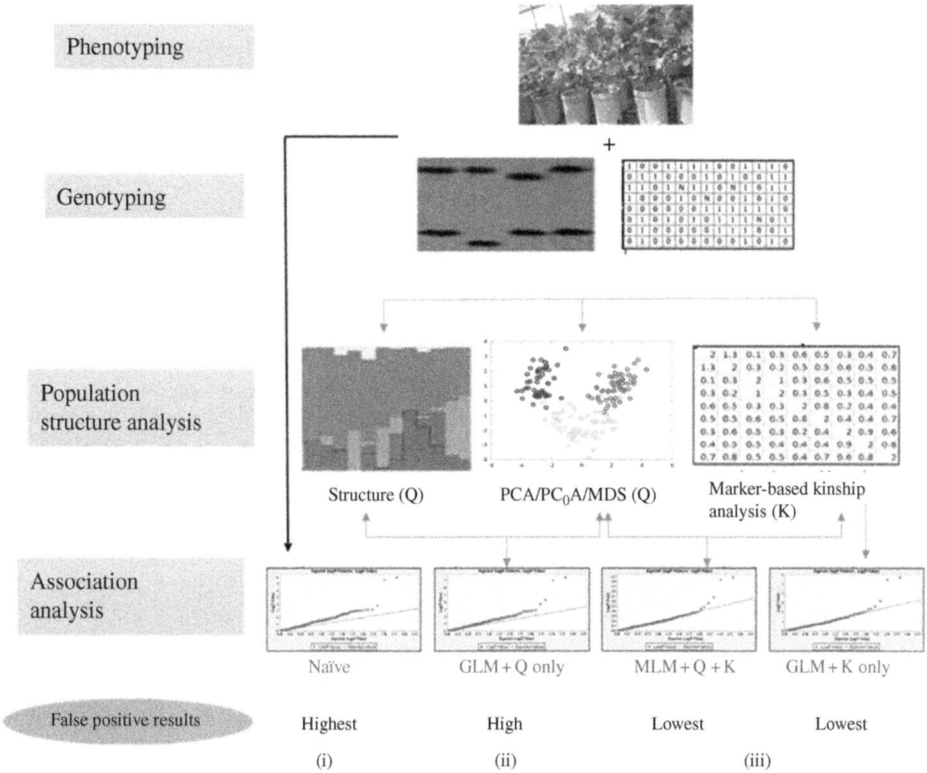


**Figure 10.4** Schematic representations of steps involved in association mapping and linkage disequilibrium analyses.

- A large sample size is required.
- Population which experienced bottleneck in the recent past, since bottleneck leads to an increase in LD that reduces number of markers and sample size.

CG Approach

- Analysis restricted to genomic regions having genes/QTLs for the trait(s) of interest.
- Information about CGs ascertained from several different sources like genome QTL analysis, sequence annotation, comparative genomics, transcript profiling.
- Genotyping focused on specific genomic regions.
- The sample size and the number of markers requirement are less.
- Previous knowledge regarding traits of interest is necessary.
- Identified from loss of function mutations in laboratory strains therefore difficult for analyzing the trait in natural populations.

Today, advanced analytical methods for investigating genome-wide association and cheap genotyping techniques have helped in understanding complex genetic traits in soybean (Zhang et al. 2016). In the past, several attempts were made to decipher the genetic nature of these complex characters in soybean (Hao et al. 2012; Hwang et al. 2014; Song et al. 2015; Zhang et al. 2015, 2016). The GWAS was found to be a robust approach to identify important genes coding complex traits in crops with self-pollinating mating systems (Lorenz et al. 2010). However, Hyten et al. in 2007 reported the pattern of LD in soybean populations is highly variable among the different populations and genomic regions (Hyten et al. 2007). They have also reported that a large number of molecular markers (SNPs) will be essential for performing whole-genome association analysis even in a self-pollinated species with fairly high LD.

The basic requirement for performing GWAS is the availability of thousands to millions of markers for attaining maximum genome coverage and getting the highest resolution in mapping. This has been greatly achieved by the advent of NGS techniques (Bhatia et al. 2013). NGS platforms can provide several molecular markers at a low cost by sequencing SNP sites in a controlled genomic subset. While many array-based molecular techniques demanding prior knowledge about the target sequence, in contrast, a method such as GBS can skip the sequence discovery and explores SNP variation in DNA fragments after restriction digestion (Biazzi et al. 2019). GBS approach was first developed in the Buckler lab by Elshire and coworkers in 2011 using a restriction enzyme, ApeKI as a "frequent cutter" and combining the ligation of both adapters into one step. It is a simple approach for creating reduced representation libraries for the NGS sequencing platform like Illumina (Elshire et al. 2011), and it aims to generate large numbers of SNPs that can be used in genetic analyses and genotyping (Beissinger et al. 2013). This system became popular because of some key features like low cost, less sample handling, less PCR and purification steps, no size fractionation, no limit of the reference sequence, effective barcoding, and easier to perform at a large scale (Davey et al. 2011). He et al. (2014) briefly outlined the general steps involved in the GBS protocol for plant breeding as shown in Figure 10.5.

GWAS accelerates molecular breeding for the improvement of various complex traits in soybean. While GWAS in combination with the GBS approach was used to provide dense genome-wide marker coverage for a large panel of diverse lines or germplasms. It saves both cost and time of plant breeder and helps in the study of genetic nature of various agronomically important traits. There are several GBS-based GWAS analysis tools in soybean which ease study of complex characters.

GBS-based association mapping has been successfully used to map *sclerotinia* stem rot in a set of 130 diverse lines of soybean (Bastien et al. 2014). In this study, 7864 SNPs were used to perform the LD analyses. Four regions associated with *sclerotinia* stem rot resistance were identified using MLM analysis and one of these genomic regions was validated in a bi-parental population. In another similar study of association mapping of QTLs for *sclerotinia* stem rot resistance in a diverse soybean plant introduction led to the identification of three genomic regions significantly associated with resistance to stem rot (Iquira et al. 2015).

Sonah et al. (2015) have demonstrated that combining GBS with GWAS can be used as a powerful complementary system to classical bi-parental mapping for studying complex traits in soybean. The study was performed using 139 diverse lines and was phenotyped for eight complex traits (flower color, pubescence color, hilum color, maturity, PH, SW, seed oil, and protein). Nearly 25 SNP loci associated with these complex traits were identified, interestingly most of these GWAS loci identified were located within previously reported

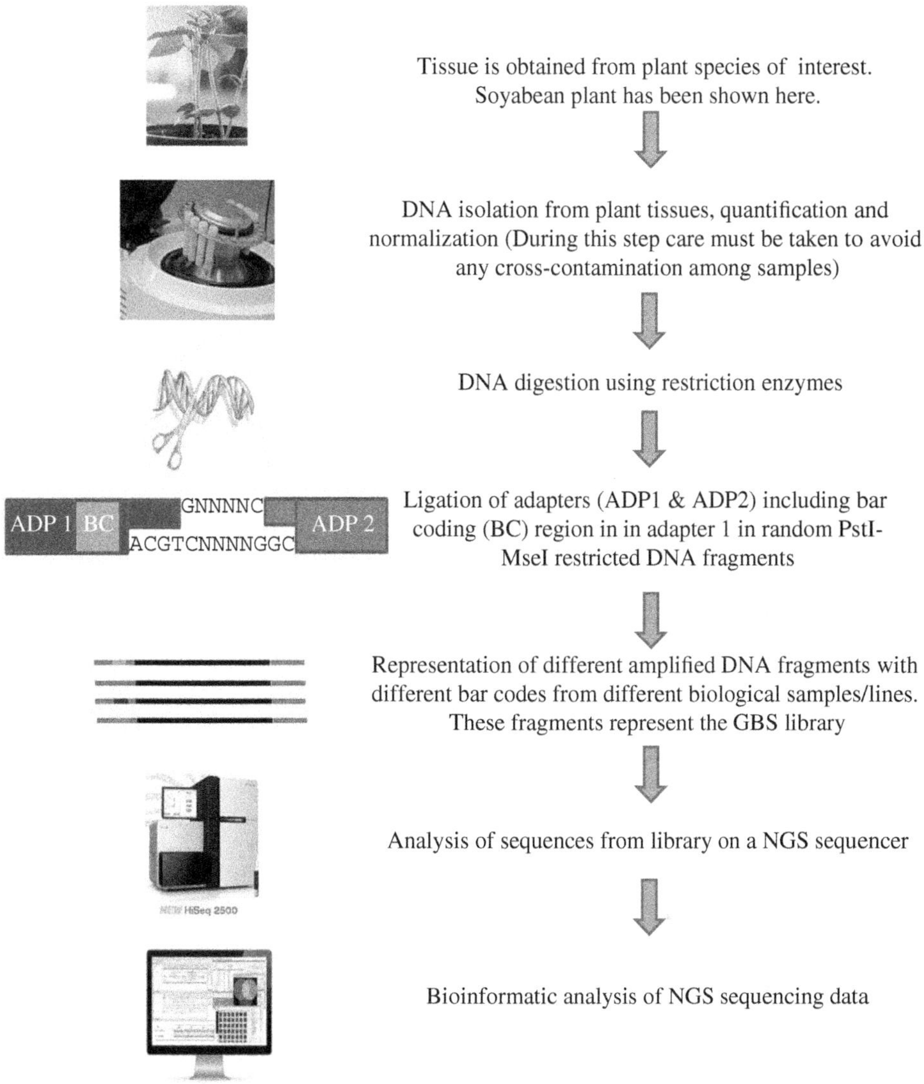


**Figure 10.5** General steps involved in the GBS protocol for plant breeding. *Source:* The figure is adapted and modified from He et al. (2014) which is available under a Creative Commons Attribution (CC-By) International License

genomic regions identified through bi-parental mapping. More recently, QTLs related to soybean protein and oil content have been identified with the help of an alternate approach called SLAF-seq approach. It has been used for whole-genome genotyping of a population consisting of 185 soybean accessions. The study found that 31 SNPs located on 12 of the 20 soybean chromosomes were associated with seed protein and oil content (Li et al. 2019). The list of GBS-based GWAS studies in soybean has been given in Table 10.4 which depicts the number of germplasms used, platform/technique used in the study, most significantly associated SNP, and salient feature of the study.

**Table 10.4** List of GBS-based GWAS studies in soybean.

| Trait | Number of genotypes used | Platform/ technique used | Most significantly associated SNP | Salient feature of the study | References |
|---|---|---|---|---|---|
| Tolerance to iron deficiency chlorosis | 138 advanced soybean breeding lines developed for the north-central region of the United States | Illumina HiSeq 2000 | 7 | 12 candidate genes associated with metabolism of iron were mapped in proximity to the QTL, *NAS3* gene which encodes the enzyme nicotianamine synthase which synthesizes the nicotianamine an iron chelator also maps to the same QTL region | Mamidi et al. (2014) |
| Sclerotinia stem rot resistance | 130 private breeding program (Semences Prograin Inc.) in Eastern Canada | Illumina HiSeq 2000 | 4 | Four regions associated with sclerotinia stem rot resistance were identified using mixed linear model analysis and one of these genomic regions was validated in a bi-parental population | Bastien et al. (2014) |
| Sclerotinia stem rot resistance | 101 diverse collection most of which composed of plant introductions | Illumina HiSeq 2000 | 3 | Three QTLs were identified which exhibited a significant association at a stringent threshold ($q = 0.10$) and all of the most highly resistant plant i ntroductions showed common alleles at these three QTLs | Iquira et al. (2015) |
| Agronomic traits like flower color, pubescence color, hilum color, maturity, plant height, seed weight, seed oil, and protein | 139 "core" genotypes spanning the genetic diversity | Illumina HiSeq 2000 | 25 | Most of these GWAS loci were found within regions identified by previously reported QTLs for these traits | Sonah et al. (2015) |

| | | | | | |
|---|---|---|---|---|---|
| Seed protein and oil content | 185 germplasms lines | SLAF-seq | 31 | Among 31 SNPs found to be associated with the these traits, 31 beneficial alleles were identified. Two SNP markers, namely rs15774585 and rs15783346 on Chr 07, were found to be correlated with seed oil content during 2015 and 2016 | Li et al. (2019) |
| Symbiotic nitrogen fixation (SNF) | 297 African cultivated soybean lines | Illumina HiSeq 2000 | 25 | Identified 25 QTL regions comprising of 40 putative candidate genes which included 20 genes with no prior identified role in SNF. A line with a large size deletion (164 kb), containing a QTL region having a candidate gene (CASTOR), exhibited a marked decrease in SNF | Torkamaneh et al. (2020) |
| Maturity | Set A = 53 NILs contrasting for maturity loci, Set B = 91 Plant introduction lines obtained from the National Plant Germplasm System, and Set C = 305 lines representative of the Eastern Canadian germplasm | Illumina HiSeq2000 | 6 (haplotypes) | The functional allele *E3Ha* and its loss of function form, *e3-tr*, were the two most prevalent in the Eastern Canadian germplasm | Tardivel et al. (2014) |

## 10.8 GBS-based Genomic Selection Efforts in Soybean

The world's population is projected to rise by 2050 which will increase demand for food crops. However, the changing climates pose a challenge to crop production and global food security (Reynolds 2010; Tester and Langridge 2010). Conventional breeding has helped to increase the potential yield of crops but is not adequate to satisfy the needs of the increasing world population (Oury et al. 2012; Fischer et al. 2014). Additionally, this method of breeding has many limitations, such as time-consuming, laborious, ineffective, costly, needs large population size, less accurate and reliable, solely relay on phenotypic selection, less effective for low heritable multigenic traits that could be restricted by environment and genotype interactions (Tester and Langridge 2010; Tuberosa 2012; Bhat et al. 2015). Molecular breeding has helped to overcome the limitations of conventional breeding. It involves the use of molecular markers linked to traits based on functional polymorphism to interrogate chromosomal region of interest or identification of new chromosomal region associated with a particular trait. Molecular breeding involves various techniques such as marker assisted-backcross breeding (MABB), MAS, and genomic selection (GS). GS uses genome-wide markers with phenotypic selection data to predict the breeding value in an early generation that accelerates breeding with accuracy (Heffner et al. 2010). GS has proven to be more effective than conventional MAS in improving QTL-controlled traits with minor effects and implementation of genomic selection is expected to have important consequences for genetic testing systems and genetic enhancement programs (Bernardo and Yu 2007; Heffner et al. 2010; Zhang et al. 2017; Goddard and Hayes 2007).

Genomic selection has been used successfully in multiple crop improvements including soybean (Zhao et al. 2012; Poland and Rife 2012; Lorenz et al. 2010; Crossa et al. 2013; Jarquin et al. 2016; Shu et al. 2013; Bao et al. 2014). Genomic selection is a promising method that utilizes molecular markers to design novel breeding programs and create new marker-based genetic assessment models. Applications of GBS are highlighted in Figure 10.6. The important consideration for GS to function in crop plants is the availability of low cost, versatile and high-density marker systems that increase the genetic benefit of complex traits per unit time and cost (Bhat et al. 2016). Genomic selection would increase the genetic advantage per year compared to the only phenotypic selection, with realistic selection accuracy assumptions, reduced breeding cycle times, time gains per unit, and selection intensities, which allow improved crop plant varieties to grow more quickly (Zhong et al. 2009; Heffner et al. 2010; Desta and Ortiz 2014). The precision of Genomic selection is influenced by various factors such as trait architecture and heritability, size and composition of training sets (TSs), marker density, and the statistical model used (Jannink et al. 2010).

GS uses two types of datasets: a TS and a validation set (VS) as it is important to base materials. In short, GS involves creating a predictive model by associating marker information with phenotypic information in which model refers to a selection set of candidates that have been genotyped but not phenotypically tested and called the TS, and the genomic breeding value of new genotypes are predicted based on the marker effect estimated based on the TS. To test the efficacy of GS, a method known as cross-validation involving bisecting a collection of genotyped and phenotyped lines into TS and VS. The genomic estimated breeding value (GEBV) calculated from TS, correlated with the observed phenotypic values

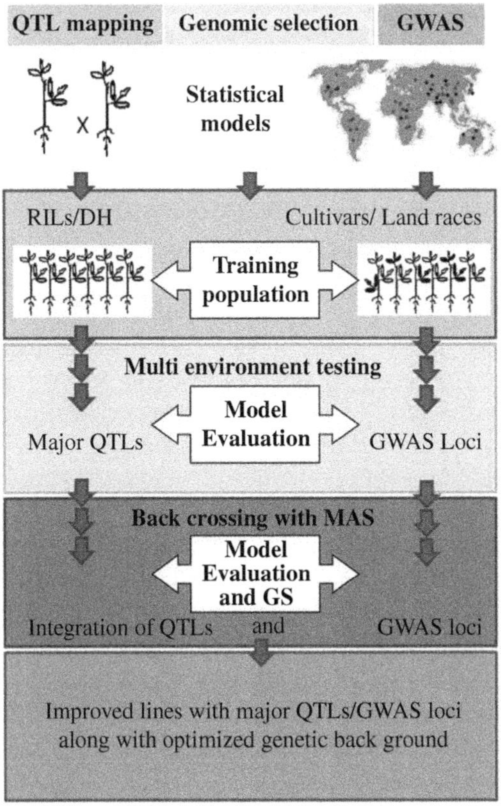


**Figure 10.6** Diagrammatic representation of application of genotyping by sequencing in genomic selection, GWAS and QTL mapping. *Source:* The figure is reproduced from Deshmukh et al. (2014) which is available under a Creative Commons Attribution (CC-By) International License.

of the VS that determines predictive capacity (Combs and Bernardo 2013; Jacobson et al. 2014; Stewart-Brown et al. 2019). Several factors, including GS models, marker density, LD level, QTL number, population size-specific, the relationship between TS and VS, and trait heritage affect GS performance (Jarquin et al. 2016; Zhang et al. 2016; Duhnen et al. 2017). GS predicts GEBVs in the target species depending on the phenotype and genotype effects on the TS can be used in integrated breeding programs (Desta and Ortiz 2014). GS relies on the degree of genetic similarity between TS and VS in the LD between marker and trait loci which is more precise and faster with the aid of NGS, high-throughput phenotyping, and computational algorithms for statistical analysis (Jannink et al. 2010; Cabrera-Bosquet et al. 2012; Dhondt et al. 2013).

A well-annotated reference genome will accelerate subsequent analysis such as identification of SNPs, variability in copy numbers, structural variability, and even genomics research in crops (Li et al. 2017). The GS sequencing approach by genotyping (GBS) is a significant complement to conventional crop improvement and it is very important to transfer genomics-assisted breeding into commercial crops (Poland and Rife 2012). GBS makes use of polymorphism recognition using NGS technologies followed by bioinformatics tools (Elshire et al. 2011). GBS has many advantages including reduced sample handling, fewer PCR and purification steps, low cost, no need for preliminary sequence details, no size fractionation, and an efficient barcoding technique that helps to conduct direct GS for crop improvement (Davey et al. 2011; Poland and Rife 2012; Narum et al. 2013; He et al. 2014). SLAF-seq, a variation of the GBS method determined to be cost-effective in genotyping soybean (Sun et al. 2013; Qi et al. 2014b).

Three hundred and one experimental soybean lines have been used to evaluate prospects for genomic selection using GBS and to evaluate the impact of selection and imputation of GBS markers on genomic prediction accuracy. Genotyping by sequencing scored total 16 502 SNPs with minor allele frequency (MAF) >0.05 and missing values > 5% were detected by GBS and predictive accuracy for grain yield, measured using cross-validation, was estimated at 0.64, effective for the use of soybean grain yield genomic selection. This is the first kind of research that shows an application of GBS for soybean genomic prediction that has strong potential to boost soybean genetics and indicates that standard additive G-BLUP models can be used without loss of accuracy on unfiltered, imputed GBS data (Jarquín et al. 2014).

A reference panel, with more than 10 million SNPs produced by the resequencing of 99 soybean lines and these SNPs genotyped in experimental lines after alignment with the soybean reference genome. This research highlights the potential of low coverage sequencing with imputation as an economic method for obtaining knowledge about the high-density SNP genotype in soybean. This research highlights the potential of low coverage sequencing with imputation as an economic method for obtaining knowledge about the high-density SNP genotype in soybean (Happ et al. 2019).

Based on 249 soybean accessions, genome-wide association analysis and genomic selection of seed amino acid content were conducted and accessions with high amino acid concentrations in seeds, and molecular markers associated with individuals and amino acid groups were identified. Twenty-five SNP markers associated with multiple amino acids can be used to concurrently increase multiamino acid concentration in soybean. Genomic selection for amino acid concentration revealed that the selection efficiency of amino acids based on the markers associated with 15 amino acids is important. The use of markers significantly linked to multiple amino acids is likely to help breeders efficiently select soybean varieties with enhanced amino acid content (Qin et al. 2019).

Total 483 elite breeding lines, genotyped with BARCSoySNP6K iSelect BeadChips to study the potential of GS for soybean and this was the first report indicating the successful application of GS to achieve higher heritability traits such as protein and oil content in soybean. During cross-validation using the RR-BLUP model, predictive abilities (rMP) of 0.81, 0.71, and 0.26 for protein, oil, and yield were achieved, respectively. Comparative analysis of different marker densities showed inflation in rMP was due to population structure. In another two approaches, a bi-parental population which includes a TS of full-sibs of the VS and across populations which includes a TS of all remaining breeding lines except for full-sibs of the VS were used to predict breeding values for lines of four bi-parental populations within the GS dataset. The across populations method was found to be more practical over bi-parental population as it reduces the breeding cycle and ignores smaller TSs (Stewart-Brown et al. 2019).

## References


Akkaya, M.S., Bhagwat, A.A., and Cregan, P.B. (1992). Length polymorphisms of simple sequence repeat DNA in soybean. *Genetics* 132 (4): 1131–1139.

Akkaya, M.S., Shoemaker, R.C., Specht, J.E. et al. (1995). Integration of simple sequence repeat DNA markers into a soybean linkage map. *Crop Science* 35 (5): 1439–1445.

Akond, M., Liu, S., Schoener, L. et al. (2013). A SNP-based genetic linkage map of soybean using the SoySNP6K Illumina Infinium BeadChip genotyping array. *Plant Genetics, Genomics, and Biotechnology* 1 (3): 80–89.

Andersen, S.B. and Torp, A.M. (2002). QTL mapping in crop plants. In: *Molecular Techniques in Crop Improvement* (eds. S.M. Jain, D.S. Brar and B.S. Ahloowalia), 541–562. Dordrecht: Springer.

Bandillo, N., Jarquin, D., Song, Q. et al. (2015). A population structure and genome wide association analysis on the USDA soybean germplasm collection. *The Plant Genome* 8 (3): 1–13.

Bao, Y., Vuong, T., Meinhardt, C., Tiffin, P., Denny, R., Chen, S. Y. Nguyen, H. T. (2014). Potential of association mapping and genomic selection to explore PI 88788 derived soybean cyst nematode resistance. Plant Genome, 7, doi:10.3835/plantgenome2013.3811.0039.

Bastien, M., Sonah, H., and Belzile, F. (2014). Genome wide association mapping of *Sclerotinia sclerotiorum* resistance in soybean with a genotyping-by-sequencing approach. *The Plant Genome* 7 (1): plantgenome2013-10.

Beissinger, T.M., Hirsch, C.N., Sekhon, R.S. et al. (2013). Marker density and read depth for genotyping populations using genotyping-by-sequencing. *Genetics* 193 (4): 1073–1081.

Bernardo, R. (2008). Molecular markers and selection for complex traits in plants: learning from the last 20 years. *Crop Science* 48 (5): 1649–1664.

Bernardo, R. and Yu, J. (2007). Prospects for genomewide selection for quantitative traits in maize. *Crop Science* 47 (3): 1082–1090.

Bhat, T.A., Ahmad, L., and Kotru, R. (2015). Relation between agrometeorological indices, crop phenology and yield of rice genotypes as influenced by real time N management. *Journal of Agrometeorology* 17 (1): 90.

Bhat, J.A., Ali, S., Salgotra, R.K. et al. (2016). Genomic selection in the era of next generation sequencing for complex traits in plant breeding. *Frontiers in Genetics* 7: 221.

Bhatia, D., Wing, R.A., and Singh, K. (2013). Genotyping by sequencing, its implications and benefits. *Crop Improvement* 40 (2): 101–111.

Biazzi, E., Nazzicari, N., Pecetti, L. et al. (2019). GBS based genome wide association and genomic selection for alfalfa (*Medicago sativa*) forage quality improvement. In: *The Model Legume Medicago truncatula* (ed. F. de Bruijn), 923–927. Wiley.

Bradbury, P.J., Zhang, Z., and Kroon, D.E. (2007). TASSEL: software for association mapping of complex traits in diverse samples. *Bioinformatics* 23 (19): 2633–2635.

Cabrera-Bosquet, L., Crossa, J., von Zitzewitz, J. et al. (2012). High-throughput phenotyping and genomic selection: the frontiers of crop breeding converge. *Journal of Integrative Plant Biology* 54 (5): 312–320.

Cao, Y., Li, S., Wang, Z. et al. (2017). Identification of major quantitative trait loci for seed oil content in soybeans by combining linkage and genome-wide association mapping. *Frontiers in Plant Science* 8: 1222.

Catchen, J., Hohenlohe, P.A., and Bassham, S. (2013). Stacks: an analysis tool set for population genomics. *Molecular Ecology* 22 (11): 3124–3140. https://doi.org/10.1111/mec.12354.

Chung, W.H., Jeong, N., Kim, J. et al. (2014). Population structure and domestication revealed by high-depth resequencing of Korean cultivated and wild soybean genomes. *DNA Research* 21: 153–167.

Collard, B.C. and Mackill, D.J. (2008). Marker-assisted selection: an approach for precision plant breeding in the twenty-first century. *Philosophical Transactions of the Royal Society B: Biological Sciences* 363 (1491): 557–572.

Combs, E. and Bernardo, R. (2013). Accuracy of genomewide selection for different traits with constant population size, heritability, and number of markers. *The Plant Genome* 6 (1): plantgenome2012-11.

Cregan, P.B., Jarvik, T., Bush, A.L. et al. (1999). An integrated genetic linkage map of the soybean genome. *Crop Science* 39 (5): 1464–1490.

Crossa, J., Beyene, Y., Kassa, S. et al. (2013). Genomic prediction in maize breeding populations with genotyping-by-sequencing. *G3: Genes, Genomes, Genetics* 3 (11): 1903–1926.

Davey, J.W., Hohenlohe, P.A., Etter, P.D. et al. (2011). Genome-wide genetic marker discovery and genotyping using next-generation sequencing. *Nature Reviews Genetics* 12 (7): 499–510.

Desta, Z.A. and Ortiz, R. (2014). Genomic selection: genome-wide prediction in plant improvement. *Trends in Plant Science* 19 (9): 592–601.

Deshmukh, R., Sonah, H., Patil, G. et al. (2014). Integrating omic approaches for abiotic stress tolerance in soybean. *Frontiers in Plant Science* 5: 244.

Dhondt, S., Wuyts, N., and Inzé, D. (2013). Cell to whole-plant phenotyping: the best is yet to come. *Trends in Plant Science* 18 (8): 428–439.

Duhnen, A., Gras, A., Teyssèdre, S. et al. (2017). Genomic selection for yield and seed protein content in soybean: a study of breeding program data and assessment of prediction accuracy. *Crop Science* 57 (3): 1325–1337.

Elshire, R.J., Glaubitz, J.C., Sun, Q. et al. (2011). A robust, simple genotyping-by-sequencing (GBS) approach for high diversity species. *PLoS One* 6 (5).

Ferreira, A.R., Foutz, K.R., and Keim, P. (2000). Soybean genetic map of RAPD markers assigned to an existing scaffold RFLP map. *Journal of Heredity* 91 (5): 392–396.

Fischer, R.A., Byerlee, D., and Edmeades, G. (2014). *Crop Yields and Global Food Security*, 8–11. Canberra: ACIAR.

Glaubitz, J.C., Casstevens, T.M., and Lu, F. (2014). TASSEL-GBS: a high capacity genotyping by sequencing analysis pipeline. *PLoS One* 9 (2): e90346. https://doi.org/10.1371/journal.pone.0090346.

Goddard, M.E. and Hayes, B.J. (2007). Genomic selection. *Journal of Animal Breeding and Genetics* 124 (6): 323–330.

Grunvald, A.K., Torres, A.R., de Lima Passianotto, A.L. et al. (2018). Identification of QTLs associated with biological nitrogen fixation traits in soybean using a genotyping by sequencing approach. *Crop Science* 58: 2523–2532.

Guo, B., Guo, Y., Hong, H. et al. (2016). Identification of genomic insertion and flanking sequence of G2-EPSPS and GAT transgenes in soybean using whole genome sequencing method. *Frontiers in Plant Science* 7: 1009.

Gupta, P.K., Kulwal, P.L., and Jaiswal, V. (2014). Association mapping in crop plants: opportunities and challenges. *Advances in Genetics* 85: 109–147.

Han, J., Han, D., Guo, Y. et al. (2019). QTL mapping pod dehiscence resistance in soybean (*Glycine max* L. Merr.) using specific-locus amplified fragment sequencing. *Theoretical and Applied Genetics* 132: 2253–2272.

Hao, D., Cheng, H., Yin, Z. et al. (2012). Identification of single nucleotide polymorphisms and haplotypes associated with yield and yield components in soybean (*Glycine max*) landraces across multiple environments. *Theoretical and Applied Genetics* 124 (3): 447–458.

Happ, M.M., Wang, H., Graef, G.L. et al. (2019). Generating high density, low cost genotype data in soybean [*Glycine max* (L.) Merr.]. *G3: Genes, Genomes, Genetics* 9: 2153–2160.

He, J., Zhao, X., Laroche, A. et al. (2014). Genotyping-by-sequencing (GBS), an ultimate marker-assisted selection (MAS) tool to accelerate plant breeding. *Frontiers in Plant Science* 5: 484.

Heffner, E.L., Lorenz, A.J., Jannink, J.L. et al. (2010). Plant breeding with genomic selection: gain per unit time and cost. *Crop Science* 50 (5): 1681–1690.

Horvath, A.A. (1926). Changes in the blood composition of rabbits fed on raw soy beans. *Journal of Biological Chemistry* 68 (2): 343–355.

Hwang, E.Y., Song, Q., Jia, G. et al. (2014). A genome-wide association study of seed protein and oil content in soybean. *BMC Genomics* 15 (1): 1.

Hyten, D.L., Choi, I.Y., Song, Q. et al. (2007). Highly variable patterns of linkage disequilibrium in multiple soybean populations. *Genetics* 175 (4): 1937–1944.

Hyten, D.L., Cannon, S.B., Song, Q. et al. (2010). High-throughput SNP discovery through deep resequencing of a reduced representation library to anchor and orient scaffolds in the soybean whole genome sequence. *BMC Genomics* 11: 38.

Iquira, E., Humira, S., and Francois, B. (2015). Association mapping of QTLs for sclerotinia stem rot resistance in a collection of soybean plant introductions using a genotyping by sequencing (GBS) approach. *BMC Plant Biology* 15 (1): 5.

Jacobson, A., Lian, L., Zhong, S. et al. (2014). General combining ability model for genomewide selection in a biparental cross. *Crop Science* 54 (3): 895–905.

Jannink, J.L., Lorenz, A.J., and Iwata, H. (2010). Genomic selection in plant breeding: from theory to practice. *Briefings in Functional Genomics* 9 (2): 166–177.

Jarquín, D., Kocak, K., Posadas, L. et al. (2014). Genotyping by sequencing for genomic prediction in a soybean breeding population. *BMC Genomics* 15: 740.

Jarquin, D., Specht, J., and Lorenz, A. (2016). Prospects of genomic prediction in the USDA Soybean Germplasm Collection: Historical data creates robust models for enhancing selection of accessions. *G3: Genes, Genomes, Genetics* 6 (8): 2329–2341.

Jennings, H.S. (1917). The numerical results of diverse systems of breeding, with respect to two pairs of characters, linked or independent, with special relation to the effects of linkage. *Genetics* 2 (2): 97.

Keim, P., Schupp, J.M., Travis, S.E. et al. (1997). A high density soybean genetic map based on AFLP markers. *Crop Science* 37 (2): 537–543.

Kim, M.Y., Lee, S., Van, K. et al. (2010). Whole-genome sequencing and intensive analysis of the undomesticated soybean (*Glycine soja* Sieb. and Zucc.) genome. *Proceedings of the National Academy of Sciences* 107 (51): 22032–22037.

Kong, L., Lu, S., Wang, Y. et al. (2018). Quantitative trait locus mapping of flowering time and maturity in soybean using next-generation sequencing-based analysis. *Frontiers in Plant Science* 9: 995.

Kumar, S., Banks, T.W., and Cloutier, S. (2012, 2012). SNP discovery through next-generation sequencing and its applications. *International Journal of Plant Genomics* https://doi.org/10.1155/2012/831460.

Lam, H.M., Xu, X., Liu, X. et al. (2010). Resequencing of 31 wild and cultivated soybean genomes identifies patterns of genetic diversity and selection. *Nature Genetics* 42: 1053–1059.

Lark, K.G., Weisemann, J.M., Matthews, B.F. et al. (1993). A genetic map of soybean (*Glycine max* L.) using an intraspecific cross of two cultivars: 'Minosy'and 'Noir 1'. *Theoretical and Applied Genetics* 86 (8): 901–906.

Lee, S., Freewalt, K.R., McHale, L.K. et al. (2015). A high-resolution genetic linkage map of soybean based on 357 recombinant inbred lines genotyped with BARCSoySNP6K. *Molecular Breeding* 35 (2): 58.

Lemay, M.A., Torkamaneh, D., and Rigaill, G. (2019). Screening populations for copy number variation using genotyping-by-sequencing: a proof of concept using soybean fast neutron mutants. *BMC Genomics* 20: 634.

Li, Y., Zhao, S., Ma, J. et al. (2013). Molecular footprints of domestication and improvement in soybean revealed by whole genome re-sequencing. *BMC Genomics* 14: 1–12.

Li, B., Tian, L., Zhang, J. et al. (2014). Construction of a high-density genetic map based on large-scale markers developed by specific length amplified fragment sequencing (SLAF-seq) and its application to QTL analysis for isoflavone content in *Glycine max*. *BMC Genomics* 15 (1): 1086.

Li, B., Fan, S., Yu, F. et al. (2017). High-resolution mapping of QTL for fatty acid composition in soybean using specific-locus amplified fragment sequencing. *Theoretical and Applied Genetics* 130 (7): 1467–1479.

Li, D., Zhao, X., Han, Y. et al. (2019). Genome-wide association mapping for seed protein and oil contents using a large panel of soybean accessions. *Genomics* 111: 90–95.

Liu, S., Ge, F., Huang, W. et al. (2019). Effective identification of soybean candidate genes involved in resistance to soybean cyst nematode via direct whole genome re-sequencing of two segregating mutants. *Theoretical and Applied Genetics* 132: 2677–2687.

Lorenz, A.J., Hamblin, M.T., and Jannink, J.L. (2010). Performance of single nucleotide polymorphisms versus haplotypes for genome-wide association analysis in barley. *PLoS One* 5 (11): e14079.

Lu, F., Lipka, A.E., and Glaubitz, J. (2013). Switchgrass genomic diversity, ploidy, and evolution: novel insights from a network-based SNP discovery protocol. *PLoS Genetics* 9 (1): e1003215. https://doi.org/10.1371/journal.pgen.1003215.

Ludke, W.H., Schuster, I., da Silva, F.L. et al. (2019). SNP markers associated with soybean partial resistance to *Phytophthora sojae*. *Crop Breeding and Applied Biotechnology* 19: 31–39.

Mackay, T.F.C., Stone, E.A., and Ayroles, J.F. (2009). The genetics of quantitative traits: challenges and prospects. *Nature Reviews Genetics* 10: 565–577.

Mamidi, S., Lee, R.K., Goos, J.R. et al. (2014). Genome-wide association studies identifies seven major regions responsible for iron deficiency chlorosis in soybean (*Glycine max*). *PLoS One* 9 (9): e107469.

Mansur, L.M., Orf, J.H., Chase, K. et al. (1996). Genetic mapping of agronomic traits using recombinant inbred lines of soybean. *Crop Science* 36 (5): 1327–1336.

Maughan, P.J., Maroof, M.S., Buss, G.R. et al. (1996). Amplified fragment length polymorphism (AFLP) in soybean: species diversity, inheritance, and near-isogenic line analysis. *Theoretical and Applied Genetics* 93 (3): 392–401.

Moose, S.P. and Mumm, R.H. (2008). Molecular plant breeding as the foundation for 21st century crop improvement. *Plant Physiology* 147 (3): 969–977.

Narum, S.R., Buerkle, C.A., Davey, J.W. et al. (2013). Genotyping-by-sequencing in ecological and conservation genomics. *Molecular Ecology* 22 (11): 2841.

Oury, F.X., Godin, C., Mailliard, A. et al. (2012). A study of genetic progress due to selection reveals a negative effect of climate change on bread wheat yield in France. *European Journal of Agronomy* 40: 28–38.

Palmer, R.G. and Kilen, T.C. (1987). Qualitative genetics and cytogenetics. In: *Soybeans: Improvement, Production, and Uses*, Agronomy Monographs 16, 2e (ed. J.R. Wilcox), 135–209. Madison, WI: ASA, CSSA, and SSSA.

Paterson, A.H., Bowers, J.E., Burow, M.D. et al. (2000). Comparative genomics of plant chromosomes. *The Plant Cell* 12 (9): 1523–1539.

Patil, G., Vuong, T.D., Kale, S. et al. (2018). Dissecting genomic hotspots underlying seed protein, oil, and sucrose content in an interspecific mapping population of soybean using high-density linkage mapping. *Plant Biotechnology Journal* 16: 1939–1953.

Patil, G.B., Lakhssassi, N., Wan, J. et al. (2019). Whole-genome re-sequencing reveals the impact of the interaction of copy number variants of the rhg1 and Rhg4 genes on broad-based resistance to soybean cyst nematode. *Plant Biotechnology Journal* 17: 1595–1611.

Pawlowski, M.L., Vuong, T.D., Valliyodan, B. et al. (2020). Whole-genome resequencing identifies quantitative trait loci associated with mycorrhizal colonization of soybean. *Theoretical and Applied Genetics* 133: 409–417.

Phansak, P., Soonsuwon, W., Hyten, D.L. et al. (2016). Multi-population selective genotyping to identify soybean [*Glycine max* (L.) Merr.] seed protein and oil QTLs. *G3: Genes, Genomes, Genetics* 6: 1635–1648.

Ping, J., Fitzgerald, J.C., Zhang, C. et al. (2016). Identification and molecular mapping of Rps11, a novel gene conferring resistance to *Phytophthora sojae* in soybean. *Theoretical and Applied Genetics* 129 (2): 445–451.

Poland, J.A. and Rife, T.W. (2012). Genotyping-by-sequencing for plant breeding and genetics. *The Plant Genome* 5: 92–102.

Priolli, R.H.G., Carvalho, C.R.L., Bajay, M.M. et al. (2019). Genome analysis to identify SNPs associated with oil content and fatty acid components in soybean. *Euphytica* 215: 54.

Qi, X., Li, M.-W., Xie, M. et al. (2014a). Identification of a novel salt tolerance gene in wild soybean by whole-genome sequencing. *Nature Communications* 5: 1–11.

Qi, Z., Huang, L., Zhu, R. et al. (2014b). A high-density genetic map for soybean based on specific length amplified fragment sequencing. *PLoS One* 9 (8): e104871.

Qin, J., Shi, A., Song, Q. et al. (2019). Genome wide association study and genomic selection of amino acid concentrations in soybean seeds. *Frontiers in Plant Science* 10: 1445.

Rehman, H.M., Nawaz, M.A., Shah, Z.H. et al. (2018). Functional characterization of naturally occurring wild soybean mutant (sg-5) lacking astringent saponins using whole genome sequencing approach. *Plant Science* 267: 148–156.

Reynolds, M.P. (ed.) (2010). *Climate Change and Crop Production*, vol. 1. CABI International.

Schmutz, J., Cannon, S.B., Schlueter, J. et al. (2010). Genome sequence of the palaeopolyploid soybean. *Nature* 463: 178–183.

Semagn, K., Bjornstad, A., and Ndjiondjop, M.N. (2006). Principles, requirements and prospects of genetic mapping in plants. *African Journal of Biotechnology* 5 (25): 2569–2587.

Shoemaker, R.C. and Specht, J.E. (1995). Integration of the soybean molecular and classical genetic linkage groups. *Crop Science* 35 (2): 436–446.

Shu, Y.J., Yu, D.S., Wang, D. et al. (2013). Genomic selection of seed weight based on low-density SCAR markers in soybean. *Genetics and Molecular Research* 12 (3): 2178–2188.

Singh, B.D. and Singh, A.K. (2015). *Marker-Assisted Plant Breeding: Principles and Practices*, 77–122. New Delhi: Springer.

Sonah, H., Bastien, M., and Iquira, E. (2013). An improved genotyping by sequencing (GBS) approach offering increased versatility and efficiency of SNP discovery and genotyping. *PLoS One* 8 (1): e54603. https://doi.org/10.1371/journal.pone.0054603.

Sonah, H., O'Donoughue, L., Cober, E. et al. (2015). Identification of loci governing eight agronomic traits using a GBS GWAS approach and validation by QTL mapping in soya bean. *Plant Biotechnology Journal* 13 (2): 211–221.

Song, Q., Hyten, D.L., Jia, G. et al. (2013). Development and evaluation of SoySNP50K, a high-density genotyping array for soybean. *PLoS One* 8: e54985.

Song, Q., Hyten, D.L., Jia, G. et al. (2015). Fingerprinting soybean germplasm and its utility in genomic research. *G3: Genes, Genomes, Genetics* 5 (10): 1999–2006.

Song, Q., Jenkins, J., Jia, G. et al. (2016). Construction of high-resolution genetic linkage maps to improve the soybean genome sequence assembly Glyma1.01. *BMC Genomics* 17 (1): 33.

Stewart-Brown, B.B., Song, Q., Vaughn, J.N. et al. (2019). Genomic selection for yield and seed composition traits within an applied soybean breeding program. *G3: Genes, Genomes, Genetics* 9 (7): 2253–2265.

Sun, X., Liu, D., Zhang, X. et al. (2013). SLAF-seq: an efficient method of large-scale de novo SNP discovery and genotyping using high-throughput sequencing. *PLoS One* 8 (3): e58700.

Tanksley, S.D., Young, N.D., Paterson, A.H. et al. (1989). RFLP mapping in plant breeding: new tools for an old science. *Bio/Technology* 7 (3): 257–264.

Tardivel, A., Sonah, H., Belzile, F. et al. (2014). Rapid identification of alleles at the soybean maturity gene E3 using genotyping by sequencing and a haplotype-based approach. *The Plant Genome* 7 (2): 1–9.

Tester, M. and Langridge, P. (2010). Breeding technologies to increase crop production in a changing world. *Science* 327 (5967): 818–822.

Torkamaneh, D., Laroche, J., and Belzile, F. (2016). Genome-wide SNP calling from genotyping by sequencing (GBS) data: a comparison of seven pipelines and two sequencing technologies. *PLoS One* 11 (8): e0161333. https://doi.org/10.1371/journal.pone.0161333.

Torkamaneh, D., Laroche, J., and Bastien, M. (2017). Fast-GBS: a new pipeline for the efficient and highly accurate calling of SNPs from genotyping-by-sequencing data. *BMC Bioinformatics* 18: 5.

Torkamaneh, D., Chalifour, F.P., Beauchamp, C.J. et al. (2020). Genome-wide association analyses reveal the genetic basis of biomass accumulation under symbiotic nitrogen fixation in African soybean. *Theoretical and Applied Genetics* 133 (2): 665–676.

Tuberosa, R. (2012). Phenotyping for drought tolerance of crops in the genomics era. *Frontiers in Physiology* 3: 347.

Valliyodan, B., Qiu, D., Patil, G. et al. (2016). Landscape of genomic diversity and trait discovery in soybean. *Scientific Reports* 6: 1–10.

Varala, K., Swaminathan, K., Li, Y. et al. (2011). Rapid genotyping of soybean cultivars using high throughput sequencing. *PLoS One* 6 (9): e24811.

Verma, S., Gupta, S., Bandhiwal, N. et al. (2015). High-density linkage map construction and mapping of seed trait QTLs in chickpea (*Cicer arietinum* L.) using Genotyping-by-Sequencing (GBS). *Scientific Reports* 5: 1–14.

Vuong, T.D., Sleper, D.A., Shannon, J.G. et al. (2010). Novel quantitative trait loci for broad-based resistance to soybean cyst nematode (*Heterodera glycines* Ichinohe) in soybean PI 567516C. *Theoretical and Applied Genetics* 121: 1253–1266.

Watanabe, S., Shimizu, T., Machita, K. et al. (2018). Development of a high-density linkage map and chromosome segment substitution lines for Japanese soybean cultivar Enrei. *DNA Research* 25 (2): 123–136.

Wu, X., Ren, C., Joshi, T. et al. (2010). SNP discovery by high-throughput sequencing in soybean. *BMC Genomics* 11: 469.

Xia, Z., Tsubokura, Y., Hoshi, M. et al. (2007). An integrated high-density linkage map of soybean with RFLP, SSR, STS, and AFLP markers using a single F2 population. *DNA Research* 14 (6): 257–269.

Xu, Y. (2010). *Molecular Plant Breeding*, 1–734. Wallingford, UK: CABI International.

Zhang, J., Song, Q., Cregan, P.B. et al. (2015). Genome-wide association study for flowering time, maturity dates and plant height in early maturing soybean (Glycine max) germplasm. *BMC Genomics* 16 (1): 217.

Zhang, J., Song, Q., Cregan, P.B. et al. (2016). Genome-wide association study, genomic prediction and marker-assisted selection for seed weight in soybean (Glycinemax). *Theoretical and Applied Genetics* 129 (1): 117–130.

Zhang, X., Pérez-Rodríguez, P., Burgueño, J. et al. (2017). Rapid cycling genomic selection in a multiparental tropical maize population. *G3: Genes, Genomes, Genetics* 7 (7): 2315–2326.

Zhang, Y., Li, W., Lin, Y. et al. (2018). Construction of a high-density genetic map and mapping of QTLs for soybean (*Glycine max*) agronomic and seed quality traits by specific length amplified fragment sequencing. *BMC Genomics* 19 (1): 641.

Zhao, Y., Gowda, M., Liu, W. et al. (2012). Accuracy of genomic selection in European maize elite breeding populations. *Theoretical and Applied Genetics* 124 (4): 769–776.

Zhong, S., Dekkers, J.C., Fernando, R.L. et al. (2009). Factors affecting accuracy from genomic selection in populations derived from multiple inbred lines: a barley case study. *Genetics* 182 (1): 355–364.

Zhong, C., Sun, S., Li, Y. et al. (2018). Next-generation sequencing to identify candidate genes and develop diagnostic markers for a novel Phytophthora resistance gene, RpsHC18, in soybean. *Theoretical and Applied Genetics* 131: 525–538.

Zhou, Z., Jiang, Y., Wang, Z. et al. (2015). Resequencing 302 wild and cultivated accessions identifies genes related to domestication and improvement in soybean. *Nature Biotechnology* 33 (4): 408–414.

Zuo, J.F., Niu, Y., Cheng, P. et al. (2019). Effect of marker segregation distortion on high density linkage map construction and QTL mapping in soybean (*Glycine max* L.). *Heredity* 123 (5): 579–592.

# 11

# Advances and Applicability of Genotyping Technologies in Cotton Improvement

*Shubham Bhardwaj*[1,4,*], *Vikas Devkar*[2,*], *Amit Kumar*[1], *Alish Alisha*[3], *Shivani Sharma*[1], *Rupesh K. Deshmukh*[1], *and Gunvant B. Patil*[2]

[1]*National Agri-Food Biotechnology Institute (NABI), Mohali, Punjab, India*
[2]*Department of Plant and Soil Science, Institute of Genomics for Crop Abiotic Stress Tolerance (IGCAST), Texas Tech University, Lubbock, TX, USA*
[3]*Department of Gene Expression, Faculty of Biology, Adam Mickiewicz University, Poznań, Poland*
[4]*National Institute of Plant Genome Research (NIPGR), New Delhi, India*
*Both authors contributed equally to this work.*

## 11.1 Introduction

Cotton is one of the most commercially important natural fiber crops that is majorly used as raw material in textile industries globally. Cotton fibers are hollow cellulose fibers and that makes it soft, cool, absorbent, breathable, and comfortable textile material. Additionally, cotton seedcakes are used in livestock feed and its seed oil is used for food and lubricant purposes. The diploid cotton species have 13 haploid chromosomes although the genome size varies between the species. The *Gossypium* genus phenotypically and genotypically is very diverse and has eight diploid genome groups (A to G and K) and one allopolyploid group (AD). It further includes approximately 52 species, including 45 diploid and 7 allotetraploids (Yang et al. 2020). Among all *Gossypium* species, currently, only four are cultivated commercially for their spinnable fiber. These species are two diploids, *Gossypium arboreum,* and *Gossypium herbaceum* (A–G and K genomes, $2n = 2x = 26$), and two tetraploids (AD genome, $2n = 4x = 52$); *Gossypium hirsutum* L. (Upland cotton) and *Gossypium barbadense* L. (Pima, Sea Island, or Egyptian cotton) (Campbell et al. 2010). The D-genome-like American diploid (*Gossypium raimondii*) together with an A-genome-like ancestral African diploid (*G. arboreum* or, *G. herbaceum*) is the ancestor of the two most economically important cotton-fiber producing species, *G. hirsutum* and *G. barbadense*. The tetraploid cotton species (AD) are thought to be formed by allopolyploidization events (1–2 million years ago) through intercontinental dispersal of a D-genome species as the male (pollen) parent hybridizing with an A-genome species as the maternal parent (Wendel and Cronn 2003). *G. hirsutum* (AADD, AD1 genome) is known for its high yield and requires shorter growing seasons, while *G. barbadense* (AADD, AD2 genome) is known for its superior quality fibers and better comprehensive resistance but requires relatively longer

*Genotyping by Sequencing for Crop Improvement,* First Edition.
Edited by Humira Sonah, Vinod Goyal, S.M. Shivaraj, and Rupesh K. Deshmukh.

growing season. Annually, more than 90% of the fiber produced comes from these two allotetraploid species of cotton.

The upland or Mexican cotton (*G. hirsutum*) alone contributes >90% of total world production while the remaining production comes from *G. barbadense* 3–4%, *G. arboreum*, and *G. herbaceum* (together, 2–6%). Cotton is grown commercially in around 80 countries in diverse climatic conditions. The annual world production of cotton is estimated at around 25 to 26.7 million tons per year with $600 billion yearly economic impacts worldwide, accounting for nearly 2.5% of the world's arable land. The major producers are India, the United States, China, Brazil, and Pakistan [Source: *ICAC-01/2019* (The International Cotton Advisory Committee); Khan et al. 2020]. Upland cotton is considered an excellent model crop for studying polyploidy, crop domestication, evolutions, and transgenic improvement. Additionally, cotton fiber is extensively studied for physiological and molecular processes associated with cell elongation and cellulose biosynthesis (Wang et al. 2012). However genetic improvement of cotton fiber yield, disease resistance, and biotic stress tolerance is a complex process because of its genetic architecture and genome complexity. This chapter describes the recent progress in genetic mapping and genomics, identifies knowledge gaps, and highlights the need for integrated approaches.

## 11.2 Challenges due to Polyploidy in Cotton

Genomic and molecular breeding studies in polyploid crops have fallen behind due to the large genome size, whole-genome duplication, and complexity of genome with highly repetitive sequence, while the diploid crops studies have achieved relatively great success in molecular mapping, marker-assisted selection (MAS), and trait discovery studies (Malik et al. 2014; Patil et al. 2016; Yang et al. 2020). Genotyping of polyploidy species is often challenging due to limitations in the identification of false polymorphism. Furthermore, sub-genome-specific primer/probe designing becomes very complicated due to the high similarity between the homoeologous region and target (Mason 2015). Secondly, the development of high-quality (gold standard) reference genome assembly in *Gossypium* is challenging, and therefore the intergenic DNA including repeat-rich regions, telomeres, and centromeres are poorly represented. Nevertheless, recent advances in sequencing technology and *de novo* assembly tools have accelerated the cotton genomics research and over the last 10 years, several reference genomes for cotton have been developed (www.cottongen.org). Despite several disadvantages of polyploidy, it offers several cytogenetical, molecular, and biological advantages. For example, duplicated genes can be lost, retained, or maintained as duplicates, leading to gene subfunctionalization or neo-functionalization (Comai 2005; Patil et al. 2015).

## 11.3 Applications of Genomics and Genotyping for Cotton Breeding Programs

The genetic improvement of several economically important crops including cotton has largely been the result of conventional breeding efforts. However, these yield gains through conventional breeding have been aided and accelerated by innovative breeding strategies. The success of the molecular breeding program depends upon the availability of genomic and genotyping technologies. Figure 11.1 describes the detailed strategies used in

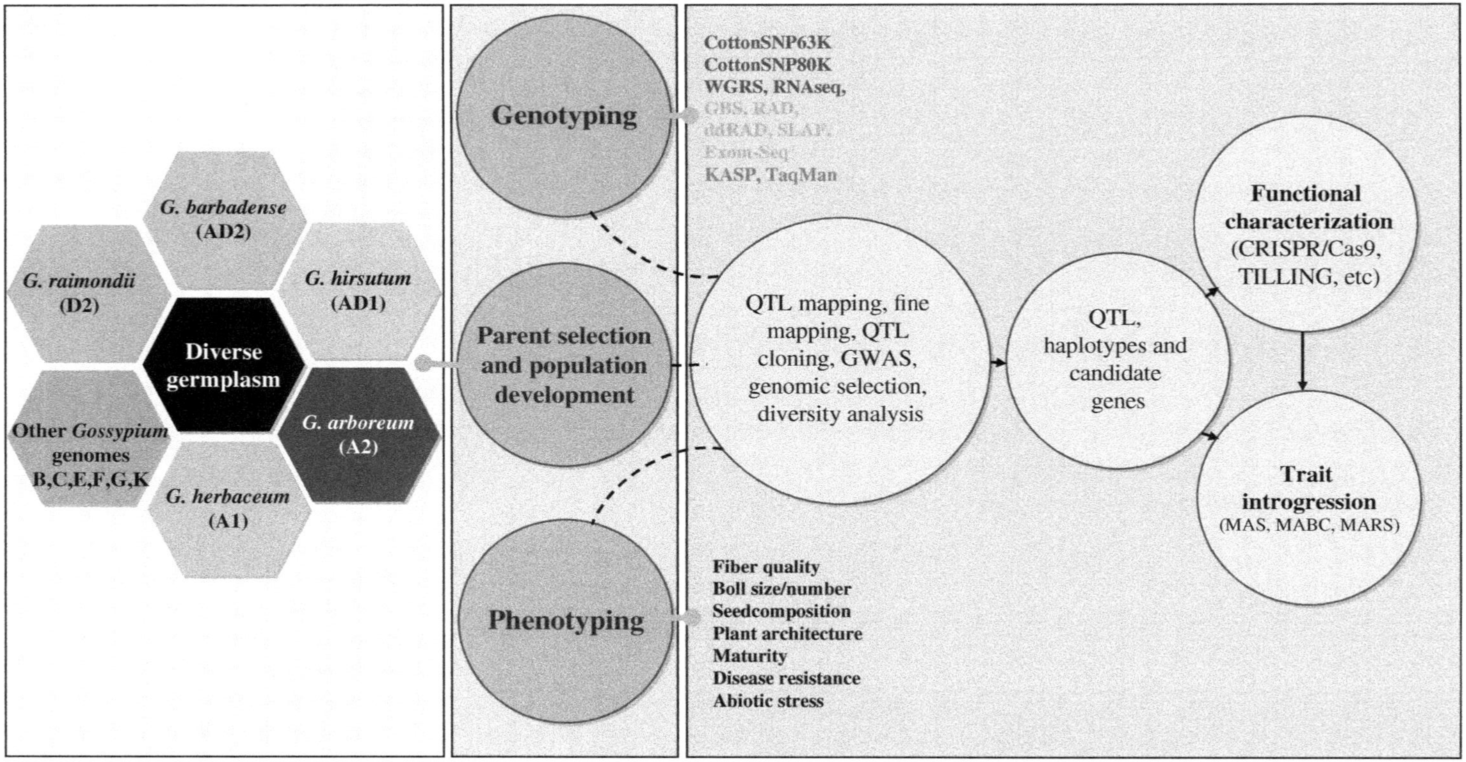


**Figure 11.1** Integrated genomics and breeding approaches for cotton improvement. This figure illustrates the systematic approach for selection and development of cotton germplasm, phenotyping for important traits, and genotyping using available platforms. Various genotyping platforms can be used to identify and map QTLs, haplotypes, and candidate genes. Furthermore, the identified QTL or genomic loci can be used in functional validation and breeding program. *Abbreviations*: QTL, quantitative trait locus; GWAS, genome-wide association study; TILLING, targeted induced local lesions in the genome; KASP, Kompetitive allele-specific PCR; GBS, genotyping by sequencing.

genomics-assisted crop improvement. Genomics and genotyping-based applications are required to improve the narrow genetic base by introgressing novel alleles from wild and exotic germplasm (Asekova et al. 2016). The need for genomic resources, such as reference genome, high-density single-nucleotide polymorphic (SNP) marker, and their linkage map, could tremendously benefit the discovery of novel genes and quantitative trait loci (QTL). Several low- to high-throughput and cost-effective genotyping platforms are now available to facilitate MAS and the development of improved germplasm in cotton (Reviewed by Rasheed et al. 2017). Among different molecular markers, SNPs are considered the marker of choice due to their frequency in the wild or recombinant population, genome-wide distribution, flexibility to convert in various cost-effective genotyping assays (KASP, CAPS, TaqMan, etc.), robustness, high-throughput, and accuracy (Rasheed et al. 2017; Patil et al. 2017; Malik et al. 2014).

Cotton has extreme morphological, cytogenetic, and genomic diversity, for example, cytogenetically cotton chromosomes in polyploid (AADD) differ in chromosome size but not in chromosome number. Therefore, researchers first sequenced the D-genome relative of allotetraploid cotton, *G. raimondii* that has small, and less complex genomic features (Paterson et al. 2012; Wang et al. 2012). Following that the draft genome of diploid *G. arboreum* was published in 2014 (Li et al. 2014). Soon after, both A and D genomes were assembled, and these resources were utilized for sequencing and assembly of *G. hirsutum* and *G. barbadense* reference genomes. Subsequently, several studies were carried out to develop high-quality *de novo* assemblies and resequencing of hundreds of diverse accessions including, wild relatives, landraces, and elite cultivars. These valuable genomic resources laid the foundation for studying genomic diversity, domestication, and application in MAS (Ulloa et al. 2020).

## 11.4 Genotyping Efforts in Cotton

Previously, several moderate to high-density linkage map has been generated using restriction fragment length polymorphism (RFLP), random amplified polymorphic DNA (RAPD), simple sequence repeat (SSR), and SNP molecular markers for genetic mapping and diversity analysis of *Gossypium* species (Ulloa et al. 2020). Among these, SNPs are the most widely used genetic variants in MAS, developing genotype-phenotype association, recombination studies, and trait discovery. Due to their high prevalence in the genome, availability of user-friendly bioinformatics tools for their detection, and compatibility with genotyping assays, SNP markers are widely used in crop breeding. Considering these aspects, this chapter specifically focuses on SNPs and other structural variants (SVs) and their use in cotton breeding and improvement. Genotyping by sequencing (GBS) is (considerably) low-cost and high-throughput SNP identification platform and offers several advantages such as simple library construction, identification of SNP as well as small InDels and microsatellites, and it has higher allele dosage detection in polyploidy species, low ascertainment bias, and no size selection and DNA shearing (Sonah et al. 2013). Like GBS technology, restriction site-associated DNA sequencing (RAD-Seq) involves restriction digestion of genomic DNA but both methods differ in substantial ways (Reviewed by Davey et al. 2011; Andrews et al. 2016). RAD-seq uses restriction enzymes to shear genomic DNA, which was size fractioned and then sequenced on the next-generation sequencing (NGS) platform. These methods are well adopted in crops including cotton, it requires high-quality DNA in large amount that is used for the size selection that restricts the wide use of RAD-seq in crop breeding. Nevertheless, several cotton researchers used this methodology for genetic mapping studies (Wang et al. 2015b; Sudakara et al. 2017). The SNP array technology is another revolutionary method to detect polymorphism within diverse germplasm in high-throughput setting (Rai et al. 2013). These arrays are capable of large-scale genotyping in a parallel fashion to assay thousands of loci in a short time which later can be used for various genomic applications including linkage mapping, genome-wide association study (GWAS), the pattern of genetic diversity, genomic selection (GS), and dissection of a complex trait. In cotton, the National Botanical Research Institute, India in collaboration with Affymetrix developed Axiom® Cotton Genotyping Array that includes 35 550 markers that were discovered in *G. hirsutum* and *G. barbadense* (Rai et al. 2013). Among 35 550 markers, 28 158 are intra-specific markers discovered through gene-enriched genomic sequences of *G. hirsutum* and 7 392 markers identified using genome reduction on restriction site method in which 5 286 markers are from inter-specific assembly of *G. hirsutum* and *G. barbadense* and 2 106 markers are intra-specific to *G. hirsutum* (Rai et al. 2013; http://www.affymetrix.com/support/technical/byproduct.affx?product=axiom_cotton).

Similarly, Cai et al. (2017) developed CottonSNP80K array that includes 77 774 SNPs and among that 59 502 were polymorphic loci. The CottonSNP80K represents high-density SNP markers distributed all over the genome and provides more precise loci which avoid false polymorphism due to homoeologous/paralogous gene. Furthermore, they have utilized these SNP arrays in association mapping for salt stress traits and identified eight significant SNPs associated with three salt stress-related traits (Cai et al. 2017). In another study, Hulse-Kemp et al. (2015) developed 63K Cotton SNP array by integrating genomic information from 13 independent data sets. Among these, nine were intra-specific and four were inter-specific populations (illustrated in Figure 11.2). This array represents different *G. hirsutum* germplasm and five other species, containing 45K intra-specific SNPs and 18K inter-specific SNPs that are relevant to a wide range of tetraploid cotton germplasms including its wild germplasm. Furthermore, they have also developed two high-density genetic maps containing a total of 22 829 SNPs. This is one of the key resources for the cotton researcher to genotype their diverse or inbred populations.

In cotton, earlier studies were limited to a couple of thousands of SNPs (Byers et al. 2012; Islam et al. 2015). But with the advent of NGS, availability of reference genomes and array technologies, the genotyping efforts in cotton increased exponentially and were used for uncovering genetic relationships and domestication patterns in *G. hirsutum*. Gore et al. (2014) first time attempted GBS study in cotton and identified over 412 SNP and 429 SSR, which covered nearly half of the tetraploid cotton genome. Furthermore, they evaluated 10 traits on TM-1 X NM24016 population and identified transgressive segregation for fiber traits. In another study, Islam et al. (2015) identified 4441 and 1176 polymorphic SNP from 11 diverse Upland cotton cultivars and their random-mated recombinant inbred progeny by using two different GBS methodologies. Additionally, the informative SNPs were utilized to develop a functional Kompetitive allele-specific PCR (KASP) genotyping assay. Similarly, Byers et al. (2012) have identified 11 834 and 1 679 nongenic SNPs by intra-specific comparisons between accessions of *G. hirsutum* and *G. barbadense* by genome

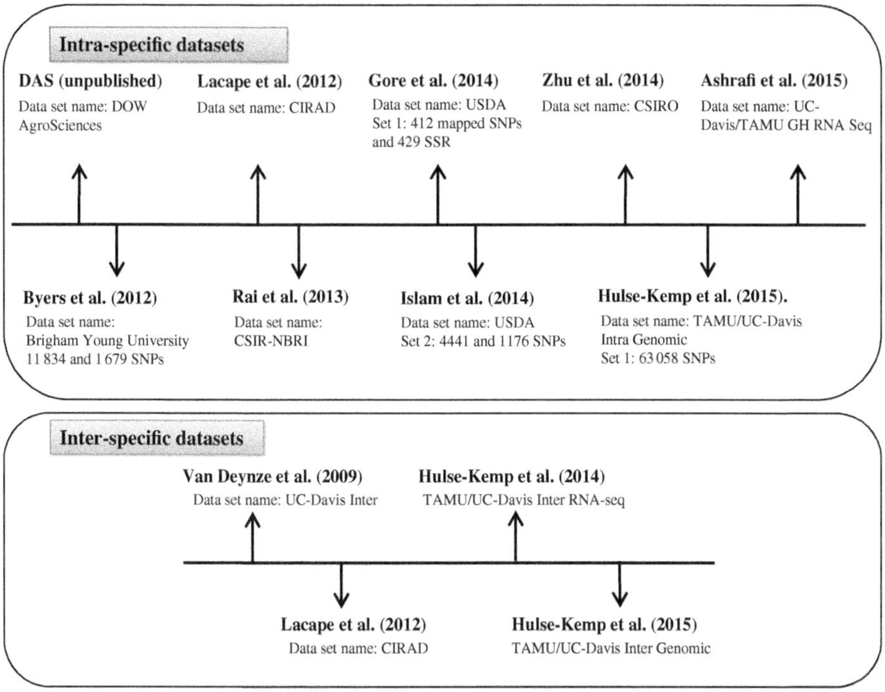


**Figure 11.2** Name of the nine intra-specific and four inter-specific datasets utilized in the development of Cotton63KSNP array by Hulse-Kemp et al. (2015) (Ashrafi et al. 2015; Byers et al. 2012; Gore et al. 2014; Hulse-Kemp et al. 2014; Rai et al. 2013; Van Deynze et al. 2009; Zhu et al. 2014). *Source*: Adapted from Hulse-Kemp et al. (2015).

reduction on restriction site method. Following the advancement in GBS workflow and computational analysis, several research groups used this technology for cotton improvement (summarized in Table 11.1).

## 11.5 High-Density Linkage Maps in Cotton

Molecular markers are the key to the development of high-resolution genetic linkage maps and fine mapping of QTL. Genetic linkage maps play a pivotal role in understanding the genome structure, explaining genome evolution, facilitating QTL mapping, MAS, and map-based cloning. A conventional marker-based genetic map cannot fully achieve the desired resolution for fine QTL mapping or positional cloning (Jia et al. 2016; Wu et al. 2014). As plant biotechnology advances, it contributes day by day to the gradual replacement of dominant DNA markers with co-dominant markers which include SNPs and SSRs. Microsatellites are second-generation markers; they are more informative and have higher transferability. However, due to its low efficacy and relatively small numbers, it has restrictions on its implementation in fine-scale QTL mapping and related subjects (Wang et al. 2019b). SNP markers have gained wider acceptance and applicability due to their abundance in genome, robustness, and suitability to automation

**Table 11.1** List of genomic resources available in cotton.

| Genotyping platform | Genotype (number of samples) | Number of SNPs/InDels | References |
|---|---|---|---|
| Sanger sequencing of ESTs | *G. hirsutum* (16) and G. *barbadense* (3) | >1000 SNPs and 279 InDels | Van Deynze et al. (2009) |
| Roche 454 pyrosequencing | *G. hirsutum* (2), *G. barbadense* (2) | 11 834 and 1 679 nongenic SNPs<br>4 327 genic SNPs in ESTs assembly | Byers et al. (2012) |
| GBS | *G. hirsutum* (1) and TM-1 X NM24016 (RIL) | 429 SSR, 412 SNPs | Gore et al. (2014) |
| GBS | *G. hirsutum* (11) and their RIL population | 4 441 and 1 176 | Islam et al. (2015) |
| GBS | F2:3 RIL population (277) | 5 178 | Diouf et al. (2018) |
| GBS | RIL population (143) | 3 557 SNPs | Fan et al. (2018) |
| RAD-seq | DH962 and Jimian5 (*G. hirsutum* parents) | 121 InDels, 441 SNPs | Wang et al. (2015a) |
| GBS | F2 RIL population (188) | 3 187 SNPs | Qi et al. (2017) |
| ddRAD | DS-28 X SBYF-425 (interspecific population) | 759 SNP, 21 InDels | Sudakara et al. (2017) |
| SLAF-seq | Five tetraploid species | 1 32 880 SNPs and 6 296 InDels | Shen et al. (2017) |
| Array-based (Affymetrix) | *G. hirsutum* and *G. Barbadense* | 35 550 SNPs | Rai et al. (2013), Affymetrix, Inc. |
| Array-based (CottonSNP63K) | *G. hirsutum* and five other species | 45 104 intraspecific and 17 954 interspecific SNPs | Hulse-Kemp et al. (2015) |
| Array-based (CottonSNP80K) | *G. hirsutum* (100) | 77 774 SNPs | Cai et al. (2017) |

(Hiremath et al. 2012; Kadam et al. 2016; Joshi et al. 2019). Previously, the majority of genetic maps were generated by using PCR-based markers which have low density, asymmetrical distribution of mapped markers, laborious and time consuming, and these aspects negatively impact the development of high-density linkage maps. Particularly in cotton, the first linkage map was developed based on the RFLP marker (Reinisch et al. 1994) and following that linkage mapping in cotton was improved by integrating various molecular markers, including RFLP (Jiang et al. 1998; Reinisch et al. 1994; Rong et al. 2004; Shappley et al. 1998; Ulloa et al. 2002), amplified fragment length polymorphism (AFLP) (Lacape et al. 2003), RAPD (Kohel et al. 2001), sequence-tagged sites (Rong et al. 2004), SSRs (Guo et al. 2007; Lacape et al. 2003; Liu et al. 2017; Wang et al. 2020; Yu et al. 2011), and SNPs (Blair et al. 2016; Diouf et al. 2018; Qi et al. 2017; Ulloa et al. 2020).

Sequencing the genomes of *G. barbadense* (Liu et al. 2015; Yuan et al. 2015), *G. arboretum* (Du et al. 2018; Li et al. 2014), *G. raimondii* (Paterson et al. 2012; Wang et al. 2012), and *G. hirsutum* (Li et al. 2015; Wang et al. 2019a; Zhang et al. 2015) has accelerated the development of genetic map construction and provided a platform for trait discoveries in cotton. Furthermore, interspecific linkage maps have been developed from a cross between *G. barbadense* and *G. hirsutum* (Chandnani et al. 2018; Guo et al. 2007; Han et al. 2006; Jiang et al. 1998; Lacape et al. 2010; Lin et al. 2009; Paterson et al. 2003; Rong et al. 2004; Wang et al. 2016; Yu et al. 2011). Similarly, interspecific maps were also developed for *G. herbaceum* × *G. arboretum*, *G. trilobum* × *G. raimondii* (Brubaker et al. 1999; Desai et al. 2006; Rong et al. 2004), and *G. nelsonii* × *G. austral* (Brubaker and Brown 2003). Besides interspecific maps, some intra-hirsutum maps were also generated (Liu et al. 2017; Lin et al. 2009; Shang et al. 2015; Tang et al. 2015; Zhang et al. 2019; Zhiyuan et al. 2014) (Table 11.2).

## 11.6 Whole-Genome Sequencing of Cotton Germplasm

In addition to the agronomic and economical importance of cotton, it is also a valuable resource for studying plant polyploidization (Du et al. 2018). The *Gossypium* genus consists of 45 diploid and 7 tetraploid species (Wang et al. 2012). Many rearrangements and structural differences during the process of diversification caused extensive chromosomal evolution in the diploid species (Yang et al. 2020). *G. raimondtii* was selected as the first species for whole-genome sequencing (WGS) and its genome assembly was reported in 2012 (Wang et al. 2012). Two years later in 2014, the draft genome of *G. arboreum* was released (Li et al. 2014). After sequencing of both diploid progenitors, the genome sequences of cultivated allotetraploid cotton species, *G. hirsutum* TM-1, and *G.barbadense* were published in 2015 (Yuan et al. 2015). The genome sequences of the economically important cotton species have provided a new set of valuable resources for better understanding the dynamics of the cotton genome. The available genome sequences further aid in the identification of the candidate genes, which play a significant role in improving the fiber quality and productivity of the cotton. These resources, allowed to resequence cotton genomes extensively to explore the genomic variations of diverse germplasms and alleles associated with quality and yield-related traits (Du et al. 2018; Ma et al. 2019a). Du et al. (2018) sequenced diverse germplasm of *G. arboreum*, resequencing 243 *G. arboreum* and *G. herbaceum* accessions (Du et al. 2018). The study revealed the findings that *G. arboreum* accessions equally diverged from *G. raimondii*. Due to the domestication-related traits, most of the accessions have experienced geographic isolation and adapted to that region (Du et al. 2018). Using a large number of SNP markers, GWAS has identified many significant associations for agronomically important traits. For example, a nonsynonymous cysteine-to-arginine substitution in *G. arboreum* KASIII is associated with changes in substantial fatty acid composition in the seeds (Du et al. 2018). In another study, resequencing of the parents (9053 and GK9708) of elite hybrid cotton CCRI63, revealed 8649 SNPs and 629 InDels (Song et al. 2018). Subsequently, 68 genes were identified related to the fiber development trait after these SNPs and InDels were co-localized with the reported QTLs of lint percentage (LP) and boll weight (BW). Further, this study concluded that accumulation of favorable variations from the parents leads to outstanding traits of LP and BW in CCRI63 hybrid (Song et al. 2018). Similarly, the WGS approaches were used to dissect novel traits involved in fiber quality and yield improvement (Table 11.3).

**Table 11.2** Development of various interspecific and intraspecific linkage (grey shade) maps.

| Interspecific crosses | Type of marker | Number of marker/loci | References |
|---|---|---|---|
| F2 (*G. nelsonii* × *G. austral*) | AFLP | *G. austral* (449) and *G. nelsonii* (397) | Brubaker and Brown (2003) |
| F2 (*G. arboreum* × *G. herbaceum*) | RFLP | 275 loci, spanned 1147 cM, avg. dist. 4.2 cM | Desai et al. (2006) |
| F2 (*G. hirsutum* × *G. barbadense*) | RFLP | 261 RFLP markers, 27 linkage groups, spanned 3767 cM, avg. dist. 14.4 cM | Jiang et al. (1998) |
| F2 of (*G. hirsutum* and *G. barbadense* acc. K101) | Sequence-tagged sites (STS) | 3347 loci, spanned 5941.2 cM, avg. dist. 1.78 cM | Rong et al. (2004) |
| BC1 (*G. hirsutum* L. TM-1 × *G. barbadense* L. cv. Hai7124) | SSR, eSSR, REMAP, CAPs, BAC-end, SNP, InDel | 3414 loci, spanned 3667.62 cM, avg. dist. 1.08 cM | Zhao et al. (2012) |
| F2 (*G. hirsutum* accession) DH962 × Jimian5 | (SRAP) combined with SSRs, RAPDs, and RGAPs | 471 loci to 51 linkage group, spanned 3070.2 cM, avg dist. 6.5 cM | Lin et al. (2009) |
| *G. raimondii* (CCRI 35 × Yumian 1) | SSR | 2051 loci, spanned 3508.29 cM with an avg. dist. of 1.71 cM | Liu et al. (2017) |
| *G. hirsutum* (GX1135 9 GX100-2) RIL | SSR | 581 loci, spanned 3714.4 cM. avg. dist. 6.39 cM | Shang et al. (2015) |
| F2:3 population *G. hirsutum* (CCRI35 × Nan Dan) | SNP | 5178 SNP markers, spanned 4768.098 cM, avg. dist. of 0.92 cM | Diouf et al. (2017) |
| RIL population (*G. hirsutum* accessions) (NC05AZ06 × NC11–2091) | SNP | 3129 SNPs mapped to 26 linkage group, avg. dist. 1.41 cM, map spanned 4422.44 cM | Zhang et al. (2019) |
| *G. hirsutum* accession (CCRI36 × G2005) | SNP | 6434 loci average distance of 0.63 cM, map spanned 4071.98 cM | Jia et al. (2016) |
| *G. hirsutum* (LU28 and ZHONG213) | SNP | 3978 SNP markers average genetic distance 0.62 cM | Li et al. (2017) |

**Table 11.3** List of genome wide association studies (GWAS) in cotton.

| Aim of study | Number of genotypes | Platform/technique used | Significant trait association | References |
|---|---|---|---|---|
| GWAS for 11 domestication related traits | 243 | PacBio long-reads and Hi-C technologies | fatty acid content, FOV resistance, seed fuzz development | Du et al. (2018) |
| GWAS for 13 fiber and yield-related traits | 419 | IlluminaHiSeq | Flowering day; fiber length; fiber strength | Ma et al. (2018) |
| GWAS for 10 traits related to yield | 318 | IlluminaHiSeq 2500 platform | Lint yield | Fang et al. (2017) |
| 12 different agronomic traits | 2 upland cotton | Illumina HiSeq4000 | Lint percentage, boll weight | Song et al. (2018) |
| Fiber quality | 26 | IlluminaHiSeq 4000 | Lint percentage | Ma et al. (2019b) |
| Evolution of allotetraploid | 19 | Illumina HiSeq X | — | Hu et al. (2019) |

## 11.7 Application of GBS Technology in Cotton Research

GBS is a robust and widely used genotyping approach in crops and it utilizes next-generation sequencing Illumina NGS platforms. Elshire et al. (2011) first time demonstrated the GBS in maize and barley. However, before GBS technology, the researcher used RAD-seq (restriction-associated DNA sequencing), which also utilizes NGS platforms to genotype large populations (Baird et al. 2008). In addition to these technologies, several other platforms for high-throughput genotyping such as multiplex shotgun genotyping (MSG), double-digested RAD-seq, ion torrent GBS, sequence-based genotyping (SBG), restriction fragment sequencing (REST-seq) and restriction enzyme sequence comparative analysis have been utilized in several crops (Baird et al. 2008; da Cunha and do Nascimento 2009; Elshire et al. 2011; Monson-Miller et al. 2012; Poland et al. 2012; Stolle and Moritz 2013; Truong et al. 2012). Typically, these GBS protocols utilize restriction enzymes to achieve the reduced genome representation through recruiting the genomic sequence flanking RE sites (Elshire et al. 2011; Poland et al. 2012). The steps that included GBS protocols are (i) isolation of genomic DNA, (ii) library preparation and sequencing, (iii) processing of raw sequence data, and SNP calling, and (iv) alignment of SNP and their functional validation (Islam et al. 2015). The *ApeKI* is a methylation-sensitive restriction enzyme used in the Institute for Genome Diversity (IGD) approach while, D2B (Data2Bio) used both methylation-sensitive *BfuCI* as well as methylation insensitive *NspI* enzyme (Elshire et al. 2011; Islam et al. 2015). Gore et al. (2014) developed and mapped 412 SNPs in bi-parental mapping upland cotton using GBS technology. In another study, Islam et al. (2015) identified 5617 SNP markers in 11 upland cotton cultivars. Furthermore, these markers were successfully validated using KASP genotyping assay (Islam et al. 2015). In addition to sequencing the population, bioinformatics pipelines are crucial for the identification of SNP markers in GBS. These bioinformatics pipelines are largely categorized into two categories, reference-based and *de novo* based that are being used in GBS Analysis. In the reference-based category, reads from reduced genome representation can be mapped on the reference genome, and SNP can be called (Li and Durbin 2009; Nielsen et al. 2011). Up to date, various reference-based GBS analysis pipelines have been developed that include TASSEL-GBS, Stacks, FAST-GBS, and IGST, and are the most widely used reference-based GBS analysis pipelines in plant genomics (Bradbury et al. 2007; Catchen et al. 2013; Glaubitz et al. 2014; Sonah et al. 2013; Torkamaneh et al. 2017). But in the absence of a reference genome, the most widely used pipelines in the *de novo*-based approach are UNEAK and Stacks (Catchen et al. 2013; Lu et al. 2013). Recent GBS-based cotton studies used the TASSEL-GBS where reference genome sequence availability can be utilized to guide the stacking of short reads from the same genomic site.

## 11.8 GBS-based Bi-Parental QTL Mapping and Association Mapping in Cotton

In cotton, majority of QTLs were identified for fiber quality, seed composition, disease resistance, abiotic stress tolerance, plant architecture, and plant maturity (www.cottongen.org). Discovering novel QTLs/genes for cotton fiber quality is the long-term objective of cotton improvement. However, many factors determine the quality of the cotton fiber including fiber strength, length, micronaire (air permeability of compressed cotton fibers), fiber composition, and color. In addition to fiber quality, fiber yield is an important target for cotton improvement. In cotton multiple QTLs for each trait have been identified but very few have been employed in MAS breeding. Most of the QTLs are localized over a wide range of genomic regions and the maps generated were derived from interspecific hybridization. Hence, to reduce the wide genomic gap and to overcome the inefficiency of maps generated by interspecific hybridization, the researcher developed intraspecific linkage maps to identify large-effect QTLs and additionally the incorporated SNP markers to improve mapping resolutions (Tables 11.2 and 11.4). Besides interspecific and intraspecific populations (Table 11.2), recombinant inbred lines (RILs) and backcross populations are also employed to generate maps for quantitative traits (Table 11.4) (Diouf et al. 2018; Fan et al. 2018; Magwanga et al. 2018). Previous studies have identified various QTLs for fiber quality-related traits (Shang et al. 2015). In a study conducted by Diouf et al. (2018), a total of 30 QTLs consistent in at least two environments were identified for fiber quality and yield-related traits in cotton (Diouf et al. 2018). Their positions within the cotton genome were identified and the genes tightly linked to these QTLs were also identified by GBS. Cotton also faces major yield losses due to abiotic stresses mainly by drought and salinity. There have been several studies to identify QTLs by GBS during stress at seedling stages and detected QTLs were further utilized in the breeding programs (Abdelraheem et al. 2018; Diouf et al. 2017; Magwanga et al. 2018) (Table 11.4). Hence, the data generated for quantitative traits by GBS could be employed in the cotton breeding programs to enhance selection and develop superior cultivars for traits, yield and fiber quality.

In contrast to bi-parental mapping, genome-wide association mapping utilizes diverse germplasm that carries more genetic recombination and provides the statistical power to detect QTLs (Do et al. 2019). The utilization of GBS for GWAS is becoming increasingly popular in genetic research, as they provide alternative cost-effective and high-throughput

**Table 11.4** List of GBS-based QTL mapping studies in cotton.

| Trait | Source | Chromosome | Genetic position/interval | QTL | LOD | References |
|---|---|---|---|---|---|---|
| Fiber quality and yield | 277 $F_{2:3}$ Intra-specific cross (CCRI35 × Nan Dan (NH)) | Chr. 17-D03 and Chr. 26-D12 | 4768.098 cM, avg. dist. 0.92 cM | 110 QTLs for11 traits | 2.5 | Diouf et al. (2018) |
| Plant height | $F_2$ (Z571 × CCRI49) | Chr. 3, 4, 5, 7, 9, 17, 19, 23, and 25 | 3828.551 cM, avg. dist. 0.687 cM | 17 QTLs | 3 | Qi et al. (2017) |
| Salinity | 277 $F_{2:3}$ Pop. from Intra-specific cross (CCRI35 × Nan Dan) | Chr. A02, A06, A12, D01, D03, D06, D08, and D13 | 4768.098 cM, avg. dist. 0.92 cM | 66 QTLs | 2.5 | Diouf et al. (2017) |
| Fiber quality | (MAGIC) Pop. of Random mating | A07 | 71 to 77 Mb | 86 QTLs | $P \leq 0.01$ | Islam et al. (2016) |
| Drought stress | $BC_2F_2$ Pop. from (donor*G. tomentosum* × recurrent *G. hirsutum*) | A01, A02, A04, A07, A08, A09, D01, D05, D06, D08, D10, D12, and D13 | 4191.3 cM, avg. dist. 0.3849 cM | 66 434 genes were mined | 2.5 | Magwanga et al. (2018) |
| Plant height | RIL (*G. hirsutum* × *G. barbadense*) | At3, At5, At10, Dt11, Dt13, Dt1, Dt3, Dt9, and Dt12 | 3433.24 cM, avg. dist. 0.67 cM | 10 PH QTLs | 2.5 | Ma et al. (2019a) |
| Fiber traits | RIL (*G. barbadense* cultivar 5917 × Pima S-7) | Chr. 5, 7 and16 | 3076.23 cM, avg. dist. 1.09 cM | 42 QTLs | $P < 0.05$ | Fan et al. (2018) |
| Early-maturity traits | F2 Pop. (LU28 × Zhong213) | D03 | 2480 cM, avg. dist. 0.62 cM | 47 QTLs | 2.5 | Li et al. (2017) |
| Fiber quality | RIL (*G. hirsutum* line TM-1 × NM24016) | AT and DT chromosomes | 2061 cM | 28 QTLs | 5 | Gore et al. (2014) |
| Fiber length (FL) | Two BILs, i.e. NMGA-062 and NMGA-105 | A08 and D03 | 3462.85 cM, avg. dist. 0.49 cM | 2 QTL | 6 | Liu et al. (2019) |
| Drought and salt stress tolerance | TM-1/NM24016 | Multiple Chr. | 2221.28 cM | 165 QTLs | 4 | Abdelraheem et al. (2018) |

options (Sonah et al. 2015). Previously, researchers have successfully used GBS data to perform GWAS in several organisms including major crops; soybean (Bastien et al. 2014; Sonah et al. 2015), rice (Bandillo et al. 2013), maize (Pace et al. 2015), and flax (Luo et al. 2019). In cotton, association mapping was carried out for a number of traits (Abdurakhmonov et al. 2008; Cai et al. 2014; Huang et al. 2017; Kantartzi and Stewart 2008; Mei et al. 2013; Su et al. 2016; Sun et al. 2017; Zeng et al. 2009; Zhang et al. 2013) (Table 11.3). Islam et al. (2016) used markers derived from the GBS study to identify genes associated with fiber quality. In this study, a multiparent advanced generation inter-cross (MAGIC) population was used to conduct a genome-wide association study (GWAS). They designed InDel marker CFBid0004 from 18bp deletion in GhRBB1_A07 gene, which was significantly found to be associated with superior fiber quality line in MAGIC RILs and NCVT cultivars. In another study, Islam et al. (2019) performed GS in cotton and evaluated the performance of the GS prediction using 550 cotton breeding RILs of the MAGIC population derived from 11 diverse parents. In a similar study, Gapare et al. (2018) used CottonSNP63K markers on historical breeding lines to perform GS for fiber length and fiber strength. Altogether, these studies show that the GBS platform can be used effectively in cotton breeding programs and it provides a cost-effective and high-throughput alternative to several other genotyping platforms.

## 11.9 Summary and Outlook

Cotton is an economically important global commodity crop and prime source of high-quality fiber as well as seed oil for human and animal consumption and industry applications. Genetic improvement efforts continue to advance superior fiber quality, yield, and resistance to biotic and abiotic stresses. Increasing the fiber quality and yield be the most important selection criteria for cotton researchers. Molecular breeding strategies are highly competent in the introgression of novel traits in several crops including cotton. The molecular breeding or MAS strategies primarily rely on cost-effective and high-throughput breeder-friendly genotyping technologies. Compared to other major food crops (rice, corn, and soybean), the genomic resources for cotton lag behind due to genomics complexities as well as research investment. Nevertheless, advancement in genotyping technologies has led to steady improvement in cotton research that facilitated the understanding of species evolution, domestication, and identification of QTLs for several traits. However, only a few QTLs reported to date have been functionally validated and used in the cotton breeding program. A combination of conventional breeding strategy, molecular breeding (introgression of exotic loci), functional genomics, and high-throughput phenotyping platforms will be helpful to dissect novel traits and use them in cotton breeding programs.

## References

Abdelraheem, A., Fang, D.D., and Zhang, J. (2018). Quantitative trait locus mapping of drought and salt tolerance in an introgressed recombinant inbred line population of Upland cotton under the greenhouse and field conditions. *Euphytica* 214: 8.

Abdurakhmonov, I., Kohel, R., Yu, J. et al. (2008). Molecular diversity and association mapping of fiber quality traits in exotic *G. hirsutum* L. germplasm. *Genomics* 92: 478–487.

Andrews, K.R., Good, J.M., Miller, M.R. et al. (2016). Harnessing the power of RADseq for ecological and evolutionary genomics. *Nature Reviews Genetics* 17 (2): 81.

Asekova, S., Kulkarni, K.P., Patil, G. et al. (2016). Genetic analysis of shoot fresh weight in a cross of wild (*G. soja*) and cultivated (*G. max*) soybean. *Molecular Breeding* 36 (7): 103.

Ashrafi, H., Hulse-Kemp, A.M., Wang, F. et al. (2015). A long-read transcriptome assembly of cotton (*Gossypium hirsutum* L.) and intraspecific single nucleotide polymorphism discovery. *The Plant Genome* 8 (2): 1–14.

Baird, N.A., Etter, P.D., Atwood, T.S. et al. (2008). Rapid SNP discovery and genetic mapping using sequenced RAD markers. *PLoS One* 3: e3376.

Bandillo, N., Raghavan, C., Muyco, P.A. et al. (2013). Multi-parent advanced generation inter-cross (MAGIC) populations in rice: progress and potential for genetics research and breeding. *Rice* 6 (1): 11.

Bastien, M., Sonah, H., and Belzile, F. (2014). Genome wide association mapping of *Sclerotinia sclerotiorum* resistance in soybean with a genotyping-by-sequencing approach. *The Plant Genome* 7 (1): 1–13.

Blair, M.W., Wu, X., Bhandari, D. et al. (2016). Genetic dissection of ICP-detected nutrient accumulation in the whole seed of common bean (*Phaseolus vulgaris* L.). *Frontiers in Plant Science* 7: 219.

Bradbury, P.J., Zhang, Z., Kroon, D.E. et al. (2007). TASSEL: software for association mapping of complex traits in diverse samples. *Bioinformatics* 23: 2633–2635.

Brubaker, C.L. and Brown, A.H.D. (2003). The use of multiple alien chromosome addition aneuploids facilitates genetic linkage mapping of the *Gossypium* G genome. *Genome* 46: 774–791.

Brubaker, C.L., Paterson, A.H., and Wendel, J.F. (1999). Comparative genetic mapping of allotetraploid cotton and its diploid progenitors. *Genome* 42: 184–203.

Byers, R.L., Harker, D.B., Yourstone, S.M. et al. (2012). Development and mapping of SNP assays in allotetraploid cotton. *Theoretical and Applied Genetics* 124: 1201–1214.

Cai, C., Ye, W., Zhang, T. et al. (2014). Association analysis of fiber quality traits and exploration of elite alleles in Upland cotton cultivars/accessions (*Gossypium hirsutum* L.). *Journal of Integrative Plant Biology* 56: 51–62.

Cai, C., Zhu, G., Zhang, T. et al. (2017). High-density 80 K SNP array is a powerful tool for genotyping *G. hirsutum* accessions and genome analysis. *BMC Genomics* 18: 654.

Campbell, B.T., Saha, S., Percy, R. et al. (2010). Status of the global cotton germplasm resources. *Crop Science* 50 (4): 1161–1179.

Catchen, J., Hohenlohe, P.A., Bassham, S. et al. (2013). Stacks: an analysis tool set for population genomics. *Molecular Ecology* 22: 3124–3140.

Chandnani, R., Kim, C., Guo, H. et al. (2018). Genetic analysis of *Gossypium* fiber quality traits in reciprocal advanced backcross populations. *The Plant Genome* 11: 170057.

Comai, L. (2005). The advantages and disadvantages of being polyploid. *Nature Reviews Genetics* 6 (11): 836–846.

da Cunha, K.P.V. and do Nascimento, C.W.A. (2009). Silicon effects on metal tolerance and structural changes in maize (*Zea mays* L.) grown on a cadmium and zinc enriched soil. *Water, Air, and Soil Pollution* 197: 323.

Davey, J.W., Hohenlohe, P.A., Etter, P.D. et al. (2011). Genome-wide genetic marker discovery and genotyping using next-generation sequencing. *Nature Reviews Genetics* 12: 499–510.

Desai, A., Chee, P.W., Rong, J. et al. (2006). Chromosome structural changes in diploid and tetraploid A genomes of *Gossypium*. *Genome* 49: 336–345.

Diouf, L., Pan, Z., He, S.-P. et al. (2017). High-density linkage map construction and mapping of salt-tolerant QTLs at seedling stage in Upland cotton using genotyping by sequencing (GBS). *International Journal of Molecular Sciences* 18: 2622.

Diouf, L., Magwanga, R.O., Gong, W. et al. (2018). QTL mapping of fiber quality and yield-related traits in an intra-specific Upland cotton using genotype by sequencing (GBS). *International Journal of Molecular Sciences* 19: 441.

Do, T.D., Vuong, T.D., Dunn, D. et al. (2019). Identification of new loci for salt tolerance in soybean by high-resolution genome-wide association mapping. *BMC Genomics* 20 (1): 318.

Du, X., Huang, G., He, S. et al. (2018). Resequencing of 243 diploid cotton accessions based on an updated A genome identifies the genetic basis of key agronomic traits. *Nature Genetics* 50: 796–802.

Elshire, R.J., Glaubitz, J.C., Sun, Q. et al. (2011). A robust, simple genotyping-by-sequencing (GBS) approach for high diversity species. *PLoS One* 6: e19379.

Fan, L., Wang, L., Wang, X. et al. (2018). A high-density genetic map of extra-long staple cotton (*Gossypium barbadense*) constructed using genotyping-by-sequencing based single nucleotide polymorphic markers and identification of fiber traits-related QTL in a recombinant inbred line population. *BMC Genomics* 19: 489.

Fang, L., Wang, Q., Hu, Y. et al. (2017). Genomic analyses in cotton identify signatures of selection and loci associated with fiber quality and yield traits. *Nature Genetics* 49: 1089.

Gapare, W., Liu, S., Conaty, W. et al. (2018). Historical datasets support genomic selection models for the prediction of cotton fiber quality phenotypes across multiple environments. *G3: Genes, Genomes, Genetics* 8 (5): 1721–1732.

Glaubitz, J.C., Casstevens, T.M., Lu, F. et al. (2014). TASSEL-GBS: a high capacity genotyping by sequencing analysis pipeline. *PLoS One* 9: e90346.

Gore, M.A., Fang, D.D., Poland, J.A. et al. (2014). Linkage map construction and quantitative trait locus analysis of agronomic and fiber quality traits in cotton. *The Plant Genome* 7: 1–10.

Guo, W., Cai, C., Wang, C. et al. (2007). A microsatellite-based, gene-rich linkage map reveals genome structure, function and evolution in *Gossypium*. *Genetics* 176: 527–541.

Han, Z., Wang, C., Song, X. et al. (2006). Characteristics, development and mapping of *Gossypium hirsutum* derived EST-SSRs in allotetraploid cotton. *Theoretical and Applied Genetics* 112: 430–439.

Hiremath, P.J., Kumar, A., Penmetsa, R.V. et al. (2012). Large-scale development of cost-effective SNP marker assays for diversity assessment and genetic mapping in chickpea and comparative mapping in legumes. *Plant Biotechnology Journal* 10 (6): 716–732.

Hu, Y., Chen, J., Fang, L. et al. (2019). *Gossypium barbadense* and *Gossypium hirsutum* genomes provide insights into the origin and evolution of allotetraploid cotton. *Nature Genetics* 51 (4): 739–748.

Huang, C., Nie, X., Shen, C. et al. (2017). Population structure and genetic basis of the agronomic traits of Upland cotton in China revealed by a genome-wide association study using high-density SNPs. *Plant Biotechnology Journal* 15: 1374–1386.

Hulse-Kemp, A.M., Ashrafi, H., Zheng, X. et al. (2014). Development and bin mapping of gene-associated interspecific SNPs for cotton (*Gossypium hirsutum* L.) introgression breeding efforts. *BMC Genomics* 15 (1): 1–14.

Hulse-Kemp, A.M., Lemm, J., Plieske, J. et al. (2015). Development of a 63K SNP array for cotton and high-density mapping of intraspecific and interspecific populations of *Gossypium* spp. *G3: Genes, Genomes, Genetics* 5: 1187–1209.

Islam, M.S., Thyssen, G.N., Jenkins, J.N. et al. (2015). Detection, validation, and application of genotyping-by-sequencing based single nucleotide polymorphisms in Upland cotton. *The Plant Genome* 8: 1940–3372.

Islam, M.S., Thyssen, G.N., Jenkins, J.N. et al. (2016). A MAGIC population-based genome-wide association study reveals functional association of GhRBB1_A07 gene with superior fiber quality in cotton. *BMC Genomics* 17: 903.

Islam, M.S., Fang, D.D., Jenkins, J.N. et al. (2019). Evaluation of genomic selection methods for predicting fiber quality traits in Upland cotton. *Molecular Genetics and Genomics* 295 (1): 67–79.

Jia, X., Pang, C., Wei, H. et al. (2016). High-density linkage map construction and QTL analysis for earliness-related traits in *Gossypium hirsutum* L. *BMC Genomics* 17: 909.

Jiang, C., Wright, R.J., El-Zik, K.M. et al. (1998). Polyploid formation created unique avenues for response to selection in *Gossypium* (cotton). *Proceedings of the National Academy of Sciences of the United States of America* 95: 4419–4424.

Joshi, V., Shinde, S., Nimmakayala, P. et al. (2019). Haplotype networking of GWAS hits for citrulline variation associated with the domestication of watermelon. *International Journal of Molecular Sciences* 20 (21): 5392.

Kadam, S., Vuong, T.D., Qiu, D. et al. (2016). Genomic-assisted phylogenetic analysis and marker development for next generation soybean cyst nematode resistance breeding. *Plant Science* 242: 342–350.

Kantartzi, S. and Stewart, J.M. (2008). Association analysis of fibre traits in *Gossypium arboreum* accessions. *Plant Breeding* 127: 173–179.

Khan, M.A., Wahid, A., Ahmad, M. et al. (2020). World cotton production and consumption: an overview. In: *Cotton Production and Uses* (eds. S. Ahmad and M. Hasanuzzaman), 1–7. Singapore: Springer.

Kohel, R.J., Yu, J., Park, Y.-H. et al. (2001). Molecular mapping and characterization of traits controlling fiber quality in cotton. *Euphytica* 121: 163–172.

Lacape, J.M., Nguyen, T.B., Thibivilliers, S. et al. (2003). A combined RFLP–SSR–AFLP map of tetraploid cotton based on a *Gossypium hirsutum* × *Gossypium barbadense* backcross population. *Genome* 46: 612–626.

Lacape, J.-M., Llewellyn, D., Jacobs, J. et al. (2010). Meta-analysis of cotton fiber quality QTLs across diverse environments in a *Gossypium hirsutum* x *G. barbadense* RIL population. *BMC Plant Biology* 10: 132.

Li, H. and Durbin, R. (2009). Fast and accurate short read alignment with Burrows–Wheeler transform. *Bioinformatics* 25: 1754–1760.

Li, F., Fan, G., Wang, K. et al. (2014). Genome sequence of the cultivated cotton *Gossypium arboreum*. *Nature Genetics* 46: 567–572.

Li, F., Fan, G., Lu, C. et al. (2015). Genome sequence of cultivated Upland cotton (*Gossypium hirsutum* TM-1) provides insights into genome evolution. *Nature Biotechnology* 33: 524–530.

Li, L., Zhao, S., Su, J. et al. (2017). High-density genetic linkage map construction by F2 populations and QTL analysis of early-maturity traits in Upland cotton (*Gossypium hirsutum* L.). *PLoS One* 12: e0182918.

Lin, Z., Zhang, Y., Zhang, X. et al. (2009). A high-density integrative linkage map for *Gossypium hirsutum*. *Euphytica* 166: 35–45.

Liu, X., Zhao, B., Zheng, H.-J. et al. (2015). *Gossypium barbadense* genome sequence provides insight into the evolution of extra-long staple fiber and specialized metabolites. *Scientific Reports* 5: 14139.

Liu, X., Teng, Z., Wang, J. et al. (2017). Enriching an intraspecific genetic map and identifying QTL for fiber quality and yield component traits across multiple environments in Upland cotton (*Gossypium hirsutum* L.). *Molecular Genetics and Genomics* 292: 1281–1306.

Liu, G., Pei, W., Li, D. et al. (2019). A targeted QTL analysis for fiber length using a genetic population between two introgressed backcrossed inbred lines in Upland cotton (*Gossypium hirsutum*). *The Crop Journal* 7: 273–282.

Lu, F., Lipka, A.E., Glaubitz, J. et al. (2013). Switchgrass genomic diversity, ploidy, and evolution: novel insights from a network-based SNP discovery protocol. *PLoS Genetics* 9: e1003215.

Luo, Z., Tomasi, P., Fahlgren, N. et al. (2019). Genome-wide association study (GWAS) of leaf cuticular wax components in *Camelina sativa* identifies genetic loci related to intracellular wax transport. *BMC Plant Biology* 19 (1): 187.

Ma, Z., He, S., Wang, X. et al. (2018). Resequencing a core collection of Upland cotton identifies genomic variation and loci influencing fiber quality and yield. *Nature Genetics* 50: 803–813.

Ma, J., Pei, W., Ma, Q. et al. (2019a). QTL analysis and candidate gene identification for plant height in cotton based on an interspecific backcross inbred line population of *Gossypium hirsutum*×*Gossypium barbadense*. *Theoretical and Applied Genetics* 132: 2663–2676.

Ma, X., Wang, Z., Li, W. et al. (2019b). Resequencing core accessions of a pedigree identifies derivation of genomic segments and key agronomic trait loci during cotton improvement. *Plant Biotechnology Journal* 17: 762–775.

Magwanga, R.O., Lu, P., Kirungu, J.N. et al. (2018). GBS mapping and analysis of genes conserved between *Gossypium tomentosum* and *Gossypium hirsutum* cotton cultivars that respond to drought stress at the seedling stage of the BC2F2 generation. *International Journal of Molecular Sciences* 19: 1614.

Malik, W., Ashraf, J., Iqbal, M.Z. et al. (2014). Molecular markers and cotton genetic improvement: current status and future prospects. *The Scientific World Journal* 2014: 607091.

Mason, A.S. (2015). Challenges of genotyping polyploid species. In: *Plant Genotyping* (ed. J. Batley), 161–168. New York, NY: Humana Press.

Mei, H., Zhu, X., and Zhang, T. (2013). Favorable QTL alleles for yield and its components identified by association mapping in Chinese Upland cotton cultivars. *PLoS One* 8: e82193.

Monson-Miller, J., Sanchez-Mendez, D.C., Fass, J. et al. (2012). Reference genome-independent assessment of mutation density using restriction enzyme-phased sequencing. *BMC Genomics* 13: 72.

Nielsen, R., Paul, J.S., Albrechtsen, A. et al. (2011). Genotype and SNP calling from next-generation sequencing data. *Nature Reviews Genetics* 12: 443–451.

Pace, J., Gardner, C., Romay, C. et al. (2015). Genome-wide association analysis of seedling root development in maize (*Zea mays* L.). *BMC Genomics* 16 (1): 47.

Paterson, A.H., Saranga, Y., Menz, M. et al. (2003). QTL analysis of genotype × environment interactions affecting cotton fiber quality. *Theoretical and Applied Genetics* 106: 384–396.

Paterson, A.H., Wendel, J.F., Gundlach, H. et al. (2012). Repeated polyploidization of *Gossypium* genomes and the evolution of spinnable cotton fibres. *Nature* 492: 423–427.

Patil, G., Valliyodan, B., Deshmukh, R. et al. (2015). Soybean (*Glycine max*) SWEET gene family: insights through comparative genomics, transcriptome profiling and whole genome re-sequence analysis. *BMC Genomics* 16 (1): 520.

Patil, G., Do, T., Vuong, T.D. et al. (2016). Genomic-assisted haplotype analysis and the development of high-throughput SNP markers for salinity tolerance in soybean. *Scientific Reports* 6 (1): 1–13.

Patil, G., Chaudhary, J., Vuong, T.D. et al. (2017). Development of SNP genotyping assays for seed composition traits in soybean. *International Journal of Plant Genomics* 2017: 6572969.

Poland, J.A., Brown, P.J., Sorrells, M.E. et al. (2012). Development of high-density genetic maps for barley and wheat using a novel two-enzyme genotyping-by-sequencing approach. *PLoS One* 7: e32253.

Qi, H., Wang, N., Qiao, W. et al. (2017). Construction of a high-density genetic map using genotyping by sequencing (GBS) for quantitative trait loci (QTL) analysis of three plant morphological traits in Upland cotton (*Gossypium hirsutum* L.). *Euphytica* 213: 83.

Rai, K.M., Singh, S.K., Bhardwaj, A. et al. (2013). Large-scale resource development in *Gossypium hirsutum* L. by 454 sequencing of genic-enriched libraries from six diverse genotypes. *Plant Biotechnology Journal* 11 (8): 953–963.

Rasheed, A., Hao, Y., Xia, X. et al. (2017). Crop breeding chips and genotyping platforms: progress, challenges, and perspectives. *Molecular Plant* 10 (8): 1047–1064.

Reinisch, A.J., Dong, J.M., Brubaker, C.L. et al. (1994). A detailed RFLP map of cotton, *Gossypium hirsutum x Gossypium barbadense*: chromosome organization and evolution in a disomic polyploid genome. *Genetics* 138: 829–847.

Rong, J., Abbey, C., Bowers, J.E. et al. (2004). A 3347-locus genetic recombination map of sequence-tagged sites reveals features of genome organization, transmission and evolution of cotton (*Gossypium*). *Genetics* 166: 389–417.

Shang, L., Liang, Q., Wang, Y. et al. (2015). Identification of stable QTLs controlling fiber traits properties in multi-environment using recombinant inbred lines in Upland cotton (*Gossypium hirsutum* L.). *Euphytica* 205: 877–888.

Shappley, Z.W., Jenkins, J.N., Meredith, W.R. et al. (1998). An RFLP linkage map of Upland cotton, *Gossypium hirsutum* L. *Theoretical and Applied Genetics* 97: 756–761.

Shen, C., Jin, X., Zhu, D. et al. (2017). Uncovering SNP and Indel variations of tetraploid cottons by SLAF-seq. *BMC Genomics* 18 (1): 1–13.

Sonah, H., Bastien, M., Iquira, E. et al. (2013). An improved genotyping by sequencing (GBS) approach offering increased versatility and efficiency of SNP discovery and genotyping. *PLoS One* 8: e54603.

Sonah, H., O'Donoughue, L., Cober, E. et al. (2015). Identification of loci governing eight agronomic traits using a GBS-GWAS approach and validation by QTL mapping in soya bean. *Plant Biotechnology Journal* 13 (2): 211–221.

Song, C., Li, W., Wang, Z. et al. (2018). Genome resequencing reveals genetic variation between the parents of an elite hybrid Upland cotton. *Agronomy* 8: 305.

Stolle, E. and Moritz, R.F.A. (2013). RESTseq–efficient benchtop population genomics with RESTriction Fragment SEQuencing. *PLoS One* 8: e63960.

Su, J., Pang, C., Wei, H. et al. (2016). Identification of favorable SNP alleles and candidate genes for traits related to early maturity via GWAS in Upland cotton. *BMC Genomics* 17: 687.

Sudakara, R.S., Katageri, I.S., Kumar, N.V. et al. (2017). Discovery of single nucleotide polymorphism in *Gossypium hirsutum* and *G. barbadense* through next generation sequencing approach. *Indian Journal of Genetics* 77 (1): 126–133.

Sun, Z., Wang, X., Liu, Z. et al. (2017). Genome-wide association study discovered genetic variation and candidate genes of fibre quality traits in *Gossypium hirsutum* L. *Plant Biotechnology Journal* 15: 982–996.

Tang, S., Teng, Z., Zhai, T. et al. (2015). Construction of genetic map and QTL analysis of fiber quality traits for Upland cotton (*Gossypium hirsutum* L.). *Euphytica* 201: 195–213.

Torkamaneh, D., Laroche, J., Bastien, M. et al. (2017). FAST-GBS: a new pipeline for the efficient and highly accurate calling of SNPs from genotyping-by-sequencing data. *BMC Bioinformatics* 18: 5.

Truong, H.T., Ramos, A.M., Yalcin, F. et al. (2012). Sequence-based genotyping for marker discovery and co-dominant scoring in germplasm and populations. *PLoS One* 7: e37565.

Ulloa, M., Meredith, W.R. Jr., Shappley, Z.W. et al. (2002). RFLP genetic linkage maps from four F2.3 populations and a joinmap of *Gossypium hirsutum* L. *Theoretical and Applied Genetics* 104: 200–208.

Ulloa, M., De Santiago, L.M., Hulse-Kemp, A.M. et al. (2020). Enhancing Upland cotton for drought resilience, productivity, and fiber quality: comparative evaluation and genetic dissection. *Molecular Genetics and Genomics* 295: 155–176.

Van Deynze, A., Stoffel, K., Lee, M. et al. (2009). Sampling nucleotide diversity in cotton. *BMC Plant Biology* 9: 125.

Wang, K., Wang, Z., Li, F. et al. (2012). The draft genome of a diploid cotton *Gossypium raimondii*. *Nature Genetics* 44: 1098–1103.

Wang, H., Huang, C., Guo, H. et al. (2015a). QTL mapping for fiber and yield traits in Upland cotton under multiple environments. *PLoS One* 10: e0130742.

Wang, H., Jin, X., Zhang, B. et al. (2015b). Enrichment of an intraspecific genetic map of Upland cotton by developing markers using parental RAD sequencing. *DNA Research* 22: 147–160.

Wang, H., Huang, C., Zhao, W. et al. (2016). Identification of QTL for fiber quality and yield traits using two immortalized backcross populations in Upland cotton. *PLoS One* 11: e0166970.

Wang, M., Tu, L., Yuan, D. et al. (2019a). Reference genome sequences of two cultivated allotetraploid cottons, *Gossypium hirsutum* and *Gossypium barbadense*. *Nature Genetics* 51: 224–229.

Wang, X., Liu, H., Pang, M. et al. (2019b). Construction of a high-density genetic linkage map and mapping of quantitative trait loci for growth-related traits in silver carp (*Hypophthalmichthys molitrix*). *Scientific Reports* 9: 17506.

Wang, F., Zhang, J., Chen, Y. et al. (2020). Identification of candidate genes for key fibre-related QTLs and derivation of favourable alleles in *Gossypium hirsutum* recombinant inbred lines with *G. barbadense* introgressions. *Plant Biotechnology Journal* 18: 707–720.

Wendel, J.F. and Cronn, R.C. (2003). Polyploidy and the evolutionary history of cotton. *Advances in Agronomy* 78: 139.
Wu, J., Li, L.-T., Li, M. et al. (2014). High-density genetic linkage map construction and identification of fruit-related QTLs in pear using SNP and SSR markers. *Journal of Experimental Botany* 65: 5771–5781.
Yang, Z., Qanmber, G., Wang, Z. et al. (2020). *Gossypium* genomics: trends, scope, and utilization for cotton improvement. *Trends in Plant Science* 25: 488–500.
Yu, Y., Yuan, D., Liang, S. et al. (2011). Genome structure of cotton revealed by a genome-wide SSR genetic map constructed from a BC1 population between *Gossypium hirsutum* and *G. barbadense*. *BMC Genomics* 12: 15.
Yuan, D., Tang, Z., Wang, M. et al. (2015). The genome sequence of Sea-Island cotton (*Gossypium barbadense*) provides insights into the allopolyploidization and development of superior spinnable fibres. *Scientific Reports* 5: 17662.
Zeng, L., Meredith, W.R., Gutiérrez, O.A. et al. (2009). Identification of associations between SSR markers and fiber traits in an exotic germplasm derived from multiple crosses among *Gossypium* tetraploid species. *Theoretical and Applied Genetics* 119: 93–103.
Zhang, T., Qian, N., Zhu, X. et al. (2013). Variations and transmission of QTL alleles for yield and fiber qualities in Upland cotton cultivars developed in China. *PLoS One* 8: e57220.
Zhang, T., Hu, Y., Jiang, W. et al. (2015). Sequencing of allotetraploid cotton (*Gossypium hirsutum* L. acc. TM-1) provides a resource for fiber improvement. *Nature Biotechnology* 33: 531–537.
Zhang, K., Kuraparthy, V., Fang, H. et al. (2019). High-density linkage map construction and QTL analyses for fiber quality, yield and morphological traits using CottonSNP63K array in Upland cotton (*Gossypium hirsutum* L.). *BMC Genomics* 20 (1): 889.
Zhao, L., Yuanda, L., Caiping, C. et al. (2012). Toward allotetraploid cotton genome assembly: integration of a high-density molecular genetic linkage map with DNA sequence information. *BMC Genomics* 13: 539.
Zhiyuan, N., Chen, H., Mei, H. et al. (2014). Molecular tagging of QTLs for fiber quality and yield in the Upland cotton cultivar Acala-Prema. *Euphytica* 195: 143–156.
Zhu, Q.H., Spriggs, A., Taylor, J.M. et al. (2014). Transcriptome and complexity-reduced, DNA-based identification of intraspecies single-nucleotide polymorphisms in the polyploid *Gossypium hirsutum* L. *G3: Genes, Genomes, Genetics* 4 (10): 1893–1905.

# 12

# Recent Advances and Applicability of GBS, GWAS, and GS in Millet Crops*

*Pankaj S. Mundada*[1,2,*], *Swapnil B. Kadam*[1,*], *Anupama A. Pable*[3], *and Vitthal T. Barvkar*[#,1]

[1]*Department of Botany, Savitribai Phule Pune University, Pune, Maharashtra, India*
[2]*Department of Biotechnology, Yashavantrao Chavan Institute of Science, Satara Maharashtra, India*
[3]*Department of Microbiology, Savitribai Phule Pune University, Maharashtra, India*
**Have contributed equally to this work.*
*[#]Corresponding Author Dr. Vitthal T. Barvkar Email: bvitthal@unipune.ac.in*

## 12.1 Introduction

Global food demand is increasing with the rapid increase in the world's population. Environmental damage along with rising population imparts adverse effects on global food production (Etesami and Jeong 2018). This will lead to land degradation, a shift in cultivation to more marginal areas and soil types, and the need for an increase in agricultural productivity. According to Food and Agriculture Organization (FAO) of the United Nations, food security is defined as "one of the fundamental need of society is said to be achieved when all the population has physical, social and economic access to the ample amount of nutritious food to meet the dietary needs of the individual for an active and healthy life."

Millets are composed of a varied group of grasses cultivated for food, feed, and forage characterized by a small grain size (Lata 2015). These belong to the family Gramineae comprising 10 genera and more than 14 species including *Sorghum bicolor* (L.) (sorghum), *Pennisetum glaucum* (L.) (pearl millet), *Eleusine coracana* (L.) (finger millet), *Setaria italica* (foxtail millet), *Paspalum scrobiculatum* (L.) (kodo millet), *Panicum miliaceum* (L.) (proso millet/broomcorn millet), *Echinochloa* spp. (L.) (barnyard millet), and *Panicum sumatrense* (L.) (little millet) (Basavaraj et al. 2010). These are known for their exceptional traits and nutritional values and hence have the potential to provide nutritional security (Kumar et al. 2016a, 2018). Millets are mostly cultivated in semiarid and tropical regions due to their ability to sustain with very low water and hence can be cultivated under nonirrigated conditions or in low rainfall regimes (200–500 mm), especially in Asia and Africa. The most important feature of the millets is their ability to adapt to adverse environmental conditions with minimal requirement of inputs (Lata et al. 2013). Their short life cycle attributes to grow in all seasons and higher nutritional value make them the crops of choice (Figure 12.1). Millets exhibit high-fiber content, gluten-free proteins, low glycemic index, and are enriched with minerals (calcium, iron, copper, magnesium, etc.) as well as

*Genotyping by Sequencing for Crop Improvement*, First Edition.
Edited by Humira Sonah, Vinod Goyal, S.M. Shivaraj, and Rupesh K. Deshmukh.

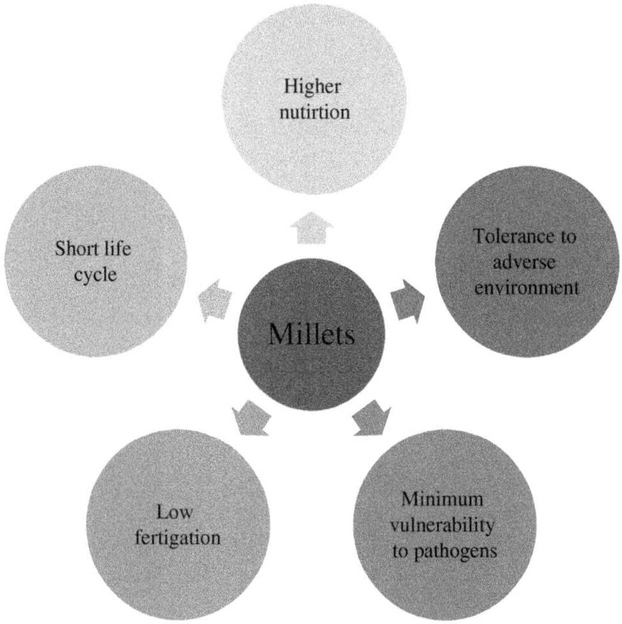


**Figure 12.1** Schematic representation of the important characteristics of millets.

antioxidants (Singh and Raghuvanshi 2012). Despite these multiple attributes, millets are sometimes regarded as orphan crops due to their less global production compared to other crops (Gupta et al. 2017).

Genetic resources play an important role in plant genomic studies to advance the scientific understanding of agriculture systems. A huge amount of diverse genetic resources of millets is collected and preserved at different national and international germplasm banks across the world. In India, International Crop Research Institute for the Semi-Arid Tropics (ICRISAT) alone has the collection of 41 816 accessions for sorghum, 24 373 accessions for pearl millet, 7519 accessions for finger millet, 1542 accessions for foxtail millet, 842 accessions for proso millet, 749 accession for barnyard millet, 473 accessions for little millet, and 665 accessions for kodo millet (www.icrisat.org).

The enormous genetic potential of millets serves as a vital resource for understanding the genomics of stress tolerance and nutritional potential (Kumar et al. 2015). The use of DNA markers for the selection of cultivars has been proven to be successful for crop improvement in cross-breeding programs. Progress in plant genomics studies provides insights into crop diversity at genetic levels and accelerates the process of genetic improvement (Muthamilarasan et al. 2014). Plant breeders select accessions based on phenotypic traits and molecular markers for the breeding program with the objective of germplasm improvement. Genetic maps along with quantitative trait loci (QTL) and association mapping (AM) populations have dichotomized the genetic control of agricultural traits, thereby allowing breeding programs to select high-value genotypes with the help of marker-assisted selection (MAS) and genomic selection (GS) (Kulwal et al. 2011). However, the use of AM and GS requires maker coverage at high density. The developments in next-generation sequencing (NGS) in the past few years have assisted the discovery of large number of genetic markers such as single-nucleotide polymorphism (SNP) and simple sequence repeats (SSRs).

## 12.2 GBS Efforts in Millet Crops

Genotyping-by-sequencing (GBS) is one of the reduced representations of genome sequencing methods used to discover SNPs. The technique uses restriction enzymes to cleave DNA coupled to DNA-barcoded adapters and reduces genome complexity. By choosing a suitable restriction enzyme, high SNP coverage can be achieved in gene-rich regions. This method has several advantages such as simplicity in the protocol, cost-effectiveness, high specificity, reproducibility, and high throughput (Elshire et al. 2011). This tool has been employed for linkage mapping and genomic diversity studies (Chung et al. 2017).

GBS has been carried out in different millet crops such as foxtail millet, sorghum, pearl millet, and finger millet (Table 12.1). Bai et al. resequenced the foxtail millet genome for cultivar Shi–Li–Xiang (SLX) in 2013. Using a comparative genomics approach against the existing genome data of Yugu1 cultivar (Bennetzen et al. 2012) and Zhang gu cultivar (Zhang et al. 2012) they reported 762 082 SNPs and 26 802 insertions or deletion (indels) in SLX versus Yugu1 genome and 915 434 SNPs and 28 546 indels in SLX versus Zhang gu genome. Out of this SNP data, they validated 465 SNPs with >90% accuracy, which can be used as markers in whole-genome mapping and marker-assisted breeding in foxtail millet. Later, Zhang et al. (2017) used genotypes A2 and Zhang gu to obtain 439 recombinant inbred lines (RILs) by cross-breeding and reported 33 579 SNPs from 542 322 SNPs to 33 587 small indels in between them. Foxtail millet cultivars, Hongmiaozhangu (P1), and Changnong 35 (P2) were crossed and 124 $F_2$ progenies were obtained. Using the restriction site-associated DNA sequencing (RAD-Seq) method, 2 668 587 SNP loci were detected, out of which 2 629 567 were located on nine chromosomes, and the remaining 39 020 were found on scaffolds (Wang et al. 2017). Jaiswal et al. (2019a) carried out genome-wide association study with 142 eco-geographically diverse cultivars of foxtail millet using 10 367 SNPs developed using GBS-ddRAD approach.

In sorghum, 971 diverse accessions from worldwide collections were used to facilitate the discovery of genetic markers (Morris et al. 2013). They reported 265 487 SNPs among the selected accessions and generated genome-wide SNP map to characterize the population structure according to the geographical origin across Africa and Asia. In 2015, 1943 geo-referenced landraces were characterized using GBS to obtain 404 627 SNPs (Lasky et al. 2015). Their findings proposed the role of genomic signatures of environmental adaptations for crop improvement and marker-assisted breeding. Later, Brenton et al. (2016) carried out GBS among 343 accessions to obtain 232 303 SNPs. They used this data for the development of a comparative understanding of bioenergy traits in sorghum. A proof-of-concept study regarding the integration of genomic predictions into a broad germplasm evaluation process was given by Yu et al. (2016). Initially, they selected 962 sorghum accessions as a reference set and carried out GBS to obtain 340 496 SNPs. Out of this, a set of 299 accessions was chosen as the training set to represent the diversity of reference set. For this, they also performed phenotypic characterization for traits such as biomass and yield. Using multiple analytical methods they cross-validated the data with high prediction accuracy.

**Table 12.1** High-throughput genomic platforms used for genotyping of millet crops.

| Crop | Number of accession used in study | Genotyping platform | Number of SNPs/indels | Multiplex capacity | References |
|---|---|---|---|---|---|
| Foxtail millet | 3 Accessions | Solexa sequencing using Illumina Genome Analyzer II | 762 082 SNPs, 26 802 (indels) in SLX versus Yugu1genome and 915 434 SNPs, 28 546 indels in SLX versus Zhang gu genome | — | Bai et al. (2013) |
| Foxtail millet | 439 Recombinant inbred lines | Illumina HiSeq 2000 | 33 579 SNPs, 33 587 small insertions and deletions (indels) | Multiplexed sequencing and paired-end strategy | Zhang et al. (2017) |
| Foxtail millet | 124 $F_2$ progeny | Illumina HiSeq X Ten | 2 668 587 SNPs | 96 Plex | Wang et al. (2017) |
| Foxtail millet | 142 Accessions | Illumina HiSeq 4000 | 10 367 SNPs | | Jaiswal et al. (2019) |
| Sorghum | 971 Accessions | Illumina Genome Analyzer IIx/HiSeq | 265 487 SNPs | 96- or 384-Plex GBS | Morris et al. (2013) |
| Sorghum | 1943 Accessions | Illumina Genome Analyzer IIx/HiSeq | 404 627 SNPs | 96- or 384-Plex GBS | Lasky et al. (2015) |
| Sorghum | 343 Accessions | Illumina HiSeq 2000 | 232 303 SNPs | 95-Plex | Brenton et al. (2016) |
| Sorghum | 962 Accessions | Genome Analyzer II | 340 496 SNPs | 96 Plex | Yu et al. (2016) |
| Sorghum | 399 Accessions | Illumina MiSeq | 690 SNPs | 96 Plex | Kong et al. (2018) |
| Sorghum | 1425 Accessions | Illumina HiSeq 2500 | 72 190 SNPs | 96-Plex | Girma et al. (2019) |
| Pearl millet | 48 Wild | Illumina HiSeq 2000 | 262 928 SNPs | 95-Plexlibrary | Berthouly-Salazar et al. (2016) |
| Pearl millet | 184 Accessions | Illumina HiSeq 2000 | 16 650 SNPs | 96-Plex | Punnuri et al. (2016) |

*(Continued)*

**Table 12.1** (Continued)

| Crop | Number of accession used in study | Genotyping platform | Number of SNPs/indels | Multiplex capacity | References |
|---|---|---|---|---|---|
| Pearl millet | 994 Accessions | Illumina HiSeq 2000 | 3 117 056 SNPs | 384-Plex and 96-plex | Varshney et al. (2017) |
| Pearl millet | 93 Accessions | Illumina HiSeq 2000 | 3321 SNPs | 95-Plex | Moumouni et al. (2015) |
| Pearl millet | 252 Global accessions and 248 Senegalese landraces | Illumina HiSeq 2500 | 83 875 SNPs | 96-Plex | Hu et al. (2015) |
| Pearl millet | $F_2$ mapped 190 plants | Illumina HiSeq 2000 | 460 SNPs | 96-Plex level | Pucher et al. (2018) |
| Pearl millet | 188 Accessions | HiSeq 2500 and a NextSeq 500 sequencing system (Illumina) | 392 493 SNPs | 96-Plex libraries | Debieu et al. (2018) |
| Pearl millet | 12 Accession | Single end ion proton sequencing | 15 306 RAD-seq SNPs and 32 463 tGBS SNPs | 384-Plex | Liang et al. (2018) |
| Finger millet | 146 Accessions | Illumina NextSeq platform with paired end | 17 245 SNPs | — | Qi et al. (2018) |
| Finger millet | 113 Accessions | Illumina True Seq version 3 HiSeq 2000 | 23 000 SNPs | 96-Plex libraries | Kumar et al. (2016) |
| Finger millet | 89 Accessions | Illumina Hi-Seq 2000 | 23 285 SNPs | — | Gimode et al. (2016) |

This project illustrated the possible global cost-effective strategy for the assessment of a huge amount of germplasm (Yu et al. 2016). In the other study, *S. bicolor* BTx623 to *S. bicolor* IS3620C was crossed to obtain 393 RILs. The GBS analysis revealed 690 SNPs which are further used for genetic and QTL mapping (Kong et al. 2018). Similarly, using GBS, among the 1425 Ethiopian landrace accessions, a total of 72 190 SNPs were selected from 879 407 SNPs by stringent quality control (Girma et al. 2019).

To facilitate breeding work in pearl millet, the GBS technique was used to construct genetic map composed of SNP markers. For mapping, 93 $F_2$ plants were obtained from nine wild cultivars. The study reported 3321 SNPs (Moumouni et al. 2015). Similarly, 83 875 SNPs were reported in another study among 500 accessions of pearl millet (Hu et al. 2015). Berthouly-Salazar et al. (2016) proposed an alternate strategy using transcriptome for the identification of SNPs using GBS in the crops without reference genome. Their study reported 262 928 SNPs among 48 wild pearl millet genotypes (Berthouly-Salazar et al. 2016). Punnuri et al. (2016) carried out GBS for 186 RILs obtained by breeding of "Tift 99D2B1" and "Tift 454" lines of pearl millet and reported 16 650 SNPs among them. A breakthrough came with the whole-genome sequencing of pearl millet by Varshney et al. (2017). They also performed GBS of 994 pearl millet lines for the establishment of association between a marker and phenotypic traits (Varshney et al. 2017). GBS-based linkage map was developed for mapping of male fertility restoration locus in pearl millet by cross-breeding of "ICMA 02777" and "ICMR 08888" accessions (Pucher et al. 2018). In another study, a potential mechanism for regulating response to drought stress in pearl millet was revealed, thereby contributing to the knowledge of breeding of pearl millet lines with improved yield and drought stress tolerance (Debieu et al. 2018). Liang et al. (2018) illustrated the modern approach of GBS to enable genomic selection in pearl millet. They evaluated the two genotyping strategies, RAD-seq (use of single restriction enzyme) (Miller et al. 2007) and tGBS (use of two restriction enzymes) (Ott et al. 2017) for characterization of inbred pearl millet lines. These studies revealed that the SNPs identified using RAD-seq exhibited low minor allele frequencies, whereas tGBS SNPs showed higher minor allele frequencies (Liang et al. 2018).

In the finger millet, GBS was used for analyzing the population structure of diverse genotypes. The study used genetically diverse 113 finger millet accessions. GBS revealed a set of 23 000 SNPs distributed across all the accessions (Kumar et al. 2016b). Gimode et al. (2016) carried out GBS in 89 diverse cultivars of finger millet and reported 23 285 SNPs among them. In addition to the existing tools or the analysis of GBS data, a new UGbS-Flex pipeline was reported, especially for the genotyping of heterozygous and polyploidy crops for which whole-genome sequence (WGS) is not available (Qi et al. 2018). This study was validated by identifying 17 245 SNPs from 146 genotypes of finger millet. Researchers also demonstrated the feasibility of cross-species application of GBS among closely related species.

## 12.3 High-density Linkage Maps in Millet Crops

In the past few decades, genetic linkage maps have been developed with the help of DNA markers for different plant species. Various types of DNA markers such as RFLP, RAPD, AFLP, SNPs, and SSR have been used by researchers (Young 2001). Marker analysis and phenotyping of agronomical traits are essential factors for the acceleration of breeding programs (Araus et al. 2018). The prior availability of genome sequence is not needed for construction of linkage maps using molecular markers. The high-density linkage maps play an important role in the field of genetics and genomics, thereby facilitating the discovery of genomic regions underlying a trait of interest. However, most of the maps are obtained using hundreds of makers which will not cover the genome completely leading to low density maps (Liu et al. 2014). Modern techniques such as GBS are used to discover highly abundant SNP markers which aided the development of high-density linkage maps in millet crops.

In the foxtail millet, using genome-wide analysis 28 342 microsatellite markers were identified with the frequency of 69 microsatellites per Mb (Pandey et al., 2013). Out of 28 342 markers, they developed 21 492 primers and mapped 15 573 markers physically on nine chromosomes spanning 405.3 Mb genome. Out of these, researchers validated 159 markers in eight different accessions and proved a high percentage of genetic transferability in millet and nonmillet crops. Fang et al. (2016) screened 10 598 SSR covering 1318.8 cM genome. The average distance between the markers is 1.27 cM; however, the distribution of these markers is uneven on different chromosomes. Wang et al. (2017) developed high-density linkage map using 9968 SNP markers for the construction of a genetic map. These markers generated the map of length 1648.8 cM with an average distance of 0.17 cM per marker, dividing them into nine linkage groups (Wang et al. 2017). Liu et al. (2020) represented the markers in the form of a bin map. They used 3413 bin markers with 3963 recombinant breakpoints developed using SNPs spanned 1222.26 cM distance with an average distance of 0.36 cM for all the nine chromosomes (Liu et al. 2020).

Several reports are available describing the genetic maps of pearl millet developed using different molecular markers; however, these maps suffer from inconsistent marker coverage and large gaps of greater than 20–30 cM between the adjacent markers. Moumouni et al. (2015) using the GBS approach developed a genetic map covering a genetic distance of 640 cM. The map was constructed using 3321 SNPs with a minor allele frequency (MAF) of $\geq$0.3 for 2809 high-quality SNPs. The SNPs were distributed evenly among the seven linkage groups having sizes 62–123 cM, and the average density was 0.51 SNP/cM. The average distance between the adjacent SNPs was 2.1 ($\pm$0.6) cM. Along with SNPs, 19 SSR markers were assessed to establish bridge among the linkage groups (Moumouni et al. 2015). Using a similar approach of GBS markers, a high-density linkage map covering the genetic distance of 716.7 cM was developed by Punnuri et al. (2016). The map was developed using 16 650 SNPs and 333 567 sequence tags distributed across seven chromosomes. The final map exhibited a density of 23.23 SNP per cM (Punnuri et al. 2016).

Ji et al. (2017) constructed high-density linkage map using specific-locus-amplified fragment (SLAF) markers in sorghum. The linkage map of 2158.1 cM covering 10 chromosomes was developed using 2246 SLAF markers. The average gap on each chromosome was 10.2 cM with 98.1% gaps of less than 5 cM. The number of markers on each chromosome ranged from 123 to 315 with an average value of 224.6 and an average distance of 0.98 cM (Ji et al. 2015).

## 12.4 GBS-based QTL Mapping Efforts in Millet Crops

Most of the traits in plants are polygenic and quantitative in nature, namely, crop yield, biomass, and biotic as well as abiotic stress tolerance. Advancement in plant breeding, molecular marker development, and ease in sequencing technology has accelerated QTL identification in an unprecedented manner. Initially, establishment of genetic linkage maps played a crucial role in plant breading. However, these methods have inherent flaws such as comparatively less number of recombinants generated from two parents in a few generations results in low resolution in linkage mapping and very large genetic and physical distance with several candidate genes for a trait of interest. These limitations were superseded by an alternative approach to the QTL mapping known as the GWAS or AM approach (Gómez et al. 2011) based on the principle of linkage disequilibrium (LD) to detect a substantial association between DNA markers and target trait using genotyping by sequencing (Gupta et al. 2005). Genetic linkage is established using extensive genotyping of a panel of germplasm or breeding populations (BPs) showing contrasting phenotypes under various environmental conditions for the trait of interest.

There have been multiple reports on the use of GBS technology for QTL analysis in millet crops. Mostly the GBS-based QTL analysis has been reported for pearl millet for various traits. Punnuri et al. (2016) reported the use of high-density linkage map using GBS for identification of *Pyricularia* leaf spot resistance and flowering time QTLs. They developed 186 RIL mapping populations from a cross between the lines Tift $99D_2B_1$ (rust and *Pyricularia* leaf spot resistant) and Tift 454 (nematode resistance and pollen fertility restorer capability) and precisely identified flowering time and leaf spot resistant QTL locations (Punnuri et al. 2016). The QTL controlling biomass production under drought stress in pearl millet has been identified by Debieu et al. (2018). They used 188 inbred lines and landraces with phenotypic data for GBS analysis and found SNPs on chromosomes 6 and 2, 3 and 6 contributing significantly for stay-green and aerial biomass, respectively. Further, they identified SNPs located in genes for sirohaem and cuticular wax biosynthesis and later considered as the main transpiration barriers in leaves conferring drought tolerance (Debieu et al., 2018). Moreover, there are attempts made in pearl millet to understand QTL associated with male-fertility restoration, an important trait with respect to the production of commercial hybrid seed (Pucher et al. 2018). They used F2 mapping population of 190 plants derived from a cross made between ICMA 02777 and ICMR 08888 for GBS-based genotyping and identified SNPs contributing significantly to pollen production and selfed seed set (traits indicating male-fertility restoration of the A4 CMS system) and plant height.

Similarly, attempt has been made in foxtail millet to utilize GBS technology to understand QTLs associated with yield and various agronomical traits in plants grown under long- and short-day photoperiods (Zhang et al. 2017). A total of 59 QTLs have been reported, out of which nine major QTLs (qpl2, qph5, qph6, qcgn6, qnl5, qnl6, qtn5, qfll9, and qflw5-1) for panicle length (PL), plant height (PH), code grain number (CGN), neck length (NL), flag-leaf length (FLL), tiller number (TN), and flag-leaf width (FLW) were detected consistently under both the photoperiods, suggesting that some important agronomic traits shared the similar genetic basis under different photoperiods (Table 12.2).

**Table 12.2** List of GBS-based QTL mapping studies in millet crops.

| Crop | Trait | Cross | Population | Chromosome | Genetic position/ marker interval | QTL | LOD | References |
|---|---|---|---|---|---|---|---|---|
| Pearl millet | Leaf spot resistance | Tift 99D $_2B_1$× Tift 454 | 186 Recombinant inbred line | LG1 | 32.3cM/S1_1423 to S1_3590 | Flowering time | 2.61 | Punnuri et al. (2016) |
| | | | | LG2 | 23.3cM/S2_1896 to S2_2803 | | 4.86 | |
| | | | | LG5 | 0.0cM/S5_0012 to S5_1669 | | 2.38 | |
| | | | | LG7 | 14.4cM/S7_0244 to S7_2067 | | 2.48 | |
| | | | | LG2 | 85.0cM/S2_7773 to S2_8331 | Resistance to *Pyricularia grisea* leaf spot | 1.78 | |
| | | | | LG3 | 114.2cM/S3_0019 to S3_4763 | | 1.82 | |
| | | | | LG5 | 30.5cM/S5_2145 to S5_4145 | | 4.83 | |
| | | | | LG7 | 30.5cM/S7_0738 to S7_3864 | | 5.05 | |
| Pearl millet | Early drought stress tolerance | Inbred lines and landraces | 188 Pearl millet inbred lines developed at ICRISAT(Niger) from landrace and ICMB88004 as a control | 6 | S6_111262313 | Stay-green | NA | Debieu et al. (2018)0 |
| | | | | 2 | S2_17262004 | Aerial biomass | | |
| | | | | 3 | S3_216179591 | | | |
| | | | | 6 | S6_64437341 | | | |
| | | | | | S6_157567763 | | | |
| Pearl millet | Male-fertility restoration | ICMA 02777 × ICMR 08888 | F2 mapping population of 190 plants | LG2 | 44cM/S2_110825781 to S2_195649011 | Pollen product ion | 7.86 | Pucher et al. (2018) |
| | | | | LG2 | 44cM/S2_110825781 to S2_195649011 | Selfed seed set | 5.39 | |
| | | | | LG4 | 38cM/S4_48328719 to S4_70429741 | Plant height | 11.52 | |

| | | | | | | | |
|---|---|---|---|---|---|---|---|
| Foxtail millet | 14 Agronomic traits including yield for plants grown under long- and short-day photoperiods | Zhang gu × A2 | 439 Recombinant inbred lines | 59 QTL | | Heading data (HD) panicle length (PL) | Zhang et al. (2017) |
| | | | | | | Tiller number (TN) | |
| | | | | | | Panicle weight (TN) | |
| | | | | | | Panicle diameter (PD) | |
| | | | | | | Flag-leaf length (FLL) | |
| | | | | | | Flag-leaf width (FLW) | |
| | | | | | | Plant height (PH) | |
| | | | | | | Stem diameter (SD) | |
| | | | | | | Stem node number (SNN) | |
| | | | | | | Code number (CN) | |
| | | | | | | Code grain number (CGN) | |
| | | | | | | Thousand-grain weight (TGW) | |
| | | | | | | Neck length (NL) | |

## 12.5 Genome Sequencing and Resequencing of Millet Crops

Nucleotide sequencing has witnessed many technological milestones since its development. Sanger sequencing, which is one of the gold standard techniques for DNA sequencing, could produce only a few bases of reads, while today's NGS technologies, such as Illumina-Solexa sequencing, are capable to read the whole genome in a very short time. The advancement in sequencing techniques has boosted the WGS analysis, though millets received attention quite late for whole-genome sequencing compared to model crops (Yu et al. 2002; Goff et al. 2002). Foxtail millet was the first millet for which WGS assembly was released (Zhang et al. 2012).

Platforms available for whole-genome sequencing are Illumina-Solexa, Roche 454 pyrosequencing, Ion Torrent, ABI SOLiD, Helicos, MinION, and single-molecule real-time (SMRT) sequencing. Illumina-Solexa is based on the sequencing-by-synthesis (SBS) principle. The whole-genome sequencing for millets started recently. WGS of foxtail millet was followed by pearl millet, finger millet, and most recently, broomcorn millet was sequenced. The reason for such delay in advance genomics in millets is partly because of their orphan crop status. Ploidy level is another hurdle (e.g. allotetraploid genome of finger millet) behind the delay of WGS. Whole-genome duplication (WGD) is one of the mechanisms through which polyploidy arises. Thus, polyploidy has a high percent of repetitive sequence. Millet genomes consist of high percentage of such repetitive elements. Pearl millet genome has more than 75% repetitive genome. A high percentage of repetitive sequences creates problems during genome assembly (Kyriakidou et al. 2018). But as the techniques for whole-genome sequencing are evolving at a high rate, such as the third-generation sequencing which includes Hi-C technology, solutions to address genomic complexity-related issues are emerging. Thus, WGS data for more millets including minor millets and pseudo-millets are expected to be available in the near future.

Despite the drought tolerance exhibited by millets, other traits such as low yield and susceptibility to diseases such as downy mildew and rust make the economy of the millets unimpressive. Many important traits are polygenic in nature. Hence, single-candidate gene study is not sufficient to understand the genotype–phenotype association of these traits. WGS information is useful to understand the events that happen during evolution at the genomic level. Gene family expansion and contraction analysis from genome sequence data provide molecular explanations for the specific traits shown by the plants. Finger millet WGS study revealed the variable composition of transposable elements (TEs), which is one of the important components of genomic architecture across the grasses. Here, WGS provides a strong association between variable compositions of these TE and diversifications observed across the grasses. In foxtail millet, WGD is predicted to have happened 27 million years ago based on WGS data, which is an important outcome for evolutionary study (Zou et al. 2019). Gene family study analysis from WGS data of broomcorn millet pointed expansion of MATH-BTB-BACK proteins in the crops such as little millet and foxtail millet, which play an important role in abiotic stress tolerance through ABA signaling and thus may contribute to the abiotic stress tolerance trait exhibited by these two plants. The significances of WGS in millet are summarized in Figure 12.2.

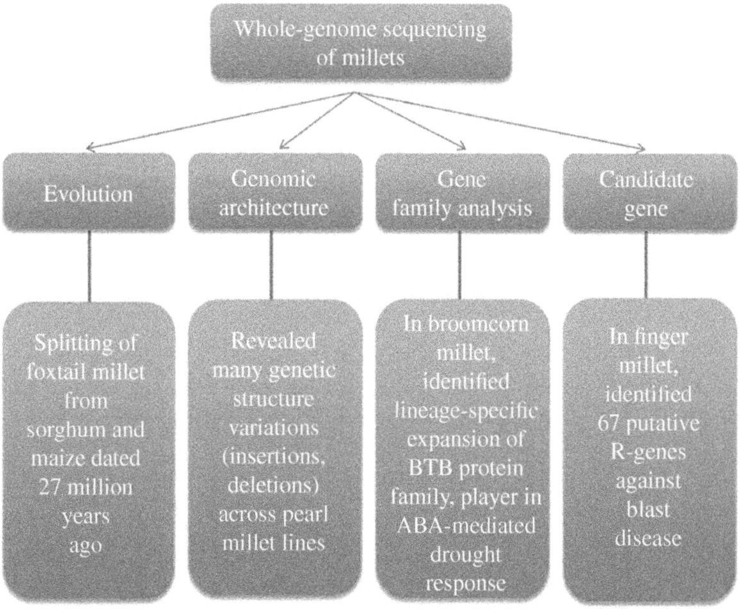


**Figure 12.2** Applications of whole-genome sequence (WGS) of millets.

### 12.5.1 Pearl Millet

Pearl millet (*P. glaucum* (L.)) is cultivated in the arid and semiarid regions of Africa and Asia. It withstands in low fertile and highly acidic soil with rough environmental conditions including high temperature, salinity, and low rainfall. The genome of pearl millet was sequenced using reference genotype Tift 23D2B1-P1-P5 in 2017 (Varshney et al. 2017).

The genome size of pearl millet is estimated to be 1.76 Gb, out of which approximately 90% of genome, i.e. 1.58 Gb genome, was reconstructed from 1.49 Tb sequencing data. The assembled genome contained 77.2% repeat elements. Gene annotations identified 38 579 protein-coding genes. The gene family study across the species showed 15 887 gene families common between pearl millet and foxtail millet, while pearl millet and barley had 15 887 common gene families in between them. Interestingly, wax biosynthesis genes and trans-membrane transporters of secondary metabolites are one of those mostly expanded gene families which may contribute to heat and drought tolerance exhibited by pearl millet. Resequencing (with 1.68× coverage) of 994 lines was done to understand the population structure, genetic diversity, and journey of domestication. The resequencing population included 963 inbred lines and 31 wild accessions. Results indicated the central part of West Africa as an origin of pearl millet. About 88 256 SSR and 29 542 173 SNPs were identified between Pearl Millet Inbred Germplasm Association Panel lines and hybrid parental lines. Gene families related to the biosynthesis of terpenoid biosynthesis are found expanded in the genome, which may impart a higher degree of heat and drought tolerance in pearl millet (Varshney et al. 2017).

### 12.5.2 Broomcorn Millet

Broomcorn millet (*P. miliaceum* L.) is the earliest domesticated crop, highly nutritious with high contents of protein, minerals, and antioxidants. It is also considered to have the highest water use efficiency (Lu et al. 2009; Washburn et al. 2015). Broomcorn millet, also known as proso millet, is an allotetraploid with an estimated genome size of approximately 923 Mb. The genome is sequenced using 132 individual plants from the F6 population of RILs (Zou et al. 2019).

Out of an estimated 923 Mb genome, 855 Mb genome was sequenced. The genome contains 58.2% repeat elements. The majority of this repeated content is composed of TEs (92.1%) from Gypsy and Copia superfamilies. The gene annotation study identified 55 930 protein-coding genes, 339 microRNA genes, 1420 transfer RNAs, 1640 ribosomal RNAs, and 2302 small nuclear RNAs. Phylogenetic analysis was done to understand the evolutionary history of the broomcorn millet. A close relationship between *Setaria italica*, *Zea mays*, *S. bicolor*, and *P. miliaceum* was confirmed by 511 single-copy orthologous genes. The gene family study classified 47 142 genes to 20 374 gene families. Out of these families, 4.2% of genes were specific to broomcorn millet, while 74.5% of gene families were common with grasses foxtail millet, maize sorghum, and rice. From the data, 221 787 SNPs and 112 158 SSR were identified. The study also identified the presence of multiple decarboxylation mechanisms which may have made broomcorn competent in dynamic and fluctuating environmental conditions (Zou et al. 2019).

### 12.5.3 Finger Millet

Finger millet (*E. coracana* (L.) Gaertn.) is a self-pollinating, drought-resistant crop grown in the semiarid tropics of Asia and Africa. It is considered the best source of minerals including calcium, iron, zinc, and manganese. It also provides enough amount of vitamins and amino acids (Chandra et al. 2016). WGS of two genotypes of finger millet is reported. Genotype ML-365 is sequenced in 2017, and sequence assembly of another genotype PR-202 (IC: 479099) is released in 2018 (Hittalmani et al. 2017; Hatakeyama et al. 2018).

In genotype ML-365, out of an estimated 1453 Mb genome size, 1196 Mb (82%) genome was sequenced. The genome contains 49.92% repetitive elements. Gene annotation study identified 85 243 genes in the genome; 2866 drought-responsive genes were identified. An abundant number of SSR makers, i.e. 11 40 833, are predicted. Transcriptome study revealed 53 300 unigenes at well-water condition and 100 046 at low-moisture condition. Further, 67 candidate R-genes against blast disease were identified. In genotype PR-202, the estimated genome size is 1500 Mb, out of which 1189 Mb genome is sequenced. Gene annotation identified 62 348 genes. Interestingly, 96.5% of genes are identified as single-copy genes (Hatakeyama et al. 2018). The genome of ML-365 found more collinear with rice and foxtail millet as compared to the other members from the Poaceae species. Based on the KAAS server, a major portion of the genes was related to carbohydrate metabolism; lipid metabolism; and folding, sorting, and degradation pathways. A close evolutionary relationship between the finger millet and other grasses including foxtail millet, sorghum, rice, and maize is marked from the intergenomic collinearity analysis (Hittalmani et al. 2017).

### 12.5.4 Foxtail Millet

Foxtail millet (*S. italica*) is an annual grass domesticated around 8700 years ago. It is cultivated in semiarid areas and is closely related to biofuel grasses such as switchgrass and napier grass. *Setaria* is getting promoted as model grass due to its short life cycle and small size (Hu et al. 2018). In 2012, foxtail millet WGS for two genotypes, i.e. Zhang gu and Yugu1, was completed (Bennetzen et al. 2012; Zhang et al. 2012). In the next year, another important landrace SLX was also resequenced (Bai et al. 2013).

The genome size estimated for Zhang gu is 490 Mb, out of which 423 Mb genome was sequenced, which is almost 86% of the total genome. The repeat region was 46%. In total, 38 801 protein-coding genes were annotated, out of which more than 80% of genes were found expressed in transcriptome study. In total, 542 322 SNPs and 33 587 insertion/deletions were observed in between A2 and Zhang gu (Zhang et al. 2012). A close evolutionary relationship between foxtail millet and other grasses (rice, sorghum, maize, and *Brachypodium*) was revealed based on highly conserved colinearity between them (Bennetzen et al. 2012; Zhang et al. 2012). Resequencing in foxtail millet was done on RIL derived from a cross between Zhang gu and A2. It resulted in the identification of 59 QTLs for 14 agronomical traits (Zhang et al. 2017). Out of the 510 Mb estimated genome, 396.7 Mb genome was sequenced from Yugu1. The genome was found to be composed of 40% TEs. Around 24 000–29 000 genes were annotated. It included six drought-responsive genes when compared with the genomes of drought-susceptible rice and maize. In total, 992 SNP markers were identified. Extensive synteny was observed between *Setaria*, sorghum, and rice genomes. Foxtail millet was found to possess gene families related to stress response. This includes a family of 586 genes annotated to trait "response to water." The study also proposed that *Ft_CA1* gene is important for C4 photosynthesis pathway in foxtail millet (Zhang et al. 2012).

In foxtail millet, 762 082 SNPs, 26 802 indel polymorphisms, and 10 109 structural variations between SLX and Yugu1 genomes were observed, while 915 434 SNPs, 28 546 indels, and 12 968 were identified between SLX and Zhang gu genomes. Interestingly, when compared with Yugu1 genome, 40% of SNPs were detected inside NB-ARC domain containing genes, protein kinase, or leucine-rich repeats. This finding suggests the role of pathogen pressure in the diversification of plant disease-resistance proteins. SNPs in SLX were validated with more than 90% accuracy and hence can be utilized as SNP markers with greater confidence (Bai et al. 2013).

### 12.5.5 Sorghum

*S. bicolor* is a staple food for nearly half a billion people. It requires low agricultural inputs, and apart from human consumption, it is widely used as animal fodder. It is also considered as a potential biofuel crop. The assembly of the sorghum genome was published by Paterson et al. (2009). They used shotgun sequencing approach and Sanger sequencing chemistry. In 2013, sorghum genome was resequenced. In total, 44 sorghum genotypes were used to sequence the genome. It included landraces, improved inbreds, and wild and weedy sorghum lines. Paired-end sequencing libraries of size 500 bp were sequenced on the HiSeq 2000 platform (Mace et al. 2013). WGS resulted in the assembly of 730 Mb, which contains 55% retrotransposons. Out of the 34 496 sorghum gene models, 27 640 protein-coding genes were predicted using *ab initio* gene prediction methods. Orthologues for 94% sorghum genes were identified in rice, Arabidopsis, and poplar. For defense-related traits, 211 nucleotide-binding-site–leucine-rich-repeat (NBS-LRR) coding genes were identified. In total, 71 000 SSR were identified (Paterson et al., 2009).

Resequencing yielded 7.97 billion 90-bp paired-end reads compromising 717 Gb high-quality sequencing raw data. A large number of SNPs, i.e. 4.9 million, were identified. In total, 1 982 971 indel mutations were identified. Resequencing demonstrated the diversity in wild lines. It also underlines the importance of wild sorghum germplasm for sorghum-domesticated genotype improvement (Mace et al. 2013). The details of WGS efforts in millets are summarized in Table 12.3. Table 12.4 highlights the statistics of millet genome assemblies.

**Table 12.3** Details of efforts performed for whole-genome resequencing and reference genome in millet crops.

| Crop | Type of germplasm/ population | Number of lines used | Sequencing depth | Sequencing platform | Number of SNPs | References |
|---|---|---|---|---|---|---|
| Pearl millet | Cultivar | 994 | 296× | Illumina HiSeq 2000 | 29 542 173 | Varshney et al. (2017) |
| Broomcorn millet | Cultivar | 136 | 87× | PacBio RS II | 221 787 | Zou et al. (2019) |
| Finger millet | Cultivar ML-365 | | | HiSeq 4000 and NextSeq 500 | | Hittalmani et al. (2017) |
| Finger millet | Cultivar PR-202 | | 96.5× | NextSeq 500, PacBio RS II system, and MiSeq | | Hatakeyama et al. (2018) |
| Foxtail millet | Cultivar Zhang gu, A2 | | A2-10× | Illumina Genome Analyzer II and HiSeq 2000 | A2-542 322 | Zhang et al. (2012) |
| Foxtail millet | Cultivar Li-Xiang | | 11× | Illumina Genome Analyzer II | 915 434 SNPs with respect to Zhang gu, 762 082 SNPs with respect to Yugu1 | Bai et al. (2013) |
| Sorghum | Cultivar BTx623 | | 8.5× | ABI-3730 | | Paterson et al. (2009) |
| Sorghum | Landraces, improved inbreds, wild, and weedy | 44 Lines (18 landraces, 17 improved inbreds, and 7 wild + weedy) | 16–45× | Illumina HiSeq 2000 | 4 946 038 | Mace et al. (2013) |

**Table 12.4** Summarized millet genome assembly statistics.

| Millet | Number of scaffold | N50 scaffold (bp) | GC content (%) | Protein-coding gene | Repetitive sequence (%) | References |
|---|---|---|---|---|---|---|
| Pearl millet | 25241 | 884945 | 47.9 | 38579 | 77.2 | Varshney et al. (2017) |
| Broomcorn millet | 1309 | 46662 | 46.8 | 55930 | 58.20 | Zou et al. (2019) |
| Finger millet: ML-365 | 525759 | 23732 | 44.7 | 85243 | 49.92 | Hittalmani et al. (2017) |
| Finger millet: PR-202 | 2387 | 905318 | 43.0 | 62348 | 49.2 | Hatakeyama et al. (2018) |
| Foxtail millet: Zhang gu | 336 | 4667 | 46.1 | 40599 | 46.3 | Zhang et al. (2012) |

## 12.6 GBS-based GWAS Efforts in Millet Crops

WGS data have brought significant changes in population genetics studies. GBS is one of the WGS-based genotyping approaches. It emerged as a promising tool by reducing the required time and cost of the analysis with high confidence output. It has widened the options available to geneticists including GWAS and GS. This section focuses GBS-based GWAS with respect to millets.

A set of genotypes showing phenotypic variation for a particular trait should possess a polymorphic genome loci linked with the observed variation. GWAS uses WGS to analyze the association between phenotype and genome. GWAS screens the complete genome to identify all variable genomic factors (such as SNPs) responsible for (or associated with) the target genotypic trait. Hence, GWAS is considered as noncandidate-driven approach (as it does not depend only on specific candidate loci). Many of the agronomical phenotypic traits are polygenic, i.e. controlled by more than one gene/loci. The power of GWAS is specifically evident in such traits. There are two possibilities about the inheritance of these multiple genomic factors controlling a single trait: they inherit either together or independently. LD represents the possibility of coinheritance of these multiple independent loci. GWAS uses LD value to predict the association between the trait and the genetic variants such as SNPs.

The major advantage of GWAS is that it does not require any parental cross, which reduces the time required. At a time, a large number of contrasting accessions and phenotypic variations can be studied in a single experiment. On the other hand, GWAS is associated with few limitations such as it requires large size population, low allele frequency found to affect the output, and possibility of exclusion of rare allele (rare allele may contribute to natural variations). The GWAS study requires a population of appropriate size with phenotypic variations for the target trait. The size of the population plays a crucial role to correct the rare allele effect. Any abnormal phenotypic data must be eliminated using tools such as boxplot analysis. Genotyping is done using WGS data and SNP arrays. Quality check for SNP data is important and is done based on factors including missing SNP, MAF, heterozygosity rate, and relatedness. For geno- and phenotype association, the models such as general linear model (GLM) and mixed linear model (MLM) are generally used. Validation of the analysis is done by the false discovery rate (FDR) or Bonferroni correction (BC). Various software including TASSEL, GenSTAT, and PLINK are available to perform GWAS. Manhattan plot is most frequently used to present the data.

In foxtail millet, GWAS was carried out to identify markers associated with the nutrients such as potassium, phosphorus, calcium, sulfur, and magnesium and macronutrients nickel, boron, iron, zinc, and manganese. The study further conducted the pyramiding effect, suggesting the contribution of multiple marker trait associations (MTAs) for a single nutrient trait (Jaiswal et al. 2019a). Another GWAS study was conducted to understand the association between 10 important agronomical traits (including days to flowering, plant height, tiller number, FLL, FLW, peduncle length, panicle length, tiller maturity, grain yield, and grain weight) and 10 367 SNPs. The study identified 81 MTAs for the above mentioned traits. For three important agronomic traits including grain yield, thousand-grain weight, and FLW, seven high confidence level MTAs were identified (Jaiswal et al. 2019b). In another high-throughput study conducted by Jia et al. (2013), 916 foxtail varieties were sequenced. A haplotype map was created using 2.58 million SNPs. Five hundred and twelve associated loci were identified for 47 agronomic traits including flowering time, plant height, and inflorescence architecture. The study also identified inter- and intraspecies variants by sequencing *Setaria viridis*, which is a wild progenitor of *S. italica* (Jia et al. 2013). Upadhyaya et al. (2015) genotyped 181 accessions of foxtail millet. They identified high-quality 17 714 SNPs across the genome. The study concluded with the identification of eight MTAs for pigmentation and seven for days to flower. The study included diversified genotypes having variations for biotic and abiotic stress tolerance (Upadhyaya et al. 2015). In pearl millet, the GWAS study was coupled with WGS and resequencing project. GWAS was conducted across 288 testcross progenies of PMiGAP lines for 20 traits and identified 1054 strongly significant marker trait associations for 15 traits including grain number per panicle, grains per square meter, panicle yield, grain yield, and plant height (Varshney et al. 2017). Genetic association related to drought stress response was studied using 392 493 SNPs. There were 10 SNPs significantly associated with stay-green and biomass phenotypes. In well water conditions, four MTAs were reported for stay-green phenotype, while six MTAs for biomass production in drought conditions (Debieu et al. 2018).

In an experiment, the population structure of finger millet germplasm of diverse origins was studied. Across 113 finger millet accessions, 23 000 SNPs were identified spread across all accessions. Three subpopulations were identified using a model-based program – Markov chain Monte Carlo. The SNP data was significant and phylogenetic analysis (Kumar et al. 2016b). A set of 385 945 heterozygous and 2 999 053 homozygous SNPs from this experiment was used to analyze its association with the traits related to grain yield. Three models were used to study the association. The single locus single trait (SLST) model identified 20 MTAs for five traits. Advanced models such as multilocus mixed model (MLMM) and multitrait mixed model (MTMM) identified more associations, i.e. 36 and 53 MTAs, respectively (Sharma et al. 2018). Further, using the same SNP set, 7 seed protein content (SPC), 10 days of maturity (DM), and 9 grain maturity (GY) MTAs were identified using the GLM model. MLM model produced 4 SPC, 9 DM, and 7 GY MTAs. This study further proposed aspartyl protease as a candidate gene for SPC (Tiwari et al. 2020).

With the availability of reference genome sequence for sorghum, large number of GBS-based GWAS studies have been undertaken in this crop. The 102 SNPs that are significantly associated with plant height, pericarp color, presence or absence of awns, glume cover, panicle exsertion, panicle compactness and shape, smut resistance, and male sterility have been identified in sorghum. The association with 61 candidate genes for different traits has been uncovered in the study (Girma et al. 2019). GWAS was used to find out MTA for plant height components and inflorescence architecture (Morris et al. 2013). Another interesting study came up with the finding that seed mass is influenced by drought stress and identified the genetic signature for this association (Wang et al. 2020). Using a public database for whole genome and SNP database, GWAS identified MTAs for biotic stress resistance in sorghum mini core lines against anthracnose, downy mildew, and head smut (Ahn et al. 2019) (Table 12.5).

## 12.7 GBS-based Genomic Selection (GS) Efforts in Millet Crops

Genomic selection is a method to predict the superior line from the testing pool based on its genetic value derived from genome-wide markers data. The method was suggested by Meuwissen et al. (2001) (Yabe and Iwata 2020). Although traditional MAS methods have played important role in crop improvement, they struggle in some scenarios, especially for polygenic traits, traits involving epistasis, and traits influenced by the environment. Genomic selection provides a promising alternative to the traditional approaches of MAS with respect to their aforesaid limitations. GS considers complete marker data across the genome, which eliminates the limitations aroused due to consideration of only the specific loci. GS counts the effect of all markers simultaneously, which includes minor or small gene effects otherwise which get escaped in traditional MAS studies.

GS employs training population (TP) for which both phenotyping and genotyping are performed. Another population for which only genotyping is done is called as breeding population (BP). High-density markers are required for the success of GS. From the geno- and phenotype data of TP, "marker effects" of all markers are considered without applying any threshold parameters to generate a statistical model (Robertsen et al. 2019). To this model, genotypic data from BP are applied to calculate the genome-estimated breeding values (GEBV) for the BP. GEBV does not express the function of loci, but it acts as a prediction criterion that can be applied for breeding. Cross-validation (CV) is done to calculate the productive ability of the model. For this, a subset from TP is used as a validation set to predict the accuracy of the model. Once the model passed through CV, it is applied to BP (Shamshad and Sharma 2018). The general scheme for the GS is presented in Figure 12.3.

In pearl millet, the GS model was implemented to study the grain yield trait. In this experiment, genotyping was conducted through two different platforms i.e. conventional GBS, RAD-seq, and tunel GBS (tGBS). tGBS is a modification of RAD-seq and overcomes the problem RAD-seq faces at heterozygous sites. The study validated the prediction model using three different CV approaches to mimic the actual field conditions and concluded that the addition of phenotypic data of parental environment and inclusion of genotype–environment interaction enhance the hybrid prediction accuracy of the GS model (Jarquin et al. 2019). Another GS study on pearl millet evaluated the two genotyping techniques, i.e. RAD-seq and tGBS. The number of SNPs generated by RAD-seq was higher than tGBS. But tGBS data were

**Table 12.5** The list of GBS-based GWAS studies in millets.

| Crop | Trait | Number of genotypes analyzed | Technique used/platform | Most significantly associated SNP | Salient feature of the study | References |
|---|---|---|---|---|---|---|
| Foxtail millet | Micronutrients (potassium, phosphorus, calcium, sulfur, and magnesium) and macronutrients (nickel, boron, iron, zinc, and manganese) | 93 | ddRAD-seq; Illumina; HiSeq 4000 | 10 SNPs associated with B, Mg, Zn, and Fe | Pyramiding effect identified multiple MTA contributing to single trait | Jaiswal et al. (2019a) |
| Foxtail millet | Agronomical traits (days to flowering, plant height, tiller number, flag leaf length, flag leaf width, peduncle length, panicle length, tiller maturity, grain yield, and grain weight) | 142 | ddRAD-seq; Illumina HiSeq 4000 | 7 SNPs associated with flag leaf width, grain weight, and 1000 grain weight | Stringent Bonferroni correction overcome the small size of testing population | Jaiswal et al. (2019b) |
| Foxtail millet | 47 Agronomic traits (including flowering time, plant height, and inflorescence architecture) | 916 | RAD-seq; Illumina Genome Analyzer IIx and Illumina HiSeq 2000 | 512 Loci associated with 47 agronomic traits | Identified inter- and intraspecies variants, impact of environmental factors on marker – trait association observed | Jia et al. (2013) |
| Foxtail millet | Pigmentation and days to 50% flowering | 181 | RAD-seq; HiSeq 2000 | 8 SNPs for pigmentation and 7 SNPs for days to 50% flowering | Study conducted on population including core collection having wide genetic variation for resistance to blast, salinity tolerance, agronomic, and grain nutritional traits | Upadhyaya et al. (2015) |
| Pearl millet | 15 agronomical traits related to grain and stover yield | 288 | RAD-seq; HiSeq 2000 | 1054 SNPs associated with 15 traits | Sequencing reference genome and large-scale resequencing | Varshney et al. (2017) |

| | | | | | | |
|---|---|---|---|---|---|---|
| Pearl millet | Drought stress and biomass production | 175 | RAD-seq; HiSeq 2500 and NextSeq 500 | Six SNPs associated with stay green phenotype and four SNPs associated with biomass production | Proposed potential mechanism for drought response and stay green phenotype | Debieu et al. (2018) |
| Finger millet | 14 Agromorphological traits | 113 | RAD-seq; HiSeq 2500 | For five traits: SLST-20 MTAs MLMM-36 MTAs<br>MTMM-53 MTAs | Study simultaneously applied three models for association analysis | Kumar et al. (2016b) and Sharma et al. (2018) |
| Finger millet | Seed protein content and days of maturity and grain maturity | 113 | RAD-seq; HiSeq 2500 | GLM model:<br>SPC-7 MTA<br>DM-10 MTA<br>GY-9 MTA<br>GLM model:<br>SPC-4 MTA<br>DM-9 MTA<br>GY-7 MTA | | Kumar et al. (2016b) and Tiwari et al. (2020) |
| Sorghum | Awns, glume cover, pericarp color, panicle compactness and shape, panicle exsertion, smut resistance, and male sterility | 1425 | RAD-seq; HiSeq 2500 | 102 SNPs | Study conducted on field-based pheno- and genotyping of a large and diverse germplasm | Girma et al. (2019) |
| Sorghum | Plant height components and inflorescence architecture | 336 | RAD-seq; Illumina Genome Analyzer *IIx*/HiSeq 2000 | Inflorescencearchitecture-22 SNPs | Study mapped loci for plant height and candidate genes for inflorescence architecture | Morris et al. (2013) |
| Sorghum | Seed mass adaptation to precipitation gradients | 336 | Public genotyping-by-sequencing (GBS) data – https://doi.org/10.5061/dryad.jc3ht | 13 SNPs | | Wang et al. (2020) |

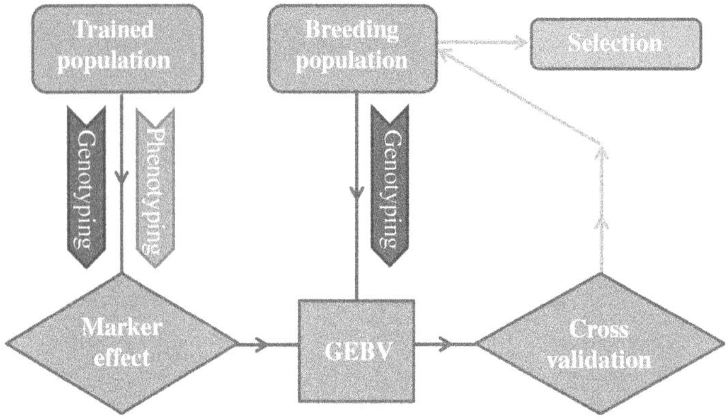


**Figure 12.3** General steps involved in genomic selection.

more efficient as they produced more numbers of MAF. The study also concluded that inbred line data can reduce the accuracy of the breeding predictions for traits with significant heterosis. With the availability of the genome sequence, the researchers undertake similar GS studies in different millets to reap the benefits of the advanced breeding technology.

## 12.8 Summary

Millets provide a sustainable solution to the enhanced food demand of the growing human population. Recently, the nutritional properties of millets popularized them among the larger population. Also, millets are the big source of fodder for domesticated animals. Most importantly, millets offer these benefits at very minimum agricultural inputs. On the contrary, limitations such as low yield have restricted the large-scale cultivation of millet crops. Since the development of whole-genome sequencing technologies, millet genomics has fascinated researchers. The WGS and WGRS outcomes of foxtail millets, finger millets, pearl millets, and broomcorn millets are made available at different databases for further genomic studies. High-density linkage maps are being generated using GBS and GWAS for millets. These tools are promising, especially in the identification of genetic loci governing the polygenic traits in millet genomes with higher ploidy levels. Millets occupy a key position in plant evolution; hence, their genomic studies with modern techniques are going to benefit the entire grass family. Psuedo-millets and small millets are yet to get the advantages of GBS, GWAS, and GS. Thus, the genomic resources available in millets will enrich the current understanding of the phenotype–genotype correlation and set a new paradigm for millet breeding programs.

## References

Ahn, E., Hu, Z., Perumal, R. et al. (2019). Genome wide association analysis of sorghum mini core lines regarding anthracnose, downy mildew, and head smut. *PLoS One* **14** (5).

Araus, J.L., Kefauver, S.C., Zaman-Allah, M. et al. (2018). Translating high-throughput phenotyping into genetic gain. *Trends in Plant Science* **23** (5): 451–466.

Bai, H., Cao, Y., Quan, J. et al. (2013). Identifying the genome-wide sequence variations and developing new molecular markers for genetics research by re-sequencing a landrace cultivar of foxtail millet. *PLoS One* **8** (9): e73514.

Basavaraj, G., Rao, P.P., Bhagavatula, S., and Ahmed, W. (2010). Availability and utilization of pearl millet in India. *SAT eJournal* **8**.

Bennetzen, J.L., Schmutz, J., Wang, H. et al. (2012). Reference genome sequence of the model plant Setaria. *Nature Biotechnology* **30** (6): 555.

Berthouly-Salazar, C., Thuillet, A.C., Rhoné, B. et al. (2016). Genome scan reveals selection acting on genes linked to stress response in wild pearl millet. *Molecular Ecology* **25** (21): 5500–5512.

Brenton, Z.W., Cooper, E.A., Myers, M.T. et al. (2016). A genomic resource for the development, improvement, and exploitation of sorghum for bioenergy. *Genetics* **204** (1): 21–33.

Chandra, D., Chandra, S., and Sharma, A.K. (2016). Review of finger millet (*Eleusine coracana* (L.) Gaertn): a power house of health benefiting nutrients. *Food Science and Human Wellness* **5** (3): 149–155.

Chung, Y.S., Choi, S.C., Jun, T.H., and Kim, C. (2017). Genotyping-by-sequencing: a promising tool for plant genetics research and breeding. *Horticulture, Environment and Biotechnology* **58** (5): 425–431.

Debieu, M., Id, B.S., Passot, S. et al. (2018). Response to early drought stress and identification of QTLs controlling biomass production under drought in pearl millet. *PLoS One* **13** (10).

Elshire, R.J., Glaubitz, J.C., Sun, Q. et al. (2011). A robust, simple genotyping-by-sequencing (GBS) approach for high diversity species. *PLoS One* **6** (5).

Etesami, H. and Jeong, B.R. (2018). Silicon (Si): review and future prospects on the action mechanisms in alleviating biotic and abiotic stresses in plants. *Ecotoxicology and Environmental Safety* **147** (2017): 881–896.

Fang, X., Dong, K., Wang, X. et al. (2016). A high density genetic map and QTL for agronomic and yield traits in foxtail millet [*Setaria italica* (L.) *P. Beauv.*]. *BMC Genomics* **17** (1): 336.

Gimode, D., Odeny, D.A., De Villiers, E.P. et al. (2016). Identification of SNP and SSR markers in finger millet using next generation sequencing technologies. *PLoS One* **11** (7): 1–23.

Girma, G., Nida, H., Seyoum, A. et al. (2019). A large-scale genome-wide association analyses of Ethiopian sorghum landrace collection reveal loci associated with important traits. *Frontiers in Plant Science* **10**: 691.

Goff, S.A., Ricke, D., Lan, T. et al. (2002). A draft sequence of the rice genome (*Oryza sativa* L. ssp. japonica). *Science* **296** (5565): 92–100.

Gómez, G., Álvarez, M.F., and Mosquera, T. (2011). Association mapping, a method to detect quantitative trait loci: statistical bases. *Agron Colombia* **29** (3): 367–376.

Gupta, P.K., Rustgi, S., and Kulwal, P.L. (2005). Linkage disequilibrium and association studies in higher plants: present status and future prospects. *Plant Molecular Biology* **57** (4): 461–485.

Gupta, S.M., Arora, S., Mirza, N. et al. (2017). Finger millet: a “certain” crop for an “uncertain” future and a solution to food insecurity and hidden hunger under stressful environments. *Frontiers in Plant Science* **8**: 643.

Hatakeyama, M., Aluri, S., Balachadran, M.T. et al. (2018). Multiple hybrid de novo genome assembly of finger millet, an orphan allotetraploid crop. *DNA Research* **25** (1): 39–47.

Hittalmani, S., Mahesh, H.B., Shirke, M.D. et al. (2017). Genome and transcriptome sequence of finger millet (*Eleusine coracana* (L.) Gaertn.) provides insights into drought tolerance and nutraceutical properties. *BMC Genomics* 18 (1): 465.

Hu, Z., Mbacké, B., Perumal, R. et al. (2015). Population genomics of pearl millet (*Pennisetum glaucum* (L.) R. Br.): comparative analysis of global accessions and *Senegalese landraces*. *BMC Genomics* **16** (1): 12.

Hu, H., Mauro-herrera, M., Doust, A.N., and Nelson, M.N. (2018). Domestication and improvement in the model C4 grass, Setaria. *Frontiers in Plant Science* **9**: 719.

Jaiswal, V., Bandyopadhyay, T., Gahlaut, V. et al. (2019a). Genome-wide association study (GWAS) delineates genomic loci for ten nutritional elements in foxtail millet (*Setaria italica* L.). *Journal of Cereal Science* **85**: 48–55.

Jaiswal, V., Gupta, S., Gahlaut, V., and Muthamilarasan, M. (2019b). Genome-wide association study of major agronomic traits in foxtail millet (*Setaria italica* L.) using ddRAD sequencing. *Scientific Reports* 9 (1): 1–11.

Jarquin, D., Howard, R., Liang, Z. et al. (2019). Enhancing hybrid prediction in pearl millet using genomic and/or multi-environment phenotypic information of inbreds. *Frontiers in Genetics* **10**: 1294.

Ji, G., Zhang, Q., Du, R. et al. (2015). Construction of a high-density genetic map using specific length amplified fragment markers and identification of a quantitative trait locus for anthracnose resistance in walnut (*Juglans regia* L.). *BMC Genomics* 16 (1): 614.

Guisu Ji, Qingjiang Zhang, Ruiheng Du, et al. (2017) “Construction of a high-density genetic map using specific-locus amplified fragments in sorghum.” *BMC Genomics* 18.1 (2017): 1–10.

Jia, G., Huang, X., Zhi, H. et al. (2013). A haplotype map of genomic variations and genome-wide association studies of agronomic traits in foxtail millet (*Setaria italica*). *Nature Genetics* **45** (8): 957–961.

Kong, W.Q., Kim, C., Zhang, D. et al. (2018). Genotyping by sequencing of 393 *Sorghum bicolor* BTx623× IS3620C recombinant inbred lines improves sensitivity and resolution of QTL detection. *G3 Genes, Genomes, Genetics* **8** (8): 2563–2572.

Kulwal, P.L., Thudi, M., and Rajeev, K. (2011). *Genomics Interventions in Crop Breeding for Sustainable Agriculture*. Springer.

Kumar, A., Gaur, V.S., Goel, A., and Gupta, A.K. (2015). De novo assembly and characterization of developing spikes transcriptome of finger millet (*Eleusine coracana*): a minor crop having nutraceutical properties. *Plant Molecular Biology Reporter* **33** (4): 905–922.

Kumar, A., Metwal, M., Kaur, S. et al. (2016a). Nutraceutical value of finger millet [*Eleusine coracana* (L.) Gaertn.], and their improvement using omics approaches. *Frontiers in Plant Science* 7: 934.

Kumar, A., Sharma, D., Tiwari, A. et al. (2016b). Genotyping-by-sequencing analysis for determining population structure of finger millet germplasm of diverse origins. *Plant Genome* **9** (2) https://doi.org/10.3835/plantgenome2015.07.0058.

Kumar, A., Tomer, V., Kaur, A. et al. (2018). Millets: a solution to agrarian and nutritional challenges. *Agricultural Food Security* 7 (1): 31.

Kyriakidou, M., Tai, H.H., Anglin, N.L., Ellis, D., and Strömvik, M. V (2018) Current strategies of polyploid plant genome sequence assembly. *Frontiers in Plant Science*, **9**, 1660.

Lasky, J.R., Upadhyaya, H.D., Ramu, P. et al. (2015). Genome-environment associations in sorghum landraces predict adaptive traits. *Science Advances* **1** (6): e1400218.

Lata, C. (2015). Advances in omics for enhancing abiotic stress tolerance in millets. *Proceedings of the Indian National Science Academy* **81**: 397–417.

Lata, C., Gupta, S., and Prasad, M. (2013). Foxtail millet: a model crop for genetic and genomic studies in bioenergy grasses. *Critical Reviews in Biotechnology* **33** (3): 328–343.

Liang, Z., Gupta, S.K., Yeh, C. et al. (2018). Phenotypic data from inbred parents can improve genomic prediction in pearl millet hybrids. *G3 Genes, Genomes Genetics* **8** (7): 2513–2522.

Liu, D., Ma, C., Hong, W. et al. (2014). Construction and analysis of high-density linkage map using high-throughput sequencing data. *PLoS One* **9** (6): e98855.

Liu, T., He, J., Dong, K. et al. (2020). QTL mapping of yield component traits on bin map generated from resequencing a RIL population of foxtail millet (*Setaria italica*). *BMC Genomics* **21** (1): 141.

Lu, H., Zhang, J., Liu, K. et al. (2009). Earliest domestication of common millet (*Panicum miliaceum*) in East Asia extended to 10,000 years ago. *Proceedings of the National Academy of Sciences of the United States of America* **106** (18): 7367–7372.

Mace, E.S., Tai, S., Gilding, E.K. et al. (2013). Whole-genome sequencing reveals untapped genetic potential in Africa's indigenous cereal crop sorghum. *Nature Communications* **4** (1): 1–9.

Meuwissen, T.H.E., Hayes, B.J., and Goddard, M.E. (2001, 2001). Prediction of total genetic value using genome-wide dense marker maps. *Genetics* 157: 1819–1829.

Miller, M.R., Atwood, T.S., Eames, B.F. et al. (2007). RAD marker microarrays enable rapid mapping of zebrafish mutations. *Genome Biology* 8 (6): R105.

Morris, G.P., Ramu, P., Deshpande, S.P. et al. (2013). Population genomic and genome-wide association studies of agroclimatic traits in sorghum. *Proceedings of the National Academy of Sciences of the United States of America* **110** (2): 453–458.

Moumouni, K.H., Kountche, B.A., Jean, M. et al. (2015). Construction of a genetic map for pearl millet, *Pennisetum glaucum* (L.) R. Br., using a genotyping-by-sequencing (GBS) approach. *Molecular Breeding* 35 (1): 5.

Muthamilarasan, M., Venkata Suresh, B., Pandey, G. et al. (2014). Development of 5123 intron-length polymorphic markers for large-scale genotyping applications in foxtail millet. *DNA Research* **21** (1): 41–52.

Ott, A., Liu, S., Schnable, J.C. et al. (2017). tGBS® genotyping-by-sequencing enables reliable genotyping of heterozygous loci. *Nucleic Acids Research* **45** (21): e178–e178.

Pandey, G., Misra, G., Kumari, K. et al. (2013). Genome-wide development and use of microsatellite markers for large-scale genotyping applications in foxtail millet. *DNA Research* **20** (2): 197–207.

Paterson, A.H., Bowers, J.E., Dubchak, I. et al. (2009). The *Sorghum bicolor* genome and the diversification of grasses. *Nature* **457** (7229): 551–556.

Pucher, A., Hash, C.T., Wallace, J.G. et al. (2018). Mapping a male-fertility restoration locus for the A 4 cytoplasmic-genic male-sterility system in pearl millet using a genotyping-by-sequencing-based linkage map. *BMC Plant Biology* 18 (1): 65.

Punnuri, S.M., Wallace, J.G., Knoll, J.E. et al. (2016). Development of a high-density linkage map and tagging leaf spot resistance in pearl millet using genotyping-by-sequencing markers. *Plant Genome* **9** (2).

Qi, P., Gimode, D., Saha, D. et al. (2018). UGbS-Flex, a novel bioinformatics pipeline for imputation-free SNP discovery in polyploids without a reference genome: finger millet as a case study. *BMC Plant Biology* 18 (1): 117.

Robertsen, C.D., Hjortshøj, R.L., and Janss, L.L. (2019). Genomic selection in cereal breeding. *Agronomy* **9** (2): 95.

Shamshad, M. and Sharma, A. (2018). The usage of genomic selection strategy in plant breeding. *Next Generation Plant Breeding* 26: 93–108.

Sharma, D., Tiwari, A., Sood, S. et al. (2018). Genome wide association mapping of agro-morphological traits among a diverse collection of finger millet (*Eleusine coracana* L.) genotypes using SNP markers. *PLoS One* 13 (8): e0199444.

Singh, P. and Raghuvanshi, R. (2012). Finger millet for food and nutritional security. *African Journal of Food Science* **6** (4): 77–84.

Tiwari, A., Sharma, D., Sood, S. et al. (2020). Genome-wide association mapping for seed protein content in finger millet (*Eleusine coracana*) global collection through genotyping by sequencing. *Journal of Cereal Science* **91**: 102888.

Upadhyaya, H.D., Vetriventhan, M., Deshpande, S.P. et al. (2015). Population genetics and structure of a global foxtail millet germplasm collection. *Plant Genome* **8** (3): 1–13.

Varshney, R.K., Shi, C., Thudi, M. et al. (2017). Pearl millet genome sequence provides a resource to improve agronomic traits in arid environments. *Natural Public Gr* **35** (10): 969.

Wang, J., Wang, Z., Du, X. et al. (2017). A high-density genetic map and QTL analysis of agronomic traits in foxtail millet [*Setaria italica* (L.) P. Beauv.] using RAD-seq. *PLoS One* **12**: 6.

Wang, J., Hu, Z., Upadhyaya, H.D. et al. (2020). Genomic signatures of seed mass adaptation to global precipitation gradients in sorghum. *Heredity (Edinb).* **124** (1): 108–121.

Washburn, J.D., Schnable, J.C., Davidse, G. et al. (2015). Phylogeny and photosynthesis of the grass tribe Paniceae. *American Journal of Botany* **102** (9): 1493–1505.

Yabe, S. and Iwata, H. (2020). Genomics-assisted breeding in minor and pseudo-cereals. *Breeding Science* **70** (1): 19–31.

Young, N.D. (2001). *Constructing a Plant Genetic Linkage Map with DNA Markers*, 31–Dordrecht, 47. Springer.

Yu, J., Hu, S., Wang, J. et al. (2002). A draft sequence of the rice genome (*Oryza sativa* L. ssp. indica). *Science* 296 (5565): 79–92.

Yu, X., Li, X., Guo, T. et al. (2016). Genomic prediction contributing to a promising global strategy to turbocharge gene banks. *Nature Plants* **2** (10): 1–7.

Zhang, G., Liu, X., Quan, Z. et al. (2012). Genome sequence of foxtail millet (*Setaria italica*) provides insights into grass evolution and biofuel potential. *Nature Biotechnology* **30** (6): 549–554.

Zhang, K., Fan, G., Zhang, X. et al. (2017). Identification of QTLs for 14 agronomically important traits in *Setaria italica* based on SNPs generated from high-throughput sequencing. *G3* **7** (5): 1587–1594.

Zou, C., Li, L., Miki, D. et al. (2019). The genome of broomcorn millet. *Nature Communications* **10** (1): 1–11.

# 13

# Recent Advances and Applicability of GBS, GWAS, and GS in Pigeon Pea

*Anuradha Singh and Nisha Singh*

*Department of Genomics, ICAR – National Institute on Plant Biotechnology, New Delhi, India*

## 13.1 Introduction

Pigeon pea (*Cajanus cajan* L. Millsp.) is one of the major legumes. It is a crop of great nutritional value with a protein content of 20–23%. It serves as an important dietary protein source for people living in Asia. Its leaves, seed husks, and pod husks are used for animal food. Pigeon pea is also used as green manure as fallen leaves from the plant provide vital nutrients to the soil, the standing crop in the field also helps to increase soil fertility by fixing atmospheric nitrogen. It is cultivated in many parts of Asia and Africa including India, China, Kenya, Myanmar, and Nepal. Among these countries, India stands first in both areas under cultivation (5.06 Mha) and production (3.29 Mt) of pigeon pea (http://faostat3.fao.org/home/, as of August 2016); however, since from last six decades, no improvement has been made in crop yield and productivity.

Pigeon pea production is affected by various biotic and abiotic stress. In the Asiatic region, it is planted from mid-June to July, during the summer season when the temperature is very high, and also during the monsoon season. During summer, the seedlings cannot withstand heat due to high temperature leading to loss of seedlings affecting the yield. In the monsoon season, the heavy rainfall causes flooding hindering the establishment of the pigeon pea seedlings. Soil salinity is another major abiotic factor that causes osmotic stress leading to physiological and biochemical changes in the plant. In addition to abiotic stresses, the biotic stresses also cause huge yield loss in pigeon pea. Many cultivated varieties are susceptible to sterility mosaic disease (SMD) and Fusarium wilt (FW). SMD is caused due to pigeon pea sterility mosaic virus (PPSMV) that is transmitted through a mite (*Aceria cajani*) and FW is caused by a fungus *Fusarium udum*. Other biotic factors like pod borer (*Helicoverpa armigera*), pod fly (*Melanoagromyza obtusa*), spotted borer (*Maruca*

*Genotyping by Sequencing for Crop Improvement*, First Edition.
Edited by Humira Sonah, Vinod Goyal, S.M. Shivaraj, and Rupesh K. Deshmukh.

*vitrata*) also affect the yield. The innate characters of the crop such as the long duration, tall stature, low harvest index, spreading plant type, photosensitivity, nonsynchronous maturity also hinder their productivity. Additionally, the majority of varieties bear open type of flowers which lead to outcrossing and causes difficulty in maintaining the purity of seeds. Hence, there is an urgent need of rapid breeding to develop varieties that can overcome the above hindering factors to increase pigeon pea production and also can increase the farmer's income.

## 13.2 Pigeon Pea Sequencing and Resequencing

The estimated size of the pigeon pea genome packed in 11 chromosomes is 858 Mbp (Greilhuber and Obermayer 1998). The first draft of pigeon pea genome (popular variety Asha) was sequenced by Singh et al. (2012) under the Pigeon pea Genomics Initiative (PGI) commenced by the Indian Council of Agricultural Research (ICAR) under the Indo-US Agricultural Knowledge Initiative (AKI). Assembly of 511 Mbp of high-quality genome sequence was carried out by using a long sequence reads obtained from 454 GS-FLX chemistry with mean read lengths of >550 bp and more than 10× genome coverage. The short reads were arranged into sequence contigs later into scaffolds. Thus obtained 59 681 scaffolds were subsequently anchored to 11 chromosomes of pigeon pea. Further, the genome-wide prediction showed the presence of 47 004 protein-coding genes and 12 511 genes encoding transposable elements. Among the protein-coding genes, 1213 were disease resistance genes, and 152 were abiotic stress tolerance genes. About 347 genic-single-nucleotide polymorphism (SNP) markers of an intra-species reference genetic map derived from Asha/UPAS120. The linkage map was developed based on two Illumina multiplex SNP assays of 1536-plex and 768-plex SNPs identified by comparing the transcriptome assemblies of the varieties Asha and UPAS 120 (Dutta et al. 2011; Kumawat et al. 2012). In 2012, another draft genome sequence of pigeon pea has been released by Varshney et al. (2012). The comparative genomics analysis between pigeon pea and soybean showed that the pigeon pea genome is having a low but significant synteny with the soybean genome. An attempt to identify SSR markers from the genome sequence revealed large number of hypervariable "Arhar" simple sequence repeat (HASSR) markers which were further validated experimentally in different pigeon pea varieties (Dutta et al. 2011, 2013). In 2017, an improved version of this genome with 648.2 Mbp coverage is the highest available genome coverage of pigeon pea to date (Mahato et al. 2018). In addition to the nuclear genome, efforts have been made to sequence mitochondrial genome by Tuteja et al. (2013) and chloroplast genome by Kaila et al. (2016) in pigeon pea.

The availability of the genome sequences and resequencing data of pigeon pea genotypes together with help of bioinformatics software led to the discoveries of genetic markers like SSR and SNPs. Further, these markers aided the development of high-density genotyping platforms (Table 13.1). Transcriptome sequencing has also been carried out in pigeon pea which can be used for SNP marker discoveries (Dutta et al. 2011; Rathinam et al. 2019; Saxena et al. 2020a, 2020b, 2020c). These genome sequence resources can help us in understanding the pigeon pea origin and diversity.

**Table 13.1** Whole-genome resequencing studies in pigeon pea.

| Number of genotypes used | Number and types of significant SNP markers discovered | References |
|---|---|---|
| Two genotypes (ASHA and UPAS) | 1536 SNP plex and a 768 SNP plex | Dutta et al. (2011), Kumawat et al. (2012), and Arora et al. (2017) |
| 256 | 48 SNP | Roorkiwal et al. (2013) |
| 104 | 51 201 SNPs | Saxena et al. (2018a) |
| 45 | 62 053 SNPs | Singh et al. (2020a, 2020b, 2020c) |

## 13.3 Development of Pigeon Pea High-density Genotyping Platforms

In 2011, SNPs were identified between contings of pigeon pea varieties Asha (NCBI accession no. EZ647865-EZ683068) and UPAS 120 (NCBI accession no. EZ617718-EZ674864) derived from high-quality transcriptome shotgun assembly (TSA) (Singh et al. 2012). The multiplex assays of 1536-plex and 768-plex were developed using Illumina's GoldenGate assay design tool (http://www.illumina.com) These multiplexed assays were used for SNP genotyping by Kumawat et al. (2012) and Arora et al. (2017) in $F_2$ mapping populations. In 2020, 62 K SNP chip array "CcSNPnks" of pigeon pea was designed after filtering 645 662 genic-SNPs identified by the resequencing of 45 diverse genotypes suitable for Affymetrix GeneTitan® platform. This chip comprises of 62 053 SNPs located in 9629 genes which belong to five different categories. These five categories include 4328 single-copy genes conserved between soybean and pigeon pea, 4314 single-copy genes specific to pigeon pea genome, 156 genes characterized for the agronomical importance, 746 disease resistance genes, and 85 multicopy genes of pigeon pea. This completely genic chip is comprised of SNPs from different regions of gene, 28.94% exonic, 33.04% intronic, 27.56% 5′ UTR, and 10.46% 3′UTR. Further, this 62 K SNP chip was used effectively to analyze of genetic diversity and population structure of 95 pigeon pea varieties. Further, this chip has also been used for high-resolution mapping of 11 yield-related quantitative trait loci (QTLs) using a bi-parental RIL population (Pusa Dwarf/H2001-4) (Singh et al. 2020a, 2020b, 2020c). Another pigeon pea 56K Axiom SNP chip array has been developed by Saxena et al. (2018b), which has 70.53% of the SNPs from intergenic regions of the genome, 5166 of the 56 512 SNPs in this chip belong to exonic regions. These recent high-density platforms were designed to make the genotyping more cost-effective per data point basis so it can aid in, gene mapping studies, marker–trait association, rapid breeding, and physiological studies, etc. To date, the 62 K SNP array seems to be the most efficient and cost-effective platform (Table 13.2).

## 13.4 Development of High-density Linkage Maps in Pigeon Pea

The next-generation sequencing (NGS)-based genotyping-by-sequencing (GBS) has been widely used in several crops for genetic diversity studies and in developing highly dense genetic maps. The availability of genomic resources in pigeon pea helped in the

**Table 13.2** List of high-density genotyping platforms developed for pigeon pea.

| High-density genotyping platforms | Number of SNP/index | References |
|---|---|---|
| Illumina BeadXpress, Golden gate array | 1536 SNP plex and a 768 SNP plex | Dutta et al. (2011), Kumawat et al. (2012), and Arora et al. (2017) |
| Illumina BeadXpress, Golden gate assay | 48 SNPs | Roorkiwal et al. (2013) |
| 50 K Axiom *Affymetrix* Gene Titan array | 56 512 SNPs | Saxena et al. (2018b) |
| 62 K SNP Axiom *Affymetrix* Gene Titan array | 62 053 SNPs | Singh et al. (2020a, 2020b, 2020c) |

**Table 13.3** High-density linkage maps generated in pigeon pea by using the recent advances of genotyping.

| S. no. | Mapping populations | Genotyping method | Population type | Average marker interval (cM) | References |
|---|---|---|---|---|---|
| 1 | Pusa Dwarf × HDM04-1 | SSR, Illumina GoldenGate bead-arrays (1536-plex and 768-plex) | F2 | | Kumawat et al. (2012) |
| 2 | Asha × UPAS 120 | SSR, Illumina GoldenGate bead-arrays (1536-plex and 768-plex), GBS-RAD | F2 | 1.39 | Arora et al. (2017) |
| 3 | ICP 8863 × ICPL 87119 | GBS | F2 | 1.6 | Saxena et al. (2017) |
| 4 | ICPL 85063 × ICPL 87119 | GBS | F2 | 2.6 | Saxena et al. (2017) |
| 5 | ICP 5529 × ICP 11605 | GBS | F2 | 0.54 | Saxena et al. (2017) |
| 6 | ICPA 2039 × ICPL 87119 | GBS | F2 | 0.3 | Saxena et al. (2018b) |
| 7 | ICPL 20096 × ICPL 332 | GBS | RIL | 0.84 | Saxena et al. (2017) |
| 8 | ICPL 20097 × ICP 8863 | GBS | RIL | 1.65 | Saxena et al. (2017) |
| 9 | ICPB 2049 × ICPL 99050 | GBS | RIL | 1.16 | Saxena et al. (2017) |
| 10 | ICPL 99010 × ICP 5529 | 50 K Axiom SNP array | RIL | 0.1 | Yadav et al. (2019) |
| 11 | Pusa Dwarf/H2001-4 | 62 k Chip SNP array | RIL | 0.88 | Singh et al. (2020a, 2020b, 2020c) |

development of high-density linkage genetic maps using SSR and SNP markers (Table 13.3). Previously the genetic maps in pigeon pea were developed using SSR markers. In 2011, an SSR-based genetic map of pigeon pea with 239 loci was developed for an inter-species F2 population [*C. cajan* acc. ICP 28 ×*Cajanus scarabaeoides* (L.) Thouars acc. ICPW 94] (Bohra et al. 2011). The F2 : 3 population derived from a cross between inbred lines "Pusa Dwarf" and "HDM04-1" was used for the construction of a linkage map comprising of 296 genic SNP and SSR markers covering a total map length of 1520.22 cM representing 11 chromosomes of the pigeon pea. This is the first dense intra-specific linkage map of pigeon pea comprising of SNP markers of pigeon pea and from this 13 QTLs for the six agronomic traits were identified (Kumawat et al. 2012). Arora et al. (2017) developed a dense intraspecific linkage map of pigeon pea with 932 markers which span a total map length of 1411.83 cM. The consensus map is based on three different linkage maps of F2 populations derived from the cross between Asha × UPAS 120, Pusa Dwarf × H2001-4, and Pusa Dwarf × HDM04-1. These SNP markers were derived using Illumina GoldenGate bead arrays, and restriction site-associated DNA (RAD) sequencing. The integrated map comprised of 547 SNP derived from bead-array, 319 SNP derived from RAD-seq, and 65 simple sequence repeat (SSR) marker loci.

## 13.5 QTL Analysis Using High-density Genotyping Platforms and GBS

The high-genotyping platforms and NGS-based GBS methods enabled us to find significant QTLs. The significant QTLs discovered using the recent genotyping methods are given in Table 13.4. As described earlier pigeon pea has limitations due to abiotic stress, attack of pathogens, photosensitivity, long duration, cleistogamy, tall height, etc., there is a need for rapid breeding programs. These advances will be useful for mapping particular trait loci/gene which can be used for the development of superior variety with higher yields.

## 13.6 GWAS Efforts in Pigeon Pea

GWAS enables us to discover marker–trait association of phenotypes in a set of distantly related germplasm. In pigeon pea limited GWAS have been reported. In one study, 94 pigeon pea genotypes with 6144 DArT and 768 SNP markers helped in associating the determinacy trait with 19 SNPs and 6 DArTs markers (Mir et al. 2013). The analysis for candidate gene from 142 pigeon pea verities in combination with the biparental QTL (ICPA 2039 × ICPR 2447) and gene expression study indicated a strong association of determinacy trait with CcTFL1 (Mir et al. 2014). An association study using 89 pigeon pea genotypes for fusarium wilt showed association of six SSR markers, HASSRs 8, 18, 30, 121 128, 174 with FW resistance (Patil et al. 2017), further, two of these markers were validated through the analysis of F2 individuals derived from Bahar (susceptible) × KPL 43 (resistant) (Patil et al. 2017). Varshney et al. (2017) analyzed marker-trait associations (MTAs) using 286 accessions with phenotyping data for eight different agronomic traits collected for two years, 2012–2013 and 2013–2014. A total of 241 significant MTAs were determined for eight traits, i.e. plant height, primary branches per plant, secondary branch per plant, days to 50% flowering, 100-seed weight, days to 75% maturity, pods per plant, and seeds per pod. These genes are associated with self-fertilization and disease resistance. The genome-wide linkage disequilibrium (LD) decay rate was 70 kb estimated by employing 446 568 high-quality SNPs for cultivated pigeon pea (Varshney et al. 2017). The pangenome of 89 pigeon pea accessions was developed, these accessions were mainly collected from India and the Philippines. To date, not very significant association mapping has been found using bi-parental population, and also, not very vast diversity has been reported yet in pigeon pea. However, the development of the high throughput genotyping arrays such as the 62K "CcSNPnks" SNP chip with the combination of vast phenotyping is expected to boost high-resolution GWAS in pigeon pea.

**Table 13.4** List of significant QTLs identified using GBS and high-density SNP platforms.

| Mapping population | Population type | Trait | QTL name | Marker interval | Reference | Genotyping platform used |
|---|---|---|---|---|---|---|
| Pusa Dwarf × HDM04-1 | F2 | Plant height | *qPH4.1* | ASNP1310–ASNP2099 | Kumawat et al. (2012) | Illumina GoldenGate bead-array technology SNPs |
| | | | *qPH5.1* | ASSR100–ASSR206 | | |
| | | Primary branches per plant | *qPB4.1* | ASNP1664–ASSR295 | | |
| | | | *qPB5.1* | ASSR206–ASNP242 | | |
| | | Secondary branches per plant | *qSB5.1* | ASSR100–ASSR206 | | |
| | | Number of pods per plant | *qPD3.1* | ASNP64–ASNP882 | | |
| | | | *qPD4.1* | ASNP2099–ASNP269 | | |
| | | | *qPD5.1* | ASSR100–ASSR206 | | |
| | | Days to flowering | *qFL4.1* | ASNP1310–ASNP2099 | | |
| | | | *qFL5.1* | ASSR100–ASSR206 | | |
| | | Days to maturity | *qMT4.1* | ASNP2099–ASNP269 | | |
| | | | *qMT5.1* | ASSR100–ASSR206 | | |
| | | | *qMT10.1* | ASNP2262–ASNP2402 | | |
| ICPL 20096 × ICPL 332 | RIL | Resistance to SMD | *qSMD11.1* | S11_30004779–S11_36027138 | Saxena et al. (2017) | GBS |
| ICP 8863 × ICPL 87119 | F2 | | *qSMD3.1* | S3_18837756–S3_5324938 | | |
| | | | *qSMD7.1* | S7_14725598–S7_7547477 | | |
| | | | *qSMD11.3* | S11_16365686–S11_5757417 | | |
| ICPB 2049 × ICPL 99050 | RIL | Resistance to FW | *qFW3.1* | S3_18695411–S3_17153283 | Saxena et al. (2017) | GBS |
| | | | *qFW6.1* | S6_22726005–S6_23553522 | | |
| | | | *qFW11.1* | S11_37262913–S11_37133265 | | |
| | | | *qFW11.2* | S11_20607023–S11_16809228 | | |
| ICPL 20096 × ICPL 332 | RIL | | *qFW7.1* | S7_6202998–S7_4645510 | | |
| | | | *qFW11.4* | S11_7119684–S11_10698013 | | |

| | | | | | | |
|---|---|---|---|---|---|---|
| ICPL 99010 × ICP 5529 | RIL | Cleistogamous flower | *qCl3.2* | Affx-123306383–Affx-123348752 | Yadav et al. (2019) | 50 K SNP array |
| | | | *qCl6.1* | Affx-123348721–Affx-123312002 | | |
| | | | *qCl3.2* | Affx-123306383–Affx-123348752 | | |
| | | Shriveled seed | *qShS6.2* | Affx-123332664–Affx-123310260 | | |
| | | Seed size | *qSS6.1* | Affx-123348721–Affx-123312002 | | |
| Pusa Dwarf × HDM04-1 | RIL | Number of primary branches per plant | *qPB3.1* | SCP-3333_576–CSCSP-1895_352 | Singh et al. (2020a, 2020b, 2020c) | 62 K *Affymetrix* SNP |
| | | Number of primary branches per plant | *qPB5.1* | CSCSP-14348_476–SCP-14083_1762 | | |
| | | Pod bearing length on main axis | *qPBLm3.1* | DRDRP-209_577–DRDRP-159_1581 | | |
| | | Pod bearing length on primary branch | *qPBLp2.1* | SCP-11875_4471–SCP-6166_1773 | | |
| | | Pod bearing length on primary branch | *qPBLp7.1* | SCP-6058_3471–CSCSP-248_411 | | |
| | | Pod bearing length on primary branch | *qPBLp7.2* | SCP-3492_834–SCP-15353_1301 | | |
| | | Pod bearing length on primary branch | *qPBLp7.3* | SCP-1543_308–SCP-15324_117 | | |
| | | Pod bearing length on primary branch | *qPBLp10.1* | CSCSP-14172_2899–DRDRP-710_2488 | | |
| | | Number of seeds per pod | *qSP3.1* | CSCSP-12895_48–CSCSP-14596_3162 | | |
| | | Number of seeds per pod | *qSP5.1* | CSCSP-15851_436–CSCSP-6821_4504 | | |
| | | Number of seeds per pod | *qSP10.1* | CSCSP-10917_684–DRDRP-720_1470 | | |

## 13.7 Genomic Selection (GS) Efforts in Pigeon Pea

Meuwissen et al. (2001) developed the concept of genomic selection (GS) in animals for the estimation of breeding values of phenotypes based on genome-wide marker data. The GS helps in the selection of superior individuals based on genome-wide marker information and evades the need for a phenotypic analysis of those individuals. Genotype selection is carried out based on genomic estimated breeding value (GEBV) can be predicted using a several GS models (Crossa et al. 2017). To date, there is no report available on GS models designed to calculate the breeding efficiency in pigeon pea. This is mainly due to the lack of funding support from both public and industrial sides made the pigeon pea research sluggish. Another important reason could be that this crop is mainly consumed and grown in developing countries thus, developed countries do not focus on pigeon pea research and development.

Although, the application of GS in other legumes such as chickpea, pea, and soybean has shown moderate to high accuracies in prediction for traits such as yield (0.59: Jarquín et al. 2014; 0.47–0.62: Duhnen et al. 2017) and protein content (0.50–0.55: Duhnen et al. 2017). Annicchiarico et al. (2017) found that the GS is superior to phenotypic selection (PS) w.r.t prediction of yield under drought stress. Recently, in chickpea, the GS was applied in 320 lines for yield and other traits (Roorkiwal et al. 2016). They have reported different prediction accuracies ranging from 0.13 (seed yield) to 0.91 (100 seed weight). The prediction accuracy of GS was found to change depending on the statistical method used and the marker density (Tayeh et al. 2015a, 2015b; Roorkiwal et al. 2016; Duhnen et al. 2017). It is also recommended that prediction accuracy of GS can be improved by using data collected from different environments and by using G × E interaction factor into GS models (multiplicative reaction norm model: Jarquín et al. 2014; Roorkiwal et al. 2018). The addition of epistasis interactions in GS models is also shown to play a significant role in improved prediction accuracy (Duhnen et al. 2017). For GS, it is critical when we are using a subset of SNPs selected from GWAS instead of whole-genome resequencing SNPs datasets. For example, Li et al. (2018) showed that the use of subset of SNP from GWAS improved accuracy of GS prediction compared to use of the entire WGRS-SNPs.

Presently there is a need for using the available pigeon pea genomics data for QTL mapping, GWAS, and GS. The integration of genomics with phenomics data will aid in the precise breeding of elite cultivars for the desired trait (Figure 13.1).

## 13.8 Summary

During the last few years, huge amount of the genomic resources in the form of genome-wide markers, genetic maps, genome sequence, high-throughput genotyping assays, marker–trait associations, and germplasm resequencing has been developed in pigeon pea. Deploying QTLs controlling important traits into elite cultivars aids faster development of the improved verities. In some cases, GWAS has been employed in breeding pigeon pea to enhance the yield and other valuable traits. The introduction of advanced breeding techniques like GS and speed breeding will improve the genetic gains in pigeon pea. The utilization of available genomic resources along with improved breeding methods will help to develop high-yielding pigeon pea cultivars with agronomically important traits.

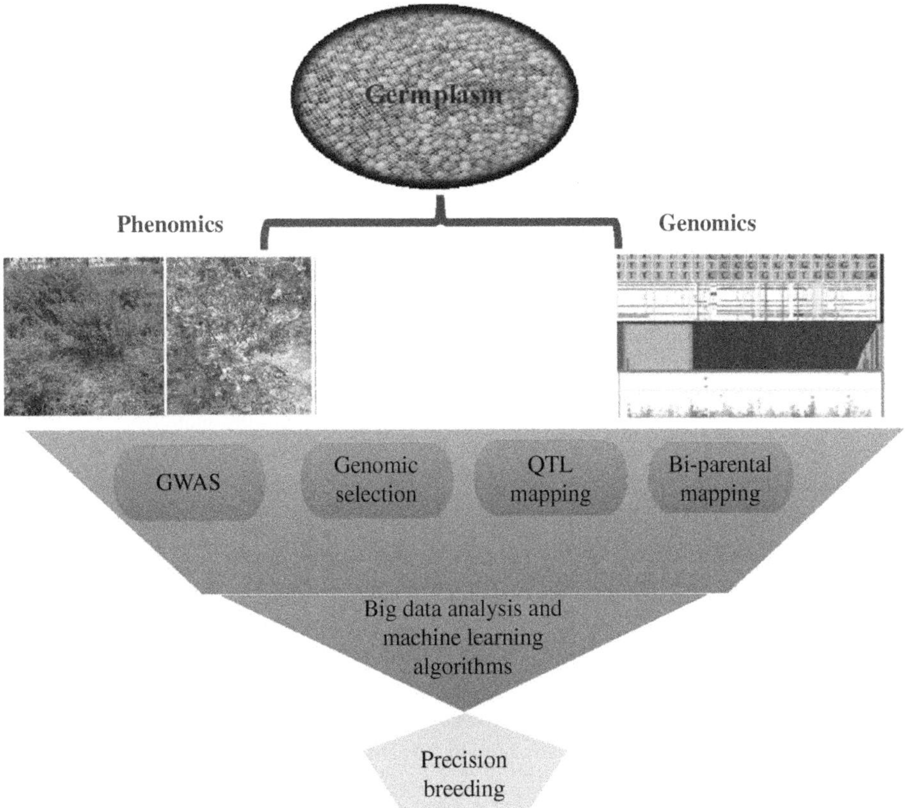


**Figure 13.1** Depicting the genomics and phenomics for the exploitation of pigeon pea genetic resources in investigating genome-wide association studies (GWASs), genomic selection, quantitative trait loci (QTLs) analysis leading to precision breeding.

## References

Annicchiarico, P., Nazzicari, N., Pecetti, L. et al. (2017). GBS-based genomic selection for pea grain yield under severe terminal drought. *Plant Genome* 10: 2.

Tayeh, N., Klein, A., Le Paslier, M.C. et al. (2015a). Genomic prediction in pea: effect of marker density and training population size and composition on prediction accuracy. *Frontiers in Plant Science* 6: 941.

Arora, S., Mahato, A.K., Singh, S. et al. (2017). A high-density intraspecific SNP linkage map of pigeonpea (*Cajanas cajan* L. Millsp.). *PLoS One* 12 (6): e0179747.

Bohra, A., Dubey, A., Saxena, R.K. et al. (2011). Analysis of BAC-end sequences (BESs) and development of BES-SSR markers for genetic mapping and hybrid purity assessment in pigeonpea (*Cajanus* spp.). *BMC Plant Biology* 11: 56. https://doi.org/10.1186/1471-2229-11-56.

Crossa, J., Pérez-Rodríguez, P., Cuevas, J. et al. (2017). Genomic selection in plant breeding: methods, models, and perspectives. *Trends in Plant Science* 22: 961–975.

Duhnen, A., Gras, A., Teyssèdre, S. et al. (2017). Genomic selection for yield and seed protein content in soybean: a study of breeding program data and assessment of prediction accuracy. *Crop Science* 57: 1–13.

Dutta, S., Kumawat, G., Singh, B.P. et al. (2011). Development of genic-SSR markers by deep transcriptome sequencing in pigeonpea (*Cajanus cajan* (L.) Millspaugh). *BMC Plant Biology* 11: 17.

Dutta, S., Mahato, A.K., Sharma, P. et al. (2013). Highly variable 'Arhar' simple sequence repeat markers for molecular diversity and phylogenetic studies in pigeonpea [*Cajanus cajan* (L.) Millisp.]. *Plant Breeding 132* (2): 191–196.

Greilhuber, J. and Obermayer, R. (1998). Genome size variation in *Cajanus cajan* (*Fabaceae*): a reconsideration. *Plant Systematics and Evolution* 212: 135–141. https://doi.org/10.1007/BF00985225.

Jarquín, D., Crossa, J., Lacaze, X. et al. (2014). A reaction norm model for genomic selection using high-dimensional genomic and environmental data. *Theoretical and Applied Genetics* 127: 595–607.

Kaila, T., Chaduvla, P.K., Saxena, S. et al. (2016). Chloroplast genome sequence of pigeon pea (*Cajanus cajan* (L.) Millspaugh) and *Cajanus scarabaeoides* (L.) Thouars: genome organization and comparison with other legumes. *Frontiers in Plant Science* 7: 1847.

Kumawat, G., Raje, R.S., Bhutani, S. et al. (2012). Molecular mapping of QTLs for plant type and earliness traits in pigeonpea (*Cajanus cajan* L. Millsp.). *BMC Genetics* 13: 84. https://doi.org/10.1186/1471-2156-13-84.

Li, Y., Ruperao, P., Batley, J. et al. (2018). Investigating drought tolerance in chickpea using genome-wide association mapping and genomic selection based on whole-genome resequencing data. *Frontiers in Plant Science* 9: 190.

Mahato, A.K., Sharma, A.K., Sharma, T.R., and Singh, N.K. (2018). An improved draft of the pigeonpea (*Cajanus cajan* (L.) Millsp.) genome. *Data in Brief* 16: 376–380.

Meuwissen, T.H.E., Hayes, B.J., and Goddard, M.E. (2001). Prediction of total genetic value using genome wide dense marker maps. *Genetics* 157: 1819–1829.

Mir, R.R., Saxena, R.K., Saxena, K.B. et al. (2013). Whole-genome scanning for mapping determinacy in pigeon pea (*Cajanus* spp.). *Plant Breeding* 132 (5): 472–478.

Mir, R.R., Kudapa, H., Srikanth, S. et al. (2014). Candidate gene analysis for determinacy in pigeonpea (*Cajanus* spp.). *Theoretical and Applied Genetics* 127 (12): 2663–2678.

Patil, P.G., Dubey, J., Bohra, A. et al. (2017). Association mapping to discover significant marker-trait associations for resistance against Fusarium wilt variant 2 in pigeonpea [*Cajanus cajan* (L.) Millspaugh] using SSR markers. *Journal of Applied Genetics* 58 (3): 307–319.

Rathinam, M., Mishra, P., Vasudevan, M. et al. (2019). Comparative transcriptome analysis of pigeonpea, *Cajanus cajan* (L.) and one of its wild relatives *Cajanus platycarpus* (Benth.) Maesen. *PLoS One* 14 (7): e0218731.

Roorkiwal, M., Sawargaonkar, S.L., Chitikineni, A. et al. (2013). Single nucleotide polymorphism genotyping for breeding and genetics applications in chickpea and pigeonpea using the BeadXpress platform. *The Plant Genome* 6 (2): 1–10.

Roorkiwal, M., Jarquin, D., Singh, M.K. et al. (2018). Genomic-enabled prediction models using multi-environment trials to estimate the effect of genotype × environment interaction on prediction accuracy in chickpea. *Scientific Reports* 8: 11701.

Roorkiwal, M., Rathore, A., Das, R.R. et al. (2016). Genome enabled prediction models for yield related traits in chickpea. *Frontiers in Plant Science* 7: 1666.

Saxena, S., Sahu, S., Kaila, T. et al. (2020a). Transcriptome profiling of differentially expressed genes in cytoplasmic male-sterile line and its fertility restorer line in pigeon pea (*Cajanus cajan* L.). *BMC plant biology 20* (1): 1–24.

Saxena, R.K., Hake, A., Hingane, A.J. et al. (2020b). Translational pigeonpea genomics consortium for accelerating genetic gains in pigeonpea (*Cajanus cajan* L.). *Agronomy 10*: 1289.

Saxena, S., Sahu, S., Kaila, T. et al. (2020c). Transcriptome profiling of differentially expressed genes in cytoplasmic male-sterile line and its fertility restorer line in pigeon pea (*Cajanus cajan* L.). *BMC Plant Biology 20* (1): 1–24.

Saxena, R.K., Singh, V.K., Kale, S.M. et al. (2017). Construction of genotyping-by-sequencing based high-density genetic maps and QTL mapping for fusarium wilt resistance in pigeonpea. *Scientific Reports 7* (1): 1–11.

Saxena, R.K., Patel, K., Kumar, C.S. et al. (2018a). Molecular mapping and inheritance of restoration of fertility (Rf) in A4 hybrid system in pigeonpea (*Cajanus cajan* (L.) Millsp.). *Theoretical and Applied Genetics 131* (8): 1605–1614.

Saxena, R.K., Rathore, A., Bohra, A. et al. (2018b). Development and application of high-density Axiom Cajanus SNP array with 56K SNPs to understand the genome architecture of released cultivars and founder genotypes. *The Plant Genome* 11 (3): 1–10.

Singh, N.K., Gupta, D.K., Jayaswal, P.K. et al. (2012). The first draft of the pigeonpea genome sequence. *Journal of Plant Biochemistry and Biotechnology 21* (1): 98–112.

Singh, N., Rai, V., and Singh, N.K. (2020a). Multi-omics strategies and prospects to enhance seed quality and nutritional traits in pigeonpea. *Nucleus* 63: 249–256.

Singh, N., Bhatt, V., Rana, N. et al. (2020b). Advances of next-generation sequencing (NGS) technologies to enhance the biofortifications in crops. In: *Advances in Agri-Food Biotechnology* (eds. T.R. Sharma, R. Deshmukh and H. Sonah). Singapore: Springer https://doi.org/10.1007/978-981-15-2874-3_17.

Singh, S., Mahato, A.K., Jayaswal, P.K. et al. (2020c). A 62K genic-SNP chip array for genetic studies and breeding applications in pigeonpea (*Cajanus cajan* L. Millsp.). *Scientific Reports* 10 (1): 4960. https://doi.org/10.1038/s41598-020-61889-0.

Tayeh, N., Klein, A., Le Paslier, M.C. et al. (2015b). Genomic prediction in pea: effect of marker density and training population size and composition on prediction accuracy. *Frontiers in Plant Science 6*: 941.

Tuteja, R.E., Saxena, R.K., Davila, J.A. et al. (2013). Cytoplasmic male sterility-associated chimeric open reading frames identified by mitochondrial genome sequencing of four Cajanus genotypes. *DNA Research* 20 (5): 485–495.

Varshney, R.K., Chen, W., Li, Y. et al. (2012). Draft genome sequence of pigeonpea (*Cajanus cajan*), an orphan legume crop of resource-poor farmers. *Nature Biotechnology 30* (1): 83–89.

Varshney, R., Saxena, R., Upadhyaya, H. et al. (2017). Whole-genome resequencing of 292 pigeonpea accessions identifies genomic regions associated with domestication and agronomic traits. *Nature Genetics* 49: 1082–1088. https://doi.org/10.1038/ng.3872.

Yadav, P., Saxena, K.B., Hingane, A. et al. (2019). An "Axiom Cajanus SNP Array" based high density genetic map and QTL mapping for high-selfing flower and seed quality traits in pigeonpea. *BMC Genomics* 20 (1): 1–10.

# 14

# Opportunity and Challenges for High-throughput Genotyping in Sugarcane

*Prathima P. Thirugnanasambandam*[1], *Avinash Singode*[2], *Lakshmipathy Thalambedu*[1] *and Senthilkumar Shanmugavel*[1]

[1] *Crop Improvement Division, ICAR – Sugarcane Breeding Institute, Coimbatore, Tamil Nadu, India*
[2] *ICAR – Indian Institute of Millets Research, Hyderabad, Telangana, India*

## 14.1 Introduction

Civilization started with agriculture since then humankind innovated technologies for higher crop production and productivity. Enhancing crop productivity genetically is the most viable strategy to increase production for securing food and nutrition. Food security refers to both physical and economic access to sufficient food for meeting the dietary needs for a productive and healthy life. Due to shrinking natural resources, agriculture is also bearing the burden of meeting fuel demand. Sugarcane is the potential crop for fuel production, specifically biofuels (sugar-based ethanol) and cogeneration of electricity (cane bagasse). Sugarcane offers alternatives to food, such as fiber and livestock feed.

Sugarcane, as the 14th most cultivated crop, contributes 60% of world bioethanol (Central and South America), while the remaining 40% is produced from raw starch as well as crops such as maize, sugar beet, and sweet sorghum (North America and European Union). In the year 2020, countries such as India, Brazil, Thailand, China, the United States, Mexico, Russia, Pakistan, France, and Australia accounted for nearly 70% of global sugarcane production. The sugar juice mainly consists of high concentrations of soluble sugars such as sucrose, glucose, and fructose and may contain trace amounts of other sugars such as trehalose. The bagasse that remains after crushing the juice can serve as a lignocellulosic feedstock for fuels and chemicals. Sugarcane is imperative as a food and industrial crop; hence, continuous efforts are made globally to improve the yield and sucrose content and making it resilient to biotic and abiotic stress. A giant leap in sugarcane is made, which is popularly known as "nobilization of sugarcane"; however, sugarcane genetics is an enigma that needs tremendous effort to unravel.

The currently available cultivated commercial sugarcane hybrids have been bred for more than 100 years by crossing different varieties of two species (*Saccharum officinarum* and *Saccharum spontaneum*) and hybrids that have a highly complex genome comprising

*Genotyping by Sequencing for Crop Improvement*, First Edition.
Edited by Humira Sonah, Vinod Goyal, S.M. Shivaraj, and Rupesh K. Deshmukh.

10000 Mbp distributed in 100–130 chromosomes. Chromosome assortment in modern cultivars largely follows preferential pairing, the meiosis follows bivalent pairing. The genetic architecture studies of large genomes are essential to identify genomic regions linked with traits of interest. However, due to the variation in genetic material composition of sugarcane, the applications of genomic tools have been drastically delayed in this crop, in comparison to other crops that have been advanced to marker-assisted selection (MAS) and genomic selection (GS). The advent of high-throughput sequencing technologies has provided opportunities to discover molecular markers, specifically the single-nucleotide polymorphisms (SNPs) and insertion–deletion (indels), at the genome-wide scale. However, attempts of molecular studies using different DNA marker systems were made. Quantitative trait loci (QTL) analysis in sugarcane was undertaken to identify genomic regions governing important agronomic traits such as red rot resistance, sugar, and fiber content. However, these studies did not translate into the development of genomic tools for sugarcane improvement.

## 14.2 Sugarcane Genome and Genetics

Naturally occurring sugarcane genotypes of *S. officinarum* (noble cane), *S. sinensis*, and *S. barberi* were cultivated to produce sugar before the introduction of modern hybrid cultivars. The discovery of the ability of sugarcane to produce viable seeds led to the establishment of sugarcane breeding programs in Java and Barbados in the late 1800s, resulting in development of POJ 2878 and POJ 2575. Modern sugarcane cultivars ($2n$ = 100–120) are products of hybridization between *S. officinarum* ($2n = 80$) and *S. spontaneum* ($2n$ = 40–128), with minor contributions from other *Saccharum* species. Hybrids inherit $2n$ chromosomes from the female parent (*S. officinarum*) and $n$ chromosomes from the male parent (*S. spontaneum*), resulting in $2n+n$ transmission known as "nobilization" and phenomena known as female restitution (Bremer 1961a, 1961b), and it ceases to exist after the first intercross. Studies have shown that modern cultivars inherit 10–20% chromosomes from *S. spontaneum* and 80–90% chromosome from *S. officinarum* (Grivet et al. 1996; Hoarau et al. 2002).

Sugarcane posses a larger genome size with ~10 Gb (Le Cunff et al. 2008), and the average haploid genome sizes of *S. officinarum* and *S. spontaneum* are 985 and 845 Mb, respectively (Zhang et al. 2012). Due to the larger genome size and the polyploid aneuploid nature of sugarcane, so far mosaic haploid reference genome for commercial hybrid (Garsmeur et al. 2018) and allele defined haploid genome of *S. spontaneum* are available (Zhang et al. 2018). All modern cultivars of sugarcane had their origin traced to POJ 2878 and POJ 2575 (Moore 2005). Sugarcane breeding programs rely heavily on phenotypic selection for the production of sugarcane cultivars, and the typical time for the release of a new variety is 10–15 years. The very first successful sugarcane cultivar Co 205 was released by Sugarcane Breeding Institute, Coimbatore, India, in the year 1918.

The important traits in sugarcane are quantitatively inherited, and due to its aneuploidy nature, it may have more than 12 alleles, making segregation of alleles complex (Sreenivasan and Ahloowalia 1987; Le Cunff et al. 2008). The presence of multiple alleles each with a minor effect on the phenotype may be difficult to predict. Apart from the effect of multiple alleles on the phenotype, environmental interaction is also one of the important factors to be accounted for. Sugarcane also exhibits high levels of Linkage Disequilibrium (LD)

(Aitken et al. 2005), since the major cultivars available are derived from few founder parental clones which were used in the earlier breeding programs (Wei et al. 2006). The genome complexity of sugarcane and the lack of complete reference genome had led to the slow adaptation of advanced biotechnological tools such as GS, genome-wide association study (GWAS), and MAS in sugarcane breeding programs.

## 14.3 Genetic Studies and Marker Systems

The classical genetic studies were made using morphological and cytological markers. Later, biochemical markers were used to study the genetic variation. These marker systems are highly influenced by the environment and limited in number. However, they laid the foundation of genetic map construction as in the case of *Drosophila* and maize. Initially, the hybridization-based marker, restriction fragment length polymorphism (RFLP), became very popular, and many genetic maps in several organisms were developed. As per the chronology of the development of marker systems, RFLP and randomly amplified polymorphic DNA (RAPD) markers were the first-generation makers; the second generation was the simple sequence repeats (SSRs), amplified fragment length polymorphism (AFLP), and their modifications. The third generation included expressed sequence tag (EST)-based markers and SNP. The second- and third-generation markers completely relied on polymerase chain reaction (PCR) and were called PCR-based DNA markers. The most common marker systems that are used in genetic studies include RAPD-, SSRs-, and EST-based markers. The PCR-based markers became popular over the hybridization-based markers such as RFLP due to the rapidity, low cost, ease of automation, and high genome coverage. PCR-based markers are expensive, time consuming, and are limited in number when compared with the present-day next-generation sequencing (NGS) platforms based on SNP markers.

The limitation in the number and the time of genotyping was very crucial in MAS and genetic studies of complex genomes such as wheat, sugarcane, and other polyploids. However, among the marker systems, SSRs were extensively used in mapping and tagging genes/QTLs and MAS in many crop species. The SSRs remained the first choice of markers in many important crops such as rice, wheat, maize, barley, soybean, and cotton. Nevertheless, the number of markers had been a major limitation in the precise estimation of the genetic effects and inter- and intra-allelic interaction in crop genomes. Understanding the cryptic genetic structure of populations required a huge number of polymorphic markers. The DNA sequencers could detect the multiple alleles at an SSR locus with high precision, but allele detection predominantly dependent on the software package and artifacts remains a major concern. The PCR-based markers such as RAPD and AFLP lacked repeatability but, nevertheless, were found to be very useful in crops where genomic information was limited. Cordeiro et al. (2000) developed the first set of microsatellite markers in sugarcane. As expected, a high number of alleles per loci and nonspecific banding patterns outside the expected size range prevented the application of multiplexing. The hypervariable and codominant nature of SSRs makes them suitable for genetic diversity, phylogenetic studies, and fingerprinting of sugarcane genotypes (Chen et al. 2009). Though SSR markers are codominant, they are being scored as dominant in sugarcane. A single primer pair produces a large number of alleles, and this can only be meaningfully interpreted if they are scored as dominant. As low as two SSR markers are sufficient to distinguish lines within *Saccharum* species and the members from old and new world *Erianthus* (Cordeiro et al. 2003). The SSR-based genetic diversity studies were found in agreement with the evolutionary course of the sugarcane cultivars; the order of contributing species in today's accessions is *S. officinarum, S. spontaneum, S. robustum, S. barberi, S. sinense, Erianthus arundinaceus*, and *Saccharum* spp. hybrids (Ali et al. 2019).

The study with SSRs showed consistent amplification and reproducibility in sugarcane, which encouraged researchers to use them in mapping; however, it required a large number of progenies as the ploidy level is high. The SSR markers also showed cross-species amplification between *S. officinarum* and *S. spontaneum*. The species-specific SSRs can be helpful in interspecific cross-identification, backcrossing programs, and interspecific introgression of genes/genomic regions (Wu et al. 2019). In the case of SSRs, the "nulls" are also recorded as polymorphism; however, it is necessary to optimize and characterize them before recording as a null allele.

The basic function of markers is to map and tag genes for MAS. The MAS methods are suitable for qualitative traits governed by one or few major effect genes. The complex polygenic traits, with small and cumulative effects, exhibit a very low probability of success in MAS. The primary limitations in the use of markers for QTL mapping of polygenic traits are (i) limiting the estimation of allelic effects to two in a biparental mapping population, generally used in QTL mapping and (ii) statistical models have not evolved to estimate empirically the small QTL effects dispersed in the crop genome (Heffner et al. 2009). Genome-wide high-density genotyping with a coverage of less than 1 cM between markers was found effective in identifying minor effect QTLs. Integration of markers information, QTL positions, gene density, and the whole-genome sequence is suggested to have better practical applications in plant breeding (Mace and Jordan 2011).

Utilization of historical LD between markers in natural and unrelated populations for associating markers to traits (association mapping [AM]) is effective not only for QTL identification but also to identify causal polymorphism for variation in a trait. The AM largely depends on the diversity in the population, a number of markers, and the historical recombination that might have taken place for thousands of years unlike biparental mapping populations with one or two rounds of recombination. The high density of markers such as SNP markers generated through genotyping-by-sequencing (GBS) is ideal for genome-wide association studies (GWASs) to deliver high-resolution map, establishing a marker–trait association, and its application in breeding program (Flint-Garcia et al. 2003). High-throughput genotyping such as GBS has opened avenue for GS. GS relies on high-density of markers across the genome for precise identification of small-effect QTLs and assigns breeding value to individual SNP genotype/haplotype.

## 14.4 Genotyping-by-Sequencing (GBS)

Prior to next-generation sequencing (NGS), hybridization and PCR-based markers were popular. Rapid analysis of genomes had been a great limitation in crop improvement and MAS. The quest for rapid, high-throughput, and breakthrough solid-state computation led to the development of novel chemistry in DNA amplification, multiplexing, and reading DNA sequences. From 1977 to 2016, three generations of sequencing technologies have emerged. Second- and third-generation sequencing technologies are referred commonly to as NGS. These technologies have significantly increased the sequencing speed, and there is a drastic reduction in sequencing cost since the inception of NGS in 2004. Without NGS-based genotyping, the concepts such as marker-assisted recurrent selection (MARS), AM, and GS could not come into existence since they are heavily dependent on high marker density.

Elshire et al. (2011) developed genome-wide SNP genotyping using NGS sequencing, which is simple, quick, extremely specific, and highly reproducible. The technique uses methylation-sensitive restriction enzymes so that the repetitive regions or the heterochromatinized regions of genomes can be avoided and lower copy regions could be targeted with higher efficiency. The restriction site-associated genomic DNA (RAD) tags for high-density SNP discovery and genotyping were first verified by Baird et al. (2008). The GBS adapters, primer sequencing, and library preparation are depicted in Figures 14.1 and 14.2. The GBS workflow includes the use of restriction enzymes that leave 2–3 bp overhangs by design. The GBS protocol using *Pst*1 and *Msp*1 in detail is published by Poland et al. (2012). Elshire et al. (2011) described two different types of adapters: one is "barcode adapter" that

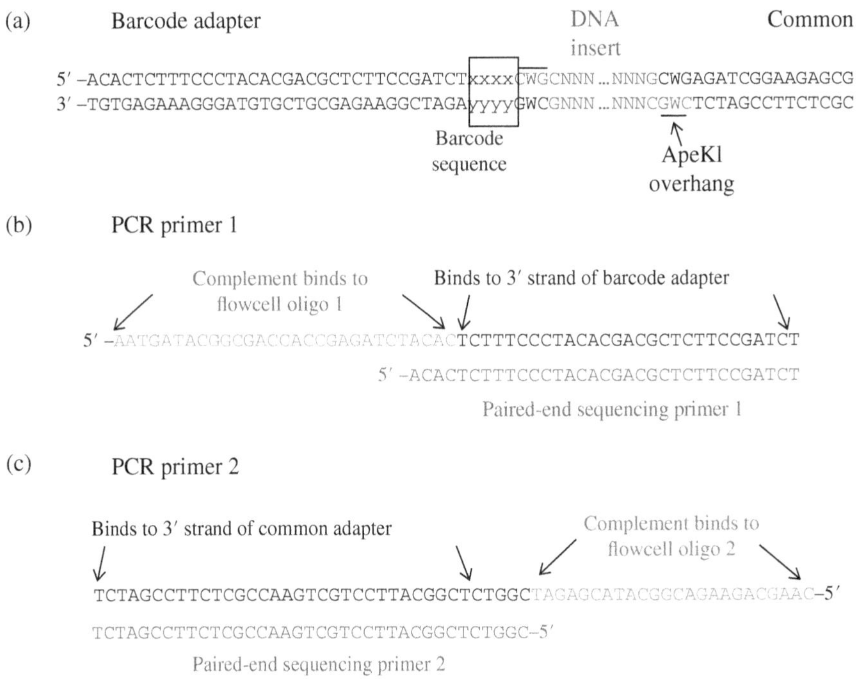


**Figure 14.1** GBS adapters, PCR and sequencing primers. (a) Sequences of double-stranded barcode and common adapters. Adapters are shown ligated to *Ape*KI-cut genomic DNA. Positions of the barcode sequence and *Ape*KI overhangs are shown relative to the insert DNA; (b) sequences of PCR primer 1 and paired end sequencing primer 1 (PE-1). Binding sites for flow cell oligonucleotide 1 and barcode adapter are indicated; (c) sequences of PCR primer 2 and paired end sequencing primer 2 (PE-2). Binding sites for flow cell oligonucleotide 2 and common adapter are indicated. *Source:* (Elshire et al., 2011)

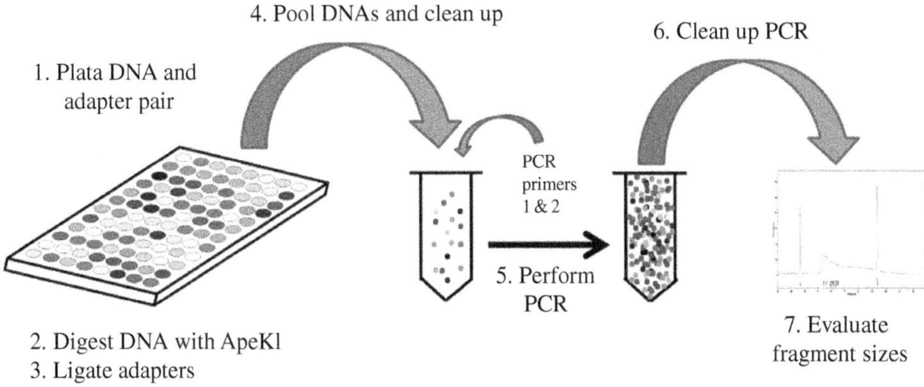


**Figure 14.2** Steps in GBS library construction. Note: Up to 96 DNA samples can be processed simultaneously. (1) DNA samples, barcode, and common adapter pairs are plated and dried; (2, 3) samples are then digested with *ApeKI* and adapters are ligated to the ends of genomic DNA fragments; (4) T4 ligase is inactivated by heating and an aliquot of each sample is pooled and applied to a size exclusion column to remove unreacted adapters; (5) appropriate primers with binding sites on the ligated adapters are added and PCR is performed to increase the fragment pool; (6, 7) PCR products are cleaned up and fragment sizes of the resulting library are checked on a DNA analyzer (BioRad Experion® or similar instrument). Libraries without adapter dimers are retained for DNA sequencing. *Source:* Elshire et al., 2011

terminates with 4–8 bp barcode at 3′ end of its top strand, a 3 bp overhang on the 5′ end of its bottom strand being complementary to the "sticky" end generated by the restriction enzyme *Ape*KI. The barcode is unique to one genotype and maintains its specificity in multiplexing. The barcodes are designed such that the possibility of misidentifying samples because of sequencing or adapter synthesis is minimum. All pair-wise combinations of the barcodes differed by a minimum of three mutational steps for correctly assigning samples with single-base barcode sequencing errors (Hamady et al. 2008). The second is the "common adapter" with only an *Ape*KI-compatible sticky end. The adapters were designed such that *Ape*KI recognition sites do not occur in any adapters and do not regenerate after ligation to genomic DNA.

The next step is the preparation of libraries for NGS. The samples along with the barcode adapter and common adapters were digested with *Ape*KI. The sticky ends thus produced will be ligated with T4 ligase. The restricted genomic DNA fragments ligated with adapter sequences are amplified with primers complementary to adapter sequences. The PCR products of the sample pool constitute the sequencing "library." The library is purified from primer dimers and is evaluated for fragment size. As per the protocol followed by Elshire et al. (2011), any fragment ~128 bp was considered as adapter dimer, which is not expected in the library or kept below 0.5%. The expected size of the DNA fragments is in the range of 170–350 bp, which is ready for DNA sequencing. The sequencing generates "fastq" files that are filtered using the following criteria: (i) one of the barcodes that matched perfectly and the expected four-base remnant of the *Ape*KI cut site (CWGC), (ii) sequence reads that are not adapter/adapter dimers, and (iii) sequences with no missing nucleotide "Ns" in their first 72 bases. These sequences were separated into different files according to their barcode. The sequences corresponding to the barcode are removed, and the remaining sequence is trimmed to 64 bases (including the initial CWGC). These filtered sequences are aligned to the reference sequence. In maize, Elshire et al. (2011) aligned the filtered sequences to B73 RefGen V1 using Burrows–Wheeler alignment tool.

## 14.5 SNP Calling Using GBS Pipelines

NGS has enabled millions of sequence reads per run; however, using these sequence reads in MAS requires extensive bioinformatic resources. The complexity of data analysis for SNP calling multiplies with every step of sample preparation (using restriction enzymes and sample multiplexing), filtering sequences (different fragment length and variable read depth), and alignment with reference sequence, which cumulatively is called a pipeline. These pipelines have enabled SNP calling even in nonmodel species such as sugarcane where the genome is complex, and genomic resources did not exist *a priori* (Davey et al. 2011). The GBS pipeline includes programs to filter out poor-quality reads, classify reads based on the individual barcode, align to reference genome sequence, and finally discover polymorphic loci (SNP). There are two types of GBS pipelines developed: (i) *de novo* based and (ii) reference based. Reference-based GBS pipelines such as TASSEL-GBS (v1 and v2), Stacks, IGST, and Fast-GBS use reference genome sequence for alignment and SNPs calling (Bradbury et al. 2007; Catchen et al. 2013; Sonah et al. 2013; Glaubitz et al. 2014). In the crop genomes where the reference sequence is not available, GBS is done using pairs of nearly identical reads in the sequence reads to identify polymorphism at a locus; however, the sequence reads are aligned precisely in all the samples. The most used *de novo*-based pipelines are UNEAK and Stacks (Catchen et al. 2013; Lu et al. 2013). Some of the *de novo*- and reference-based pipelines are briefly explained below.

***"Fast-GBS"***: The Fast-GBS analysis pipeline (available at https://bitbucket.org/jerlar73/fast-gbs) can handle data from different sequencing platforms and can identify SNPs from GBS data very efficiently using standard programming languages such as Python. It has a mix of different packages with seven core functions of processing of raw reads such as demultiplexing and cleaning, quality assessment of raw reads and mapping to references, filtering of mapped and unmapped reads, realigning and construction of local haplotypes, estimation of population frequencies, and individual haplotypes, variant calling, and identification of highly consistent variants.

***IGST (IBIS GBS tool)***: IGST is developed in Perl and Python scripts, utilizing publicly available software tools (FASTX toolkit, BWA, SAMtools, and VCFtools) and the tools developed by Li and Durbin (2009), Li et al. (2009), and Danecek et al. (2011) to process Illumina sequence reads. IGST demultiplexes and cleans barcoded reads using Sabre (https://github.com/najoshi/sabre). The pipeline can perform heterozygous corrections.

***TASSEL-GBS (versions 1 and 2)***: It is a java-based pipeline, and two versions are released till now, v1 (TASSEL 3.0; Bradbury et al. 2007) and v2 (TASSEL 5.0; Glaubitz et al. 2014). Version 1 works on 64-bp trimmed reads, while version 2 can work on longer reads up to 92 bp, which improves the accuracy of alignment to reference genome and SNP discovery.

***UNEAK (Universal Network Enabled Analysis Kit)***: The UNEAK is the nonreference GBS SNP calling pipeline, which is an extension of the Java program of TASSEL. The general design of UNEAK is described by Torkamaneh et al. (2016) as follows: UNEAK commands are run as TASSEL plugins via the command line.

***Stacks:*** *Stacks* is implemented by component programs written in C++ and Perl. It uses sequenced DNA fragments from any restriction enzyme-based GBS protocol. This pipeline is suitable for both reference genome alignment and *de novo* SNP discovery (Kumar et al. 2012). *Stacks* is accessible as open-source software under the GPL license and available for download at http://creskolab.uoregon.edu/stacks/. *Stacks* utilizes a catalog to document all loci identified from the population and matches individuals to that catalog to find out which haplotype alleles are there at different locus in each individual.

New molecular approaches such as GWAS and GS are now being applied to sugarcane, which is based on the high level of LD observed within sugarcane cultivars. GBS for mapping studies in sugarcane was first made by Balsalobre et al. (2017). The study established a pipeline for identifying SNPs, developed the first GBS-based integrated genetic map with additional SSR and TRAP markers, and identified QTLs governing the yield-related traits based on the integrated genetic map. It has also performed annotation of the sequences linked with QTLs to search putative candidate genes controlling yield traits in sugarcane.

## 14.6 Sugarcane Genome Sequencing

Sugarcane is by far the complex plant genome and is the recent genome to be sequenced because almost every other plant crop genome is sequenced years before sugarcane. Sugarcane is an autopolyploid with variable ploidy levels (frequently occurring aneuploidy) and a large genome. Several other crop genomes such as wheat and maize that are relatively larger were, however, sequenced well ahead of sugarcane. The major problem in sequencing sugarcane was its complex dynamics of the genome organization rather than its size. Every cycle of meiosis leads to a variable chromosome number. Nevertheless, the ancestry of present-day sugarcane is also complex that the authors believe that sequencing its progenitors will not deliver much insight into the genetic plasticity of sugarcane. Probably, sugarcane genome complexity is much greater than any other important crop species (D'Hont and Glaszmann 2001; Grivet and Arruda 2002). For comparison, the wheat genome contains 17 000 Mbp organized into 46 chromosomes, while the human genome has a mere 3200 Mbp, also organized into 46 chromosomes. Organisms may exhibit complex geno- and phenotypic plasticity, but the regular nature of meiosis and segregation of chromosomes like diploids make them easy for sequencing and other genomic studies. The sugarcane genome has ~300 000 genes, with humans just at 23 000, although it is well known that the complexity in organism and C-value is a paradox. Although sugarcane's monoploid genome is about 1 Gb, which is nearly about sorghum, rice, and other crops that were sequenced, its highly polymorphic nature presented a significant challenge for obtaining a close representation of the genome. In comparison to sugarcane, sorghum, which is the closest relative of sugarcane outside of the *Saccharum* complex, is a diploid with 10 chromosomes and a genome of about 730 Mb. Both the crops were reported to share a common ancestor about eight to nine million years ago.

A large-scale BAC library construction for the sugarcane variety R570 with 115 chromosomes funded by International Consortium for Sugarcane Biotechnology (ICSB) was made possible through an international collaboration consisting of Joint Genome Institute (JGI), Keygene, Queensland Alliance for Agriculture and Food Innovation, Commonwealth Scientific and Industrial Research Organization (CSIRO), South African Sugarcane Research Institute, University of Sao Paulo, and the Joint BioEnergy Institute. For sequencing, R570, a cultivar from Reunion island having one of the best characterized genomes through several molecular and genomics studies (Grivet et al. 1996; Asnaghi et al. 2000; Hoarau et al. 2001; Raboin 2006), was utilized. The BAC library consisting of 103 296 clones representing 14× monoploid genome coverage and 1.3× total genome coverage of this cultivar was developed. The genomic DNA of sugarcane was fragmented and cloned into BAC libraries. These BAC clones were then aligned with the sorghum genome sequence. A whole-genome profiling approach (Van Oeveren et al. 2011) was used for obtaining a high-quality single tiling path assembly of 382 Mb size from a minimum tiling path of 4660 sugarcane BACs covering the gene-rich part of the sorghum genome. The BAC clones of the minimum tiling path were selected based on the sorghum genome (as a bait to filter out those clones hybridized with the sorghum genome). Those BAC clone sequences that were aligned to the sorghum genome were selected based on hybridization studies, and the sequences were assembled to form a draft sequence of sugarcane, representing just one copy of the total genome (Garsmeur et al. 2018). This draft monoploid sugarcane genome sequence opened several comparative studies with other sugarcane, sorghum, and related grasses. BAC screening was undertaken using a fast and efficient 3D pool approach which was developed at CNRGV (The National Center for Plant Genomic Resources), France. The Sugarcane Genome Sequencing Initiative (SUGESI) was formed to integrate the efforts to produce a reference genome sequence of a sugarcane cultivar (*Saccharum hybrid*) and its ancestors (*S. officinarum* and *S. spontaneum*) using a combination of different approaches but mainly focused on BAC sequencing. The SUGESI Consortium comprises researchers from Australia, France, Brazil, South Africa, and the United States.

The other variety, SP80-3280, was chosen for sequencing because more data were available about this variety and is the top variety grown and also used for sugarcane breeding in Brazil. It was used for Project Sugarcane Genome wherein SUCEST, the largest collection of sugarcane EST-based library, was constructed during 1999–2002 by a consortium of Brazilian researchers. About 238,000 ESTs were sequenced from 26 diverse cDNA libraries (Vettore et al. 2003). A functional genomics part of the project undertook after the EST sequencing. Subsequently, integrated database was created with the EST sequences, gene expression data, gene categories, and data mining tools, providing a comprehensive portal to sugarcane genomics resources. The sequencing of SP80-3280 took 20 years of planning and execution of the research, though the sequencing technology available in 2008 could produce smaller fragments/reads, which further required assembling to form contigs. In 2014, Microsoft Research Institute impressed and intrigued by the crop's complexity collaborated with São Paulo Research Foundation, FAPESP, to tackle the computational challenges in assembling the sugarcane genome. A proposal was made in collaboration with FAPESP (The Fundação de Amparo à Pesquisa do Estado de São Paulo), Brazil, which took the form of the project "Development of an algorithm for the assembly of the sugarcane polyploid genome," with Glaucia Mendes Souza as the principal investigator. Assembling the genome with the highly fragmented sequences with multiple copies required significant computing power, which was supplied by Microsoft. This, however, led to the development of the monoploid draft genome of sugarcane which was publicly available. This served as a basis for the genetic improvement of sugarcane according to the UN Food and Agriculture Organization (FAO). The sequencing effort is one of Brazil's important contributions in sugarcane genomic research. Large-scale research involving global partnership on sugarcane is one of the factors that enabled Brazil to achieve 40% of its total energy demands from renewables, especially sugarcane with low-carbon emissions.

Recently, an international team from China, the United States, Brazil, and Canada sequenced the genome of *S. spontaneum*, the progenitor species corresponding to 10–15% of the commercial sugarcane genome. The study was published in Nature Genetics, 2018 (Zhang et al. 2018). Having initiated the genome sequencing projects, the researchers are now aiming to alter carbon pathways to produce new compounds and bioproducts from sugar, lignin, fiber, and other plant metabolites. It is expected that the knowledge of the sugarcane genome will allow editing metabolic pathways to increase sugar content, biosynthesis of different molecules, and bring in higher bioenergy potential.

## 14.7 Linkage and QTL Mapping in Sugarcane

Development of a commercially successful sugarcane variety takes 12–15 years and is a continuous process. The early selection and evaluation take huge resources and time. Considering the difficulty in the prediction of performance in the progenies derived from interspecific crosses, the on-farm and multi-location trials are the only way to select clones for commercial release. In other important crops where the genome complexity was not a challenge, enormous information is generated through genetic and genomic studies which can be used in MAS.

Before DNA-based markers, biochemical and molecular markers in sugarcane were the only means for monitoring unifactorial inheritance (Glaszmann et al. 1989; D'Hont et al. 1994). The challenge for developing a genetic map based on DNA markers were the multiple copies, polysomy, paralogous loci, and aneuploidy. Many genetic maps were generated in sugarcane using RFLP probes; however, marker order could not be arranged due to the small population. Moreover, the RFLP probes were borrowed from the maize studies due to high synteny found between the three Andropogoneae tribe members, maize, sugarcane, and sorghum (Grivet et al. 1996).

The use of single-dose markers in sugarcane and similar genomes was helpful in the mapping (Wu et al. 1992). The first QTL mapping study in sugarcane was performed by Sills et al. (1995). Integration of single-dose markers such as SNPs, SSR, and AFLP was attempted to generate a robust linkage map in sugarcane. These markers were expected to segregate in 1 : 1 and 1 : 2 : 1 ratio. The dosage and ploidy of single-dose SNPs were estimated using "SuperMASSA" to generate the first genetic map in which SNP markers were segregating in a 1 : 2 : 1 fashion (Costa et al. 2016). This method was refined using GBS for single-dose SNP genotyping. This refinement has led to the identification of additive and dominance effects; in addition, it also estimated the segregation patterns for all mapped QTLs (Balsalobre et al. 2017). These insights into mapping have led to the understanding that it is necessary to develop statistical approaches to enable the inclusion of markers at multiple doses to enhance the coverage of single-dose markers. Moreover, polyploids invariably have multiple doses; hence, the inclusion of multiple dose markers will permit the understanding of phenotype governed by the dosage effect. In sugarcane, some of the disease-resistant major QTLs were identified such as brown rust resistance *Bru1* (Daugrois et al. 1996) and *Bru2* (Raboin 2006), smut resistance (Wei et al. 2006), yellow spot resistance (Aljanabi et al. 2007), and sugarcane yellow leaf resistance (Costet et al. 2012). Diagnostic markers were successfully developed for brown rust resistance, *Bru1,* and are the most successful diagnostic markers. The impact of this finding is such that the frequency of *Bru1* in sugarcane clones increased from 15% (1975–1985) to 47% (2002–2012) after brown rust was introduced to Florida (Yang et al. 2018b). The sugar-related QTLs were validated through AM, which has led to the identification of functional alleles associated with sucrose biosynthesis (Ukoskit et al. 2019).

QTL mapping in sugarcane involves generating a biparental mapping population by crossing parents contrasting for the traits to be mapped. At times, the same mapping population can be used for mapping different traits, provided parents differed for the traits under question. DNAs of individuals in mapping population are then digested with restriction enzymes (such as *Pst*I) and genotyped with Illumina Hiseq 2000 platform (Illumina® Inc., San Diego, CA, USA), and the SNP variants are called by various methods as described in Yang et al. (2017), and then only single-dose markers and rarely double-dose markers are selected for mapping. These markers are used to develop linkage maps for each parent separately, and then QTLs are detected using software such as WinQTLCart 2.5 (Wang et al. 2012). The positions of QTLs, LOD values, and their contribution to phenotypic variation are determined through effective phenotyping. The linkage groups formed through this procedure are aligned to form homoeologous groups. Heterozygous marker genotype at a particular QTL locus may have a positive or negative effect on the phenotype; accordingly, the selections need to be made. For instance, Yang et al. detected three QTLs for resistance to orange rust in sugarcane in which qORR4 had a negative contribution to disease resistance, while the other two contributed positively. With the combination of the three positive QTLs (heterozygote for QTL qORR109 and qORR102, and homozygote for QTL qORR4), disease severity reduced by 56.4%. Besides, already available PCR-based genic markers can be used to validate the QTLs identified using GBS-based markers.

Mapping QTLs in sugarcane is challenging; however, its complexity is an opportunity to understand the dosage effect and dynamics of interspecific genomes in the manifestation of a phenotype. The marker technologies can never be obsolete, rather a conscious use of technology(ies) and complementation can be very useful in better understanding of genomes and phenotypes (Table 14.1).

**Table 14.1** Studies on linkage and QTL mapping in sugarcane through GBS-based markers.

| Cross | Trait | Pop | Chr | Genetic position/ marker interval | QTL | LOD | Reference |
|---|---|---|---|---|---|---|---|
| CP95-1039 (Resistant) X CP88-1762 (Susceptible), biparental cross | Resistance to orange rust | F1 | LG 109 (CP95-1039) | 27.3 | qORR109 | 19.3 | Yang et al. (2018a) |
| | | | LG 4 (CP88-1762) | 29.5 | qORR4 | 5.4 | |
| | | | LG 102 (CP88-1762) | 2.2 | qORR102 | 3.4 | |
| CP95-1039 (Sus) X CP88-1762 (Res), biparental cross | Resistance to sugarcane brown rust | F1 | | | qBRR143 | 6.9 | Yang et al. (2017) |
| | | | | | qBRR73 | 13.8 | |
| SP80-3280 (female) and RB835486 (male) | Brix | F1 | LG 4 | 43.32 | B1 | 5.75 | Balsalobre et al. (2017) |
| | | | LG 47 | 40 | B2 | 4.79 | |
| | POL% | | LG 4 | 43.32 | P1 | 4.17 | |
| | Stalk diameter | | LG 29 | 2.00 | SD1 | 6.02 | |
| | Fiber content | | LG 46 | 171.00 | FIB1 | 4.77 | |
| Resistant cultivar LCP 85–384 (female) and a susceptible cultivar L 99–226 (male), biparental cross | Leaf scald resistance | F1 | LG 37 | 41 | qLSR37.1 | 4.90 | Gutierrez et al. (2018) |
| | | | LG 77 | 61.3 | qLSR77.1 | 7.63 | |
| | | | LG 104 | 18.4 | qLSR104.1 | 3.67 | |
| | | | LG 156 | 271.71 | qLSR156.1 | 3.95 | |
| | | | LG 247 | 19.30 | qLSR247.1 | 3.46 | |
| | | | LG 250 | 316.41 | qLSR250.1 | 3.25 | |
| | | | LG 262 | 89.70 | qLSR262 | 2.96 | |

## 14.8 GWAS in Sugarcane

GWAS was proposed as a potential alternative to biparental mapping population-based QTL studies in crops. This approach is based on LD between markers in the genome. The LD refers to the nonrandom association between alleles at different loci. AM uses LD between associated markers (individual SNPs or SNP haplotypes) for the trait of interest in a group of unrelated individuals or populations (Weir 2008). The historical recombination events that are fixed in populations over many generations are used to determine the LD between markers/SNPs. In a chromosomal region, the LD spanning over loci is called LD blocks. The statistical association between marker and trait implies that the loci are physically close within the LD block. Unlike maize and rice, the LD decay in sugarcane is slow up to 5 cM (Debibakas et al. 2014). This large extent of LD in sugarcane is attributed to the limited number of progenitors involved in the nobilization of cane (Janoo et al. 1999; Wei et al. 2006). The other reason is few generations of recombination events in the development of cultivars. Genomic *in situ* hybridization suggested that *S. officinarum* and *S. spontaneum* contribute ~80% and 10–20%, respectively, to the genome of modern cultivars, while ~10% are recombinant chromosomes (Fickett et al. 2019).

Conventionally, unstructured biparental mapping populations are used in QTL mapping. In contrast, AM harnesses the natural variation fixed in populations, which forms LD blocks in the genome. AM has many advantages over conventional QTL mapping. QTL mapping requires the development of specialized populations usually biparental ($F_2$, RILs, NILs, DHs, etc.). The variation accounted by QTLs is based on two alleles, whereas nature may harbor more than two alleles. The minor effect QTLs are usually ignored in MAS. Moreover, the QTLs identified in one population cannot be applied across all populations. The QTL analysis is based on variation created through few generations of recombination which miss the historical recombination events that are fixed in nature. GWAS on the other hand can be carried out on the existing populations, and it does not require any prior information about the genome and can identify QTLs that have a minor effect on the trait. The QTL mapping and AM in a biparental mapping population may not give the same result. There could be several reasons for such discrepancies (Ukoskit et al. 2019).

The very first account of GWAS in sugarcane was reported by McIntyre et al. (2005) utilizing a biparental mapping population and RFLP markers. Later on, several studies have been reported in sugarcane using different marker systems such as AFLP, SSR, SNPs, and DArT arrays (Gouy et al. 2014; Banerjee et al. 2015). A host of studies have been carried out in sugarcane utilizing the GWAS approach (Gouy et al. 2014; Banerjee et al. 2015; Racedo et al. 2016; Siraree et al. 2017). GWAS requires dense marker coverage in the genome. SNP-based marker systems such as GBS are highly useful in sugarcane. GWAS studies utilizing GBS were relatively low in sugarcane due to complexities discussed before. There is also a lack of software tools specifically designed for complex polyploids such as sugarcane, which is not the case in crops with a simpler genome. The only software tool available for polyploid was designed for tetrapolyploid potato called "GWASpoly" (Rosyara et al. 2016). So, it is necessary to develop software tools and methods that were specifically designed for crops like sugarcane.

Controlling the errors related to population size, marker number and population structure must be addressed in sugarcane. The lower population and few markers may result in false positives. The use of highly diverse genetic material may have a strong population structure, which could result in lower QTL/genes detection power using "Structure" which is allele frequency-based population structure analysis program. Most of the assumptions laid in the "Structure" algorithm are not met while working with sugarcane. Principal component analysis (PCA) is found to be a better alternative to the Q-matrix generated by "Structure" (Racedo et al. 2016). The structure is likely to generate subpopulations arbitrarily with clear-cut distinguishing the populations or k-value (Wei et al. 2010). Ukoskit et al. (2019) in contrast demonstrated that the information on pedigree relationships passport data of the sugarcane clones will help in better understanding of the structure in the sugarcane populations. A subset of representative markers for population structure analysis was found useful in distinguishing populations than random markers (Racedo et al. 2016). The studies also show that the use of single-dose markers is more useful than SNP markers in population structure analysis and further GWAS (Barreto et al. 2019). Association analysis with GBS-based SNP markers shows a high number of associated markers; however, validation of these markers must be done before using them in MAS (Fickett et al. 2019).

GWAS in sugarcane is done for almost all the important traits, namely, fiber components (Yang et al. 2019), cane yield, and related traits (Fickett et al. 2019). However, the traits that have low heritability and show complex inheritance such as cane yield and sugar content are less likely to produce similar results across populations and phenotyping locations. The AM in sugarcane can yield associated markers; however, using this information in MAS across populations remains unclear. GWAS is a robust tool to find functional putative SNPs (quantitate trait nucleotides), which can be validated by integrating other techniques in biotechnology, namely, cloning and characterizing in a model organism, transcription factor activity, and gene editing (Table 14.2).

**Table 14.2** Studies on GWAS for various traits in sugarcane using GBS markers

| Trait | Number of genotypes analyzed | Platform/ technique used | Most significantly associated SNP | Salient feature of the study | Reference |
|---|---|---|---|---|---|
| Yield-related traits | 97 | Illumina HiSeq 2500 | SNP5806, SNP5501, SNP0938, SNP2578, SNP1190, SNP3258, SNP4204, SNP1571, SNP1190, and SNP5851 | Structured association mapping | Fickett et al. (2019) |
| Fiber composition | 229 | Illumina HiSeq 2000 | 4P6213973, 8P58249945, 1P56412057, 1ind60871784, 1P19803608, 10P5522707, etc. | | Yang et al. (2019) |

## 14.9 Genomic Selection in Sugarcane

Genomic selection was first conceived in animal breeding. The idea was to enhance the accuracy of MAS of complex traits for realizing a higher rate of genetic gains and reducing the number of selection cycles (Meuwissen 2003). Meuwissen and Goddard (1996) theorized that the QTL effects can be estimated with certainty using best linear unbiased prediction (BLUP), provided that the region is not undergoing recombination from one generation to other. They demonstrated that early generation selection using haplotype surrounding QTL has better use in MAS than the conventional single-marker QTL analysis. The emphasis in this scheme was given on the selection response during early generations of selection, and new QTL will be continuously detected in ongoing MAS schemes. The selection of individual-based on genomic-estimated breeding values (GEBVs) is called GS. The fundamental feature of GS is the use of high-density markers across the genome, ensuring that every trait locus has the probability of being in LD with a minimum of one marker locus in the entire target population (Heffner et al. 2009). The GS model estimates all marker effects genome-wide; even the small genetic effects are captured. These models are trained and optimized on a subset of the entire population. The individuals in the subset are phenotyped and genotyped, which is called the "training population" or "training set." (Figure 14.3). The model parameters in the training set are used to calculate the GEBV of individuals that are only genotyped. The selection is entirely based on the predictions made by the models using genotypic data alone. The prediction accuracy and genetic gain from selection depend on the robustness of the model in the estimation of marker effects. The accuracy of GEBV depends on the training set; it must represent the individuals under selection. This situation is less likely to be present in a crop such as sugarcane, where the purity of genotype is maintained by clones, and one sexual cycle will change the entire chromosomal structure. The training population should also be large enough to capture the maximum variations possible. A previous experiment by Daetwyler et al. (2010) showed an improvement of 50% prediction accuracy when the training population was increased from 500 to 1000. The number of markers utilized is also an important factor of GS studies,

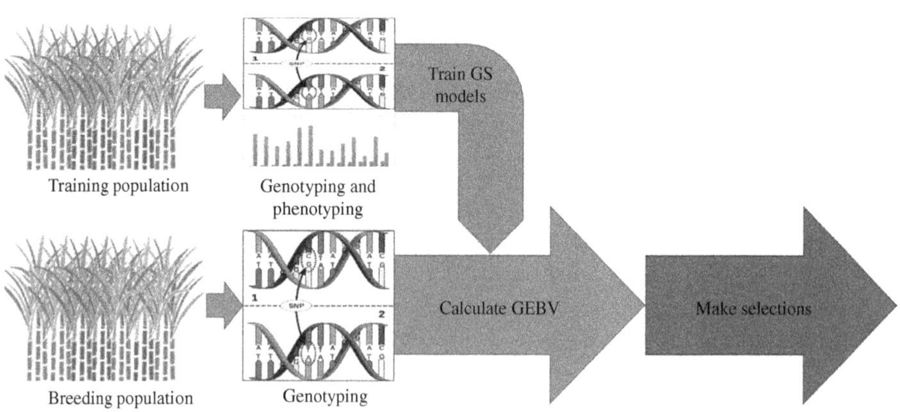


**Figure 14.3** Schematic representation of genomic selection processes from training population to selection.

which is decided by the linkage decay of the population as it varies between plant species and population employed. So, it is necessary to account for LD decay and the structure of the population.

GS is a new approach in which the entire genome was covered using genetic markers rather than using a limited number of markers as in QTL studies, leading to the detection of all QTLs that are linked to the trait which is in LD with at least one marker (Goddard and Hayes 2007). GS was proposed to address the deficiencies of biparental population-based QTL studies where it cannot detect all the QTLs, and the statistical methods used were ill-suited for their application if the trait was controlled by multiple genes (Heffner et al. 2009). Since the sequencing costs are declining with advancements in NGS technologies, it is now possible to utilize a larger sample population in GS studies. Covering the entire sugarcane genome with markers such as SSRs and DArT markers is difficult, given the larger and complex genome of sugarcane. Recent advancements in NGS technologies such as GBS to identify SNP markers provide an opportunity to cover the sugarcane genome.

Complexity in genome size, polyaneuploid nature, and long breeding cycles makes application of MAS in sugarcane an extremely difficult task. Several studies have been conducted to identify potential QTLs for agronomically important traits in sugarcane utilizing biparental and selfed progenies of sugarcane (Ming et al. 2001, 2002a, 2002b; Hoarau et al. 2002; Aitken et al. 2006, 2008; Piperidis et al. 2008; Pinto et al. 2010a, 2010b; Singh et al. 2013), but none of them translated to application in breeding programs unlike other crop species such as maize, rice, and sorghum. Since GS can account for multiple alleles simultaneously, it is better suitable for application in sugarcane breeding programs. The first report of GS in sugarcane was by Gouy et al. (2013). Two panels of sugarcane population, each with 167 genotypes, were genotyped at two locations (Reunion Island and Guadeloupe). Both populations were evaluated for 10 traits corresponding to sugar and bagasse contents, digestibility and composition of the bagasse, plant morphology, and disease resistance and genotyped using 1499 DArT markers. For the prediction algorithm, four models were utilized, which included Bayesian LASSO, ridge regression, reproducing kernel Hilbert space, and partial least-square regression. The accuracy of the predictions was verified through correlation analysis between the observed and predicted values, both across and within the two panels. Median correlations ranging from 0.29 to 0.62 (Reunion Island) and 0.11 to 0.5 (Guadeloupe) were obtained for within-panel cross-verification, and cross-validation between panels showed correlation ranging from 0.13 (smut resistance) to 0.55 (brix). The results provided the first concrete proof for the potential for utilization of GS in polyploid crops such as sugarcane. In a similar study, three different populations in the early and advanced selection stage were used to study the accuracy of GS for two important traits, yield and commercial cane sugar (CCS). Five models, Bayes A, Bayes B, Bayesian LASSO, Bayesian GBLUP, and RKHS, were used for genomic prediction. Both pedigree and genotype data from SNP analysis were used for prediction. The prediction models for CCS were found to be highly accurate for genotypes from the advanced selection stage compared to the early selection stage, indicating that the accuracy was not lost due to intense selection for CCS, while it is not the case for cane yield where the prediction accuracies were comparably low. The accuracies of models that included genotype data were also high compared to models where the genotype data were not utilized. The results provided by the existing studies were encouraging, and further research is required to better understand the potential of GS in sugarcane.

## 14.10 Summary

Genomic mapping in polyploids is a very complex process owing to the uncertainty in the ploidy level, chromosome number, gene number, and its isoforms. The sugarcane is cytologically complex due to the high chromosome number, the complex pairing behavior, and meiotic abnormalities. In addition to chromosome number variation, the composition of chromosomes differs among the commercial genotypes. For instance, R570 is composed of 80% *S. officinarum*, 10% *S. spontaneum*, and 10% recombinant chromosomes. All these factors must be considered in setting the genetic parameters for QTL analysis and designing models for GS analysis, failing which the analysis would not be informative or applicable for the sugarcane crop improvement.

## References

Aitken, K.S., Jackson, P.A., and McIntyre, C.L. (2005). A combination of AFLP and SSR markers provides extensive map coverage and identification of homo(eo)logous linkage groups in a sugarcane cultivar. *Theoretical and Applied Genetics* 110 (5): 789–801. https://doi.org/10.1007/s00122-004-1813-7.

Aitken, K.S., Jackson, P.A., and McIntyre, C.L. (2006). Quantitative trait loci identified for sugar related traits in a sugarcane (*Saccharum* spp.) cultivar x *Saccharum officinarum* population. *Theoretical and Applied Genetics* 112 (7): 1306–1317. https://doi.org/10.1007/s00122-006-0233-2.

Aitken, K.S., Hermann, S., Karno, K. et al. (2008). Genetic control of yield related stalk traits in sugarcane. *Theoretical and Applied Genetics* 117 (7): 1191–1203. https://doi.org/10.1007/s00122-008-0856-6.

Ali, A., Pan, Y.B., Wang, Q.N. et al. (2019). Genetic diversity and population structure analysis of *Saccharum* and *Erianthus genera* using microsatellite (SSR) markers. *Scientific Reports* 9 (1): 1–10. https://doi.org/10.1038/s41598-018-36630-7.

Aljanabi, S.M., Parmessur, Y., Kross, H. et al. (2007). Identification of a major quantitative trait locus (QTL) for yellow spot (*Mycovellosiella koepkei*) disease resistance in sugarcane. *Molecular Breeding* 19 (1): 1–14. https://doi.org/10.1007/s11032-006-9008-3.

Asnaghi, C., Paulet, F., Kaye, C. et al. (2000). Application of synteny across Poaceae to determine the map location of a sugarcane rust resistance gene. *Theoretical and Applied Genetics* 101 (5–6): 962–969. https://doi.org/10.1007/s001220051568.

Baird, N.A., Etter, P.D., Atwood, T.S. et al. (2008). Rapid SNP discovery and genetic mapping using sequenced RAD markers. *PLoS One* 3 (10): 3376. https://doi.org/10.1371/journal.pone.0003376.

Balsalobre, T.W.A., da Silva Pereira, G., Margarido, G.R.A. et al. (2017). GBS-based single dosage markers for linkage and QTL mapping allow gene mining for yield-related traits in sugarcane. *BMC Genomics* 18 (1): 1–19. https://doi.org/10.1186/s12864-016-3383-x.

Banerjee, N., Siraree, A., Yadav, S. et al. (2015). Marker-trait association study for sucrose and yield contributing traits in sugarcane (*Saccharum* spp. hybrid). *Euphytica* 205 (1): 185–201. https://doi.org/10.1007/s10681-015-1422-3.

Barreto, F.Z., Rosa, J.R.B.F., Almeida Balsalobre, T.W. et al. (2019). A genome-wide association study identified loci for yield component traits in sugarcane (*Saccharum* spp.). *PLoS One* 14 (7): 1–22. https://doi.org/10.1371/journal.pone.0219843.

Bradbury, P.J., Zhang, Z., Kroon, D.E. et al. (2007). TASSEL: software for association mapping of complex traits in diverse samples. *Bioinformatics* 23 (19): 2633–2635. https://doi.org/10.1093/bioinformatics/btm308.

Bremer, G. (1961a). Problems in breeding and cytology of sugar cane – III. The cytological crossing research of sugar cane. *Euphytica* 10 (2): 229–243. https://doi.org/10.1007/BF00022215.

Bremer, G. (1961b). Problems in breeding and cytology of sugarcane. *Euphytica* 10: 59–78. https://doi.org/10.1007/BF00037206.

Catchen, J., Hohenlohe, P.A., Bassham, S. et al. (2013). Stacks: an analysis tool set for population genomics. *Molecular Ecology* https://doi.org/10.1111/mec.12354.

Chen, P.H., Pan, Y.B., Chen, R.K. et al. (2009). SSR marker-based analysis of genetic relatedness among sugarcane cultivars (*Saccharum* spp. hybrids) from breeding programs in China and other countries. *Sugar Technology* 11 (4): 347–354. https://doi.org/10.1007/s12355-009-0060-2.

Cordeiro, G.M., Taylor, G.O., and Henry, R.J. (2000). Characterisation of microsatellite markers from sugarcane (*Saccharum* sp.), a highly polyploid species. *Plant Science* 155 (2): 161–168. https://doi.org/10.1016/S0168-9452(00)00208-9.

Cordeiro, G.M., Pan, Y.-B.B., and Henry, R.J. (2003). Sugarcane microsatellites for the assessment of genetic diversity in sugarcane germplasm. *Plant Science* 165 (1): 181–189. https://doi.org/10.1016/S0168-9452(03)00157-2.

Costa, E.A., Anoni, C.O., Mancini, M.C. et al. (2016). QTL mapping including codominant SNP markers with ploidy level information in a sugarcane progeny. *Euphytica* 211 (1) https://doi.org/10.1007/s10681-016-1746-7.

Costet, L., Raboin, L.-M., Payet, M. et al. (2012). A major quantitative trait allele for resistance to the *Sugarcane yellow leaf virus* (*Luteoviridae*). *Plant Breeding* 131 (5): 637–640. https://doi.org/10.1111/j.1439-0523.2012.02003.x.

D'Hont, A. and Glaszmann, J.C. (2001). Sugarcane genome analysis with molecular markers: a first decade of research. *Proceedings of the International Society of Sugar Cane Technologists* 24: 556–559.

D'Hont, A., Hai, L.Y., Gonzalez de Leon, D. et al. (1994). A molecular approach to unraveling the genetics of sugarcane, a complex polyploid of the Andropogoneae tribe. *Genome* 37 (2): 222–230. https://doi.org/10.1139/g94-031.

Daetwyler, H.D., Pong-Wong, R., Villanueva, B. et al. (2010). The impact of genetic architecture on genome-wide evaluation methods. *Genetics* 185 (3): 1021–1031. https://doi.org/10.1534/genetics.110.116855.

Danecek, P., Auton, A., Abecasis, G. et al. (2011). The variant call format and VCF tools. *Bioinformatics* 27 (15): 2156–2158. https://doi.org/10.1093/bioinformatics/btr330.

Daugrois, J.H., Grivet, L., Roques, D. et al. (1996). A putative major gene for rust resistance linked with a RFLP marker in sugarcane cultivar "R570". *Theoretical and Applied Genetics* 92 (8): 1059–1064. https://doi.org/10.1007/BF00224049.

Davey, J.W., Hohenlohe, P.A., Etter, P.D. et al. (2011). Genome-wide genetic marker discovery and genotyping using next-generation sequencing. *Nature Reviews Genetics* https://doi.org/10.1038/nrg3012.

Debibakas, S., Rocher, S., Garsmeur, O. et al. (2014). Prospecting sugarcane resistance to sugarcane yellow leaf virus by genome-wide association. *Theoretical and Applied Genetics* 127 (8): 1719–1732. https://doi.org/10.1007/s00122-014-2334-7.

Elshire, R.J., Glaubitz, J.C., Sun, Q. et al. (2011). A robust, simple genotyping-by-sequencing (GBS) approach for high diversity species. *PLoS One* 6 (5) https://doi.org/10.1371/journal.pone.0019379.

Fickett, N., Gutierrez, A., Verma, M. et al. (2019). Genome-wide association mapping identifies markers associated with cane yield components and sucrose traits in the Louisiana sugarcane core collection. *Genomics* 111 (6): 1794–1801. https://doi.org/10.1016/j.ygeno.2018.12.002.

Flint-Garcia, S.A., Thornsberry, J.M., and Edward, S.B. IV (2003). Structure of linkage disequilibrium in plants. *Annual Review of Plant Biology* 54: 357–374. https://doi.org/10.1146/annurev.arplant.54.031902.134907.

Garsmeur, O., Droc, G., Antonise, R. et al. (2018). A mosaic monoploid reference sequence for the highly complex genome of sugarcane. *Nature Communications* 9 (1) https://doi.org/10.1038/s41467-018-05051-5.

Glaszmann, J., Fautret, A., Noyer, J.L. et al. (1989). *Biochemical Genetic Markers in Sugarcane.* Springer.

Glaubitz, J.C., Casstevens, T.M., Lu, F. et al. (2014). TASSEL – GBS: a high capacity genotyping by sequencing analysis pipeline. *PLoS One* 9 (2): e90346.

Goddard, M.E. and Hayes, B.J. (2007). Genomic selection. *Journal of Animal Breeding and Genetics* 124 (6): 323–330. https://doi.org/10.1111/j.1439-0388.2007.00702.x.

Gouy, M., Rousselle, Y., Bastianelli, D. et al. (2013). Experimental assessment of the accuracy of genomic selection in sugarcane. *Theoretical and Applied Genetics* 126 (10): 2575–2586. https://doi.org/10.1007/s00122-013-2156-z.

Gouy, M., Rousselle, Y., Thong Chane, A. et al. (2014). Genome wide association mapping of agro-morphological and disease resistance traits in sugarcane. *Euphytica* 202 (2): 269–284. https://doi.org/10.1007/s10681-014-1294-y.

Grivet, L. and Arruda, P. (2002). Sugarcane genomics: depicting the complex genome of an important tropical crop. *Current Opinion in Plant Biology* 5 (2): 122–127. https://doi.org/10.1016/S1369-5266(02)00234-0.

Grivet, L., D'Hont, A., Roques, D. et al. (1996). RFLP mapping in cultivated sugarcane (*Saccharum* spp.): genome organization in a highly polyploid and aneuploid interspecific hybrid. *Genetics* 142 (3): 987–1000. https://doi.org/10.1093/genetics/142.3.987.

Gutierrez, A.F., Hoy, J.W., Kimbeng, C.A. et al. (2018). Identification of genomic regions controlling leaf scald resistance in sugarcane using a bi-parental mapping population and selective genotyping by sequencing. *Frontiers in Plant Science* 9: 877.

Hamady, M., Walker, J.J., Harris, J.K. et al. (2008). Error-correcting barcoded primers for pyrosequencing hundreds of samples in multiplex. *Nature Methods* 5: 235–237. https://doi.org/10.1038/nmeth.1184.

Heffner, E.L., Sorrells, M.E., and Jannink, J.L. (2009). Genomic selection for crop improvement. *Crop Science* 49 (1): 1–12. https://doi.org/10.2135/cropsci2008.08.0512.

Hoarau, J.Y., Offmann, B., D'Hont, A. et al. (2001). Genetic dissection of a modern sugarcane cultivar (*Saccharum* spp.). I. Genome mapping with AFLP markers. *Theoretical and Applied Genetics* 103 (1): 84–97. https://doi.org/10.1007/s001220000390.

Hoarau, J.Y., Grivet, L., Offmann, B. et al. (2002). Genetic dissection of a modern sugarcane cultivar (*Saccharum* spp.). II. Detection of QTLs for yield components. *Theoretical and Applied Genetics* 105 (6–7): 1027–1037. https://doi.org/10.1007/s00122-002-1047-5.

Janoo, N., Grivet, L., Seguin, M. et al. (1999). Molecular investigation of the genetic base of sugarcane cultivars. *Theoritical Applied Genetics* 99: 171–184.

Kumar, S., Banks, T.W., and Cloutier, S. (2012). SNP discovery through next-generation sequencing and its applications. *International Journal of Plant Genomi* https://doi.org/10.1155/2012/831460.

Le Cunff, L., Garsmeur, O., Raboin, L.M. et al. (2008). Diploid/polyploid syntenic shuttle mapping and haplotype-specific chromosome walking toward a rust resistance gene (Bru1) in highly polyploid sugarcane (2n ~ 12x ~ 115). *Genetics* 180 (1): 649–660. https://doi.org/10.1534/genetics.108.091355.

Li, H. and Durbin, R. (2009). Fast and accurate short read alignment with Burrows-Wheeler transform. *Bioinformatics* https://doi.org/10.1093/bioinformatics/btp324.

Li, H., Handsaker, B., Wysoker, A. et al. (2009). The sequence alignment/map format and SAMtools. *Bioinformatics* 25 (16): 207–209. https://doi.org/10.1093/bioinformatics/btp352.

Lu, F., Lipka, A.E., Glaubitz, J. et al. (2013). Switchgrass genomic diversity, ploidy, and evolution: novel insights from a network-based SNP discovery protocol. *PLoS Genetics* https://doi.org/10.1371/journal.pgen.1003215.

Mace, E.S. and Jordan, D.R. (2011). Integrating sorghum whole genome sequence information with a compendium of sorghum QTL studies reveals uneven distribution of QTL and of gene-rich regions with significant implications for crop improvement. *Theoretical and Applied Genetics* 123 (1): 169–191. https://doi.org/10.1007/s00122-011-1575-y.

McIntyre, C.L., Whan, V.A., Croft, B. et al. (2005). Identification and validation of molecular markers associated with Pachymetra root rot and brown rust resistance in sugarcane using map- and association-based approaches. *Molecular Breeding* 16 (2): 151–161. https://doi.org/10.1007/s11032-005-7492-5.

Meuwissen, T. (2003). Genomic selection: the future of marker assisted selection and animal breeding. Conference 10 of the FAO Biotechnology Forum is Entitled "Molecular Marker Assisted Selection as a Potential Tool for Genetic Improvement of Crops, Forest Trees, Livestock and Fish in Developing Countries". http://www.fao.org/biotech/docs/meuwissen.pdf (accesssed 17 November to 12 December 2003).

Meuwissen, T. and Goddard, M. (1996). The use of marker haplotypes in animal breeding schemes. *Genetics, Selection, Evolution* 28 (2): 161. https://doi.org/10.1186/1297-9686-28-2-161.

Ming, R., Liu, S.C., Moore, P.H. et al. (2001). QTL analysis in a complex autopolyploid: genetic control of sugar content in sugarcane. *Genome Research* 11 (12): 2075–2084. https://doi.org/10.1101/gr.198801.

Ming, R., Wang, Y.W., Draye, X. et al. (2002a). Molecular dissection of complex traits in autopolyploids: mapping QTLs affecting sugar yield and related traits in sugarcane. *Theoretical and Applied Genetics* 105 (2–3): 332–345. https://doi.org/10.1007/s00122-001-0861-5.

Ming, R., Wang, Y.W., Draye, X. et al. (2002b). Molecular dissection of complex traits in autopolyploid: mapping QTLs affecting sugar yield and related traits in sugarcane. *Theoretical and Applied Genetics* 105: 332–345. https://doi.org/10.1007/s00122-001-0861-5.

Moore, P.H. (2005). Integration of sucrose accumulation processes across hierarchical scales: Towards developing an understanding of the gene-to-crop continuum. *Field Crops Research* 92 (2–3): 119–135. https://doi.org/10.1016/j.fcr.2005.01.031.

Pinto, L.R., Leite, D.C., Fávero, T.M. et al. (2010a). Identification of microsatellite markers associated with yield components and quality parameters in sugarcane. *Proceedings of ISSCT* 27: 1–7.

Pinto, L., Garcia, A.A.F., Pastina, M. et al. (2010b). Analysis of genomic and functional RFLP derived markers associated with sucrose content, fiber and yield QTLs in sugarcane (*Saccharum* spp.) commercial cross. *Euphytica* 172: 313–327. https://doi.org/10.1007/s10681-009-9988-2.

Piperidis, N., Jackson, P.A., D'Hont, A. et al. (2008). Comparative genetics in sugarcane enables structured map enhancement and validation of marker-trait associations. *Molecular Breeding* 21 (2): 233–247. https://doi.org/10.1007/s11032-007-9124-8.

Poland, J.A., Brown, P.J., Sorrells, M.E., and Jannink, J.L. (2012). Development of high-density genetic maps for barley and wheat using a novel two-enzyme genotyping-by-sequencing approach. *PLoS One* 7 (2).

Raboin, L.-M. (2006). *Improvement of Sugarcane Through Breeding and Genetics View Project SPAD-Systèmes de Production d'Altitude et Durabilité à Madagascar View Project*. Springer https://doi.org/10.1007/s00122-006-0240-3.

Racedo, J., Gutiérrez, L., Perera, M.F. et al. (2016). Genome-wide association mapping of quantitative traits in a breeding population of sugarcane. *BMC Plant Biology* 16 (1): 142. https://doi.org/10.1186/s12870-016-0829-x.

Rosyara, U.R., De Jong, W.S., Douches, D.S., and Endelman, J.B. (2016). Software for genome-wide association studies in autopolyploids and its application to potato. *Plant Genome* 9 (2): 1–10. https://doi.org/10.3835/plantgenome2015.08.0073.

Sills, G.R., Bridges, W., Al-Janabi, S.M., and Sobral, B.W.S. (1995). Genetic analysis of agronomic traits in a cross between sugarcane (*Saccharum officinarum* L.) and its presumed progenitor (*S. robustum* Brandes & Jesw. ex Grassl). *Molecular Breeding* 1 (4): 355–363. https://doi.org/10.1007/BF01248413.

Singh, R.K., Jena, S.N., Khan, S. et al. (2013). Development, cross-species/genera transferability of novel EST-SSR markers and their utility in revealing population structure and genetic diversity in sugarcane. *Gene* 524 (2): 309–329. https://doi.org/10.1016/J.GENE.2013.03.125.

Siraree, A., Banerjee, N., Kumar, S. et al. (2017). Identification of marker-trait associations for morphological descriptors and yield component traits in sugarcane. *Physiology and Molecular Biology of Plants* 23 (1): 185–196. https://doi.org/10.1007/s12298-016-0403-x.

Sonah, H., Bastien, M., Iquira, E. et al. (2013). An improved genotyping by sequencing (GBS) approach offering increased versatility and efficiency of SNP discovery and genotyping. *PLoS One*: 8–e54603. https://doi.org/10.1371/journal.pone.0054603.

Sreenivasan, T.V. and Ahloowalia, B.S. (1987). *Sugarcane Improvement Through Breeding* (ed. D.J. Heinz), 211–253. Amsterdam: Elsevier.

Torkamaneh, D., Laroche, J., and Belzile, F. (2016). Genome-wide SNP calling from genotyping by sequencing (GBS) data: a comparison of seven pipelines and two sequencing technologies. *PLoS One* 11 (8): 1–14. https://doi.org/10.1371/journal.pone.0161333.

Ukoskit, K., Posudsavang, G., Pongsiripat, N. et al. (2019). Detection and validation of EST-SSR markers associated with sugar-related traits in sugarcane using linkage and association mapping. *Genomics* 111 (1): 1–9. https://doi.org/10.1016/j.ygeno.2018.03.019.

Van Oeveren, J., De Ruiter, M., Jesse, T. et al. (2011). Sequence-based physical mapping of complex genomes by whole genome profiling. *Genome Research* 21 (4): 618–625. https://doi.org/10.1101/gr.112094.110.

Vettore, A.L., da Silva, F.R., Kemper, E.L. et al. (2003). Analysis and functional annotation of an expressed sequence tag collection for tropical crop sugarcane. *Genome Research* 13 (12): 2725–2735.

Wang, S., Basten, C.J., and Zeng, Z.B. (2012). *Windows QTL Cartographer 2.5*. Raleigh, NC: Department of Statistics, North Carolina State University http://statgen.ncsu.edu/qtlcart/WQTLCart.htm.

Wei, X., Jackson, P.A., McIntyre, C.L. et al. (2006). Associations between DNA markers and resistance to diseases in sugarcane and effects of population substructure. *Theoretical and Applied Genetics* 114 (1): 155–164. https://doi.org/10.1007/s00122-006-0418-8.

Wei, X., Jackson, P.A., Hermann, S. et al. (2010). Simultaneously accounting for population structure, genotype by environment interaction, and spatial variation in marker-trait associations in sugarcane 1. *Genome* 53 (11): 973–981. https://doi.org/10.1139/G10-050.

Weir, B.S. (2008). Linkage disequilibrium and association mapping. *Annual Review of Genomics and Human Genetics* 9 (1): 129–142. https://doi.org/10.1146/annurev.genom.9.081307.164347.

Wu, K.K., Burnquist, W., Sorrells, M.E. et al. (1992). The detection and estimation of linkage in polyploids using single-dose restriction fragments. *Theoretical and Applied Genetics* 83 (3): 294–300. https://doi.org/10.1007/BF00224274.

Wu, J., Wang, Q., Xie, J. et al. (2019). SSR marker-assisted management of parental germplasm in sugarcane (*Saccharum* spp. Hybrids) Breeding Programs. *Agronomy* 9 (8): 1–16. https://doi.org/10.3390/agronomy9080449.

Yang, X., Sood, S., Glynn, N. et al. (2017). Constructing high-density genetic maps for polyploid sugarcane (*Saccharum* spp.) and identifying quantitative trait loci controlling brown rust resistance. *Molecular Breeding* 37: 116.

Yang, X., Islam, M.S., Sood, S. et al. (2018a). Identifying quantitative trait loci (QTLs) and developing diagnostic markers linked to orange rust resistance in sugarcane (*Saccharum* spp.). *Frontiers in Plant Science* 9 (March): 1–10. https://doi.org/10.3389/fpls.2018.00350.

Yang, X., Song, J., Todd, J. et al. (2018b). Target enrichment sequencing of 307 germplasm accessions identified ancestry of ancient and modern hybrids and signatures of adaptation and selection in sugarcane (*Saccharum* spp.), a 'sweet' crop with 'bitter' genomes. *Plant Biotechnology Journal* 17: 1–11.

Yang, X., Sood, S., Luo, Z. et al. (2019). Genome-wide association studies identified resistance Loci to orange rust and yellow leaf virus diseases in sugarcane (*Saccharum* spp.). *Phytopathology.* 109: 623–631.

Zhang, J., Nagai, C., Yu, Q. et al. (2012). Genome size variation in three *Saccharum* species. *Euphytica* 185 (3): 511–519. https://doi.org/10.1007/s10681-012-0664-6.

Zhang, J., Zhang, X., Tang, H. et al. (2018). Allele-defined genome of the autopolyploid sugarcane *Saccharum spontaneum* L. *Nature Genetics* 50 (11): 1565–1573. https://doi.org/10.1038/s41588-018-0237-2.

# 15

# Recent Advances and Applicability of GBS, GWAS, and GS in Polyploid Crops

*Vandana Thakral, Himanshu Yadav, Gunashri Padalkar, Surbhi Kumawat, Gaurav Raturi, Virender Kumar, Rushil Mandlik, Nitika Rajora and Manipal Singh*

*Department of Agriculture Biotechnology, National Agri-Food Biotechnology Institute (NABI), Mohali, Punjab, India*

## 15.1 Introduction

The polyploid nature of plants is very prominent especially in angiosperms due to the occurrence of hybridization (allopolyploidy) and/or genome doubling (autopolyploidy) (Doyle et al. 2008). Genome duplication can result in novel traits and adaptive mechanisms by producing genes with distinct functions (Crow and Wagner 2005). Many of the cultivated crops including *Saccharum officinarum* (sugarcane), *Musa* sp. (banana), *Coffea arabica* (coffee), *Solanum tuberosum* (potato), *Brassica* species, and *Fragaria ananassa* (strawberry) are polyploid in nature. Polyploids are thought to account for around 80% of all living plants. It results in the production of novel plants in terms of genetic, genomic, and phenotypic traits (Soltis et al. 2016). It is an important mechanism and driving factor for the origin of the new plant species (Bento et al. 2011). However, due to the high complexity of polyploid genome, it becomes difficult to link genotype to phenotype (Bhat et al. 2020). Polyploids contain several alleles linked with a single locus, and as a result, segregation is more complicated than in diploids (Sonah et al. 2016; Agarwal et al. 2018; Shivaraj et al. 2019). Traditional breeding of polyploid crops is difficult because of their heterozygous character; however, techniques such as marker-assisted selection may substantially enhance breeding efficiency.

Polyploid plants are known to be superior and have increased vigor as compared to their diploid progenitors. This superiority of polyploids has been exploited by numerous plant breeders over the last century, by inducing polyploidy and/or utilizing natural polyploids in a variety of methods to create cultivars with higher yield and increased tolerance to biotic and abiotic stress. After the discovery of the property of inducing polyploidy in Colchicine, various researchers attempted to produce polyploid crops between the 1930s and 1970s, like sugar beet, red clover, watermelon, alsike clover, and grape. However, only a few of

*Genotyping by Sequencing for Crop Improvement*, First Edition.
Edited by Humira Sonah, Vinod Goyal, S.M. Shivaraj, and Rupesh K. Deshmukh.

them displayed valuable traits and became commercial. Development of triploid sugar beets is one of the examples of an induced polyploid plant that is generated by crossing sterile male diploid plants with tetraploids or through reciprocal crossing (Kinoshita and Takahashi 1969). Triploid sugar beets have larger roots than diploids, but they produce more sugar per unit area, as they have the same sugar content as diploids. Furthermore, triploid sugar beets are sterile and do not generate seeds, which is not a drawback because only their roots are of commercial significance (Dabholkar 2006). Seedless watermelon (triploid) is produced by the colchicine treatment of diploid plant to obtain tetraploid form and then pollen of another diploid plant is used to pollinate tetraploid stigma resulting in the production of triploid watermelon. However, the cost of seeds for triploid variety is much higher, at least four times than that of the seeded diploid, due to the challenging seed germination and establishment which leads to higher market price (Ho et al. 2017; Crow 1994).

The polyploid genome's complexity is an impediment to many molecular and breeding techniques. Identification of quantitative trait loci (QTL) in genomes of polyploid crops is challenging (Cai et al. 2012; Tomar et al. 2016; Mishra et al. 2021). Also, due to the low resolution of available genome-wide markers, genome-wide association studies in these crops have had minimal success yet (Sonah et al. 2015; Zargar et al. 2015). Candidate gene methods based on information from well-characterized, closely related model species are similarly challenging in wheat and oilseed rape, because numerous homologues and paralogues of any given gene of interest might obstruct a straightforward interpretation of its significance in the crop genome. These drawbacks of polyploidy, extended gene families, and abundant repetitive sequences, all hinder the identification and mapping of single-nucleotide polymorphisms (SNPs), a crucial initial step in genomics-based crop development (Edwards et al. 2013).

Next-generation sequencing (NGS) has profoundly altered the genomics landscape during the last decade by enabling the investigation of polyploid genomes (Varshney et al. 2009; Gupta et al. 2020). 2013 with the help of NGS, whole reference genomes of progenitors of polyploid species have been sequenced, resolving the issues of genome assembly (Kyriakidou et al. 2018). The bioinformatic analysis of NGS data for polyploid crops is a great challenge (Batley and Edwards 2009), however an increasing number of tools have been developed to analyze the big data. Researchers may use whole-genome sequence data from various plant species to create precise genotyping systems that can be used by breeders in their field experiments in the form of genomics-assisted breeding (GAB) (Kim et al. 2020).

## 15.2 Challenges for Genotyping in Polyploidy Crops

Genotyping involves the identification of polymorphic alleles in DNA sequences of individuals. Various molecular techniques and genomic tools can be used for identification of these allelic polymorphisms. Some of the widely used combinations of PCR-based molecular markers include randomly amplified polymorphic DNA (RAPDs), simple-sequence repeats (SSRs), Cleaved amplified polymorphic sequences (CAPS), amplified fragment length polymorphisms (AFLPs) and single nucleotide polymorphisms (SNPs). The AFLP technique identifies the alleles with the help of PCR-based amplification and restriction enzyme digestion. Direct sequencing of important genomic region including ITS (internal transcribed spacer) regions, rDNA loci, and chloroplasts, is rather simple than the DNA sequence-based genotyping.

Due to duplication of whole genome in polyploid plants, multiple copies of restriction enzyme sites and primers binding sites may be present. It is difficult to identify whether the allelic variation exists in homeologous regions or homologous regions in DNA sequence analysis (Imelfort and Edwards 2009). Homeologous intergenomic markers are frequently mistakenly identified as polymorphic by sequencing techniques for identifying allelic variation. This type of overestimation is common in approaches such as finding SNP in polyploids (Duran et al. 2009; Hayward et al. 2012). Further in PCR-based marker genotyping, polyploidy nature of plant interferes with the designing of primers. Designed primers tend to amplify the homologous region along with the target region. Up to 50% of PCR-based genotyping SNPs identified from sequencing data in modern polyploids with similar genomes, such as wheat, hexaploid oats, and canola, turn out to be attributable to polymorphisms across homeologous regions rather than between alleles at the same locus. Similarly, due to genome doubling multiple copies of restriction enzyme sites present in polyploids (Oliver et al. 2011; Hayward et al. 2012). A lack of understanding of the polyploidy event will have an impact on all inferences drawn from the genotyping data.

## 15.3 Genotyping Platforms for Barley

Barley is one of the major cereal crops having nutritional and economic importance. It is ranked fourth grain crop in terms of production by the UN's Food and Agriculture Organization (http://faostat.fao.org). In 2006, high-throughput genotyping was first introduced in barley, with the advent of two Illumina Golden Gate tests (Fan et al. 2003) each included 1572 SNP markers (Close et al. 2009). Advancements in genetic research and breeding of grain crops, genotyping with SNP markers is now widely used and can be divided into two types, i.e. semi-open and closed based on the type of data generated (Darrier et al. 2019). Closed system genotyping includes SNP arrays (Bayer et al. 2017) in which the same panel of SNPs is tested repeatedly for variance across all trials and germplasm. Whereas, RAD-seq (Mr et al. 2007), DArTseq (Kilian et al. 2012), or genotyping-by-sequencing (GBS) (Elshire et al. 2011) are included in semi-open system, which analyses new SNP variation in each individual present in a set of various genetic material. Commercial SNP assay and genotyping by sequencing are the two most prevalent types of high-throughput genotyping systems. In case of barley, because of highly robust technology and exceptional data quality with few missing values SNP-based assay are generally favored over other genotyping platforms from the last one year. Most recent iteration of SNP-based assay is 50K iSelect custom genotyping array comprising of SNPs chosen from a pool of previously relevant "legacy" Illumina platform SNPs and a large cohort obtained from complete exome capture sequence data (Bayer et al. 2017). Using Illumina iSelect SNP platform for barley genotyping leads to the representation of highly diversified range of domesticated barley with minor allele frequencies (MAFs) of >5% which is physically and genetically mapped as well as well-distributed across each of the seven barley chromosomes. However, utilising SNP assays carries a number of challenges. Firstly, it is species specific and had to be developed uniquely for each crop. Further, their development and testing may be time-consuming and expensive, and adding additional SNPs to the platform is challenging and costly. Also, cost of a single genotyping assay is higher than that of GBS (Moragues et al. 2010).

GBS is an elegant NGS technique based on reduced representative sequencing for the production of large-scale genomic information, initially reported by Elshire et al. (2011). GBS does not require any prior knowledge of the target genome and may create a huge number of SNPs at a low cost. GBS as a platform for "germplasm genomics" could be considered as an alternative to whole-genome (shotgun) sequencing (WGS), which is highly expensive for large genome sized crops such as maize (~2.5 Gbp), barley (~5 Gbp), and wheat (~17 Gbp), in contrast to rice (~370 Mbp) (Wang et al. 2018) and Arabidopsis (~135 Mbp) (Alonso-Blanco et al. 2016). Darrier et al. (2019) compared both genotyping platforms (GBS and SNP arrays), for 1000 diverse barley genotypes (Figure 15.1). Both platforms found equal numbers of robust bi-allelic SNPs and had a modest overlap of SNPs, showing that the techniques utilized selectively access relevant variation in distinct parts of the barley genome. During this comparison approximately half of the SNPs in the GBS dataset had MAFs of less than 1%, demonstrating GBS's ability to discover uncommon alleles in various germplasm collections while the extremely robust calling of alleles at the same SNPs across various populations is a benefit of the SNP array. The cost per genotyping is lower for SNP arrays than for GBS, resulting in a substantially lower cost per informative data point for the SNP array. The technology used should be carefully considered in light of the needed applications and objectives, research questions being addressed, group preferences, and their ability to collect and understand various forms of output data quickly and readily. Current commercial pricing comparisons call into doubt the commonly known belief that GBS is significantly less expensive than SNP arrays, at least in barley.

## 15.4 Long-Read Sequencing-based Genotyping in Polyploid Canola

Hybridization events between *Brassica rapa* and *Brassica oleracea* resulted in a allotetraploid *Brassica napus* whose genome showed high collinearity with their progenitor crops (Chalhoub et al. 2014). Highly homologous progenitor genomes along with genome duplication and frequent homologous recombination make SNP genotyping for canola difficult. The complexity leads to misalignment of short sequencing reads resulting in reporting of false SNPs. Numerous studies in brassica reported that 90% of SNPs variation were detected across homoeologous areas and were dubbed "hemi-SNPs" (Trick et al. 2009). Also, in other short probe-based techniques with the *Brassica* 6 K SNP array revealed that up to 62% of SNPs in a pair of doubled haploids (DHs) were heterozygous (Cai et al. 2012). When filtering SNP loci, the flanking sequence which matches to diverse regions in the reference genome, results in 3040 K SNPs removal from the 60 K array that are informative and can be uniquely matched (Mason et al. 2017).

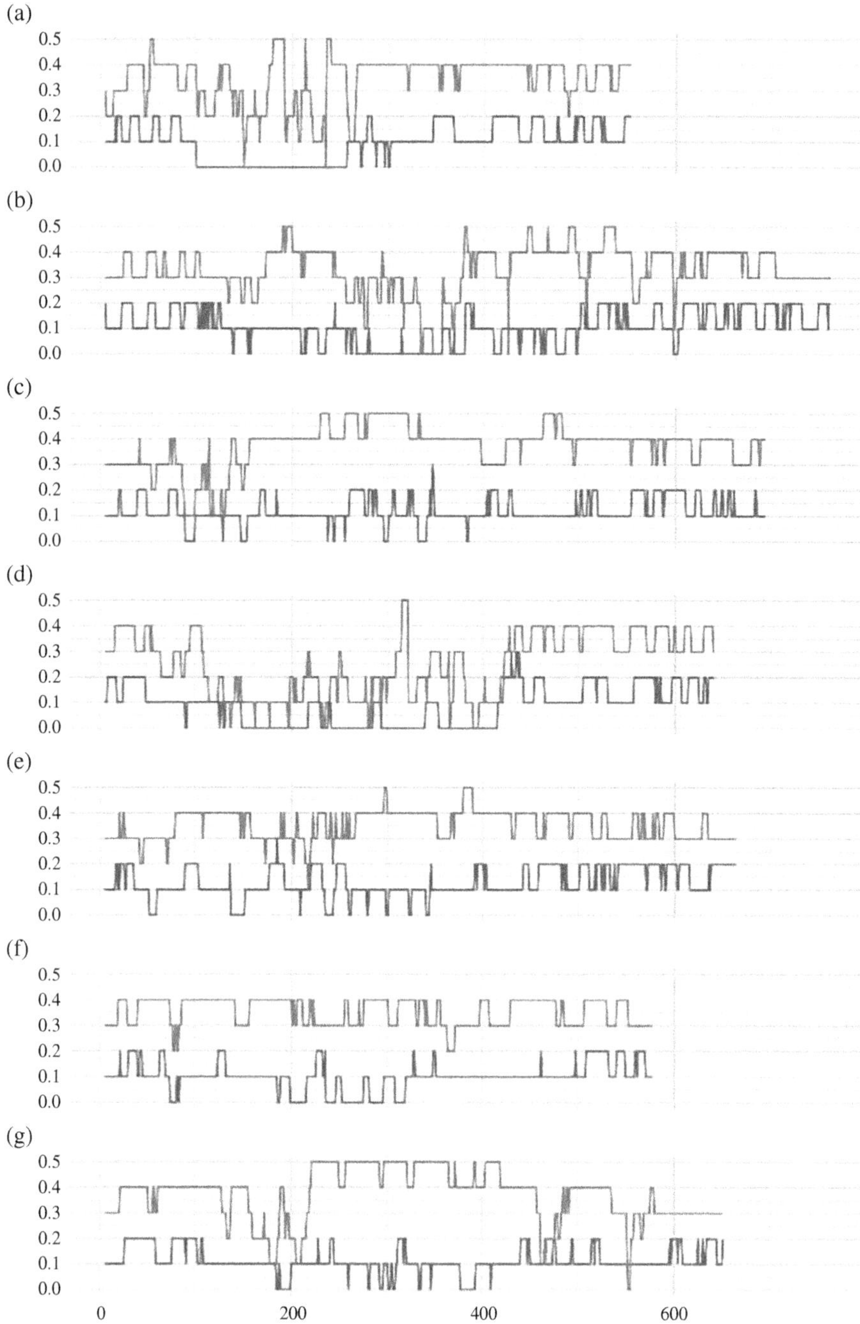


**Figure 15.1** GBS data of seven barley chromosomes 1H–7H showing genetic diversity ($\pi$) (a–g). *Source:* The figure is reproduced from Darrier et al. (2019) which is publicly available under a Creative Commons Attribution 4.0 (CC-By 4.0) International License.

Rapid technological advancements lead to the development of second- and third-generation sequencing platforms that can produce a large amount of short- or long-read data at relatively low cost. Identification of false positives in complex genomes due to multiple alignments in the reference genome can be overcome by long-read sequencing (LRS) technology. It can generate reads that can be uniquely aligned to a reference genome. Because of its capacity to traverse repeated areas and complete variants, LRS rapidly become the method of choice for de novo genome assembly and structural variant detection (Ebbert et al. 2018; Fuselli et al. 2018; Miller et al. 2018). However, the present error rate of around 10%, to ensure that the final sequence is correct, more precise short-read sequencing (SRS) is required to polish such assemblies, which has likely precluded LRS from being evaluated for regular SNP genotyping applications (Magi et al. 2017).

Skim whole genome re-sequencing (skim WGR) using LRS is a potent method for *de novo* SNP discovery along with correction or substantial sequencing depth using SRS for accuracy of the generated data (Jia et al. 2013). To be realistically utilized for genotyping, minION LRS genotype calls must include less mistakes than the overall 10% error rate, but this may be tolerated for genomic selection, which has been shown to be reasonably resilient to error rates of up to 10% (Pérez-Enciso et al. 2015). MinION LRS is free of the biases that affect other SRS systems, such as PCR bias, and GC-content, which leads to more even representation of the genome (Magi et al. 2017). Using this technique with simple bioinformatic pipeline, genotype accuracy as high as 98.4% has been achieved in canola (Malmberg et al. 2019). More advanced approaches are likely to result in accurate genotyping of even more complex germplasm in the future. At the moment, this method may be utilized to genotype known SNP locations in DHs and enhance current SNP lists in species with extensively duplicated genomes.

## 15.5 Peanut Genotyping with Targeted Amplicon Sequencing

Peanuts are a major food, feed, and oil crop that is cultivated in over 100 countries. It has oil (48–50%) and protein (25–28%) in an abundant amount in the kernels. Peanut is an allotetraploid ($2n = 2\times = 40$) comprising two diploid genomes of progenitor *Arachis duranensis* and *Arachis ipaënsis* (Bertioli et al. 2016). Because of the complexity of the peanut genome, tracing the inheritance of characteristics linked to oil components, as well as other agronomically relevant traits, is challenging (Song et al. 2016; Agarwal et al. 2018). Improved peanut cultivars with great production potential have been created and released for planting across the world. For the improvement of its oil components, various breeding projects have been proposed. SSR markers, which need time-consuming electrophoretic studies, were used to create genetic maps for such breeding schemes. Peanut has a limited genetic diversity owing to a single hybridization event between two diploid species, followed by chromosomal doubling and wild diploid species crossing barrier, makes it difficult for the improvement in important traits using genomic tools. Recent studies have reported the development of DNA markers and molecular genetics map for peanut (Pandey et al. 2012; Shirasawa et al. 2013). The discovery of QTL helps DNA marker-assisted selection to accelerate breeding (Leal-Bertioli et al. 2015).

Development of various NGS platforms contributed to the sequencing of >100 plant species including peanut (Michael and VanBuren 2015). With this discovery it is possible to locate the SNPs throughout the peanut genome with the help of whole-genome resequencing strategy. In peanut, huge amount of data generated by NGS technology helps in genome-wide genotyping techniques such as GBS and restriction site-associated DNA sequencing (RAD-Seq) (Zhou et al. 2014). High flexibility and cost-effectiveness make them a desirable tool for SNP genotyping and their wide use in plant genetics, genomics, and breeding. A genetic linkage map in peanut was constructed with the help of ddRAD-Seq analysis (Zhou et al. 2014). However, due to low genetic diversity in peanut fewer SNPs had been listed by ddRAD-Seq. This limitation can be overcome by targeted amplicon sequencing (TAS) strategy. TAS is a method of determining marker genotypes by sequencing PCR fragments generated from polymorphic loci (Mamanova et al. 2010). It provides an alternate technique for genome-wide genotyping in plants with ultralow SNP density. For TAS analysis, genome-wide SNP data are obtained using whole-genome resequencing analysis. Furthermore, a sample of SNPs encompassing the peanut genome is chosen as the target for amplicon sequencing analysis to overcome significant biases in the number and genomic locations of interspecific and intraspecific SNPs (Shirasawa et al. 2016). TGAs' substantial percentage of experimental flexibility, less DNA requirement, and high operating speed make it a better genotyping platform over SNP-chip array technologies.

## 15.6 SNP Genotyping Methods and Platforms Available for Sugarcane

Sugarcane is one of the largest produced tropical crops in the world and has a significant economic value due to its high sucrose content. It has recently received interest for ethanol and biodiesel production as an alternate for fossil fuels and might serve as a major renewable resource to alleviate the present energy crisis and greenhouse gas emissions making it a substantial dual-purpose crop (Savage 2011). Its role as a biofuel crop as well as important food crop makes it a candidate for genomic analysis to identify novel and useful alleles to incorporate into present germplasm. Genome sequencing of sugarcane is a complex process due to its extreme level of heterozygosity and polyploidy. Cultivated sugarcane is the hybrid of two progenitors *S. officinarum* and *S. spontaneum* and is thought to have ploidy level between 5× and 16× suggesting that it has 5–16 homologous copies of each gene in its genome (D'Hont 2005). Furthermore, the chromosomal inheritance ratio in a hybrid may be inconsistent, and each hybridization may result in a different genetic composition.

The sugarcane genome is becoming more tractable with technical advancements in DNA sequencing technology and genome assembly. Progress has been made to unravel the sugarcane nuclear and organellar genomes with the availability of the genome sequences of related species (Daniels and Roach 1987; Junior et al. 2004). The genome size of monoploid sugarcane spans from 760 to 926 Mb, about twice the size of the rice genome (389 Mbp), whereas the genome size of a hybrid sugarcane may reach 10 Gb (D'Hont and Glaszmann 2001; Souza et al. 2011), which is almost three times the size of the human genome.

For comparative genomics to match the level of polyploidy existing in sugarcane, *Sorghum bicolor* genome can be used because of its higher synteny with the sugarcane genome (Guimarães et al. 1997).

The creation of an appropriate reference assembly is the initial step toward producing a genomic sequence for a crop species (Okura et al. 2012; Berkman et al. 2014). Correct reference assembly can be achieved by various NGS techniques including PacBio single-molecule real-time (SMRT) sequencing (Eid et al. 2009; Souza et al. 2011), the Illumina TruSeq synthetic long-read technology. All of these technologies produce long-length reads upto several kilobases. The first-ever sugarcane genome assembly comprising of 382 Mb (Garsmeur et al. 2018) with reference sequence of the basic (monoploid) sugarcane genome was established using these technologies. These third-generation sequencing technologies switched the traditional short-read RNA-Seq technology as it improves comprehension of the complexity of isoforms for species that lack a reference. A read length of approximately 30 kb enables more accurate characterization of nucleotide variants. Transcript datasets obtained from these techniques will be a useful reference for SNP discovery, genome annotation, and splice variant identification (Jiao et al. 2017).

The advancement of NGS and subsequent discovery of SNPs has replaced the medium-throughput SSRs. Variations in the nucleotide sequence of coding regions result in an altered phenotype. SNPs present in such regions have significant value to study phenotypic polymorphism. They can be identified by aligning accession with the reference genome or utilizing de novo assembly techniques to compare reads from various genotypes (You et al. 2011). High density of SNPs in the crop genome increases their scope in genotyping techniques. Some earlier SNP-based genotyping techniques include CAPS marker (Thiel et al. 2004), TaqMan-based PCR system from Applied Biosystems (Schleinitz et al. 2011), and KASP (Semagn et al. 2014). Because of medium to high throughput, cheap cost, and sensitivity, the TaqMan and KASP assays are widely utilized. SNP genotyping based on TaqMan and KASP is possible with the real-time machines' high-resolution melting (HRM) curve analysis. Currently, commonly used mainstream SNP-based genotyping platforms are sequencing-based methods and SNP microarray. The SNP arrays have been used for genotyping of many species like rice (Chen et al. 2014; Thomson et al. 2017), maize (Unterseer et al. 2014), apple (Bianco et al. 2016), wheat (Winfield et al. 2016), and other polyploids (You et al. 2018). It is a relatively simple genotyping platform with easy data analysis without using any bioinformatic tools. Complexity of sugarcane genome with ~10 Gb size, for significant coverage of SNPs to be represented on an SNP array, a large amount of sequencing data are required. Once the data are available, analysis and genotyping can be performed without any exertion. High-density SNP arrays in sugarcane allow for the screening of a large number of genotypes. However, there is certain biasness in diversity analyses when the SNPs are selected for building arrays which can be overcome by using GBS. GBS is a suitable genotyping platform for nonmodel species because it does not require a reference genome but require computational tools for data processing. A specific bioinformatics pipeline is required to analyze allelic differences between homologous and homologous chromosomes. Both array-based and sequencing-based methods have their own pros and cons. A combination of GBS and array-based genotyping is preferable to optimize genomic selection (GS)/prediction accuracy. Targeted GBS is better for marker-assisted breeding.

## 15.7 Recent Advances and Applicability of GBS, GWAS, and GS in Polyploidy Crop Species

### 15.7.1 Linkage Maps and Mapping Software Tools for Polyploids

Linkage maps are used to describe the recombination landscapes within the species using generally one experimental cross between the parent lines of interest. As the knowledge of genetic distance is more important than physical distance, studying linkage maps is of great significance (Sonah et al. 2013, 2015, 2016). Linkage maps enable scientists to decipher the recombination frequencies as well as the linkage drag around loci of interest (de Ronne et al. 2020; Katara et al. 2010; Zargar et al. 2015). Various software used for QTL analysis also requires linkage maps of markers as it gives precise information about the co-inheritance of phenotype and markers as compared to physical maps, as physical distance does not give direct information of recombination frequencies especially in the pericentromeric regions. Another rationale for generating linkage maps is because they provide the initial genomic representation of a species, which can then be used to build more sophisticated representations (Patil et al. 2018; Agarwal et al. 2018; Singh et al. 2012). Also, they provide significant long-range linkage information throughout the whole chromosome, which is missing from short sequence reads assembly. Due to this ability of connecting and correctly orientating, the scaffolds are continuously exploited in genome assembly projects (Bartholomé et al. 2015; Fierst 2015).

Linkage maps in polyploids exhibit 1 : 1 segregation pattern, commonly known as simplex markers. These markers are having countless advantages over other segregation markers along with some distinct disadvantages. Irrespective of the mode of inheritance, in polyploid species simplex markers behave the same as in diploid species. In these species, the repulsion-phase recombination frequency estimate does not depend on ploidy level and thus has very less influence on map construction because of low LOD value. Due to this advantage, simplex markers could be used to create "assumption-free" linkage maps in unexplored polyploid species and their mode of inheritance can be studied. Although there is one exception to this is double reduction. Double reduction arises during meiosis when segment of chromosome along with its sister chromatid copy gets transmitted to an offspring which is a result of multivariant pairing (Haldane 1930). As double reduction occurs in a random manner in polysomic species and introduces only small biasness, it does not influence the recombination frequency estimates (Bourke et al. 2015). As a consequence, diploid mapping software can be used for simplex markers sets despite of any ploidy level and meiotic pairing behavior (Cheema and Dicks 2009) (Figure 15.2).

In conjunction with these advantages, simplex marker sets exhibit some limitations also. They provide very limited information about linkage in repulsion phase, chiefly at higher ploidy levels. This means that the homolog-specific maps can be generated, but they may not display integration between homologs in a single parent and do not integrate across parents (van Geest et al. 2017). Besides this, it reduces the map coverage because of limited use of markers with different segregation types. Therefore, a true polyploidy linkage mapping tool is needed which includes not only 1 : 1 segregating markers but all types of segregation markers.

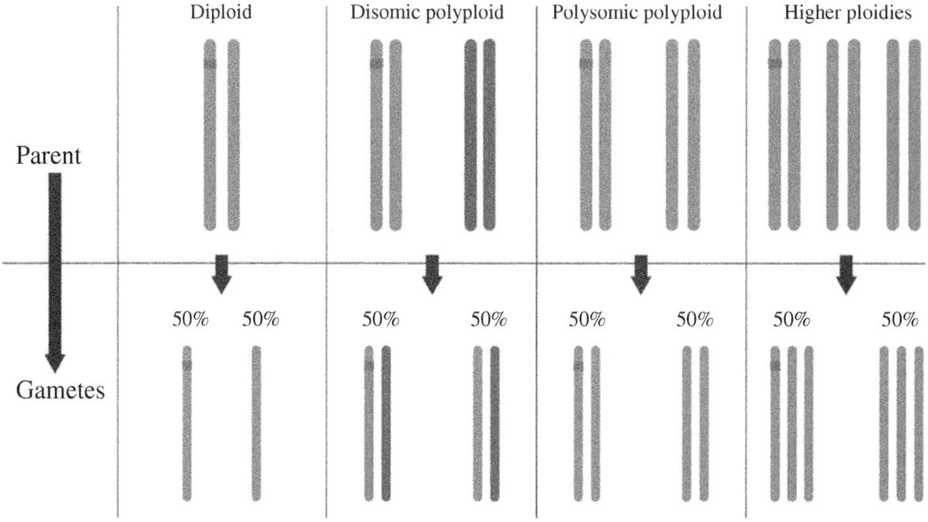


**Figure 15.2** Illustration to depict that the simplex markers show similar mode of inheritance and pairing behavior across all ploidy level, permitting usage of diploid mapping software. Here, SNPs (simplex) are shown in red color bands. *Source:* The figure is reproduced from Bourke et al. (2018) which is available under a Creative Commons Attribution 4.0 (CC-By 4.0) International License, which permits reproduction.

Linkage mapping is performed in the following order: linkage analysis, marker clustering, and ordering. There are very few software tools that can execute all the three tasks for polysomic species. The most commonly used tool for linkage mapping in polyploidy species is TetraploidMap (Goudet 1995; Hackett et al. 2007). This program also does QTL interval mapping and dosage-scored SNP data in addition to creating linkage maps for allotetraploid species (Hackett et al. 2013). More recently, TetrploidSNPMap has been developed and is publicly available from the Scottish BioSS website which uses a very sophisticated graphical user interface (GUI). Another linkage mapping package includes polymapR, TSNPM which collects data from $F_1$ population to generate maps using dosage-scored marker information and estimates recombination frequencies using maximum likelihood. They are R-based softwares and are used to analyze linkage in polysomic triploids, tetraploids, hexaploids, and segmental allotetraploid populations (Preedy and Hackett 2016). So far, they have been used to develop high-density linkage maps in tetraploid rose (Bourke et al. 2017), tetraploid potato (Bourke et al. 2016), and hexaploid chrysanthemum (van Geest et al. 2017). Another R-package-based software that is used to generate linkage map is the netgwas. It utilizes conditional dependence relationships between markers and develops maps at any ploidy level in both in-bred and out-bred populations. The algorithm used in this software automatically recognizes the linkage group and does not depend on parental dosage score. It gives output in the form of clustered and ordered marker names but does not assign genetic positions and marker phasing, imparting a major drawback to this software (Behrouzi and Wit 2019). PERGOLA is R-based software that performs all the three steps of linkage mapping. It creates maps by analyzing marker data and mode of inheritance at all ploidy levels; however, it is limited to populations descended from inbred (homozygous) founding parents such as F2 and BC1. PolyGembler, a recently developed software, uses combinatorial algorithm to develop linkage maps. It calculates recombination frequencies by counting recombination events within and between phased multimarker scaffolds or haplotypes, thereby calculating r with no associated LOD score. Missing data and mistakes related with NGS can be handled using PolyGembler. Apart from these tools, other tools that constitute to develop linkage mapping are MDSMap and LPmerge. Both these software's are R-based tools that create maps using multidimensional scaling (Endelman and Plomion 2014; van Geest et al. 2017).

## 15.8 Haplotype-based Genotyping

Haplotypes are the set of nearby SNPs (polymorphic) present in the genome, having strong linkage disequilibrium within (Maldonado et al. 2019). Haplotyping allows identifying the functional variations leading to traits of interest which can be achieved by using second-generation and third-generation sequencing. It is crucial as it enables to identify polymorphism and unlocks the genomics potential of high-throughput sequencing. In polyploid crops, identification of SNPs is quite difficult as they possess polyploid genomes which attribute to low confidence in short read alignment. SNP arrays and complexity reduction methods such as GBS and RADSeq can be used to solve this challenge (Elshire et al. 2011; Willing et al. 2011; Iquira et al. 2015; Sonah et al. 2013). These platforms provide a wide range of markers to be used in genomics. However, there are several drawbacks to these technologies including intrinsic ascertainment bias during SNP array usage for genotyping (Lachance and Tishkoff 2013). The sampling of a smaller population leads to the occurrence of this ascertainment biasness. As the SNP probes on arrays are fixed, it is quite difficult to assay the rare variants. Hence, causing the biasness in population genetics studies leading to difficulty in recognizing functional variants controlling traits of interest (Clevenger et al. 2018).

Due to continuous plunge in the cost of sequencing and increment in both efficiency and accuracy, it is now possible to generate sequence on whole population. Because of this, access to all possible polymorphism is achieved which gives better insights into genetic mapping and genome-wide association mapping studies. In polyploids, uncertain alignment of short sequences within the genome arises during the identification of SNPs. Mapping of these homeologous reads results in perplexities as their appearance *in silico* of a polymorphism lie between the accessions which is present between the subgenomes. To surpass this problem, sliding window extraction of explicit polymorphisms (SWEEP) has been designed. Besides the SWEEP, numerous haplotyping tools have been developed recently. These include polyHap (Su et al. 2008), HapCompass (Schwartz 2010), SATlotyper (Neigenfind et al. 2008), HapTree (Berger et al. 2014), SHEsisplus (Shen et al. 2016), and sDhaP (Das and Vikalo 2015). Comparison and evaluation of these tools reveal the different advantages of these software programs. For instance, HapTree was found to be suitable for producing more accurate haplotypes for triploids and tetraploids, whereas HapCompass works best for higher ploid levels (6× and higher) (Motazedi et al. 2018).

## 15.9 GBS Analytical Pipelines for Polyploids

### 15.9.1 Fast-GBS

GBS permits the simultaneous discovery of variants and exclaims genotypes within a wide range of samples (Elshire et al. 2011, p. 27; Sonah et al. 2013, 2015). It has numerous advantages over other genotyping platforms including flexibility, low cost, rapidity, and robust genotyping method. These characteristics make it a choice of technique for many applications from conservation biology to functional genomics. Fast-GBS is the second generation of GBS which offers several advantages over GBS including broad understanding of data analysis, accurate missing data imputation, improvement in bioinformatics pipelines, and active postrelease support. Also, with the help of Fast-GBS, multiple sequencing data can be run (Torkamaneh et al. 2020). Apart from these benefits, it provides a wide range of new options such as processing paired-end reads which is achieved by demultiplexing and trimming. It also allows imputation of missing data by creating a pool of read count. The samples which are having a mean number of less than 10% will automatically be removed. These properties make it a better option than other available analysis platform such as Gb-eaSy and GBS-SNP-CROP (Torkamaneh et al. 2020).

### 15.9.2 PolyRAD

PolyRAD is R package software platform for genotyping in polyploids and diploid plants. Its working principle is originally based on the design developed by Li (2011). At first, TASSEL (Glaubitz et al. 2014) or Stacks (Catchen et al. 2013) are used to discover SNP, with or without using the reference genome, afterwards the generated read depth is transferred to polyRAD or to the read counting software TagDigger (Clark and Sacks 2016). The polyRAD software can specify one or several ploidies at a time including auto and/or allopolyploidy, thus allowing variation of inheritance mode across the genome. It estimates genotype probability by using the Bayesian framework which is in accordance with mapping population design, Hardy–Weinberg equilibrium (HWE) with or without linkage disequilibrium. It is the first Bayesian genotype caller used to incorporate multiple modes of inheritance and population structure. PolyRAD allows haplotypes, as if the LD span comes in range of one RAD tag then it will be not informative for genotype imputation. Other usage of PolyRAD is in genome-wide association mapping and in genomic prediction softwares such as GAPIT (Lipka et al. 2012), TASSEL (Bradbury et al. 2007), and FarmCPU (Liu et al. 2016). It is publically available at https://github.com/lvclark/polyRAD and https://CRAN.R-project.org/package=polyRAD (Clark et al. 2019).

### 15.9.3 UGbS-Flex

UGbS-Flex is the modified version of GBS software that uses in-house Python and Perl scripts to find SNPs within genomes. It allows analysis of paired-end GBS data at different ploidy levels, polymorphism levels, breeding systems, and availability of reference genome. Through this pipeline, several thousands of SNPs have been identified within $F_3$-derived $F_2$

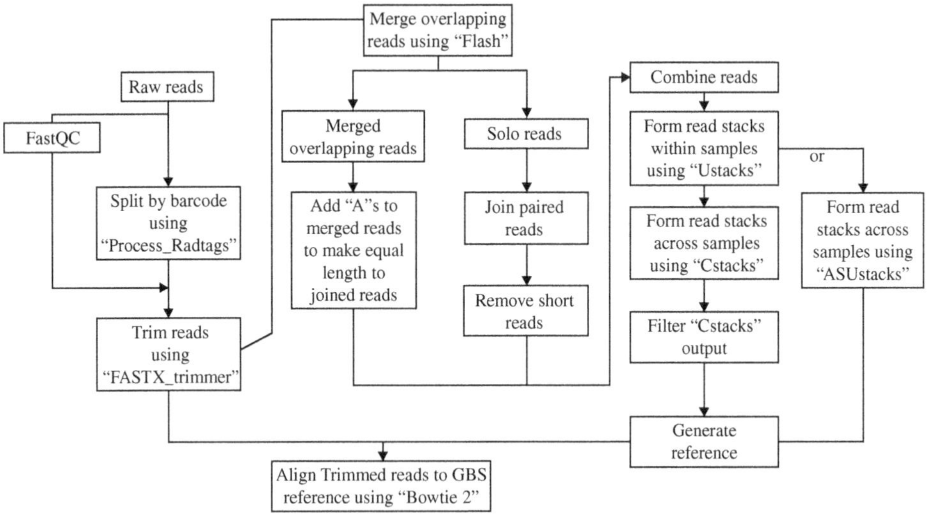


**Figure 15.3** Diagrammatic overview of UGbs-Flex Pipeline developing GBS reference. *Source:* The figure is reproduced from Qi et al. (2018) which is available under a Creative Commons Attribution 4.0 (CC-By 4.0) International License, which permits reproduction.

population in the allotetraploid finger millet and subsequent high-density genetic maps were generated using MAPMAKER. The main advantage of UGbS-Flex protocol is in identifying SNPs in outcrossing species and the species with very few genomic resources (Qi et al. 2018). Schematic overview of UGbs-Flex is shown in the Figure 15.3.

### 15.9.4 HaploTag

HaploTag is novel GBS software designed for the genotyping of self-pollinating polyploid species. With the help of tag-level haplotype within the sequenced tag, it identifies and analyzes polymorphism. It has several benefits over other GBS software, such as it can operate without reference genome, can be used at wide range of ploidy levels, offers both discovery and production mode, accounts for SNPs as well as haplotype-based genotype, and provides schematic representation of each inferred locus. Haplotag was originally designed for genotyping in self-pollinating allopolyploid species. But it can work well in self-pollinated diploid species too. If paralogous loci are present, it applies simple heterozygosity filter and builds models of haplotag loci which removes haplotypes from nonhomologous loci. This is the reason behind its best suitability for self-pollinating crops where heterozygosity is rare (Tinker et al. 2016).

## 15.10 GBS-based QTL Mapping Efforts in Polyploids

As mentioned before, GBS is a high-throughput approach to perform genotyping of organisms based on the identification of the polymorphism in allelic forms (Kim et al. 2016). The workflow of GBS involves the following steps – the isolated DNA samples are subjected to restriction enzyme digestion followed by barcode adaptor ligation and PCR amplification using barcode-specific primers. The amplified fragments are then sequenced and molecular marker identification is performed followed by the construction of a genetic map (Elshire et al. 2011). The GBS approach is cost-effective and it covers alterations in the genome sequence and therefore identification of high-density molecular markers is possible. GBS technology is known to reduce the complexity of large genomes (Davey et al. 2011), but the use of GBS for polyploid crop genomes poses challenges in the development of specific markers. Polyploid crop plants contain multiple sequence copies derived as a result of duplication that generate complications in deciphering the dosage of the alleles (Mason 2015). However, many economically important crops are polyploids and therefore, several efforts have been made to utilize the GBS approach for genotyping the polyploid crop plants as discussed in the subsequent sections.

### 15.10.1 Wheat

Wheat (*Triticum aestivum*) is one of the staple crops of the world and it is widely cultivated all around the world. In recent years, wheat production is proving incapable to provide for the rising demands (Denćić 1994). Efforts are being made towards crop improvement through breeding. Traits such as yield, crop resistance against various diseases are of considerable importance to escalate the production. However, the identification of genes in wheat becomes a difficult task because of its allohexaploid nature and complex expression of traits derived from interaction with the environment (Li et al. 2015a). Li et al. (2015b) have constructed a consensus map using 28644 GBS markers using three RIL populations generated by PBW343 × Kingbird, PBW343 × Kenya Swara, and PBW343 × Muu. High segregation variation was observed among three populations. The map was finalized by matching up against already existent rust-resistant genes in wheat and also with the reference wheat genome available. The map revealed three rust resistance genes as well as 15 QTLs associated with wheat rust disease. The QTLs identified were used to confirm the *Adult Plant Resistance (APR)* genes responsible for yellow rust, leaf rust, and root rust. Using the SNP markers developed through GBS technology, Yang et al. (2017b) have mapped the *Pis1* gene responsible for the three-pistil trait of wheat which controls the number of seeds per spike in wheat. The mapping population consisted of F2 generation derived from Chuanmai 28 (CM28) × CM28TP (A near-isogenic line of CM28.) The study would prove useful in marker-assisted breeding of wheat. Hussain et al. (2017) have constructed a high-density linkage map based on the RIL populations derived from Harry × Wesley. The map would be further helpful in performing breeding.

### 15.10.2 Potato

Uitdewilligen et al. (2013) conducted a study in which SNP genotyping assay was used to confirm the accuracy of genotyping-by-sequencing. The generated marker collection may be used to describe allele frequencies, nucleotide diversity,and population structure in potatoes, as well as to validate QTLs using association analysis. Potato crop is susceptible to a disease late blight caused by Phytophthora infestans. It has been observed in potato that choosing the right restriction enzyme for GBS library construction would help to generate appropriate SNPs and further ease the process of mapping (Bastien et al. 2018). 16 organ-specific QTL conferring resistance to late blight was reported on the basis of 83,862 SNPs obtained by 2b-restrictionsite–associated DNA and Genotyping by sequencing approaches. Overall 15putative defence genes was revealed by this study, which contribute to a better knowledge of the genetic regulation of late blight resistance in diploid potatoes and might be exploited in future potato breeding efforts (Juyo Rojas et al., 2019).

### 15.10.3 Sugarcane

Sugarcane is used for the production of table sugar and it is responsible for 80% of global table sugar contributing to its high economic value. Orange rust disease is known to adversely affect the sugarcane crop all over the world. Identification of QTLs associated with the orange rust disease has been performed. Three QTLs governing the resistance to the disease have been identified. The identified QTLs and markers would be helpful in the breeding and development of disease resistance. Despite sugarcane being a polyploid, several QTLs governing resistance to various diseases have been identified so far. It involves – Bru1 and Bru2 for brown rust resistance, Pachymetra root rot and leaf scald resistance, 11 for smut resistance, and 5 for Fiji leaf gall resistance. QTL for yellow spot resistance, QTL for sugarcane yellow leaf resistance (Yang et al. 2018). Another study by Yang et al. (2017a) has reported two major QTLs associated with brown rust resistance in sugarcane. CP95-1039 (Brown rust susceptible) and CP88-1762 (Brown rust-resistant) were crossed. The F1 population obtained was subjected to GBS. High-resolution maps were generated using SNP markers derived through pipelines. A study performed by Balsalobre et al. (2017) has mapped seven QTLs responsible for four sugarcane yield-related traits.

### 15.10.4 Canola

GBS has been utilised for marker discovery, linkage mapping, and QTL analysis in a variety of crops, including the major oilseed crops like canola. Canola (rapeseed) is an economically important oil crop. There are twelve rapeseed genomes available, covering all three ecotypes and offering useful resources for finding genetic variants (Bayer et al., 2017; Song et al.,2020). The research carried out to date focuses on the improvement of seed oil content and yield. Different QTLs responsible for seed weight, seeds per silique, silique length, and seed oil content have been identified by Chen et al. (2017). To better understand the genetic architecture of root vigour and days to flowering in canola (*B. napus*), a high density linkage map was constructed based on SNPs detected using GBS approach. QTL for both traits were identified in an F2 population resulted from a winter type and spring type cross which further can be used for marker assisted selection (ArifUzZaman et al., 2016). The genetic basis of flowering in B. napus was investigated using double digest restriction-site associated DNA sequencing (ddRAD), a GBSmethod. A hybrid between an early-flowering spring type and a late-flowering winter type resulted in an F2 mapping population. A high-density linkage map was constructed and QTL analysis was carried out for flowering time and the associated trait budding time (Scheben et al., 2020). GBS based QTL mapping using various doses of markers must be regarded as a significant step toward identifying the regions that govern quantitative characteristics in polyploid organisms.

### 15.10.5 Peanut

Peanut is an economically important crop because of its use in food preparations and for oil production. Peanut production is severely affected by early and late leaf spots (ELSs and LLSs). High-quality SNP data generated through GBS was used to identify QTLs associated with disease resistance (Han et al. 2018). Li et al. (2019) have mapped QTLs responsible for growth-related traits in peanuts by constructing a high-density genetic map. A RIL population composed of 188 individuals was used for the construction of the genetic map. Identification of QTLs would assist in breeding for the improvement of peanut crop.

### 15.10.6 Cotton

Cotton has always been a prominent crop for fiber production and has high economic importance globally because of its use in textile industries. Production of cotton fiber and its quality are major traits to focus on for improvement in cotton crop. Cotton crop worldwide is highly affected by salinity stress and the development of tolerant plant varieties is a goal for breeders. The cottonseed, although unpopular, is also an equally important product to be focused on for improvement because of its protein and oil content. The upland cotton varieties *Gossypium barbadense* and *Gossypium hirsutum* are tetraploid in nature. Magwanga et al. (2018) have used $BC_2F_2$ population derived from *Gossypium tomentosum* × *G. hirsutum* and constructed a linkage map which was used for mapping drought stress tolerance genes. Diouf et al. (2018) used the F2:3 population derived from CCRI35 × Nan Dan Ba Di Da Hua to identify SNPs using the GBS approach. A total of 110 QTLs responsible for fiber quality as well as various phenotypic traits were mapped for 11 genes. Diouf et al. (2017) have worked to map out 12 salinity stress-related genes in cotton. They generated high-density GBS-based SNP markers from the F2:3 population of CCR135 (salinity tolerant) × Nan Dan (salinity sensitive). A total of 66 QTLs responsible for 10 salinity traits were mapped out of which, 14 QTLs were stable under all environmental variations. *G. barbadense* cultivar 5917 and American Pima S-7 derived RIL population were analyzed for lint and fiber-related traits followed by GBS to identify the 3557 SNP markers which were used to construct a high-density linkage map. A total of 42 QTLs responsible for fiber quality and lint yield were identified and common as well as overlapping QTLs were studied (Fan et al. 2018).

## 15.11 GWAS and GS Using High-throughput Genotyping in Polyploidy Crops

GWAS is a novel approach to correlate a genotype to a phenotype. It simplifies the understanding of the expression of complex traits in plants (Bartoli and Roux 2017). Genomic selection uses genomic estimated breeding values (GEBV) to select the individuals having better phenotypic traits. The GEBV values denote the effect of multiple genes that control a phenotype (Heffner et al. 2010; Lorenz et al. 2011). The combined use of GWAS and GS could bypass the need to perform an actual analysis of phenotypic traits. The GWAS predicted optimal phenotypic values can be used to perform selection, thereby accelerating the breeding process (Iwata et al. 2013).

### 15.11.1 Wheat

Wheat is an important crop as it is one of the major sources of energy. Because of the rising population and increasing demands, there is a challenge to increase wheat production. To accelerate the breeding procedures for crop improvements is the need of the hour (Voss-Fels et al. 2019). A study performed on Chinese winter wheat grain yield concluded that the use of GWAS-derived dense markers to perform genomic selection would highly improve the efficiency of selection. Using high-resolution GWAS, Pang et al. (2021) identified QTLs and predicted genes associated with Stress resistance in wheat with great accuracy. Lozada et al. (2019), used GWAS and GS to evaluate the complex snow mold resistance trait in winter wheat and predicted the associated QTLs. They realized the co-localization of snow mold resistance QTLs with freeze resistance and vernalization QTLs.

### 15.11.2 Potato

Common scab disease caused by a bacterial pathogen affects potato crops worldwide. Developing resistant cultivars appears to be a potential strategy to tackle the disease. Yuan et al. (2019) have performed GWAS for the identification of the genomic regions responsible for the development of resistance in potato crops. Prodhomme et al. (2020) have identified markers associated with wart disease resistance in potatoes. Wart is caused by the fungus *Synchytrium endobioticum* and affects the yield of the potato crop. Tracing the disease resistance markers would aid in breeding efforts channeled toward the improvement of the potato crop.

### 15.11.3 Sugarcane

Sugarcane breeding is targeted toward yield improvement and improvement in biotic and abiotic stress tolerance. The polyploidy genome of sugarcane makes it difficult to use the traditional breeding methods for its improvement. NGS-based strategies such as GWAS could prove useful in simplifying the functional genomics studies of such complex polyploids by outgrowing the limitations of the conventional methods of breeding (Parameswari et al. 2021). Racedo et al. (2016) have performed GWAS to identify molecular markers associated with yield-related traits of sugarcane. The combined use of GWAS and GS would be a valuable approach for the improvement of sugarcane crop yield and conferring disease resistance so that the rising demand for sugarcane can be fulfilled.

### 15.11.4 Canola

Canola plant is affected by the fungal disease Sclerotinia stem rot (SSR). Roy et al. (2021) have identified the SNPs associated with the SSR disease resistance using GWAS. The study also depicts the importance of genomic selection studies using GWAS to speed up the process of breeding for the improvement of SSR tolerance in canola. The yellow seeded canola is considered a superior variety. The quantity of acid detergent lignin (ADL) in the seeds and the hull content (HC) of the canola plant decide the yellow color of the seeds. A study conducted by Wang et al. (2015) has identified the SNP markers associated with yellow seed using GWAS.

### 15.11.5 Peanut

Peanut crop is affected by early leaf spot (ELS) and late leaf spot (LLS). Zhang et al. (2020) have identified QTLs showing significant association with ELS and LLS resistance in peanuts. The QTLs identified would assist in breeding procedures for the development of ELS and LLS resistance in peanuts. GWAS studies have also been carried out to understand the association of genomic regions with yield-related phenotypes in peanuts. The results would be further helpful to increase the yield of peanut crop through breeding (Wang et al. 2019). Similar studies have been carried out for seed composition traits, root-knot nematode resistance which are aimed toward the betterment of the Peanut crop (Kumral 2019; Zhang et al. 2021).

### 15.11.6 Cotton

Hulse-Kemp et al. (2015) have constructed a high-density map of interspecific and intraspecific populations of cotton. Such a map would be beneficial for performing GWAS and GS studies for cotton breeding program. GWAS has been extensively performed in the cotton plant for the detection of genetic loci related to fiber quality, flowering time, early maturation, and yield-related traits (Islam et al. 2016; Li et al. 2016; Su et al. 2016; Sun et al. 2017). These studies could be employed further for the prediction of superior phenotypes in cotton for breeding purposes.

## References

Agarwal, G., Clevenger, J., Pandey, M.K. et al. (2018). High-density genetic map using whole-genome resequencing for fine mapping and candidate gene discovery for disease resistance in peanut. *Plant Biotechnology Journal* 16: 1954–1967.

Aguiar, D. and Istrail, S. (2013). Haplotype assembly in polyploid genomes and identical by descent shared tracts. *Bioinformatics* 29: i352–i360.

Alonso-Blanco, C., Andrade, J., Becker, C. et al. (2016). 1135 Genomes reveal the global pattern of polymorphism in *Arabidopsis thaliana*. *Cell* 166: 481–491.

Balsalobre, T.W.A., da Silva, P.G., Margarido, G.R.A. et al. (2017). GBS-based single dosage markers for linkage and QTL mapping allow gene mining for yield-related traits in sugarcane. *BMC Genomics* 18: 1–19.

Bartholomé, J., Mandrou, E., Mabiala, A. et al. (2015). High-resolution genetic maps of *Eucalyptus* improve *Eucalyptus grandis* genome assembly. *New Phytologist* 206: 1283–1296.

Bartoli, C. and Roux, F. (2017). Genome-wide association studies in plant pathosystems: toward an ecological genomics approach. *Frontiers in Plant Science* 8: 763.

Bastien, M., Boudhrioua, C., Fortin, G., and Belzile, F. (2018). Exploring the potential and limitations of genotyping-by-sequencing for SNP discovery and genotyping in tetraploid potato. *Genome* 61: 449–456.

Batley, J. and Edwards, D. (2009). Genome sequence data: management, storage, and visualization. *BioTechniques* 46: 333–336.

Bayer, M.M., Rapazote-Flores, P., Ganal, M. et al. (2017). Development and evaluation of a barley 50k iSelect SNP array. *Frontiers in Plant Science* 8: 1792.

Behrouzi, P. and Wit, E.C. (2019). De novo construction of polyploid linkage maps using discrete graphical models. *Bioinformatics* 35: 1083–1093.

Bento, M., Gustafson, J.P., Viegas, W., and Silva, M. (2011). Size matters in Triticeae polyploids: larger genomes have higher remodeling. *Genome* 54: 175–183.

Berger, E., Yorukoglu, D., Peng, J., and Berger, B. (2014). HapTree: a novel Bayesian framework for single individual polyplotyping using NGS data. *PLoS Computational Biology* 10: e1003502.

Berkman, P.J., Bundock, P.C., Casu, R.E. et al. (2014). A survey sequence comparison of *Saccharum genotypes* reveals allelic diversity differences. *Tropical Plant Biology* 7: 71–83.

Bertioli, D.J., Cannon, S.B., Froenicke, L. et al. (2016). The genome sequences of *Arachis duranensis* and *Arachis ipaensis*, the diploid ancestors of cultivated peanut. *Nature Genetics* 48: 438–446.

Bhat, J.A., Deshmukh, R., Zhao, T. et al. (2020). Harnessing high-throughput phenotyping and genotyping for enhanced drought tolerance in crop plants. *Journal of Biotechnology*.

Bianco, L., Cestaro, A., Linsmith, G. et al. (2016). Development and validation of the Axiom® Apple480K SNP genotyping array. *The Plant Journal* 86: 62–74.

Bourke, P.M., Voorrips, R.E., Visser, R.G. et al. (2015). The double-reduction landscape in tetraploid potato as revealed by a high-density linkage map. *Genetics* 201: 853–863.

Bourke, P.M., Voorrips, R.E., Kranenburg, T. et al. (2016). Integrating haplotype-specific linkage maps in tetraploid species using SNP markers. *Theoretical and Applied Genetics* 129: 2211–2226.

Bourke, P.M., Arens, P., Voorrips, R.E. et al. (2017). Partial preferential chromosome pairing is genotype dependent in tetraploid rose. *The Plant Journal* 90: 330–343.

Bourke, P.M., Voorrips, R.E., Visser, R.G.F. et al. (2018). Tools for Genetic Studies in Experimental Populations of Polyploids. *Frontiers in Plant Science* 9: 513.

Bradbury, P.J., Zhang, Z., Kroon, D.E. et al. (2007). TASSEL: software for association mapping of complex traits in diverse samples. *Bioinformatics* 23: 2633–2635.

Cai, G., Yang, Q., Yang, Q. et al. (2012). Identification of candidate genes of QTLs for seed weight in *Brassica napus* through comparative mapping among *Arabidopsis* and *Brassica* species. *BMC Genetics* 13: 105.

Catchen, J., Hohenlohe, P.A., Bassham, S. et al. (2013). Stacks: an analysis tool set for population genomics. *Molecular Ecology* 22: 3124–3140.

Chalhoub, B., Denoeud, F., Liu, S. et al. (2014). Early allopolyploid evolution in the post-Neolithic *Brassica napus* oilseed genome. *Science* 345: 950–953.

Cheema, J. and Dicks, J. (2009). Computational approaches and software tools for genetic linkage map estimation in plants. *Briefings in Bioinformatics* 10: 595–608.

Chen, H., Xie, W., He, H. et al. (2014). A high-density SNP genotyping array for rice biology and molecular breeding. *Molecular Plant* 7: 541–553.

Chen, J., Wang, B., Zhang, Y. et al. (2017). High-density ddRAD linkage and yield-related QTL mapping delimits a chromosomal region responsible for oil content in rapeseed (*Brassica napus* L.). *Breeding Science* 67: 296–306.

Clark, L.V. and Sacks, E.J. (2016). TagDigger: user-friendly extraction of read counts from GBS and RAD-seq data. *Source Code for Biology and Medicine* 11: 1–6.

Clark, L.V., Lipka, A.E., and Sacks, E.J. (2019). polyRAD: Genotype calling with uncertainty from sequencing data in polyploids and diploids. *G3: Genes, Genomes, Genetics* 9: 663–673.

Clevenger, J.P., Korani, W., Ozias-Akins, P. et al. (2018). Haplotype-based genotyping in polyploids. *Frontiers in Plant Science* 9: 564.
Close, T., Bhat, P., Lonardi, S. et al. (2009). Development and implementation of high-throughput SNP genotyping in barley. *BMC Genomics* 10: 582.
Crow, J.F. (1994). Hitoshi Kihara, Japan's pioneer geneticist. *Genetics* 137: 891.
Crow, K.D. and Wagner, G.P. (2005). What is the role of genome duplication in the evolution of complexity and diversity? *Molecular Biology and Evolution* 23: 887–892.
D'Hont, A. (2005). Unraveling the genome structure of polyploids using FISH and GISH; examples of sugarcane and banana. *Cytogenetic and Genome Research* 109: 27–33.
D'Hont, A. and Glaszmann, J.C. ed. (2001). International Society of Sugar Cane Technologists. Proceedings of the XXIV Congress, Brisbane, Australia, vol. 2, 17–21.
Dabholkar, A.R. (2006). *General Plant Breeding*. Concept Publishing Company.
Daniels, J. and Roach, B.T. (1987). Taxonomy and evolution. In: *Developments in Crop Science*, 7–84. Elsevier.
Darrier, B., Russell, J., Milner, S.G. et al. (2019). A comparison of mainstream genotyping platforms for the evaluation and use of barley genetic resources. *Frontiers in Plant Science* 10: 544.
Das, S. and Vikalo, H. (2015). SDhaP: haplotype assembly for diploids and polyploids via semi-definite programming. *BMC Genomics* 16: 1–16.
Davey, J.W., Hohenlohe, P.A., Etter, P.D. et al. (2011). Genome-wide genetic marker discovery and genotyping using next-generation sequencing. *Nature Reviews Genetics* 12: 499–510.
Denćić, S. (1994). Designing a wheat ideotype with increased sink capacity. *Plant Breeding* 112: 311–317.
Diouf, L., Pan, Z., He, S.-P. et al. (2017). High-density linkage map construction and mapping of salt-tolerant QTLs at seedling stage in upland cotton using genotyping by sequencing (GBS). *International Journal of Molecular Sciences* 18: 2622.
Diouf, L., Magwanga, R.O., Gong, W. et al. (2018). QTL mapping of fiber quality and yield-related traits in an intra-specific upland cotton using genotype by sequencing (GBS). *International Journal of Molecular Sciences* 19: 441.
Doyle, J.J., Flagel, L.E., Paterson, A.H. et al. (2008). Evolutionary genetics of genome merger and doubling in plants. *Annual Review of Genetics* 42: 443–461.
Duran, C., Appleby, N., Edwards, D., and Batley, J. (2009). Molecular genetic markers: discovery, applications, data storage and visualisation. *Current Bioinformatics* 4: 16–27.
Ebbert, M.T., Farrugia, S.L., Sens, J.P. et al. (2018). Long-read sequencing across the C9orf72 'GGGGCC'repeat expansion: implications for clinical use and genetic discovery efforts in human disease. *Molecular Neurodegeneration* 13: 1–17.
Edwards, D., Batley, J., and Snowdon, R.J. (2013). Accessing complex crop genomes with next-generation sequencing. *Theoretical and Applied Genetics* 126: 1–11.
Eid, J., Fehr, A., Gray, J. et al. (2009). Real-time DNA sequencing from single polymerase molecules. *Science* 323: 133–138.
Elshire, R., Glaubitz, J., Sun, Q. et al. (2011). A robust, simple genotyping-by-sequencing (GBS) approach for high diversity species. *PLoS One* 6: e19379.
Endelman, J.B. and Plomion, C. (2014). LPmerge: an R package for merging genetic maps by linear programming. *Bioinformatics* 30: 1623–1624.
Fan, J.-B., Oliphant, A., Shen, R. et al. (2003). Highly parallel SNP genotyping. *Cold Spring Harbor Symposia on Quantitative Biology* 68: 69–78.

Fan, L., Wang, L., Wang, X. et al. (2018). A high-density genetic map of extra-long staple cotton (*Gossypium barbadense*) constructed using genotyping-by-sequencing based single nucleotide polymorphic markers and identification of fiber traits-related QTL in a recombinant inbred line population. *BMC Genomics* 19: 1–12.

Fierst, J.L. (2015). Using linkage maps to correct and scaffold de novo genome assemblies: methods, challenges, and computational tools. *Frontiers in Genetics* 6: 220.

Fuselli, S., Baptista, R., Panziera, A. et al. (2018). A new hybrid approach for MHC genotyping: high-throughput NGS and long read MinION nanopore sequencing, with application to the non-model vertebrate *Alpine chamois* (*Rupicapra rupicapra*). *Heredity* 121: 293–303.

Garsmeur, O., Droc, G., Antonise, R. et al. (2018). A mosaic monoploid reference sequence for the highly complex genome of sugarcane. *Nature Communications* 9: 2638.

van Geest, G., Bourke, P.M., Voorrips, R.E. et al. (2017). An ultra-dense integrated linkage map for hexaploid chrysanthemum enables multi-allelic QTL analysis. *Theoretical and Applied Genetics* 130: 2527–2541.

Glaubitz, J.C., Casstevens, T.M., Lu, F. et al. (2014). TASSEL-GBS: a high capacity genotyping by sequencing analysis pipeline. *PLoS One* 9: e90346.

Goudet, J. (1995). FSTAT (version 1.2): a computer program to calculate F-statistics. *Journal of Heredity* 86: 485–486.

Guimarães, C.T., Sills, G.R., and Sobral, B.W. (1997). Comparative mapping of Andropogoneae: *Saccharum* L.(sugarcane) and its relation to sorghum and maize. *Proceedings of the National Academy of Sciences of the United States of America* 94: 14261–14266.

Gupta, M., Salgotra, R.K., and Chauhan, B.S. (2020). Next-generation sequencing technologies and their implications for efficient utilization of genetic resources. In: *Rediscovery of Genetic and Genomic Resources for Future Food Security* (eds. R.K. Salgotra and S.M. Zargar), 239–250. Springer Singapore: Singapore.

Hackett, C.A., Milne, I., Bradshaw, J.E. et al. (2007). TetraploidMap for Windows: linkage map construction and QTL mapping in autotetraploid species. *Journal of Heredity* 98: 727–729.

Hackett, C.A., McLean, K., and Bryan, G.J. (2013). Linkage analysis and QTL mapping using SNP dosage data in a tetraploid potato mapping population. *PLoS One* 8: e63939.

Haldane, J. (1930). Theoretical genetics of autopolyploids. *Journal of Genetics* 22: 359–372.

Han, S., Yuan, M., Clevenger, J.P. et al. (2018). A SNP-based linkage map revealed QTLs for resistance to early and late leaf spot diseases in peanut (*Arachis hypogaea* L.). *Frontiers in Plant Science* 9: 1012.

Hayward, A., Mason, A.S., Dalton-Morgan, J. et al. (2012). SNP discovery and applications in *Brassica napus*. *Journal of Plant Biotechnology* 39: 49–61.

Heffner, E.L., Lorenz, A.J., Jannink, J.L. et al. (2010). Plant breeding with genomic selection: gain per unit time and cost. *Crop Science* 50: 1681–1690.

Ho, L.-H., Khandaker, M.M., Fah, J.B.C. et al. (2017). Cultivation, common diseases and potential nutraceutical values of watermelon. In: *Agricultural Research Updates*, vol. 20 (eds. P. Gorawala and S. Mandhatri). Chapter 2. Hauppauge, NY: Nova Science Publishers Inc.

Hulse-Kemp, A.M., Lemm, J., Plieske, J. et al. (2015). Development of a 63K SNP array for cotton and high-density mapping of intraspecific and interspecific populations of *Gossypium* spp. *G3: Genes, Genomes, Genetics* 5: 1187–1209.

Hussain, W., Baenziger, P.S., Belamkar, V. et al. (2017). Genotyping-by-sequencing derived high-density linkage map and its application to QTL mapping of flag leaf traits in bread wheat. *Scientific Reports* 7: 1–15.

Imelfort, M. and Edwards, D. (2009). De novo sequencing of plant genomes using second-generation technologies. *Briefings in Bioinformatics* 10: 609–618.

Iquira, E., Humira, S., and François, B. (2015). Association mapping of QTLs for sclerotinia stem rot resistance in a collection of soybean plant introductions using a genotyping by sequencing (GBS) approach. *BMC Plant Biology* 15 (1): 1–12.

Islam, M.S., Thyssen, G.N., Jenkins, J.N. et al. (2016). A MAGIC population-based genome-wide association study reveals functional association of GhRBB1_A07 gene with superior fiber quality in cotton. *BMC Genomics* 17: 1–17.

Iwata, H., Hayashi, T., Terakami, S. et al. (2013). Potential assessment of genome-wide association study and genomic selection in Japanese pear *Pyrus pyrifolia*. *Breeding Science* 63: 125–140.

Jia, G., Huang, X., Zhi, H. et al. (2013). A haplotype map of genomic variations and genome-wide association studies of agronomic traits in foxtail millet (*Setaria italica*). *Nature Genetics* 45: 957–961.

Jiao, Y., Peluso, P., Shi, J. et al. (2017). Improved maize reference genome with single-molecule technologies. *Nature* 546: 524–527.

Junior, T.C., Carraro, D.M., Benatti, M.R. et al. (2004). Structural features and transcript-editing analysis of sugarcane (*Saccharum officinarum* L.) chloroplast genome. *Current Genetics* 46: 366–373.

Katara, J., Sonah, H., Deshmukh, R. et al. (2010). Molecular analysis of QTLs associated with resistance to brown spot in rice (*Oryza sativa* L.). *Indian Journal of Genetics* 70 (1): 17–21.

Kilian, A., Wenzl, P., Huttner, E. et al. (2012). Diversity arrays technology: a generic genome profiling technology on open platforms. In: *Data Production and Analysis in Population Genomics*, 67–89. Springer.

Kim, C., Guo, H., Kong, W. et al. (2016). Application of genotyping by sequencing technology to a variety of crop breeding programs. *Plant Science* 242: 14–22.

Kim, K.D., Kang, Y., and Kim, C. (2020). Application of genomic big data in plant breeding: past, present, and future. *Plants* 9: 1454.

Kinoshita, T. and Takahashi, M.-E. (1969). Studdies in polyploid varieties of sugar beets: XIV. Use of cytoplasmic male sterility in the production of triploid hybrids, and their performance in trials. *Journal of the Faculty of Agriculture* 56: 171–186.

Kumral, F.E. (2019). Genome Wide Association Study (GWAS) on Root-Knot Nematode Resistance in Cultivated Peanut.

Kyriakidou, M., Tai, H.H., Anglin, N.L. et al. (2018). Current strategies of polyploid plant genome sequence assembly. *Frontiers in Plant Science* 9: 1660.

Lachance, J. and Tishkoff, S.A. (2013). SNP ascertainment bias in population genetic analyses: why it is important, and how to correct it. *BioEssays* 35: 780–786.

Leal-Bertioli, S.C.M., Cavalcante, U., Gouvea, E.G. et al. (2015). Identification of QTLs for rust resistance in the peanut wild species arachis magna and the development of KASP markers for marker-assisted selection. *G3 Genes|Genomes|Genetics* 5: 1403–1413.

Li, H. (2011). A statistical framework for SNP calling, mutation discovery, association mapping and population genetical parameter estimation from sequencing data. *Bioinformatics* 27 (21): 2987–2993.

Li, A.-L., Geng, S.-F., Zhang, L.-Q. et al. (2015a). Making the bread: insights from newly synthesized allohexaploid wheat. *Molecular Plant* 8: 847–859.

Li, H., Vikram, P., Singh, R.P. et al. (2015b). A high density GBS map of bread wheat and its application for dissecting complex disease resistance traits. *BMC Genomics* 16: 1–15.

Li, C.Q., Xu, X.J., Dong, N. et al. (2016). Association mapping identifies markers related to major early-maturating traits in upland cotton (*Gossypium hirsutum* L.). *Plant Breeding* 135: 483–491.

Li, L., Yang, X., Cui, S. et al. (2019). Construction of high-density genetic map and mapping quantitative trait loci for growth habit-related traits of peanut (*Arachis hypogaea* L.). *Frontiers in Plant Science* 10: 745.

Lipka, A.E., Tian, F., Wang, Q. et al. (2012). GAPIT: genome association and prediction integrated tool. *Bioinformatics* 28: 2397–2399.

Liu, X., Huang, M., Fan, B. et al. (2016). Iterative usage of fixed and random effect models for powerful and efficient genome-wide association studies. *PLoS Genetics* 12: e1005767.

Lorenz, A.J., Chao, S., Asoro, F.G. et al. (2011). Genomic selection in plant breeding: knowledge and prospects. *Advances in Agronomy* 110: 77–123.

Lozada, D., Godoy, J.V., Murray, T.D. et al. (2019). Genetic dissection of snow mold tolerance in US Pacific Northwest winter wheat through genome-wide association study and genomic selection. *Frontiers in Plant Science* 10: 1337.

Magi, A., Giusti, B., and Tattini, L. (2017). Characterization of MinION nanopore data for resequencing analyses. *Briefings in Bioinformatics* 18: 940–953.

Magwanga, R.O., Lu, P., Kirungu, J.N. et al. (2018). GBS mapping and analysis of genes conserved between *Gossypium tomentosum* and *Gossypium hirsutum* cotton cultivars that respond to drought stress at the seedling stage of the BC2F2 generation. *International Journal of Molecular Sciences* 19: 1614.

Maldonado, C., Mora, F., Scapim, C.A., and Coan, M. (2019). Genome-wide haplotype-based association analysis of key traits of plant lodging and architecture of maize identifies major determinants for leaf angle: Hap LA4. *PLoS One* 14: e0212925.

Malmberg, M., Spangenberg, G., Daetwyler, H. et al. (2019). Assessment of low-coverage nanopore long read sequencing for SNP genotyping in doubled haploid canola (*Brassica napus* L.). *Scientific Reports* 9: 1–12.

Mamanova, L., Coffey, A.J., Scott, C.E. et al. (2010). Target-enrichment strategies for next-generation sequencing. *Nature Methods* 7: 111–118.

Mason, A.S. (2015). Challenges of genotyping polyploid species. *Plant Genotyping*: 161–168.

Mason, A.S., Higgins, E.E., Snowdon, R.J. et al. (2017). A user guide to theBrassica60K Illumina Infinium™ SNP genotyping array. *Theoretical and Applied Genetics* 130: 621–633.

Michael, T.P. and VanBuren, R. (2015). Progress, challenges and the future of crop genomes. *Current Opinion in Plant Biology* 24: 71–81.

Miller, D.E., Staber, C., Zeitlinger, J. et al. (2018). Highly contiguous genome assemblies of 15 *Drosophila* species generated using nanopore sequencing. *G3: Genes, Genomes, Genetics* 8: 3131–3141.

Mishra, A., Singh, A., Mantri, S. et al. (2021). Decoding the genome of superior chapatti quality Indian wheat variety 'C 306'unravelled novel genomic variants for chapatti and nutrition quality related genes. *Genomics* 113: 1919–1929.

Moragues, M., Comadran, J., Waugh, R. et al. (2010). Effects of ascertainment bias and marker number on estimations of barley diversity from high-throughput SNP genotype data. *Theoretical and Applied Genetics* 120: 1525–1534.

Motazedi, E., Finkers, R., Maliepaard, C., and de Ridder, D. (2018). Exploiting next-generation sequencing to solve the haplotyping puzzle in polyploids: a simulation study. *Briefings in Bioinformatics* 19: 387–403.

Mr, M., Dunham, J., Cresko WA, A.A. et al. (2007). Rapid and cost-effective polymorphism identification and genotyping using restriction site associated DNA (rAD) markers. *Genome Research* 17: 240–248.

Neigenfind, J., Gyetvai, G., Basekow, R. et al. (2008). Haplotype inference from unphased SNP data in heterozygous polyploids based on SAT. *BMC Genomics* 9: 1–26.

Okura, V., da Silva, F.R., da Silva, M.J. et al. (2012). A BAC library of the SP80-3280 sugarcane variety (*Saccharum* sp.) and its inferred microsynteny with the sorghum genome. *BMC Research Notes* 5: 1–11.

Oliver, R.E., Lazo, G.R., Lutz, J.D. et al. (2011). Model SNP development for complex genomes based on hexaploid oat using high-throughput 454 sequencing technology. *BMC Genomics* 12: 77.

Pandey, M.K., Monyo, E., Ozias-Akins, P. et al. (2012). Advances in *Arachis* genomics for peanut improvement. *Biotechnology Advances* 30: 639–651.

Pang, Y., Wu, Y., Liu, C. et al. (2021). High-resolution genome-wide association study and genomic prediction for disease resistance and cold tolerance in wheat. *Theoretical and Applied Genetics* 134: 2857–2873.

Parameswari, B., Nithya, K., Kumar, S. et al. (2021). Genome wide association studies in sugarcane host pathogen system for disease resistance: an update on the current status of research. *Indian Phytopathology*: 1–10.

Patil, G., Vuong, T.D., Kale, S. et al. (2018). Dissecting genomic hotspots underlying seed protein, oil, and sucrose content in an interspecific mapping population of soybean using high-density linkage mapping. *Plant Biotechnology Journal* 16 (11): 1939–1953.

Pérez-Enciso, M., Rincón, J.C., and Legarra, A. (2015). Sequence- vs. chip-assisted genomic selection: accurate biological information is advised. *Genetics Selection Evolution* 47: 43.

Preedy, K. and Hackett, C. (2016). A rapid marker ordering approach for high-density genetic linkage maps in experimental autotetraploid populations using multidimensional scaling. *Theoretical and Applied Genetics* 129: 2117–2132.

Prodhomme, C., Vos, P.G., Paulo, M.J. et al. (2020). Distribution of P1 (D1) wart disease resistance in potato germplasm and GWAS identification of haplotype-specific SNP markers. *Theoretical and Applied Genetics* 133: 1859–1871.

Qi, P., Gimode, D., Saha, D. et al. (2018). UGbS-Flex, a novel bioinformatics pipeline for imputation-free SNP discovery in polyploids without a reference genome: finger millet as a case study. *BMC Plant Biology* 18: 1–19.

Racedo, J., Gutiérrez, L., Perera, M.F. et al. (2016). Genome-wide association mapping of quantitative traits in a breeding population of sugarcane. *BMC Plant Biology* 16: 1–16.

de Ronne, M., Labbé, C., Lebreton, A. et al. (2020). Integrated QTL mapping, gene expression and nucleotide variation analyses to investigate complex quantitative traits: a case study with the soybean–phytophthora sojae interaction. *Plant Biotechnology Journal* 18 (7): 1492.

Roy, J., Shaikh, T., Luis, E. et al. (2021). Genome-wide association mapping and genomic prediction for adult stage sclerotinia stem rot resistance in Brassica napus (L) under field environments.

Savage, N. (2011). Fuel options: the ideal biofuel. *Nature* 474: S9–S11.

Schleinitz, D., DiStefano, J.K., and Kovacs, P. (2011). Targeted SNP genotyping using the TaqMan® assay. In: *Disease Gene Identification: Methods and Protocols* (ed. J.K. DiStefano), 77–87. Totowa, NJ: Humana Press.

Schwartz, R. (2010). Theory and algorithms for the haplotype assembly problem. *Communications in Information and Systems* 10: 23–38.
Semagn, K., Babu, R., Hearne, S. et al. (2014). Single nucleotide polymorphism genotyping using Kompetitive Allele Specific PCR (KASP): overview of the technology and its application in crop improvement. *Molecular Breeding* 33: 1–14.
Shen, J., Li, Z., Chen, J. et al. (2016). SHEsisPlus, a toolset for genetic studies on polyploid species. *Scientific Reports* 6: 1–10.
Shirasawa, K., Bertioli, D.J., Varshney, R.K. et al. (2013). Integrated consensus map of cultivated peanut and wild relatives reveals structures of the A and B genomes of *Arachis* and divergence of the legume genomes. *DNA Research* 20: 173–184.
Shirasawa, K., Kuwata, C., Watanabe, M. et al. (2016). Target amplicon sequencing for genotyping genome-wide single nucleotide polymorphisms identified by whole-genome resequencing in peanut. *The Plant Genome* 9: plantgenome2016.2006.0052.
Shivaraj, S., Deshmukh, R., Sonah, H. et al. (2019). Identification and characterization of aquaporin genes in *Arachis duranensis* and *Arachis ipaensis* genomes, the diploid progenitors of peanut. *BMC Genomics* 20: 222.
Singh, R., Singh, A.K., Sharma, T.R. et al. (2012). Fine mapping of grain length QTLs on chromosomes 1 and 7 in basmati rice (*Oryza sativa* L.). *Journal of Plant Biochemistry and Biotechnology* 21 (2): 157–166.
Soltis, D.E., Visger, C.J., Marchant, D.B. et al. (2016). Polyploidy: pitfalls and paths to a paradigm. *American Journal of Botany* 103: 1146–1166.
Sonah, H., Bastien, M., Iquira, E. et al. (2013). An improved genotyping by sequencing (GBS) approach offering increased versatility and efficiency of SNP discovery and genotyping. *PloS One* 8 (1): e54603.
Sonah, H., O'Donoughue, L., Cober, E. et al. (2015). Identification of loci governing eight agronomic traits using a GBS-GWAS approach and validation by QTL mapping in soya bean. *Plant Biotechnology Journal* 13 (2): 211–221.
Sonah, H., Zhang, X., Deshmukh, R.K. et al. (2016). Comparative transcriptomic analysis of virulence factors in *Leptosphaeria maculans* during compatible and incompatible interactions with canola. *Frontiers in Plant Science* 7: 1784.
Song, H., Wang, P., Lin, J.-Y. et al. (2016). Genome-wide identification and characterization of WRKY gene family in peanut. *Frontiers in Plant Science* 7: 534.
Souza, G.M., Berges, H., Bocs, S. et al. (2011). The sugarcane genome challenge: strategies for sequencing a highly complex genome. *Tropical Plant Biology* 4: 145–156.
Su, S.-Y., White, J., Balding, D.J., and Coin, L.J. (2008). Inference of haplotypic phase and missing genotypes in polyploid organisms and variable copy number genomic regions. *BMC Bioinformatics* 9: 1–9.
Su, J., Fan, S., Li, L. et al. (2016). Detection of favorable QTL alleles and candidate genes for lint percentage by GWAS in Chinese upland cotton. *Frontiers in Plant Science* 7: 1576.
Sun, Z., Wang, X., Liu, Z. et al. (2017). Genome-wide association study discovered genetic variation and candidate genes of fibre quality traits in *Gossypium hirsutum* L. *Plant Biotechnology Journal* 15: 982–996.
Thiel, T., Kota, R., Grosse, I. et al. (2004). SNP2CAPS: a SNP and INDEL analysis tool for CAPS marker development. *Nucleic Acids Research* 32: e5–e5.

Thomson, M., Singh, N., Dwiyanti, M. et al. (2017). Large-scale deployment of a rice 6 K SNP array for genetics and breeding applications. *Rice* 10: 40.

Tinker, N.A., Bekele, W.A., and Hattori, J. (2016). Haplotag: software for haplotype-based genotyping-by-sequencing analysis. *G3: Genes, Genomes, Genetics* 6: 857–863.

Tomar, R.S.S., Tiwari, S., Naik, B.K. et al. (2016). Molecular and morpho-agronomical characterization of root architecture at seedling and reproductive stages for drought tolerance in wheat. *PLoS One* 11: e0156528.

Torkamaneh, D., Laroche, J., and Belzile, F. (2020). Fast-GBS v2. 0: an analysis toolkit for genotyping-by-sequencing data. *Genome* 63: 577–581.

Trick, M., Long, Y., Meng, J. et al. (2009). Single nucleotide polymorphism (SNP) discovery in the polyploid *Brassica napus* using Solexa transcriptome sequencing. *Plant Biotechnology Journal* 7: 334–346.

Unterseer, S., Bauer, E., Haberer, G. et al. (2014). A powerful tool for genome analysis in maize: development and evaluation of the high density 600 k SNP genotyping array. *BMC Genomics* 15: 1–15.

Varshney, R.K., Nayak, S.N., May, G.D. et al. (2009). Next-generation sequencing technologies and their implications for crop genetics and breeding. *Trends in Biotechnology* 27: 522–530.

Voss-Fels, K.P., Cooper, M., and Hayes, B.J. (2019). Accelerating crop genetic gains with genomic selection. *Theoretical and Applied Genetics* 132: 669–686.

Wang, J., Jian, H., Wei, L. et al. (2015). Genome-wide analysis of seed acid detergent lignin (ADL) and hull content in rapeseed (*Brassica napus* L.). *PLoS One* 10: e0145045.

Wang, W., Mauleon, R., Hu, Z. et al. (2018). Genomic variation in 3,010 diverse accessions of Asian cultivated rice. *Nature* 557: 43–49.

Wang, J., Yan, C., Li, Y. et al. (2019). GWAS discovery of candidate genes for yield-related traits in peanut and support from earlier QTL mapping studies. *Genes* 10: 803.

Willing, E.-M., Hoffmann, M., Klein, J.D. et al. (2011). Paired-end RAD-seq for de novo assembly and marker design without available reference. *Bioinformatics* 27: 2187–2193.

Winfield, M.O., Allen, A.M., Burridge, A.J. et al. (2016). High-density SNP genotyping array for hexaploid wheat and its secondary and tertiary gene pool. *Plant Biotechnology Journal* 14: 1195–1206.

Yang, X., Sood, S., Glynn, N. et al. (2017a). Constructing high-density genetic maps for polyploid sugarcane (*Saccharum* spp.) and identifying quantitative trait loci controlling brown rust resistance. *Molecular Breeding* 37: 1–12.

Yang, Z., Chen, Z., Peng, Z. et al. (2017b). Development of a high-density linkage map and mapping of the three-pistil gene (Pis1) in wheat using GBS markers. *BMC Genomics* 18: 1–8.

Yang, X., Islam, M.S., Sood, S. et al. (2018). Identifying quantitative trait loci (QTLs) and developing diagnostic markers linked to orange rust resistance in sugarcane (*Saccharum* spp.). *Frontiers in Plant Science* 9: 350.

You, F.M., Huo, N., Deal, K.R. et al. (2011). Annotation-based genome-wide SNP discovery in the large and complex *Aegilops tauschii* genome using next-generation sequencing without a reference genome sequence. *BMC Genomics* 12: 1–19.

You, Q., Yang, X., Peng, Z. et al. (2018). Development and applications of a high throughput genotyping tool for polyploid crops: single nucleotide polymorphism (SNP) array. *Frontiers in Plant Science* 9: 104.

Yuan, J., Bizimungu, B., De Koeyer, D. et al. (2019). Genome-wide association study of resistance to potato common scab. *Potato Research* 63: 253–266.

Zargar, S., Raatz, B., Sona, H. et al. (2015). Recent advances in molecular marker techniques: insight into QTL mapping, GWAS and genomic selection in plants. *Journal of Crop Science and Biotechnology* 18 (5): 293–308.

Zhang, H., Chu, Y., Dang, P. et al. (2020). Identification of QTLs for resistance to leaf spots in cultivated peanut (*Arachis hypogaea* L.) through GWAS analysis. *Theoretical and Applied Genetics* 133: 2051–2061.

Zhang, H., Wang, M.L., Dang, P. et al. (2021). Identification of potential QTLs and genes associated with seed composition traits in peanut (*Arachis hypogaea* L.) using GWAS and RNA-Seq analysis. *Gene* 769: 145215.

Zhou, X., Xia, Y., Ren, X. et al. (2014). Construction of a SNP-based genetic linkage map in cultivated peanut based on large scale marker development using next-generation double-digest restriction-site-associated DNA sequencing (ddRADseq). *BMC Genomics* 15: 351.

# 16

# Recent Advances and Applicability of GBS, GWAS, and GS in Oilseed Crops

*Sanskriti Vats[1,3], Yogesh Sharma[1], Virender Kumar[1], Rushil Mandlik[1], Surbhi Kumawat[1], Himanshu Yadav[1], Pallavi Dhiman[1], Vandana Thakral[1], Md Aminul Islam[2], and Sreeja Sudhakaran[1]*

[1]*Agricultural Biotechnology, National Agri-Food Biotechnology Institute (NABI), Mohali, Punjab, India*
[2]*Department of Botany, Majuli College, Majuli, Assam, India*
[3]*Regional Centre for Biotechnology, Faridabad, Haryana (NCR Delhi), India*

## 16.1 Introduction

In the current scenario, society is heading to a greater reliance on agriculture to fulfill food, fuel, and raw material demand. India is one of the major consumers of oils and has the world's fourth-largest vegetable oil economy. The oilseed crops also contribute largely to the agricultural economy, next only to cereals within the segment of field crops. Not only economics but oilseeds are also better for evolving the human ethos of food and environmental security. For example, most oilseed crops can be grown in very wet areas, such as swamps and rainforests. The climatic diversity in India favors cultivation of nine oilseed crops. The fatty acid composition in oil seeds determines nutritional or industrial end use-value of the crop. Camelina and linseed oils, for example, are high in linolenic acid; sunflower oil has very high-oleic acid in addition to high linoleic acids; and *Brassica napus* cultivars include a wide range of oil types, including high-erucic acid and high-oleic acid. In addition, vegetable oils have advantageous minor nutrients like tocopherols. Oilseed rape extract, on the other hand, also has a lot of antinutritive fiber components like glucosinolates, phenolic acids, and phytate. Thus, current oilseed breeding projects attempt to increase oil content, quality, and yield, while lowering antinutritive components while maintaining traditional breeding objectives like biotic and abiotic tolerance.

With the rapid adoption of next-generation sequencing technology, genomics-assisted breeding has eased trait discovery, accurate association mapping, and linkage analysis (Bhat et al. 2018; Chaudhary et al. 2019a, 2019b). Reference-based genotyping-by-sequencing (GBS) approaches have contributed to the characterization of allelic variation in agronomic important genes and their concomitant incorporation in germplasm for cultivar development (André et al. 2017). Modern genomic-assisted breeding products like Tifguard HighO/L groundnut (high-oleic acid content and nematode resistance) have already shown

*Genotyping by Sequencing for Crop Improvement*, First Edition.
Edited by Humira Sonah, Vinod Goyal, S.M. Shivaraj, and Rupesh K. Deshmukh.

importance in oilseed crops (Chu et al. 2007). Thereafter, many genomic-assisted breeding varieties with improved oil quality combined with rust and leaf spot resistance had been commercialized. Upgraded computation interventions in genome annotations have expanded horizon of genomics-assisted breeding to crops without sequenced genome. Developing a draft genome on the basis of distant relatives, along with transcriptome data sets can be used for annotating gene models and developing high-density genetic maps for such species. GBS thus offers advantage to develop polymorphic markers distributed on the genetic map for crop species without reference genome (Elshire et al. 2011; Iquira et al. 2015; Sonah et al. 2013, 2015). The shortlisted genome-wide markers combined with statistical tools to associate plant genotype with phenotypic variations can be employed for association mappings and quantitative trait loci (QTLs) prediction in order to get better understanding of genetic architecture in oilseeds.

## 16.2 GBS Efforts in Oilseed Crops

GBS is a highly accurate platform for studying genomic inheritance studies for single gene or whole-genome evolution (Poland and Rife 2012). The sequenced data allows breeders to carry out GWAS, linkage analysis, developing markers, genomic selection, and genomic diversity study for a particular trait. Ease to use tags for sequence assembly in GBS allows genomic selection in crops without prior knowledge of species genome (Poland and Rife 2012; Narum et al. 2013). Recently, Malmberg et al. (2018) performed diversity analysis on Australian as well as Global *B. napus* L. germplasm using whole-genome resequencing and transcriptomics and found high degree of diversity in the Asian Brassica population specifically confined to Japan, South Korea, India, and Pakistan (Chen et al. 2008; Gyawali et al. 2013). The most polymorphic SNPs were found to be retained in the Asian population for LD analysis, suggesting these varieties as a rich source of diversity for oil breeders (Malmberg et al. 2018). GBS strategy cannot detect low SNP densities in crops such as in peanut. In such cases, whole-genome resequencing and targeted amplicon sequencing of SNPs can be a promising alternative (Shirasawa et al. 2016). In cultivated peanut, the SNP density was reported around 4.9 SNPs per 100 kb. High-resolution trait mapping in peanut had been further eased by the development of high-density SNP array, "Axiom_*Arachis*." This array has 58 K SNPs out of 163 782 total SNPs derived from DNA and RNA-sequencing of wild diploid ancestors and 41 groundnut accessions (Pandey et al. 2017). In soybean population consisting of 185 diverse soybean accessions, a total of 12 072 single-nucleotide polymorphisms (SNPs) were shortlisted using specific length amplified fragment sequencing technology (Li et al. 2019). For seed protein and oil content in soybean, 31 SNPs were found on 12 out of the 20 chromosomes. Among them, SNP on chromosome 7 was reported fundamental to oil content.

Compared to above mentioned oil crops like soybean, rapeseed, and peanut, sesame has highest oil content in dry seed (approximately 55%) (Wei et al. 2013; Chaudhary et al. 2019a, 2019b). Genomic exploration of sesame has led to greater insight in studying oil biosynthesis in the plant seed. Surprisingly, sesame has lower gene copy number of lipid-related genes than soybean and still has 30% higher seed oil content (Wang et al. 2014). Genomic analysis revealed tandem duplication event in sesame, leading to expansion of gene families like alcohol-forming fatty acyl-CoA reductase (AlcFAR) midchain alkane hydroxylase, lipid transfer protein type 1 (LTP1), and FAD4-like desaturase (FAD4-like), and the contraction of lipid acyl hydrolase-like (Sahebi et al. 2015) and lipid degradation-related families like lipoxygenase (LOX) (Wang et al. 2014). Recently, efforts had been made to understand genetic basis of fatty acid content in 601 sunflower inbred lines. GWAS study on the basis of 15 483 SNPs and the concentration of 23 fatty acids shortlisted 11 genetically associated markers in sunflower genome (Chernova et al. 2021). Two novel loci on chromosome 3 and 14 cumulatively explained 34.5% of the total variation in docosanoic acid (22 : 0).

More and more oilseed crops like linseed and safflower are gaining attention for their high oil content. Studies have been done to assign SNPs on genetic maps using various approaches like reduced representation libraries, GBS, and whole-genome resequencing. A collection of 86 accessions of linseed yielded 10 057 SNPs on whole genome (Singh et al. 2019). In another comprehensive genotyping of 350 accessions of flaxseeds, 6200 SNPs were distributed over 15 chromosomes (Hoque et al. 2020). On the basis of these SNPs, two clusters with low genetic differentiation were predicted to enhance accumulation of oil and fibers upon crossing. Another study (Kumar et al. 2012) on the population of 96 RILs from cross between safflower (*Carthamus tinctorius* L.) and its wild progenitor (*Carthamus palaestinus* Eig), mapped 2 008 196 SNPs to assembled *C. tinctorius* genome to facilitate SNP identification. These maps can be explored to study genetic variations in oil production in wild and cultivated crops and can be interogressed for genetic gain in oilseed crops for cultivar improvement.

## 16.3 High-density Linkage Maps for Oilseed Crops

### 16.3.1 High-density Linkage Maps for Peanut

The densest linkage map for peanut was developed using whole-genome resequencing approach on RIL population developed from inbred lines Tifrunner and GT-C20 which were found to be highly and moderately resistant to leaf spot diseases (Agarwal et al. 2018). Sequencing was carried out with coverage of 100×, 10×, and 2–5× for Tifrunner, GT-C20, and RILs, respectively. Around 60% of sequencing data were mapped on B subgenome whereas only 40% of data were mapped on A subgenome with exception of single RIL which showed opposite trend. A total of 16 674 polymorphic SNPs distributed on 20 chromosomes were identified out of which 10 274 SNPs were used for mapping. Genetic map with length of 3120.71 cM was developed by mapping of 8869 SNPs on 2156 loci which had mapping density of 1.45 cM/locus. The genetic map also displayed number of homologous and translocated markers. A total of 35 QTLs were identified for leaf spot disease using high-density genetic map. The phenotypic variation explained (PVE) for these QTLs ranged from 6.32% to 47.63% (Figure 16.1).

Another high-density map for peanut was built by exploiting double-digest restriction-site-associated DNA sequencing (ddRADseq) on RILs varying for resveratrol content (Luo et al. 2021). A total of 8114 SNPs were mapped at 2183 loci which gave genetic map of length 2063.55 cM. The composite interval mapping identified nine additive QTLs for resveratrol content with PVE ranging from 5.07% to 8.19%.

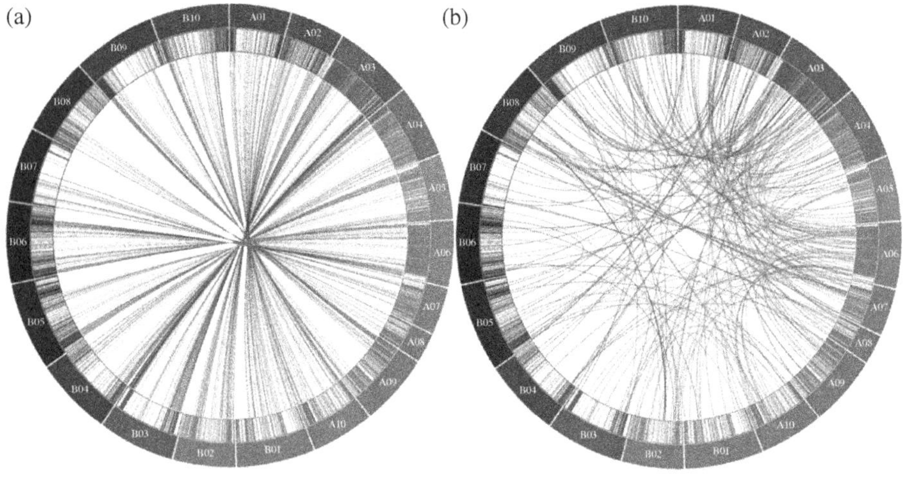


**Figure 16.1** Genomic distribution of single-nucleotide polymorphism (SNPs) mapped on different peanut chromosomes. (a) Chromosomes belong to A and B subgenome are denoted respectively as A01–A10 and B01–B10. Homeologous SNPs identified on A-chromosome are showed with connecting threads to corresponding mapped position on B- chromosome and vice versa (b) SNPs translocated to chromosomal locations other than the homeologous positions. *Source:* The figure is reproduced from Agarwal et al. (2018) which is available under a Creative Commons Attribution 4.0 (CC-By 4.0) International License, which permits reproduction.

### 16.3.2 High-density Linkage Maps for Sunflower

Construction of high-density linkage map for sunflower was carried out using specific-locus amplified fragment sequencing (SLAFseq) on F2 population (Zhou et al. 2018). A total of 39 589 polymorphic SLAFs were obtained out of which 27 261 were divided into eight segregation patterns. After further analysis, 6136 markers containing 11 980 SNP loci distributed into 17 linkage groups were used for the construction of the genetic map. This gave genetic map of length 2221.86 cM with 0.36 cM as adjacent markers average distance. The QTL analysis using the constructed map gave 8 QTLs for 4 traits *viz.* oleic acid content (OAC), plant height (PH), head diameter (HD), and stem diameter (SD).

### 16.3.3 High-density Linkage Maps of Sesamum

SLAFseq was used to construct the high-density linkage map in F2 sesamum population developed from parents with contrasting seed-related traits (Du et al. 2019). A total of 9134 polymorphic SLAF markers were obtained out of which 8159 markers were divided into eight segregation patterns. Further 6682 markers were finally selected for map construction. A total of 13 linkage groups were formed after mapping of 2159 markers, which gave genetic map of 2128.51 cM with average adjacent marker distance of 0.99 cM. QTL mapping identified around 19 seed trait-related QTLs with more than 10% of phenotypic effect. Several candidate genes for specific traits were analyzed and characterized further.

Similar SLAF-seq technique was used to construct genetic map on BC1 population developed for studying Basal Branching Habit and Flowers per Leaf Axil in Sesame (Mei et al. 2017). A total of 42 115 polymorphic SLAFs were obtained, out of which 36 592 were classified into eight segregation patterns. High-density genetic map with 9378 SLAFs and 13 linkage group was developed. Genetic map covered 1974.23 cM of distance with adjacent marker distance of 0.22 cM. The genes for flowers per leaf axil (*SiFA*) and basal branching habit (*SiBH*) were identified on basis of map constructed in study.

### 16.3.4 High-density Linkage Maps of Safflower

High-density linkage map for safflower was constructed by whole-genome resequencing of RIL population developed by cross between safflower and its wild progenitor (Bowers et al. 2016). The linkage map formed 12 linkage group matching the actual chromosome number. A total of 2 008 196 SNPs were mapped at 1178 loci with average of 1 SNP per 95.9 bp.

### 16.3.5 High-density Linkage Maps of Linseed

High-density linkage map for linseed was developed for linseed using SLAF-seq on population developed from F2 segregation progeny (Yi et al. 2017). On an average, every individual carried around 117 836 SNPs. Further analysis identified 65 845 markers which were classified into four segregation patterns. A total of 22 441 SNPs with segregation pattern of aa × bb were used for linkage map construction. After series of filters and based on MLOD values 4145 SNPs were used in the formation of 15 linkage groups. This gave linkage map of 2632.94 cM with an average length of 0.64 cM.

The F2 population developed by crossing between fiber flax and oilseed flax was used for the construction of high-density genetic map using SLAF-seq technique (Wu et al. 2018). A total of 23 115 polymorphic SLAFs were identified, out of which 10 434 were classified into eight segregation patterns. Finally, 7417 markers were used for linkage analysis which displayed aa × bb segregation pattern. A total of 2339 markers were finally mapped genome after series of filters. The genetic map of 1483.25 cM was obtained with 0.63 cM of average distance between adjacent markers. The 2339 markers gave 5106 SNP loci. QTL analysis identified 12 QTLs which were related with six fiber-related traits.

### 16.3.6 High-density Linkage Maps of Soybean

The SLAF-seq technique was used in the construction of high-density genetic map for soybean using $F_{21}$ RILs developed from cross between "Charleston" and "Dongnong594" (Qi et al. 2014). A total of 12 577 polymorphic SLAFs were obtained out of which 8381 were used in further analysis after series of filters. A total of 5308 markers showing appropriate segregation pattern were used for linkage map construction. These 5308 were mapped into 20 linkage groups forming map of 2294.433 cM with 0.43 cM as average distance between adjacent markers.

Similarly, high-density linkage mapping was performed with $F_{8:11}$ population derived from cross between semi-wild and cultivated soybean using SLAF-seq (Zhang et al. 2018). A total of 53 132 polymorphic markers were generated, out of which 30 415 were classified into 8 segregation patterns. After series of filters and analysis, 5111 markers were used for map construction. The map with length of 2909.46 cM and 0.57 cM of the average distance between adjacent markers was constructed. Mining of genes in 18 identified QTLs gave 43 potential candidate genes for different traits. Several studies have successfully used GBS for the linkage map development, QTL mapping, and GWAS in soybean (Zargar et al. 2015; Bastien et al. 2014; Sonah et al. (2015).

### 16.3.7 High-density Linkage Maps of Rapeseed

High-density SNP map was constructed studying high seed fiber content in *B. napus* L (Liu et al. 2013). The RIL population developed from cross between high seed fiber and low seed fiber content lines was used in the study. The genotyping was performed with Brassica 60 K SNP BeadChip Array. A total of 9804 homologous SNPs showing appropriate 1:1 segregation were used in linkage map construction. Finally, 9164 SNP markers were assigned into 10 linkage groups. The SNP bin map was made up consisting of 1232 bins was constructed. The genetic map had length of 1832.9 cM with average marker distance of 0.53 and 0.93 cM in A and C genomes, respectively. QTL analysis using this high-density map identified 11 QTLs related to seed color and fiber traits.

## 16.4 GBS Protocols and Analytical Pipelines

GBS is a highly multiplexed, next-generation sequencing technique useful in breeding, germplasm categorization, population studies, and trait mapping. It helps in genomic subset sequencing of species samples targeted by restriction enzymes at a low cost (Elshire et al. 2011; Sonah et al. 2013). The first step includes DNA extraction, in which DNA is extracted from the samples that GBS needs to be done. After DNA extraction, sequencing libraries are prepared in which DNA is digested with restriction enzyme (*Ape*KI, *Pst*I, or *Msp*I) followed by adapters ligation at the sticky end generated by restriction digestion of enzyme. Two different adaptors can be used in sequencing, one barcode adaptor, and a second common adaptor. These adaptor ligated restriction fragments are amplified using complementary ligated adaptor sequences, constituting to form a sequencing library. Library in which adaptor dimers are minimal (>0.5%) and other DNA fragments (170–350 bp in length) are maxima is considered suitable for sequencing. With the help of Illumina, sequencing is performed. This sequencing can be single-end sequencing or pair-end sequencing (Elshire et al. 2011; Sonah et al. 2013). After sequencing, obtained raw sequenced data are filtered. The obtained sequence reads are then aligned with the reference genome and are further scored based on presence or absence (Elshire et al. 2011). Alternatively, raw sequence reads can be processed with pipelines implemented using different bioinformatics resources such as Perl. These resources use different software tools which are publically available (SAMtools, BWA, VCFtools, and FASTX toolkit) in combination with some in-house tools. Thus obtained raw SNPs are filtered using VAFtools on the basis of read depth and genotypes missing data. By using phyton script, heterozygous corrections are performed, and with the help of fastPHASE software, missing information is imputed in the filtered set of SNPs. Further, functional annotation of SNPs can be done by co-locating SNPs with gene models or using SnpEff v3.0 program for detection of structural and functional variance.

## 16.5 GBS-based QTL Mapping Efforts in Oilseed Crops

Quantitative trait locus (QTL) mapping approaches facilitate the identification of loci associated with complex phenotypic traits. The following sections deal with the GBS-based QTL studies.

### 16.5.1 GBS-based QTL in Peanut

The abundance of simple-sequence repeats (SSRs) in the genome of peanut, its codominance, high polymorphism, presence of multiple alleles, ease of transferability from other species, and PCR-based simple analysis make it a valuable marker for genetic studies (He et al. 2003; Ferguson et al. 2004). SNPs are also distributed throughout the genome, which can be utilized for the construction of linkage map of peanut genome (Gautami et al. 2012; Nagy et al. 2012; Shirasawa et al. 2013). These linkage maps make basic genomic studies feasible, including QTL analysis. GBS facilitates the identification of SNPs and construction of high-density linkage maps, marker-assisted selection, and QTL analysis. Tseng et al. (2016) utilized SSR markers and SNPs generated via GBS of twelve lines for identifying QTLs lying under tomato spotted wilt virus (TSWV) resistance in peanut. They identified Florida-EPTM "113" which is a runner variety with superior wilt resistance (Tillman and Gorbet 2012). F2 populations were obtained by crossing between Georgia Valencia and Florida-EPTM "113." Georgia Valencia is susceptible to TSWV. They found that 329 SSR markers were polymorphic among 2431 in the peanut genome. Only SSR markers at chromosome A01 segregated with the phenotype of interest. GBS of the two parental lines and 10 F6 lines revealed 2670 SNPs, which were processed via the GBS-Tassel pipeline. Association studies based on the GBS data also revealed three loci with seven SNPs on chromosome A01 to be associated with TSWV. Three of these SNPs were flanked by markers GM 672 and AHGS 4584, and two were between markers AHGS4584 and AHGS 1646. These results were in accordance with the QTLs revealed in other varieties of peanut (Qin et al. 2012). In addition, they identified another QTL possibly associated with TSWV resistance at chromosome A09, with one SNP strongly correlating with TSWV resistance as indicated by the GBS data. Other traits, such as seed size, seed coat color, seed number per pod, seed maturity, and leaf spot resistance, also exhibited differences in this population. This highlights the utility of GBS for QTL mapping.

Another study utilizing the F9 RIL population with 192 individuals originating from a cross between Florida-07 × GP-NC WS 16 was conducted to identify QTLs underlying early and late leaf spot resistance in peanut (Han et al. 2018). GBS was exploited to construct a high-resolution map with 2753 SNPs distributed among the 20 chromosomes (linkage groups) with an average distance of 1.34 cM. Three major QTL (*qLLS2015-B05-2, qLLS2015-B05-1,* and *qLLS2016-B05*) for late leaf spot was identified on chromosome B05, with *qLLS2016-B05* anchored by two genes responsible for NBS-LRR resistance. Two QTLs (*qELS2016-B04* and *qELS2017-A03-2*) for early leaf spot resistance were mapped onto chromosome A03 and B04 (Table 16.1) (Han et al. 2018).

**Table 16.1** List of GBS-based QTL in conducted in oilseed crops.

| Crop | Trait | Cross | Population | Chromosome | QTL | LOD | PVE (%) | References |
|---|---|---|---|---|---|---|---|---|
| Peanut | TSWV resistance | Florida-EPTM "113" × Georgia Valencia | F2 and F6 | A01 | | 9 | 35.8 | Tseng et al. (2016) |
| | Late leaf spot resistance | Florida-07 × GP-NC WS 16 | F9 RIL | B05 | *qLLS2015-B05-2* | — | 11.64 | Han et al. (2018) |
| | | | | | *qLLS2015-B05-1* | | 5.03 | |
| | | | | | *qLLS2016-B05* | | 16.60 | |
| | Early leaf spot resistance | Florida-07 × GP-NC WS 16 | F9 RIL | A03; B04 | *qELS2016-B04* | — | 10.63 | Han et al. (2018) |
| | | | | | *qELS2017-A03-2* | | 11.67 | |
| Soybean | Stearic acid content | A6 × 194D and 194D × A6 | F4:5 and F5:6 | 14 | *q.14s* | 34.25/40.84 | 56.76 | Heim and Gillman (2017) |
| | Total seed oil | A6 × 194D and 194D × A6 | F4:5 and F5:6 | 9 | *q.9oi* | — | 14.43 | Heim and Gillman (2017) |
| | | | | 14 | *q.14oi* | | 20.58 | |
| | Palmitic acid | A6 × 194D and 194D × A6 | F4:5 and F5:6 | 2, 5, and 10 | *Q.2p; q.5p*; q.10p | — | 10–13 | Heim and Gillman (2017) |
| | Oleic acid | A6 × 194D and 194D × A6 | F4:5 and F5:6 | 4 | *q.4o* | — | 12.17 | Heim and Gillman (2017) |
| | | | | 14 | *q.14o* | | 28.04 | |
| | Linoleic acid | A6 × 194D and 194D × A6 | F4:5 and F5:6 | 4 | *q.4o* | — | 14.47 | Heim and Gillman (2017) |
| | | | | 14 | *q.14o* | | 9.49 | |
| | Linolenic acid | A6 × 194D and 194D × A6 | F4:5 and F5:6 | 2, 4, 8, 10, 13, and 14 | *q.2n, q.4n, q.8n, q.10n, q.13n, q.14n* | — | 7.97 to 12.16 | Heim and Gillman (2017) |
| | Plant height | A6 × 194D and 194D × A6 | F4:5 and F5:6 | 10 | *q.10h* | — | 20.90 | Heim and Gillman (2017) |
| | | | | 19 | *q.19h* | | 10.42 | |
| | Seed weight | Majesta × Hikmok sorip | — | 13 | *qHSW13* | 5.5 | 17.6 | Sonah et al. (2015) |
| | Oil content | | | 10 | *qOC10* | 3.7 | 11.8 | |
| | Protein content | | | 19 | *qPC19* | 2.95 | 10.3 | |

| | | | | | | | | |
|---|---|---|---|---|---|---|---|---|
| Rapeseed | Relative root length | GH01 × ZS9 | $BC_2 F_3$ | — | — | — | — | Ding et al. (2020) |
| | Relative hypocotyl length | | | | | | | |
| | Relative fresh weight | | | | | | | |
| | *Plasmodiophora brassicae* pathotypes 2, 3, 5, 6, and 8 resistance | T19 × ACDC | $BC_1S_1$ | A03 | *Rcr4* | 30.3–38.8 | 85–95% | Yu et al. (2017) |
| | *P. brassicae* pathotype 5x resistance | T19 × ACDC | $BC_1S_1$ | A02 | *Rcr8* | 15 | 36 | Yu et al. (2017) |
| | | | | A08 | *Rcr9* | 15.8 | 39 | |
| Canola | Root vigour | *Regent × Lagoda, Lagoda × Regent* | F2 | A01 | *NRV (Napus Root Vigor)* | 4 | 16.3 | ArifUzZaman et al. (2017) |
| | Flowering time | *Regent × Lagoda, Lagoda × Regent* | F2 | C08 | *DTF1* | 4 | 21.7 | ArifUzZaman et al. (2017) |
| | | | | C04 | *DTF2* | | 15 | |

Abbreviations: PVE, phenotypic variations explained; TSWV, tomato spotted wilt virus; LOD, log-of-odds.

### 16.5.2 GBS-based QTL in Soybean

Initially, GBS for QTL identification in soybean was utilized by Sonah et al. (2015) to map previously reported QTL for seed weight, and seed oil and protein content to prove the utility of GBS–GWAS technique. They used 141 RILs from the Majesta X Hikmok sorip (PI372415A) cross as it provides a very diverse genetic background for the traits under study. *Ape*K1 restriction enzyme was used for library preparation. Linkage map with 778 SNP markers was constructed. The details of QTLs identified have been provided in Table 16.1. Following this, Heim and Gillman (2017) identified 21 major and minor QTLs to be associated with plant height and six oil-related traits (total seed oil, stearic acid, palmitic acid, stearic acid, linoleic acid, linolenic acid, and OAC) using 173 RILs developed by crossing two mutants with different stearic acid contents. One mutant line was 194D, with approx. 11%, and the other A6 with approx. 24% stearic acid content. Four of the QTLs identified were for stearic acid content. A dense linkage map with 2977 GBS markers was constructed, and marker c14.loc58 (at chromosome 14, 58 cM, Gm14:42206409) was found to be most significantly associated with stearic acid content. Epistatic interactions between QTL linolenic acid content (*q.8n* and *q.13n*) were revealed. GBS-based approaches have been widely used in soybean breeding programs for characterization of germplasm resources and marker-based applications (Ratnaparkhe 2020, p. 84; Song 2016; Chaudhary et al. 2019a).

### 16.5.3 GBS-based QTL in Rapeseed

Oilseed rape is a vital oil crop, which is particularly sensitive to waterlogging stress. Rapeseed varieties with different waterlogging tolerance were used. GH01 (waterlogging-sensitive inbred line) and ZS9 (waterlogging-tolerance cultivar) were utilized to generate reciprocal intervarietal inbred lines (IL). The populations were named based on the recurrent parent, ZIL with ZS9, and GIL with GH01 as the recurrent parents, and consisted of 58 and 50 $BC_2$ $F_3$ families respectively. Relative root length (RLL), relative fresh weight (RFW), and relative hypocotyl length (RHL) were taken as representative traits for waterlogging. GBS was performed using *Ape*KI digestion and Illumina HiSeq 2500 producing 100-bp pair-end reads, and the SNPs were analyzed using the GBS pipeline of TASSEL v3.0.173 (Glaubitz et al. 2014). After aligning with the reference genome, 2689319 SNPs were detected in ZIL and GIL populations, but after filtering only 5485 SNPs were employed for QTL identification. A total of 5485 SNPs were distributed in all the linkage groups of rapeseed genome, except C9. Over 66 QTLs were identified for the three traits, 31 for RRL, 17 for RHL, and 18 for RFW. Six QTL clusters were also identified to have 56 candidate genes based on the physical position of SNPs and functionally annotated Arabidopsis genome. Transcriptome and resequencing data of GH01 and ZS9 further helped narrow down the number of candidate genes to twelve for six QTL clusters, highlighting the scope of such studies.

Another study revealed the QTLs responsible for clubroot disease resistance in rapeseed. Here, 1.32 million reads from the resistant parent (T19), 1.77 million reads from the susceptible parent (ACDC), and 0.07 to 1.41 million reads from their $BC_1$ generation were mapped to the reference genome, revealing 16618 SNPs, out of which 4963 were analyzed further as the rest were monomorphic. Three QTLs, *Rcr8, 4, 9*, were detected on chromosomes A02, A03, and A08, respectively. Out of these, the QTL on A03 provided resistance to *Plasmodiophora brassicae* (clubroot causing protist) pathotypes 2, 3, 5, 6, and 8 but not to 5×. *Rcr8* and *9* provided resistance to pathotype 5×. All these loci were found to be derived from the resistant parent, T19 (Yu et al. 2017). GBS has been used in canola as well highlighting the utility of this technique for QTL dissection in complex genomes (Table 16.1) (ArifUzZaman et al. 2017).

## 16.6 GBS-based GWAS Efforts in Oilseed Crops

QTL mapping is usually executed by utilizing segregating biparental populations, where low-density marker coverage is sufficient to unveil numerous QTL underlying a trait. However, despite available knowledge about multiple QTL the genes underlying multiple important agronomic traits remain unknown since a QTL spans hundreds of cM which translates into hundreds of candidate genes. The limited number of recombination events accumulating over a few generations in biparental populations utilized in traditional QTL mapping is the cause of this low resolution (Balasubramanian et al. 2009). The use of unrelated genotypes composing an association panel with large number of variations, as they have accumulated a large number of recombination events since their last common progenitor, can be used to overcome this fundamental limitation. Since their inception, such genome-wide association studies (GWASs) have been efficaciously applied in various animal as well as plant systems (Korte and Farlow 2013; Appels et al. 2014). However, due to such diverse genotypes utilized in GWAS, hence there is much-limited linkage disequilibrium between the adjecent markers, rendering much larger number of markers needed for acceptable coverage of the whole-genome (Hyten et al. 2007; Bastien et al. 2014). The development of high-throughput GBS techniques has made the availability of huge data of SNP markers conceivable (Sonah et al. 2013). Several GBS-based studies have been undertaken in oilseed crops (Table 16.2)

Since its inception, GWAS based on GBS has been utilized in model plants as well as oilseed crops (Bastien et al. 2014; Brock et al. 2020). Significantly associated markers for sclerotinia stem rot (SSR) resistance, maturity, hundred seed weight, plant height, seed oil and protein content, flower color, hilum color, pubescence color, and iron deficiency chlorosis in soybean have been retrieved using GWAS based on GBS (Bastien et al. 2014; Mamidi et al. 2014; Iquira et al. 2015; Sonah et al. 2015; Wei et al. 2017). A total of 111 resistance loci for *Leptosphaeria maculans (Lm)* were dissected in rapeseed at both seedling and adult plant stages in Canadian and Chinese canola/rapeseed accessions (Fu et al. 2020). Another study, utilizing 139 accessions from 16 countries, revealed one major QTL (at A01 at 9.66 Mbp) and 35 additional markers located on 13 chromosomes for *Lm* resistance in rapeseed (Rahman et al. 2016). A total of 40 SNPs out of a set of 37721 unfiltered SNPs were found to be significantly associated with flowering time, crude fat, and fatty acid composition in rapeseed (1 SNP for flowering time, 2 SNPs for crude fat content, and 37 for fatty acid composition). This study also revealed 21 genes involved in fatty acid content (Ryu et al. 2021). Intervarietal substitution lines (Bhat et al. 2018) from the Zhongyou 821 (ZY821) rapeseed variety as the recurring parent was used to study 43 phenotypic traits at

**Table 16.2** List of GBS-based GWAS studies in oilseed crops.

| Crop | Trait | Number of genotypes/ accessions analyzed | Platform/technique used | Total Number of SNPs | Number/location of MSS/QTL | References |
|---|---|---|---|---|---|---|
| Soybean | Iron deficiency chlorosis | | — | 79000 | One located on Gm03 at 45.03 Mbp, one on Gm05 at 8.82 Mbp, and three on Gm11 at 0.53 Mbp | Mamidi et al. (2014) |
| | SSR resistance | 130 | Illumina HiSeq2000; MLM | 12193 | 13651235 on Gm15 | Bastien et al. (2014) |
| | Seed oil | 304 | GLM | 47702 | 31114364 on Gm02; 45563060 on Gm 08; 1811225 on Gm10; 58684 on Gm14; 4183401 on Gm16; 50418986 on Gm19; 10052759 and 30728503 on Gm20 | Sonah et al. (2015) |
| | Seed protein | 304 | GLM | 47702 | 31114364 on Gm05; 45563060 on Gm08; 1335189 on Gm10; 120437 on Gm10; 4183401 on Gm16; 50418986 on Gm19; 10052759 and 30619328 on Gm20 | Sonah et al. (2015) |
| Rapeseed | *Leptosphaeria maculans (Lm)*/Blackleg disease resistance | 243 | Illumina HiSeq 2500 (PE125); GLM + Q model | 78632 | 111 MSS is mostly located on Chr. A03, A08, and A09 identified | Fu et al. (2020) |
| | | 139 | Illumina HiSeq 2500 (PE125); GLM + Q model | 42575 | One major QTL (at A01 at 9.66 Mbp) and 35 additional markers located on 13 chromosomes | Rahman et al. (2016) |
| | Flowering time | Gamma radiation generated 46 $M_9$ Mutants | FarmCPU model; fixed and random effect models | 35397 | 19120040 on A09 | Ryu et al. (2021) |
| | Crude oil content | Gamma radiation generated 46 $M_9$ Mutants | FarmCPU model; fixed and random effect models | 35397 | 17724544 on C08 | Ryu et al. (2021) |
| | Fatty acid content | Gamma radiation generated 46 $M_9$ mutants | FarmCPU model; fixed and random effect models | 35397 | 37 SNPs at various locations | Ryu et al. (2021) |
| | Electrolyte leakage | 366 | Kinship; PCA | 42575 | 36 markers and 9 QTLs located on A01 (19.846 Mbp), A02 (3.690 Mbp), A04 (11.900 Mbp), A06 (21.440 Mbp), A07 (15.780 Mbp), C02 (22.001 Mbp), C05 (42.900 Mbp), C07 (11.985 Mbp), C09 (6.480 Mbp) | Fiebelkorn et al. (2018) |
| *Camelina sativa* | 15 leaf wax traits (including major wax components) | 213 | FaST-LMM; MLM; | 6192 | 42 MSS | Luo et al. (2019) |
| Peanut | Pod size and weight | 250 | GLM | 105814 | 57 MSS | Zhou et al. (2021) |
| Sesame | γ-Tocopherol | 96 | GLM; MLM | 5962 | LG08_6621957 | He et al. (2019) |
| | β-Tocopherol | 96 | GLM; MLM | 5962 | SLG03_13104062 | He et al. (2019) |

Abbreviations: MLM, mixed linear model; GLM, general linear model; MSS, most significantly associated SNP.

12 time points, revealing more than 599 QTL (Li et al. 2020). Electrolyte leakage is another trait dissected via GBS-based GWAS in rapeseed (Fiebelkorn et al. 2018). *Camelina sativa* (Luo et al. 2019), peanut (Zhou et al. 2021), sesame (He et al. 2019) are some of the other oilseed crops where GBS has been employed for GWAS for dissecting markers associated with a variety of traits. GWAS and QTL mapping both have their merits and demerits and can at times be used to complement each other.

## References

Agarwal, G., Clevenger, J., Pandey, M.K. et al. (2018). High-density genetic map using whole-genome resequencing for fine mapping and candidate gene discovery for disease resistance in peanut. *Plant Biotechnology Journal* 16: 1954–1967.

André, L.L., Sonah, H., Dias, W.P. et al. (2017). Genome-wide association study for resistance to the southern root-knot nematode (Meloidogyne incognita) in soybean. *Molecular Breeding* 37: 148.

Appels, R., Nystrom-Persson, J., and Keeble-Gagnere, G. (2014). Advances in genome studies in plants and animals. *Functional & Integrative Genomics* 14: 1–9.

ArifUzZaman, M., Mamidi, S., McClean, P. et al. (2017). QTL mapping for root vigor and days to flowering in Brassica napus L. *Canadian Journal of Plant Science* 97: 99–109.

Balasubramanian, S., Schwartz, C., Singh, A. et al. (2009). QTL mapping in new *Arabidopsis thaliana* advanced intercross-recombinant inbred lines. *PLoS One* 4: e4318.

Bastien, M., Sonah, H., and Belzile, F. (2014). Genome wide association mapping of *Sclerotinia sclerotiorum* resistance in soybean with a genotyping-by-sequencing approach. *The Plant Genome* 20: 195.

Bhat, J.A., Shivaraj, S., Ali, S. et al. (2018). Genomic resources and omics-assisted breeding approaches for pulse crop improvement. In: *Pulse Improvement*, 13–55. Springer.

Bowers, J.E., Pearl, S.A., and Burke, J.M. (2016). Genetic mapping of millions of SNPs in safflower (*Carthamus tinctorius* L.) via whole-genome resequencing. *G3: Genes, Genomes, Genetics* 6: 2203–2211.

Brock, M.T., Rubin, M.J., DellaPenna, D. et al. (2020). A nested association mapping panel in *Arabidopsis thaliana* for mapping and characterizing genetic architecture. *G3: Genes, Genomes, Genetics* 10: 3701–3708.

Chaudhary, J., Deshmukh, R., Mir, Z.A. et al. (2019a). Metabolomics: an emerging technology for soybean improvement. In: *Biotechnology Products in Everyday Life*, 175–186. Cham: Springer.

Chaudhary, J., Shivaraj, S., Khatri, P. et al. (2019b). Approaches, applicability, and challenges for development of climate-smart soybean. In: *Genomic Designing of Climate-Smart Oilseed Crops*, 1–74. Cham: Springer.

Chen, S., Nelson, M., Ghamkhar, K. et al. (2008). Divergent patterns of allelic diversity from similar origins: the case of oilseed rape (*Brassica napus* L.) in China and Australia. *Genome* 51: 1–10.

Chernova, A.I., Gubaev, R.F., Singh, A. et al. (2021). Genotyping and lipid profiling of 601 cultivated sunflower lines reveals novel genetic determinants of oil fatty acid content. *BMC Genomics* 22: 1–15.

Chu, Y., Ramos, L., Holbrook, C. et al. (2007). Frequency of a loss-of-function mutation in oleoyl-PC desaturase (ahFAD2A) in the mini-core of the US peanut germplasm collection. *Crop Science* 47: 2372.

Ding, X.-Y., Xu, J.-S., Huang, H. et al. (2020). Unraveling waterlogging tolerance-related traits with QTL analysis in reciprocal intervarietal introgression lines using genotyping by sequencing in rapeseed (*Brassica napus* L.). *Journal of Integrative Agriculture* 19: 1974–1983.

Du, H., Zhang, H., Wei, L. et al. (2019). A high-density genetic map constructed using specific length amplified fragment (SLAF) sequencing and QTL mapping of seed-related traits in sesame (*Sesamum indicum* L.). *BMC Plant Biology* 19: 1–20.

Elshire, R.J., Glaubitz, J.C., Sun, Q. et al. (2011). A robust, simple genotyping-by-sequencing (GBS) approach for high diversity species. *PLoS One* 6: e19379.

Ferguson, M.E., Burow, M.D., Schulze, S. et al. (2004). Microsatellite identification and characterization in peanut (*A. hypogaea* L.). *Theoretical and Applied Genetics* 108: 1064–1070.

Fiebelkorn, D., Horvath, D., and Rahman, M. (2018). Genome-wide association study for electrolyte leakage in rapeseed/canola (*Brassica napus* L.). *Molecular Breeding* 38: 129.

Fu, F., Zhang, X., Liu, F. et al. (2020). Identification of resistance loci in Chinese and *Canadian canola*/rapeseed varieties against *Leptosphaeria maculans* based on genome-wide association studies. *BMC Genomics* 21: 501.

Gautami, B., Foncéka, D., Pandey, M.K. et al. (2012). An international reference consensus genetic map with 897 marker loci based on 11 mapping populations for tetraploid groundnut (*Arachis hypogaea* L.). *PLoS One* 7: e41213.

Glaubitz, J.C., Casstevens, T.M., Lu, F. et al. (2014). TASSEL-GBS: a high capacity genotyping by sequencing analysis pipeline. *PLoS One* 9: e90346.

Gyawali, S., Hegedus, D.D., Parkin, I.A. et al. (2013). Genetic diversity and population structure in a world collection of *Brassica napus* accessions with emphasis on South Korea, Japan, and Pakistan. *Crop Science* 53: 1537–1545.

Han, S., Yuan, M., Clevenger, J.P. et al. (2018). A SNP-based linkage map revealed QTLs for resistance to early and late leaf spot diseases in peanut (*Arachis hypogaea* L.). *Frontiers in Plant Science* 9: 1012.

He, G., Meng, R., Newman, M. et al. (2003). Microsatellites as DNA markers in cultivated peanut (*Arachis hypogaea* L.). *BMC Plant Biology* 3: 1–6.

He, Q., Xu, F., Min, M.-H. et al. (2019). Genome-wide association study of vitamin E using genotyping by sequencing in sesame (*Sesamum indicum*). *Genes & Genomics* 41: 1085–1093.

Heim, C.B. and Gillman, J.D. (2017). Genotyping-by-sequencing-based investigation of the genetic architecture responsible for a ~sevenfold increase in soybean seed stearic acid. *G3 Genes|Genomes|Genetics* 7: 299–308.

Hoque, A., Fiedler, J.D., and Rahman, M. (2020). Genetic diversity analysis of a flax (*Linum usitatissimum* L.) global collection. *BMC Genomics* 21: 1–13.

Hyten, D.L., Choi, I.-Y., Song, Q. et al. (2007). Highly variable patterns of linkage disequilibrium in multiple soybean populations. *Genetics* 175: 1937–1944.

Iquira, E., Humira, S., and François, B. (2015). Association mapping of QTLs for sclerotinia stem rot resistance in a collection of soybean plant introductions using a genotyping by sequencing (GBS) approach. *BMC Plant Biology* 15: 5.

Korte, A. and Farlow, A. (2013). The advantages and limitations of trait analysis with GWAS: a review. *Plant Methods* 9: 1–9.

Kumar, S., You, F.M., and Cloutier, S. (2012). Genome wide SNP discovery in flax through next generation sequencing of reduced representation libraries. *BMC Genomics* 13: 1–12.

Li, D., Zhao, X., Han, Y. et al. (2019). Genome-wide association mapping for seed protein and oil contents using a large panel of soybean accessions. *Genomics* 111: 90–95.

Li, H., Feng, H., Guo, C. et al. (2020). High-throughput phenotyping accelerates the dissection of the dynamic genetic architecture of plant growth and yield improvement in rapeseed. *Plant Biotechnology Journal* 18: 2345–2353.

Liu, L., Qu, C., Wittkop, B. et al. (2013). A high-density SNP map for accurate mapping of seed fibre QTL in *Brassica napus* L. *PLoS One* 8: e83052.

Luo, Z., Tomasi, P., Fahlgren, N. et al. (2019). Genome-wide association study (GWAS) of leaf cuticular wax components in *Camelina sativa* identifies genetic loci related to intracellular wax transport. *BMC Plant Biology* 19: 187.

Luo, H., Guo, J., Yu, B. et al. (2021). Construction of ddRADseq-based high-density genetic map and identification of quantitative trait loci for trans-resveratrol content in peanut seeds. *Frontiers in Plant Science* 12: 438.

Malmberg, M.M., Shi, F., Spangenberg, G.C. et al. (2018). Diversity and genome analysis of australian and global oilseed *Brassica napus* L. germplasm using transcriptomics and whole genome re-sequencing. *Frontiers in Plant Science*: 9, 508.

Mamidi, S., Lee, R.K., Goos, J.R. et al. (2014). Genome-wide association studies identifies seven major regions responsible for iron deficiency chlorosis in soybean (*Glycine max*). *PLoS One* 9: e107469.

Mei, H., Liu, Y., Du, Z. et al. (2017). High-density genetic map construction and gene mapping of basal branching habit and flowers per leaf axil in sesame. *Frontiers in Plant Science* 8: 636.

Nagy, E.D., Guo, Y., Tang, S. et al. (2012). A high-density genetic map of *Arachis duranensis*, a diploid ancestor of cultivated peanut. *BMC Genomics* 13: 1–11.

Narum, S.R., Buerkle, C.A., Davey, J.W. et al. (2013). Genotyping-by-sequencing in ecological and conservation genomics. *Molecular Ecology* 22: 2841.

Pandey, M.K., Agarwal, G., Kale, S.M. et al. (2017). Development and evaluation of a high density genotyping 'Axiom_ Arachis' array with 58 K SNPs for accelerating genetics and breeding in groundnut. *Scientific Reports* 7: 1–10.

Poland, J.A. and Rife, T.W. (2012). Genotyping-by-sequencing for plant breeding and genetics. *The Plant Genome* 5.

Qi, Z., Huang, L., Zhu, R. et al. (2014). A high-density genetic map for soybean based on specific length amplified fragment sequencing. *PLoS One* 9: e104871.

Qin, H., Feng, S., Chen, C. et al. (2012). An integrated genetic linkage map of cultivated peanut (*Arachis hypogaea* L.) constructed from two RIL populations. *Theoretical and Applied Genetics* 124: 653–664.

Rahman, M., Mamidi, S., del Rio, L. et al. (2016). Association mapping in *Brassica napus* (L.) accessions identifies a major QTL for blackleg disease resistance on chromosome A01. *Molecular Breeding* 36: 90.

Ryu, J., Lyu, J.I., Kim, D.-G. et al. (2021). Single nucleotide polymorphism (SNP) discovery and association study of flowering times, crude fat and fatty acid composition in rapeseed (*Brassica napus* L.) mutant lines using genotyping-by-sequencing (GBS). *Agronomy* 11: 508.

Sahebi, M., Hanafi, M.M., Siti Nor Akmar, A. et al. (2015). Importance of silicon and mechanisms of biosilica formation in plants. *BioMed Research International* 2015: 396010.

Shirasawa, K., Bertioli, D.J., Varshney, R.K. et al. (2013). Integrated consensus map of cultivated peanut and wild relatives reveals structures of the A and B genomes of Arachis and divergence of the legume genomes. *DNA Research* 20: 173–184.

Shirasawa, K., Kuwata, C., Watanabe, M. et al. (2016). Target amplicon sequencing for genotyping genome-wide single nucleotide polymorphisms identified by whole-genome resequencing in peanut. *The Plant Genome* 9: 1–8.

Singh, N., Agarwal, N., and Yadav, H.K. (2019). Genome-wide SNP-based diversity analysis and association mapping in linseed (*Linum usitatissimum* L.). *Euphytica* 215: 1–11.

Sonah, H., Bastien, M., Iquira, E. et al. (2013). An improved genotyping by sequencing (GBS) approach offering increased versatility and efficiency of SNP discovery and genotyping. *PLoS One* 8: e54603.

Sonah, H., O'Donoughue, L., Cober, E. et al. (2015). Identification of loci governing eight agronomic traits using a GBS-GWAS approach and validation by QTL mapping in soya bean. *Plant Biotechnology Journal* 13: 211–221.

Tillman, B.L. and Gorbet, D.W. (2012). Peanut cultivar UFT113. Google Patents.

Tseng, Y.-C., Tillman, B.L., Peng, Z., and Wang, J. (2016). Identification of major QTLs underlying tomato spotted wilt virus resistance in peanut cultivar Florida-EPTM '113'. *BMC Genetics* 17: 128.

Wang, L., Yu, S., Tong, C. et al. (2014). Genome sequencing of the high oil crop sesame provides insight into oil biosynthesis. *Genome Biology* 15: 1–13.

Wei, W., Zhang, Y., Lü, H. et al. (2013). Association analysis for quality traits in a diverse panel of chinese sesame (*Sesamum indicum* L.) germplasm. *Journal of Integrative Plant Biology* 55: 745–758.

Wei, W., Mesquita, A.C.O., Figueiró AdA, W.X. et al. (2017). Genome-wide association mapping of resistance to a Brazilian isolate of *Sclerotinia sclerotiorum* in soybean genotypes mostly from Brazil. *BMC Genomics* 18: 1–16.

Wu, J., Zhao, Q., Zhang, L. et al. (2018). QTL mapping of fiber-related traits based on a high-density genetic map in flax (*Linum usitatissimum* L.). *Frontiers in Plant Science* 9: 885.

Yi, L., Gao, F., Siqin, B. et al. (2017). Construction of an SNP-based high-density linkage map for flax (*Linum usitatissimum* L.) using specific length amplified fragment sequencing (SLAF-seq) technology. *PLoS One* 12: e0189785.

Yu, F., Zhang, X., Peng, G. et al. (2017). Genotyping-by-sequencing reveals three QTL for clubroot resistance to six pathotypes of *Plasmodiophora brassicae* in *Brassica rapa*. *Scientific Reports* 7: 4516.

Zargar, S.M., Raatz, B., Sonah, H. et al. (2015). Recent advances in molecular marker techniques: insight into QTL mapping, GWAS and genomic selection in plants. *Journal of Crop Science and Biotechnology* 18 (5): 293–308.

Zhang, Y., Li, W., Lin, Y. et al. (2018). Construction of a high-density genetic map and mapping of QTLs for soybean (*Glycine max*) agronomic and seed quality traits by specific length amplified fragment sequencing. *BMC Genomics* 19: 1–14.

Zhou, F., Liu, Y., Liang, C. et al. (2018). Construction of a high-density genetic linkage map and QTL mapping of oleic acid content and three agronomic traits in sunflower (*Helianthus annuus* L.) using specific-locus amplified fragment sequencing (SLAF-seq). *Breeding Science* 68: 596–605.

Zhou, X., Guo, J., Pandey, M.K. et al. (2021). Dissection of the genetic basis of yield-related traits in the chinese peanut mini-core collection through genome-wide association studies. *Frontiers in Plant Science* 12: 637284.

# Index

## a



## b



*Genotyping by Sequencing for Crop Improvement*, First Edition.
Edited by Humira Sonah, Vinod Goyal, S.M. Shivaraj, and Rupesh K. Deshmukh.

## h

## n



## o



## p



## q